MULTISCALE MODELING OF PARTICLE INTERACTIONS

MULTISCALE MODELING OF PARTICLE INTERACTIONS

Applications in Biology and Nanotechnology

Edited by

MICHAEL R. KING
Department of Biomedical Engineering
Cornell University

DAVID J. GEE
Department of Mechanical Engineering
Rochester Institute

WILEY
A JOHN WILEY & SONS, INC., PUBLICATION

Published by John Wiley & Sons, Inc., Hoboken, New Jersey.
Published simultaneously in Canada.

For general information on our other products and services or for technical support, please contact our Customer Care Department within the United States at (800) 762-2974, outside the United States at (317) 572-3993 or fax (317) 572-4002.

Wiley also publishes its books in a variety of electronic formats. Some content that appears in print may not be available in electronic formats. For more information about Wiley products, visit our web site at www.wiley.com.

Library of Congress Cataloging-in-Publication Data:

Multiscale modeling of particle interactions: applications in biology and nanotechnology / edited by Michael R. King, David J. Gee.
p. cm.
Includes index.
ISBN 978-0-470-24235-3 (cloth)
1. Nanotechnology. 2. Nanoparticles. 3. Biomedical engineering. I. King, M. R. (Michael R.),
1973–II. Gee, D. J. (David J.), 1964–
R857.N34M85 2010
610.28–dc22

2009031390

Printed in the United States of America

10 9 8 7 6 5 4 3 2 1

CONTENTS

CONTRIBUTORS

Alexander Alexeev, Ph.D., The George W. Woodruff School of Mechanical Engineering, Georgia Institute of Technology, Atlanta, Georgia

Prosenjit Bagchi, Ph.D., Department of Mechanical and Aerospace Engineering, Rutgers University, Piscataway, New Jersey

Anna C. Balazs, Ph.D., Department of Chemical and Petroleum Engineering, University of Pittsburgh, Pittsburgh, Pennsylvania

David J. Gee, Ph.D., Department of Mechanical Engineering, Rochester Institute of Technology, Rochester, New York

Nikolaos M. P. Kakalis, Ph.D., Department of Engineering Science, University of Oxford, Oxford, UK

Emmanouil D. Karagiannis, Ph.D., Department of Chemical Engineering, Massachusetts Institute of Technology, Cambridge, Massachusetts

Michael R. King, Ph.D., Department of Biomedical Engineering, Cornell University, Ithaca, New York

Dmitry I. Kopelevich, Ph.D., Department of Chemical Engineering, University of Florida, Gainesville, Florida

Petr Král, Ph.D., Department of Chemistry, University of Illinois at Chicago, Chicago, Illinois

Ian J. Laurenzi, Ph.D., Department of Chemical Engineering, Lehigh University, Bethlehem, Pennsylvania

Dan T. Lussier, Ph.D., Department of Engineering Science, University of Oxford, Oxford, UK

Feilim Mac Gabhann, Ph.D., Institute for Computational Medicine and Department of Biomedical Engineering, Johns Hopkins University, Baltimore, Maryland

Aleksander S. Popel, Ph.D., Department of Biomedical Engineering, School of Medicine, Johns Hopkins University, Baltimore, Maryland

Constantine Pozrikidis, Ph.D., Department of Chemical Engineering, University of Massachusetts, Amherst, Massachusetts

Amina A. Qutub, Ph.D., Department of Bioengineering, Rice University, Houston, Texas

Talid Sinno, Ph.D., Department of Chemical and Biomolecular Engineering, University of Pennsylvania, Philadelphia, Pennsylvania

Harry A. Stern, Ph.D., Department of Chemistry, University of Rochester, Rochester, New York

Yiannis Ventikos, Ph.D., Department of Engineering Science, University of Oxford, Oxford, UK

Boyang Wang, Ph.D., Department of Chemistry, University of Illinois at Chicago, Chicago, Illinois

1

INTRODUCTION

MICHAEL R. KING

Department of Biomedical Engineering, Cornell University, Ithaca, New York

DAVID J. GEE

Department of Mechanical Engineering, Rochester Institute of Technology, Rochester, New York

Multiscale physical problems, with important features across multiple length and time scales, are becoming increasingly prevalent in the computational sciences. The term *multiphysics* is used to describe problems in which different physical mechanisms are dominant at different length scales. For example, as one moves to the nanoscale, the roles of surface forces and thermal fluctuations become increasingly important. Multiscale problems encompass diverse fields such as mathematics, chemistry, physics, engineering, computer science, and materials science. Monoscale approaches to complex physical problems often prove to be inadequate, even with the increasing availability of high-performance supercomputers. With potentially many scales and many variables associated with problems of interest, there is a need for systematic approaches to modeling multiscale systems.

Examples of multiscale problems can be found in the field of fluid mechanics, for instance. One example is gas flow through micro- and nanochannels, where the mean free path of the gas molecules can be compared to the length scale of the flow volume via the dimensionless Knudsen number (Kn). For large-Kn problems, in which the mean free path (typical distance that a gas molecule travels between collisions) is larger than the dimensions of the channel conduit, the continuum approximation breaks down and flow must be calculated using discrete approaches such as direct

Multiscale Modeling of Particle Interactions: Applications in Biology and Nanotechnology,
Edited by Michael R. King and David J. Gee

simulation Monte Carlo or lattice Boltzmann [1]. For Kn $\ll 1$, the fluid behaves as a continuum, and the well-known Navier–Stokes equations with no-slip boundary conditions can be used to calculate the flow. For Kn $= O(1)$, gas flows behave in a transitional regime where modified continuum-level equations such as the quasi-gas dynamics or Burnett equations can be used with phenomenological wall slip boundary conditions. Another multiscale example from fluid mechanics is turbulent flow, where energy transfer occurs between self-similar vortices of varying dimension that form coherent structures in the fluid [2]. In boundary layer theory, as exemplified by the classic problem of flow past a plate, there is a thin region close to the wall in which viscosity effects are important, and this local solution must be matched to an outer solution governed by ideal flow [3]. A fourth example of a multiscale problem that arises naturally in fluid mechanics is the flow of complex fluids such as polymer melts or suspensions (see Chapters 11 and 12), in which the microstructure of polymer chains or suspended particles has a profound influence on the macroscopic rheological properties of the fluid. Some of these problems have previously been considered quite successfully using classical analytical methods or phenomenological models derived from experimental observations, whereas other problems lend themselves more readily to multiscale computational approaches.

Particle interactions and, in particular, the collective phenomena that arise from nearby discrete units, represent a hallmark of multiscale problems. In biology, for example, one finds a natural hierarchy of discrete length scales: from electrons to atoms, molecules, macromolecular assemblies, organelles, cells, multicellular aggregates, tissues, organs, organisms, and populations of organisms. The interactions between proteins and between cellular "particles" receive the most attention in Part II of the book, although a broader range, from molecular- to organ-scale phenomena important in angiogenesis, is discussed in Chapter 10. As we have written elsewhere, to gain a mechanistic rather than "black box" understanding of complex systems, it is necessary to concern oneself with the phenomena that occur at least one length scale below the scale you care most about. This is certainly true in biomedical systems, where the ultimate motivation must be the overall health of the organism.

In Part I, on nanoparticle applications, we have adopted broader definitions of interacting particles. Pozrikidis (Chapter 4) examines the mechanics of an atomic sheet, treating this as a particulate sheet connected by elastic springs. Sinno (Chapter 2) considers atomic defects in crystalline silicon as "particles" that perturb the lattice strongly up to a finite distance from the edge of the feature and only weakly or not at all farther away. Both Kopelevich (Chapter 3) and Stern (Chapter 8) are focused on the self-assembly of amphiphilic molecules into micelles and bilayers, where surfactant molecules or clusters of atoms are represented as interacting particles. Ventikos and co-workers (Chapter 6) study, among other phenomena, the rupture and coalescence of nanodroplets comprised of many interacting liquid molecules. The authors of other chapters in Part I examine the interaction between deformable capsules (Chapter 7) or water molecules (Chapter 5). A lattice of spatial nodes can be used to reduce dramatically the dimensionality of such computational problems (see Chapters 2 through 4, and 7).

Granularity refers to the extent to which a system is broken down into small parts. An important theme in multiscale modeling is coarse graining, where some of the fine detail of low-level models has been smoothed over or averaged out. Stern (Chapter 8) provides an excellent review of coarse graining in molecular dynamics (MD) simulations. Atomistic MD simulations provide the finest level of detail and, in principle, should be the most accurate. However, the time step required for numerical integration must not exceed the period of the highest-frequency motion of the system: in most cases, covalent hydrogen bond vibrations of order about 1 fs. Thus, on inexpensive computers, simulations of about 1 ns in duration can be executed: far shorter than biologically relevant molecular events such as lipid self-assembly or protein–ligand docking. In coarse-grained MD, one "particle" represents three or four atoms, effectively averaging over vibrational dynamics and resulting in an order-of-magnitude improvement in computational speed. Bagchi (Chapter 11) discusses some coarse-grained models specific to fluid dynamics. Sinno (Chapter 2) points out that with coarse-grained models it is often difficult to identify whether lack of agreement with experimental data is due to a poor choice of parameter values or to a fundamental shortcoming of the physical model and its associated assumptions, and presents strategies to address this.

Another important concept in multiscale modeling is linking between scales; that is, how one reconciles coupled fine-scale and coarser-grained models in an accurate, predictive, and unambiguous way. Sinno (Chapter 2) provides an excellent strategy for accomplishing this by first categorizing multiscale simulations into either sequential or concurrent frameworks. Sequential simulations, where parameters are passed from one scale to another, are the focus here. Atomistic simulations can be used as *in silico* experiments, and coarser-grained models tested against data generated by the atomistic simulation. This proves to be an effective approach, as mechanistic elements of the coarse-grained model are isolated and tested independent of the model parameters. Balazs (Chapter 7) describes a hybrid model that involves sequential linking between a lattice Boltzmann fluid dynamical model and a lattice spring model to simulate the elastic shell of a compliant microcapsule, although these lattices are of comparable spatial scales. In Chapter 10, Popel categorizes multiscale models of transport in biological systems as either continuous or discrete models. The continuous models, usually arising from differential balance equations, have been used at the molecular, cellular, tissue, and organ scales. Discrete models treat cells as discrete objects, and are also discussed in Chapters 11 and 12 in the context of blood cell transport and adhesion.

In this book we present a variety of approaches and methodologies for the use of multiscale computational models to study and predict the behavior of complex systems. The chapters are organized around applications in nanoscience (Part I) and biology (Part II), although other connections can be made, such as molecular dynamics (Chapters 3, 5, 6, and 8) and stochastic receptor–ligand binding (Chapters 9 and 12). Slide-show presentations to support the content in each chapter are available at the book Website at (www.Wiley.com) for instructors who wish to use this book for graduate teaching. It is our hope that the common themes and breadth of current examples will enable and inspire researchers seeking to understand multiscale physical systems and the particle interactions that lead to collective phenomena.

REFERENCES

1. Colin S. Single-phase gas flow in microchannels. In: *Heat Transfer and Fluid Flow in Minichannels and Microchannels*, ed. SG Kandlikar, S Garimella, D Li, S Colin, MR King, Elsevier, Kidlington, UK, 2006.
2. Brooke JW, Hanratty TJ. Origin of turbulence-producing eddies in a channel flow. *Phys. Fluids* **5**:1011–1022, 1993.
3. Bird RB, Stewart WE, Lightfoot EN. *Transport Phenomena*, Wiley, New York, 1960.

PART I

APPLICATIONS IN NANOTECHNOLOGY

2

MULTISCALE MODELING OF NANOSCALE AGGREGATION PHENOMENA: APPLICATIONS IN SEMICONDUCTOR MATERIALS PROCESSING

TALID SINNO
Department of Chemical and Biomolecular Engineering, University of Pennsylvania, Philadelphia, Pennsylvania

2.1 INTRODUCTION

The increasing need for an atomic and/or molecular understanding of many physiochemical phenomena tied to advanced processing of today's materials continues to drive the development of various multiscale modeling and simulation approaches. Example applications include semiconductor crystal growth and microelectronic device fabrication [1], thin-film and coatings engineering [2], and advanced metal and metal alloy processing [3]. In many of these examples, atomic diffusion enables the nucleation and growth of nano- and microstructures that strongly affect the final chemical, electronic, and mechanical properties of the material. In the case of metals, thermal treatments are crucial for tailoring the defect (grain boundaries and dislocations) and/or precipitate distributions, which strongly affect the ductility and strength of the material. In crystalline semiconductor materials such as silicon and silicon alloys, the nucleation and growth of atomic defects such as clusters and dislocations

Multiscale Modeling of Particle Interactions: Applications in Biology and Nanotechnology,
Edited by Michael R. King and David J. Gee

must be controlled carefully to minimize their effects on fabricated microelectronic devices. The inherent range of time and lengths scales that must be considered in such systems implies that some form of compound or multiscale computational approach must be used. A critical aim of multiscale modeling and simulation is the ability to capture the atomistic features of nonequilibrium dynamic phenomena at finite temperature while maintaining relevance at the experimental or process scale.

In this chapter we utilize a prototypical example of solid-state microstructure nucleation and growth: the aggregation of lattice vacancies in crystalline silicon into nanoscale cavities or voids. Vacancies are the simplest types of defects in crystals and are defined as missing lattice atoms. Their presence, which generally increases the energy of the crystal, is entropically favored at high temperature (i.e., the equilibrium concentration of vacancies increases with temperature). As a result, materials processing, which includes significant heating and cooling (such as crystal growth from the melt), can lead to the presence of large vacancy supersaturation. This supersaturation is often released by homogeneous nucleation of vacancy aggregates, or voids, whose size and density distribution is a function of the processing parameters.

In the case of silicon crystal growth, such voids, which form during silicon crystal melt growth under certain conditions [1], have long been understood to degrade the performance of microelectronic devices fabricated on crystalline silicon wafers, and therefore are of significant technological importance. Examples of void structures observed in commercial silicon crystals using transmission electron microscopy are shown in Fig. 2.1 [4,5]. Void formation in silicon is an ideal test bed for the development of multiscale modeling frameworks because of the availability of the

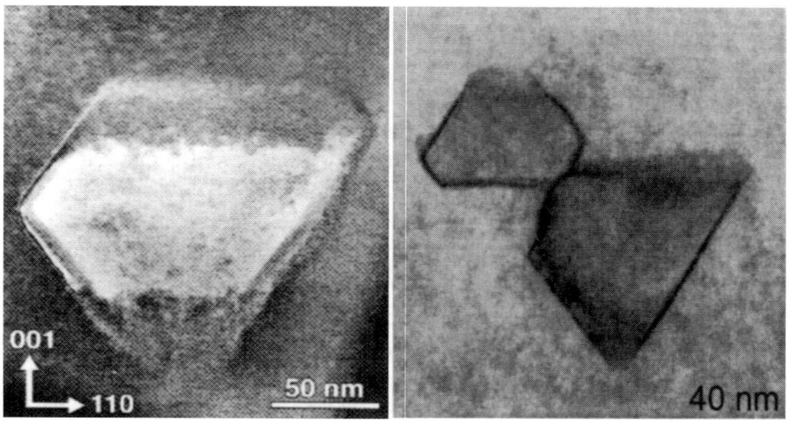

(a) (b)

Figure 2.1 TEM images of various void structures observed in crystals grown by the Czochralski (CZ) method, in which a seed crystal is pulled from a crucible containing molten silicon under highly controlled heat transfer conditions: (a) single octahedral void; (b) double void (see ref. [1] for more details). (From [4,5], with permission.)

large amount of experimental data that is necessary for model development and validation. These data are themselves multiscale in nature, ranging from total void densities averaged over an entire silicon ingot, to the atomic details of a single nanoscopic void. Moreover, the ubiquitous nature of silicon has led to the development of a large number of empirical interatomic potentials which are required for large-scale atomistic calculations. Examples of such potentials include Stillinger–Weber [6], Tersoff [7], environment-dependent interatomic potential [8], and many others. Finally, void nucleation and growth in silicon constitute essentially a problem of solid-state homogeneous nucleation, which provides a well-understood theoretical basis from which to develop detailed models.

There is currently no single simulation method that can efficiently access the large range of length and time scales required to address the types of problems of interest here. Two broad categories of multiscale simulation frameworks can be identified in general: sequential and concurrent. *Sequential multiscale modeling*, which we focus on in this chapter, refers to the use of several distinct simulation components, each representing different length and time scales, that are loosely connected, usually by parameters that are passed from one simulation scale to the other. The *concurrent approach*, which we mention here only briefly, seeks to link multiple representations into a single simulation framework, thereby greatly increasing the amount of information that can be passed from one scale to the next. Although, in principle, the concurrent approach is more powerful (i.e., more integrated across its individual components) than the sequential approach, it is generally more difficult to construct a concurrent simulation, especially one that is applicable to a broad range of problems. Examples of concurrent multiscale methods include the quasicontinuum method [9], MAAD [10], coarse-grained molecular dynamics [11], the bridging-scale method [12], and kinetic Monte Carlo–molecular dynamics hybrids [13]. In several of these methods, a fully atomistic region is centered about a feature of interest: for example, a defect, or the region immediately below a nanoscale indenter. The atomistic region is surrounded by a continuum region which is matched to the atomistic region by appropriate boundary conditions. The atomistic–continuum interface is adaptive and evolves to confine the atomistic region to the vicinity of the feature of interest.

2.1.1 Sequential Multiscale Modeling Approach for Solid-State Diffusion and Aggregation

A typical approach to executing a sequential multiscale modeling study is to use an atomistic-scale simulation technique, such as molecular dynamics (MD) or Monte Carlo, to compute model parameters that are required as input to a coarser simulation (see Fig. 2.2a). Atomistic simulations can be based either on electronic-structure theories such as density-functional theory or empirical potentials. Once parameterized, predictions of the coarse models can then be compared directly to experimental data. Examples of coarse models include continuum-rate equation models based on partial differential equations [14] and lattice kinetic Monte Carlo simulations, both of which require parametric and mechanistic input (e.g., transport and/or reaction

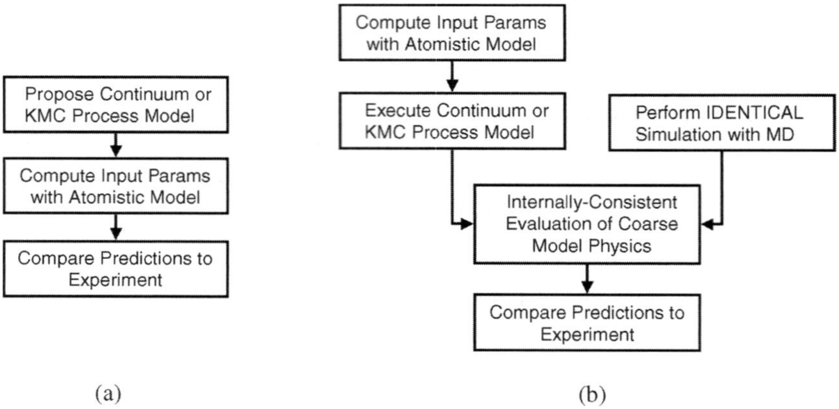

Figure 2.2 Various methods for passing information between atomistic and coarser simulation scales: (a) traditional parameter passing; (b) mechanistic information passing by direct model-on-model comparison.

mechanisms) before any model predictions can be made. A key limitation of this overall approach is that it is not possible to isolate the contributions of model parameters and model physics to errors in matching experimental data. Coarse models that cannot represent a given set of experimental data adequately are either poorly parameterized (e.g., due to a weakness in the interatomic potential description or because of insufficient experimental measurements) or simply physically inaccurate.

The increasing availability of computing power and better simulation algorithms offers another possibility for passing information between atomistic and coarse models, as shown in Fig. 2.2b. Here, a direct comparison is made between the output of large-scale atomistic simulations and coarse-grained model predictions for some reference process [15]. For the system considered in this chapter, an example reference process could be a system of supersaturated vacancies diffusing and aggregating within a silicon crystal matrix under some applied thermal and mechanical conditions. The output of this process, which is used to test model fidelity, would be various measures of the cluster size distribution as a function of time. In this approach, the atomistic simulation is applied as an *in silico* experiment and the coarse model is tested against data generated by this experiment.

There are various approaches for making such a comparison between the outputs of the two simulation scales. In one example, all the input parameters required for the coarse model are computed separately with targeted supplementary atomistic calculations based on the same interatomic potential as that used to model the reference process. The coarse model physics can then be adjusted until its prediction is consistent with the atomistically modeled process. In another approach, model-on-model regression can be performed, whereby the coarse model input parameters are regressed directly by matching the reference process data obtained from the two simulations. An example of such data could be the cluster size distribution as a function of time. Mechanistic accuracy is then ensured by comparing the regressed parameters to estimates derived from supporting atomistic calculations (again based on the same interatomic potential to guarantee parametric consistency).

A key advantage of both of these types of sequential multiscale approaches is that the mechanistic elements of the coarse model are being isolated and tested independently from the model parameters. The internal consistency provided by computing any needed input parameters with essentially no error (for the idealized material represented by the particular interatomic potential) removes any uncertainties associated with parameters and makes the agreement between the two scales almost entirely dependent on the coarse model mechanistic accuracy. In other words, these approaches allow mechanistic rather than simply parametric information to be passed between scales. The primary limitation of the approach shown in Fig. 2.2b is the time- and length-scale scope of the atomistic simulations, which in this case are based on the MD method. Microscopic processes that require length and/or time scales beyond those possible with standard MD cannot easily be considered within this framework.

The remainder of this chapter is organized as follows. In the following section, a recently developed approach for extending the (length scale) scope of large-scale direct MD simulations of atomic aggregation is discussed. This method, which is referred to as *feature-activated molecular dynamics* (FAMD), focuses the computational effort of MD simulations on the localized features of interest. In the case of vacancy aggregation within a crystalline matrix, the features are the vacancies; the lattice far away from any vacancies is essentially undisturbed and does not need to be considered dynamically. The FAMD method is applied to the direct simulation of vacancy clustering in crystalline silicon. The output of these simulations is then used to develop and test coarse-grained models of vacancy aggregation. In Section 2.3 we introduce a lattice kinetic Monte Carlo model for vacancy aggregation that provides a highly efficient representation of the process, which is improved dramatically using the approach highlighted in Fig. 2.2b. Finally, current efforts to develop novel multiscale lattice kinectic Monte Carlo simulations are discussed briefly in Section 2.4.

2.2 LARGE-SCALE MD SIMULATIONS OF MICROSTRUCTURAL EVOLUTION

The basic computational task in MD simulations is to compute the trajectory of each atom by integrating Newton's equations of motion, $\mathbf{F} = m\mathrm{a}$, where $\mathbf{F}$ represents the total force acting on each atom. The total force is a sum of internal forces (generated by other atoms) and external forces (such as those arising from applied fields). Ideally, the internal forces are computed by considering the electronic, as well as atomic, distribution within the system. The explicit consideration of quantum mechanical forces generally leads to good accuracy in the properties computed, but these simulations are limited to very small system sizes (hundreds of atoms) and simulation times (a few picoseconds). The classical MD simulation, on the other hand, relies on the specification of an empirical interatomic potential function that attempts to represent the bonding environment with a parametrically tuned function, such as the examples provided in the introduction. The accuracy of empirical potential MD simulations is variable, but the reduced computational costs allow for the simulation of millions of atoms for up to tens of nanoseconds [16].

In this section we first describe a reference process related to vacancy aggregation in crystalline silicon, which is generated with large-scale MD simulation. The data obtained from these simulations are used in subsequent sections to develop and test kinetic Monte Carlo simulations. In the second part of this section, a technique for extending the scope of large-scale MD simulations of aggregation is presented. In all cases, the environment-dependent interatomic potential (EDIP) [8] was used to describe the forces between the silicon atoms. The EDIP potential has been tested extensively and has been shown to provide a good representation of point defect and defect cluster thermodynamic properties in the crystalline silicon bulk.

The reference simulation applied here is based on a cubic periodic cell consisting of 216,000 host silicon atoms arranged in a tetrahedral lattice from which 1 000 equally spaced vacancies were removed. The simulation was carried out within the canonical NVT ensemble (constant number of particles, volume, and temperature) at a temperature of 1600 K and zero pressure. These conditions were chosen to give very high vacancy supersaturation and vacancy cluster diffusivities so that adequate nucleation and growth could be accessed over the nanosecond time scale [17]. During the course of the simulation, atomic coordinates for the entire system were stored at 1000 to 80,000 time-step intervals. Each of these coordinate files was then used as a starting point for a rapid-quench MD simulation in which the temperature was lowered from 1600 K to 100 K in 3000 time steps. The quenched configurations were then compared to a reference perfect lattice at the same density in order to locate the vacancies in the system. Vacancy clusters were then identified using the Stillinger definition [18], which connects vacancies based on their separation distances.

Using a parallel implementation of MD, the simulation above was carried out for several million time steps, corresponding to several nanoseconds of real time. Importantly, the accessible simulation time scale and system size lead to sufficient aggregation to allow for a quantitative analysis of the aggregation dynamics (see Figs. 2.4, 2.6 and 2.8 in the following sections). It should be emphasized that the very large vacancy concentrations (ca.0.5 atomic percent) that are required to accelerate the nucleation and growth dynamics of clusters to the MD time scale are not physically realistic in the context of silicon crystal growth or wafer thermal annealing [14]. The latter are generally limited to a maximum of 1×10^{-7} atomic fraction. However, the mechanisms by which vacancy aggregation proceeds, diffusion and binding, are not affected by the artificially high concentration, enabling the use of these simulations for model construction and validation. The choice of system temperature, 1600 K, which is quite close to the melting temperature of silicon, was based on the fact that the initial phases of point defect clustering during melt growth of crystalline silicon take place shortly after solidification. Relatively little is known about the properties of small defect clusters at very high temperature, and in the following sections it is shown that, in fact, defect behavior can be affected substantially by temperature. More practically, the utilization of high temperature also allows for more rapid system evolution and therefore extends the effective scope of the MD simulations described in this section.

2.2.1 Extending the Scope of Large-Scale MD Simulation:FAMD

The feature-activated molecular dynamics (FAMD) method [19,20] is designed to increase the scope of empirical MD simulations of collective phenomena such as aggregation. It can be applied in conjunction with other approaches to reduce the computational expense of MD simulations, such as parallelization, time-scale separation, and coarse graining [21]. The essential concept in FAMD is that any disturbance (or feature) within a perfect lattice, such as a point defect or defect cluster, perturbs the lattice strongly up to a finite distance from the edge of the feature, but only weakly (or not at all) farther away. In FAMD, the computational domain is divided into two types of regions: MD regions, which contain features of interest and are treated with standard MD, and static regions, which are considered to be (quasi) stationary (see Fig. 2.3). The entire system is still represented explicitly by atoms, but most of them are "placeholders" that do not contribute directly to the dynamics of the system.

An FAMD simulation begins with the specification or identification of all feature atom positions. A *feature* is broadly defined as any species that perturbs (locally) the evolution of the MD system away from perfect crystal behavior; it can be a vacancy, an interstitial, an impurity atom, or even a lattice atom that is displaced from its perfect lattice position by an external driving force. Once the individual feature positions are specified, they are grouped according to the Stillinger connectivity criterion [18]. This grouping is performed to partition the simulation system into isolated active regions, with each feature group occupying a unique active region (see Fig. 2.3). Such a region

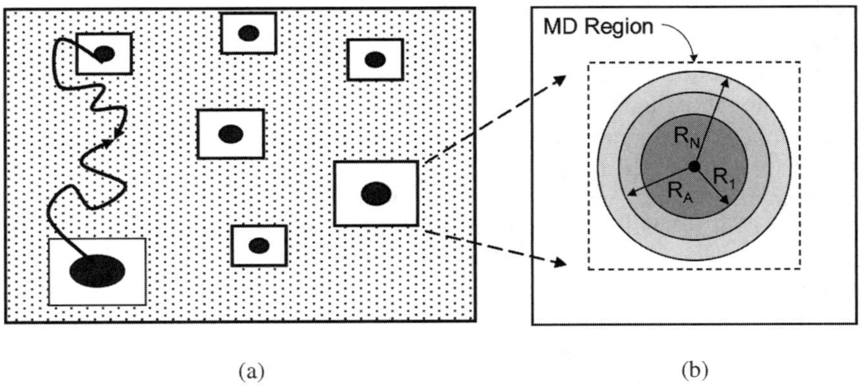

Figure 2.3 (a) Representation of active regions (open squares) surrounding defect features (solid circles) in an otherwise perfect lattice (shaded regions); (b) various components of an *MD region* construct: fully active atom region, R_1; transition region, R_A; inactive neighbor region, R_N. Atoms that are farther than R_N from the nearest feature are included in the MD region but do not contribute to the overall computational effort. (From [19], with permission.)

is an algorithmic construct containing all lattice atoms (surrounding a cluster) needed to evaluate the potential energy and forces during molecular dynamics simulation of the active atoms inside it. Active regions must be dynamic in nature; that is:

1. They must be able to move and change shape to keep the feature surrounded by a sufficient number of active atoms so that it does not interact with the edges of the active region.
2. They must be able to coalesce with other active regions as individual features diffuse to within interaction range so that aggregation can be enabled.
3. They must be able to split into multiple daughter regions when a fragmentation event takes place so that the total number of active atoms is kept at a minimum.

The structure of a cubic (for example) MD region containing a single feature is shown schematically in Fig. 2.3b. The various shells (shown here surrounding a single feature atom, for simplicity) represent different temperatures, which are specified using a temperature envelope function, T_A, that scales the simulation temperature. The value of the temperature envelope function for each atom in an active region is determined based on that atom's distance from a feature. The value of the temperature is equal to the desired simulation value up to a distance R_1, and then decreases gradually to zero in the transition interval from R_1 to R_A. Any atoms that lie beyond R_A from any feature atom are assigned a temperature of zero. The radius R_1 is large enough to accommodate any feature–feature interaction distance and includes a buffer for feature diffusion between updates. Atoms that are located within a radius R_1 of any feature atom therefore are fully active and are thermostatted to the desired simulation temperature, while atoms that are located in the region $R_1 < r < R_A$ are denoted as transition atoms. Details of the algorithms used to control the temperature and pressure across active regions in an FAMD simulation may be found in refs. [19,20].

The FAMD method was verified by comparison to the reference vacancy aggregation process described at the beginning of this section. Several aggregation metrics were monitored during both FAMD and standard (parallel) MD simulations, including the number of monomers (X_1) and dimers (X_2), and various moments of the overall size distribution, defined as $M_n = \sum_i i^n X_i$, where the index i runs over all cluster sizes. The evolution of these quantities as a function of time is shown in Fig. 2.4 for both standard MD and FAMD. The agreement between the two methods is excellent in each case, demonstrating the validity of the localized disturbance assumption in FAMD.

A slab of the FAMD cubic simulation cell after 2 ns of evolution is shown in Fig. 2.5a, in which the vacancy clusters are highlighted (large dark atoms) inside the active regions (medium-shaded atoms). Active region coalescence can be observed in multiple instances, along with the large fraction of static atoms (small light atoms). The evolution of the fraction of active atoms as a function of time during this simulation is shown in Fig. 2.5b. At the beginning of the simulation, the 1000 uniformly distributed single vacancies require a total of about 136,000 active atoms

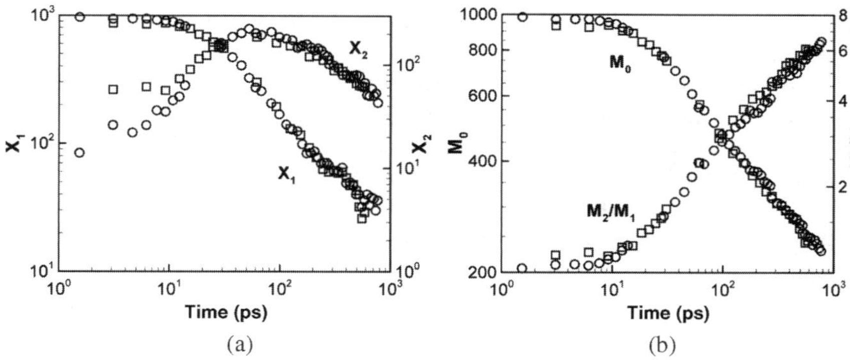

Figure 2.4 (a) Evolution of the number of single vacancies (X_1) and vacancy dimers (X_2) as a function of time for standard MD (circles) and FAMD (squares) simulations; (b) evolution of the mean cluster size (M_2/M_1) and total cluster number (M_0) of vacancy clusters as a function of time. System configuration: 1 000 vacancies in a 216,000-atom lattice at 1600 K and zero pressure. (From [19], with permission.)

(or 63% of the total system). After about 3×10^6 time steps, this number is reduced to about 34,000 atoms (16% of the overall system). In other words, the FAMD approach becomes increasingly attractive as vacancy aggregation proceeds due to active-region coalescence. The low overhead required for constructing and periodically updating the MD regions in FAMD generally results in the algorithm scaling linearly with the number of atoms activated. Depending on the feature concentration and the morphological evolution in the system, the FAMD approach can provide well over an order-of-magnitude acceleration relative to standard MD. More recently, the FAMD

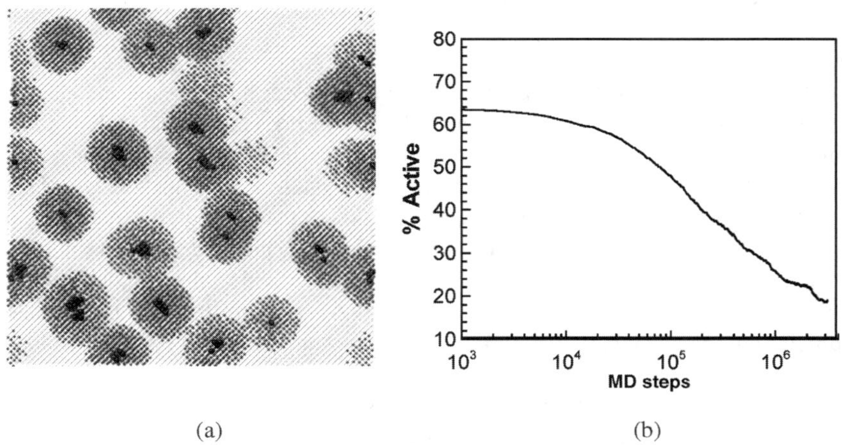

Figure 2.5 (a) Slab taken from an FAMD simulation of vacancy aggregation in silicon. Large black "atoms" are vacancies, medium-gray atoms are active atoms, and small light atoms are static. (b) Fraction of active atoms as a function of time steps for the same system. (From [19], with permission.)

approach has been parallelized to further increase the scope of large-scale MD simulation [20].

Finally, we mention briefly an extensive body of work aimed at extending the accessible *time* scale of molecular dynamics simulations (see ref. [22] for a review). These techniques attempt to accelerate the dynamics of simulations by taking advantage of the fact that many solid-state phenomena, such as atomic diffusion in crystals, are rare or infrequent events relative to the vibrational time scale of atoms, because they must overcome large barriers in the multidimensional potential-energy landscape of the system.

To summarize, the results shown in Figs. 2.4 and 2.5 represent specific examples of the *in silico* experimental data that may be used to develop and test coarse-grained simulations of the same aggregation process. Once such coarse models have been validated mechanistically, they can be applied under more realistic operating conditions to make connections with experimental measurements. In the following section, a lattice kinetic Monte Carlo simulation is developed and verified using these data, along with additional measurements taken from other MD simulations, including individual cluster diffusivities as a function of size and cluster structures. The availability of an essentially unlimited amount of data that can be collected in this way allows for extremely stringent testing of any mechanistic element of a coarse model. It will be shown that although it is quite straightforward to generate coarse-grained models that are qualitatively reasonable, quantitative accuracy requires substantially more effort, even for the relatively simple case of vacancy aggregation in single-crystal silicon.

2.3 MECHANISTICALLY VALIDATED LKMC MODELING OF ATOMIC AGGREGATION

The lattice kinetic Monte Carlo (LKMC) method is, for two principal reasons, ideally suited for the simulation of atomic diffusion and aggregation in crystalline materials. First, LKMC (and other variants of kinetic Monte Carlo) coarse-grains the details of atomic vibration, which severely limit the scope of MD simulations while retaining full atomistic resolution. Second, as applied to crystalline systems such as the one under consideration in this chapter, LKMC takes advantage of the presence of a lattice, resulting in a highly reduced problem dimensionality [23,24]. In the present application, that of vacancy diffusion and aggregation within the silicon lattice, the problem dimensionality can be reduced further by considering only the vacancies explicitly and the actual silicon atoms as a background tetrahedral grid on which the vacancies can move.

The basic elements of the LKMC method, as applied here, are as follows [25]. At each simulation step, all the possible events (e.g., diffusion or reaction) are grouped into different types according to their rates. Each type can therefore include multiple (different) events that possess the same rate, thereby reducing the number of unique event types in the simulation. The standard binary-tree search algorithm is first used to pick an event type, and then a particular particle is randomly chosen to execute the event. Once the event is executed, the local configuration and the rate database for vacancies that are possibly affected by the execution of the event are updated accordingly. The rate database only includes rate types that have been accessed thus far. If a new event type is generated by the previous move, it is added to the database. Finally, the simulation time is updated using $\Delta t = -\ln(U)/P$ (U is a uniform random number, $U \in [0,1]$), and the algorithm is repeated until the final time t_{end} is reached.

The principal drawback of most implementations of the KMC method is that mechanistic and rate information must be supplied externally (i.e., a physical model of the process must be constructed) [26]. This is in contrast to the MD simulations described in Section 2.2, which require only specification of an interatomic force field. There are, however, variants of KMC in which the potential-energy landscape generated by an interatomic force field is probed on-the-fly to locate saddle points between basins, thereby generating both an evolving database of possible events and their corresponding rates as the system evolves [27,28]. These variants of KMC, which are frequently referred to as *off-lattice* or *on-the-fly KMC*, are generally more robust than the coarse-grained lattice variant discussed in this section, but are also far more computationally demanding.

The typical input for LKMC simulations is a database of rates for the various allowable events, which are typically species hops between adjacent lattice sites and/or reaction events. In the LKMC model discussed in this section, the hopping rates between neighboring lattice sites are specified by a lattice potential, which is parameterized in terms of "bonds" between vacancies on neighboring lattice sites. These bonds are simply a parameterization of the binding (free) energy between neighboring vacancies and provide a convenient way to summarize an otherwise intractable number of different rates.

In the present application, the following approach is applied to compute the vacancy hopping rate for any vacancy in any configuration. First, all rates are cast into the framework of transition-state theory (TST) [29]. In TST, the rate of a particular transition is computed by identifying the dividing surface in the potential-energy landscape between an initial and a final state, and then analyzing the crossings across this surface. Additional approximations within the framework of TST can be made to further simplify the evaluation of rates. The most common of these is the harmonic approximation, in which the overall rate is decomposed into a temperature-independent energy barrier and a preexponential factor. Within TST, the rate for a given hop, i, is therefore

$$r_i = \nu_0 \exp\left(\frac{-\Delta E_i}{k_B T}\right), \tag{2.1}$$

where r_i is the rate for event i, ν_0 is the attempt frequency, k_B is the Boltzmann constant, and T is the simulation temperature. The energy barrier, ΔE_i, is related to the vacancy–vacancy bond energies by [30]

$$\Delta E_i = \max\left(0, \quad \Delta E_{\text{hop}} - 0.5 \sum_{j=1}^{\text{NN}} \Delta NB_j E_b^j\right), \tag{2.2}$$

where ΔE_{hop} is the (constant) energy barrier for a single isolated vacancy jump, NN is the maximum interaction shell (i.e., the range of the lattice potential), ΔNB_j is the bond number change due to a hop associated with interaction range j, and E_b^j is the corresponding bond energy. Note that all barriers, ΔE_i, are positive. If the energy difference between the initial and final states is negative and larger in magnitude than ΔE_{hop}, no energy barrier exists for the hop.

By its very nature, the KMC method (lattice or otherwise) is particularly attractive at low temperatures where thermally activated processes are slow, and therefore each particle move represents a long time increment. For example, KMC simulations of atomic diffusion and aggregation on metallic surfaces can span times of seconds and longer at temperatures below 300 K. Under these conditions, the underlying picture that the diffusion and aggregation proceed as a sequence of uncorrelated rare events that take place on lattice sites also becomes more accurate. It is also notable that this is exactly where the MD simulation becomes especially inefficient; that is, MD and KMC can be considered to be somewhat complementary tools in the analysis of solid-state diffusion, nucleation, and growth.

2.3.1 Physics of Vacancy–Vacancy Interactions in Silicon

The framework represented by Eqs. (2.1) and (2.2) is generally applicable to a broad range of lattice-based problems. To fully parameterize the lattice potential to the specific case of vacancies diffusing and aggregating in silicon at 1600 K (the temperature at which the MD simulation data was obtained), the parameters NN, ν_0, $\{E_b^j : j = 1, \text{NN}\}$ must be specified. The hopping attempt frequency and the isolated vacancy hop barrier can both be obtained directly from MD simulations of a single vacancy [15]; these were calculated as $9.5 \times 10^{13}\,\text{s}^{-1}$ and 0.3 eV, respectively, at 1600 K. The vacancy–vacancy interaction range is also relatively straightforward to measure and has been estimated to extend to the eighth-nearest-neighbor shell on the silicon lattice (i.e., NN = 8) [17]. The long range of this interaction arises from the interaction of the lattice strain fields surrounding each point defect.

The eight bonding energies (one for each neighbor shell) are more difficult to determine. First, it is important to realize that the energetics of a vacancy cluster within a lattice is determined by the surface area of the cluster, not the volume. This is because the energetic penalty for forming the cluster arises from broken bonds at the matrix–void interface; by definition, no silicon atoms are present in the bulk of the cluster. A simple bonding model in which the total binding energy is assumed to be the sum of the total number of bonds will fail to capture this trend. One approach to resolving this difficulty is to introduce interaction screening, or "line-of-sight" physics, whereby the interaction between individual vacancies can be shielded by other vacancies that lie in the path that connects them [31]. Here we assume that the interaction between two vacancies is shielded if one or more vacancies lie in the sphere circumscribed by them [25]. The screening physics leads to the correct energy scaling with cluster size, but also adds significant computational expense to the evaluation of vacancy hopping rates. Finally, it should be noted that this representation is probably not unique — it simply satisfies some basic criteria that must be met for a valid representation of vacancy aggregate energies; other bonding frameworks may be equally valid.

2.3.2 Lattice-Based Approach to Vacancy–Vacancy Bond Energies

The eight bonding energies needed to fully parameterize the LKMC model were first computed by regression using the energies of a large number of vacancy cluster configurations, each of which was computed using static energy minimization. Each cluster configuration was initialized randomly on a perfect silicon lattice by removing a number of vacancies. The locations of removed atoms were constrained so that the vacancy cluster was connected (i.e., every vacancy was within the eighth-nearest-neighbor shell of at least one other vacancy). The domain was then relaxed with conjugate gradient energy minimization (based on forces computed with the EDIP potential), and the cluster formation energy was calculated by reference to a perfect crystal system containing the same number of atoms [32]. About 60,000 configurations in total (generated for clusters with sizes $6 \leq n \leq 18$) were generated and relaxed in this manner.

The formation energy data were then used to regress the eight bond energies $\{E_b^1, E_b^2, \ldots, E_b^8\}$ according to the relationship

$$E_f(n) = 3.23n - \sum_{n} \sum_{i=1}^{NN} NB_i E_b^i \quad \text{eV}, \tag{2.3}$$

where NB_i is the number of vacancy–vacancy bonds of type i (i.e., bonds connecting a pair of vacancies i neighbor shells apart) associated with a cluster of n vacancies. The first term in Eq. (2.3) represents the formation energy for n single isolated vacancies, and the second term represents the total binding energy of the cluster. Note that the screening physics discussed earlier were taken into account during the LKMC bond energy regression. The eight regressed LKMC bond energies were found to describe almost exactly the formation energies of over 99.99% of all EDIP-relaxed structures. Note that, by construction, all cluster configurations generated in this approach are essentially on-lattice structures. In other words, the relaxation process allows only for optimization within the particular local minimum in the potential-energy landscape that corresponds to the initial configuration [32].

The parameters calculated in this and Section 2.3.1 were used to execute an LKMC simulation of the identical aggregation process described in Section 2.2.1 (1000 vacancies distributed uniformly over 216,000 lattice sites at 1600 K and zero pressure). A comparison between the LKMC model and MD results for several components of the vacancy cluster size distribution is shown in Fig. 2.6. The LKMC model clearly under-predicts the aggregation dynamics for each of the comparison metrics considered (i.e., monomer and dimer concentrations, average cluster size, and seventh-order moment of the size distribution). The bond energies in the LKMC model are such that clusters of all sizes, including dimers, are predicted to be tightly bound and therefore almost immobile. Under these conditions, the primary mechanism for further cluster growth is Oswald ripening, which proceeds by the exchange of

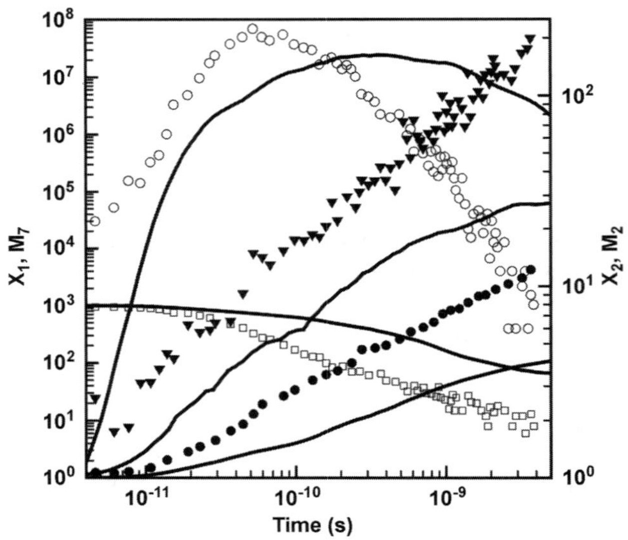

Figure 2.6 Comparison of MD and LKMC model predictions for vacancy cluster size evolution. Initial conditions consisted of 1 000 single vacancies distributed in a 216,000-atom lattice at 1600 K. Evolution profiles are symbols for MD, solid lines for LKMC; X_1, monomer (open squares); X_2, dimer (open circles); M_2, average cluster size (solid circles); M_7, seventh-order moment (solid gradients). (From [32], with permission.)

single vacancies between immobile clusters. By contrast, the MD simulation shows that small vacancy clusters are highly mobile and assume numerous, loosely bound configurations during the simulation. This is not unexpected given the high temperature of the simulation.

In summary, the LKMC model described here appears to provide a qualitatively reasonable picture of vacancy aggregation, but fails to capture the quantitative details of the aggregation process, a fact that would not have been highlighted were it not for the detailed model-to-model comparison. The structure and mobility of the clusters observed in MD simulations provides some insight into the problem. The discussion in the following section shows that many of the configurations observed in the MD simulations are highly distorted structures that do not correspond to vacancies, being clearly associated with a single lattice site. Such configurations simply cannot be represented explicitly in an LKMC model. In the following section, the source of these loosely bound, high-energy structures is discussed, and an alternative approach for parameterizing the LKMC model is presented that better represents the MD simulation data.

2.3.3 Role of Configurational Entropy in the Thermodynamics of Lattice Defects

To fully understand the source of discrepancy between the LKMC and MD results highlighted in Section 2.3.2, it is useful to consider the thermodynamics of lattice defects in detail. The conventional approach to computing the thermodynamic properties of point defects or their clusters is to search for the energetic ground-state structure of each species and compute finite temperature properties by incorporating the vibrational entropy [33,34]. Implicit in this approach is the assumption that the zero-temperature ground-state structure is the dominant configuration at all temperatures of interest. Moreover, the configurational entropy is often neglected because the ground-state structure is usually a low-symmetry, on-lattice configuration. This is, in fact, the case with vacancy clusters in silicon, whose lowest-energy configurations are usually missing lattice atoms with small displacements in the surrounding lattice atoms.

A more comprehensive approach to analyzing the thermodynamics of defects at finite temperature is to apply a framework that has been developed for supercooled liquids and glasses [35,36]. The free energy of a system in the canonical ensemble is generally given by $G = -k_B T \ln Z$, where Z is the canonical partition function. Assuming that the phase space can be divided into a distribution of energy basins, which are characterized uniquely by their minima, the partition function is given by

$$Z = \frac{1}{\Lambda^{3N}} \int G(V_\alpha) \exp(-\beta \, V_\alpha) \; dV_\alpha, \tag{2.4}$$

where $\Lambda = (h^2/2\pi \, mk_B T)^{1/2}$ is the thermal de Broglie wavelength that arises from integration of the kinetic portion of the partition function and V_α is the minimum potential energy of basin α. The density-of-states function, $G(V_\alpha)$, represents the distribution of both configurational and vibrational states; that is, $G(V_\alpha) = N^\alpha_{\text{vib}} g(V_\alpha)$, where $g(V_\alpha)$ is the configurational density of states and N^α_{vib} is the number of vibrational states in a given basin (i.e., $S^\alpha_{\text{vib}} = -k \ln N^\alpha_{\text{vib}}$). A probability distribution function for the cluster formation energies, E^f, can then be defined by [37]

$$p(E^f) = G(E^f) \exp(-\beta E^f) = g(E^f) \exp(-\beta E^f) \exp\left[\frac{S_{\text{vib}}(E^f)}{k}\right], \tag{2.5}$$

where the formation energy of a defect, E^f, is defined as $E^f = E_d - E_p(N_d/N_p)$; d and p represent defect-containing and reference defect-free domains, respectively; and N is the number of atoms in a simulation cell.

It is possible to compute the probability distribution function directly for a cluster's formation energy using standard MD simulation. The probability distribution for the six-vacancy cluster shown in Fig. 2.7 was obtained by performing NVT-ensemble MD at 1600 K on a periodic simulation cell containing a six-vacancy cluster. The atomic coordinates were quenched periodically using a conjugate gradient method, and the formation energy of the configuration, which represents a local potential energy minimum in the $3N$-dimensional phase space, was calculated. The formation energy data were collected into bins of width 0.1 eV. Note that only states that correspond to connected clusters, as defined by the Stillinger criterion, were included in the construction of the probability distribution function.

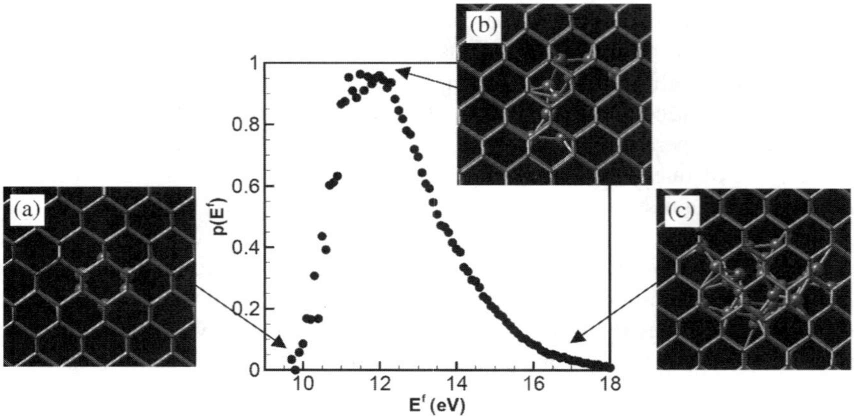

Figure 2.7 Probability distribution function for the formation energies of the six-vacancy cluster obtained by EDIP MD simulation at 1600 K. The insets are sample configurations; the large red spheres denote atoms that are displaced more than 10% of a bond length from their equilibrium positions. (From [38], with permission.) (*See insert for color representation of figure.*)

That is, the probability distribution function is a property of the cluster alone and does not include contributions from fragmented configurations (e.g., two three-interstitial fragments). A connectivity cutoff distance of 7.8 Å, corresponding to the eighth-nearest-neighbor shell distance, was employed based on the MD-derived vacancy–vacancy interaction distance.

The probability distribution of formation energies, $p(E^f)$, shows that a wide range of cluster formation energies and configurations are sampled during the MD simulation. Example structures at different formation energies are also shown in Fig. 2.7. The ground-state configuration identified by the EDIP potential [(a), $E^f = 9.8$ eV] is consistent with the predictions of electronic structure calculations and corresponds to the well-known hexagonal ring configuration [17], in which the size vacancies are arranged in a planar structure and the number of dangling bonds is minimized. On the other hand, note that the probability of observing this structure at 1600 K is almost zero, despite its very low formation energy. The most likely configurations [(b), $E^f \sim 11.8$ eV] are highly disordered and are substantially less favorable than the ground state, on an energetic basis. Note that these structures are not readily represented in a lattice-based framework because many atoms are severely displaced from their equilibrium positions. Finally, very extended structures are observed at the tail end of the distribution [(c), $E^f > 16$ eV].

The preceding observations can be explained by considering the entropic contributions implied by Eq. (2.5). Both configurational and vibrational entropy contributions increase as E^f increases. Extended, highly disordered configurations can assume numerous, almost degenerate relaxations, each of which corresponds to a unique local minimum in the potential-energy landscape. In other words, as the number of lattice atoms that are included into the defect structure increases, so does the number of small reconstructions that produce nearlydegenerate local minima. As a result, we find that the density of states for a given cluster increases exponentially with formation energy. Moreover, the vibrational entropy of each configuration also increases (approximately linearly) with formation energy [37,38]. This trend can be explained qualitatively by the fact that more extended (and disordered) structures introduce more vibrational modes into the lattice.

Both configurational and vibrational entropic contributions therefore increase the probability (i.e., lower the free energy) of higher-energy cluster configurations. The peak of the distribution is thus determined by how quickly entropic gains are balanced by energetic penalties as the formation energy increases. Clearly, as the temperature increases, the peak shifts to higher energy (and more disordered clusters). Only at low temperature (i.e., $T \leq 0.5T_m$) does the energetic ground state provide a comprehensive representation of the cluster thermodynamics and structure.

The type of broad distribution shown in Fig. 2.7 is not unique to vacancy clusters or defects in crystalline silicon. It is simply a projection of the potential-energy landscape, as defined by a particular interatomic potential, onto a formation energy axis. In the present case, the potential-energy landscape is dominated by the presence of the defect cluster. That is, the density of local minima in the energy interval shown in Fig. 2.7 is much lower when the defect is not present. Similar distributions are observed with different empirical potential functions, different types of defects (e.g., self-interstitial clusters), and different materials. These observations bring to light the idea that at least at high temperature, defects and clusters in crystals should be considered as distributions of states rather than single entities.

2.3.4 LKMC Simulation of Continuous Systems Using Direct Regression

The analysis presented in Section 2.3.3 provides an explanation for the discrepancy between the LKMC and MD predictions highlighted in Section 2.3.2. In summary, the ground-state "on-lattice" configurations used to parameterize the vacancy–vacancy bonding energies is simply not relevant at the high temperatures under consideration (i.e., 1600 K). Instead, the numerous disordered structures that dominate the distribution in Fig. 2.7 must be captured adequately in order to simulate the correct dynamics. But the solution to this problem is not straightforward. First, the relevant structures are complex off-lattice structures that cannot be mapped directly onto a lattice. Second, each cluster size possesses an almost infinite number of these configurations.

The approach we take here is to reinterpret the vacancy–vacancy bond energies introduced earlier as effective bonding free energies and the on-lattice structures in the LKMC model as coarse-grained representations of actual configurations. Although this reinterpretation is conceptually appealing, it does not suggest a direct approach for computing these bond free energies. Instead, we resort to a model-on-model regression approach, as suggested in Fig. 2.2b. The regression approach for determining the LKMC bonding parameters is as follows. A set of effective bond free energies $\{E_b^1, E_b^2, \ldots\}$ and the overall attempt frequency, ν_0, are chosen. An LKMC simulation, which is initialized in exactly the same manner as the reference MD simulation, is then executed with these parameter values. Once the LKMC run is completed, a normalized objective function is evaluated based on a least-squares comparison of the LKMC and MD size distribution components; that is,

$$\underset{\nu_0,\, E_b^j}{\mathrm{OF}} = \frac{1}{N}\sum_{i=1}^{N}\left[\frac{2\left(X_i^{\mathrm{KMC}} - X_i^{\mathrm{MD}}\right)}{X_i^{\mathrm{KMC}} + X_i^{\mathrm{MD}}}\right]^2, \tag{2.6}$$

where OF is the objective function, X_i^{KMC} and X_i^{MD} represent the comparison points of the LKMC and MD simulations, respectively, and N is the total number of comparison points used in the regression. Each LKMC simulation is executed three times and the results averaged at each objective function evaluation to further reduce the noise in the objective function.

The optimizer used in this work is based on a hybrid genetic algorithm (GA), described in detail in ref. [39], which allows for a global parametric search on a nonconvex surface. Various constraints can be imposed during the optimization process. Some of these constraints are predefined *hard constraints* on the allowable values of the bond energies and attempt frequency to physically reasonable values. Additional constraints can be implemented as *soft constraints* which add a penalty term to the objective function defined in Eq. (2.6). An example of such a constraint is the isolated vacancy monomer diffusivity, which is wellknown from MD simulation and which can be computed for the LKMC model directly from the bonding and attempt frequency parameters.

Each time the GA generates a new population of *individuals* (defined as a parameter set), the corresponding sets of parameters are passed to the LKMC simulation to obtain new objective function (fitness) values. The ability of the GA to sample local minima is enhanced by including periodic local minimization based on the simplex method. In this approach, the best $N + 1$ individuals are used to create an N-dimensional simplex every three or four generations. This simplex is evolved to the local minimum, and the new simplex vertices are then passed back to the GA to replace the worst individual in the current generation. The GA then proceeds to create the next generation using a sequence of mating and mutation operations.

A comparison of the MD data with a regressed LKMC model is shown in Fig. 2.8. A total of nine parameters were fitted in this example (eight bonding free energies and one attempt frequency). In this example, the concentration of monomers and dimers, as well as several moments of the size distribution, are included in the comparison, but no additional constraints are applied. The motivation for using high-order moments is to emphasize the large cluster contribution to the data given the relatively short evolution time of the MD data. An excellent representation of the MD dynamics is now demonstrated for both the individual component and size distribution moment metrics across the entire simulation time scale. Note that the LKMC simulations require only a few minutes to execute for a given set of input parameters, which strongly motivates the need for such coarse-grained models when the multiday cost of the MD simulation is considered.

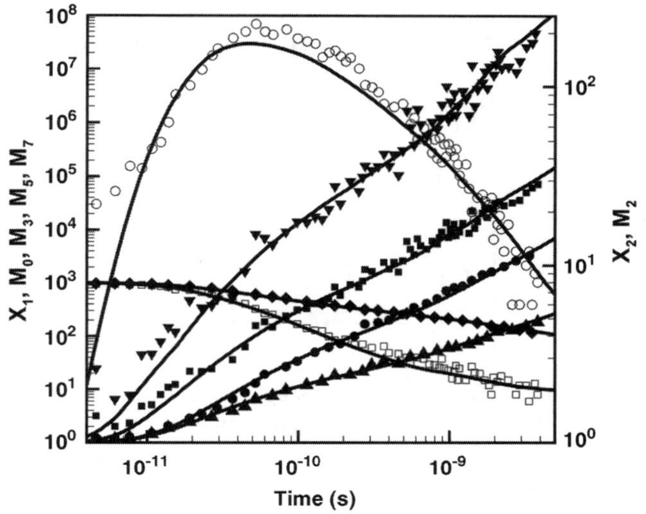

Figure 2.8 Comparison of MD and MD-regressed KMC model predictions for the vacancy cluster size evolution. Evolution profiles are symbols for MD, solid line for KMC; X_1, monomer (open squares); X_2, dimer (open circles); M_0, number of clusters (solid diamonds); M_2, average cluster size (solid circles); M_3, third-order moment (solid deltas); M_5, fifth-order moment (solid squares); M_7, seventh-order moment (solid gradients). (From [32], with permission.)

2.3.5 Uniqueness and Robustness of the Regressed LKMC Model

The robustness and uniqueness of LKMC model parameters generated by the direct regression approach must be investigated carefully. This is especially important in the case of the bonding energies, which now represent coarse-grained free energies that cannot be measured directly by any other means. The robustness of the regression was tested first by varying the assumed vacancy–vacancy interaction range. A sequence of LKMC models with different vacancy–vacancy interaction ranges were regressed to the MD data shown in Fig. 2.8. The interaction distance in these models ranged from the second-nearest-neighbor shell (2NN), up to the eighth-nearest-neighbor shell (8NN; shown in Fig. 2.8). Note that the number of regression parameters increases proportionately to the assumed vacancy–vacancy interaction distance. The overall fit quality as measured by the minimum objective function was generally found to increase somewhat with increasing interaction range, although almost all models led to good representations of the MD data, which raises some doubts regarding the robustness of the direct regression approach.

The best objective functions obtained for each LKMC model are shown in Table 2.1. Also shown in Table 2.1 for each run is the ratio of the LKMC single-vacancy diffusivity, D_{V1}^{KMC}, to the MD value, D_{V1}^{MD}. The single-vacancy diffusivity is given by $D_{V1} = \nu_0 \exp(-\Delta E_{\mathrm{hop}}/kT)$ and is proportional to the overall attempt frequency, but does not depend on any of the regressed bonding energies because $\Delta E_{\mathrm{hop}} = 0.3$ eV is externally fixed to using an independent MD simulation. Although the fit quality achieved by each of the LKMC models does not change appreciably, the

Table 2.1 Objective Function Values and Single Isolated Vacancy Diffusivity Ratios for LKMC Models with Various Interaction Ranges

Model	OF	$D_{V1}^{\mathrm{LKMC}}/D_{V1}^{\mathrm{MD}}$
2NN	0.0741	2.37
3NN	0.0257	2.27
4NN	0.0214	2.05
5NN	0.0239	1.81
6NN	0.0361	1.15
7NN	0.0157	1.25
8NN	0.0211	1.05

ratio of the LKMC and MD values for the single-vacancy diffusivity, $D_{V1}^{\mathrm{LKMC}}/D_{V1}^{\mathrm{MD}}$, decreases almost linearly from 2.4 to unity. In other words, the best objective function does not improve significantly as the interaction distance increases because the regression process increases the overall attempt frequency in the system artificially to compensate for the incorrect physics (i.e., the underestimated vacancy–vacancy interaction distance). It is notable that the correct diffusivity is reproduced almost exactly once the correct interaction distance (8NN) is specified. This information is not provided explicitly in any constraint, but rather, is present implicitly in the cluster size distribution evolution.

Additional tests of robustness are possible by considering other metrics. A comparison of the LKMC cluster diffusivities predicted and the MD values calculated for several different cluster sizes is shown in Fig. 2.9. The 8NN MD-regressed LKMC model is able to capture the correct scaling for the cluster diffusivity as a function of cluster size. On the other hand, the LKMC model described in Section 2.3.2, which was parameterized with lattice structures, severely underpredicts the mobility of clusters. The insets of Fig. 2.9 show vacancy cluster morphologies predicted by the two LKMC models in relation to the MD structure for the 35-vacancy cluster. While the lattice-cluster LKMC model predicts a compact, faceted structure (a), the MD-regressed LKMC model leads to a highly extended, loosely bound structure (b), which is much closer in morphology to the MD structure (c). These structures explain why the cluster diffusivities predicted by the MD-regressed LKMC model are much closer to the MD values than are those predicted by the lattice-cluster LKMC model. The loosely bound structures are able to reconfigure through a sequence of low-energy barrier processes, which lead collectively to enhanced center-of-mass motion. The structural comparisons can be quantified by defining an order parameter, $r_{\mathrm{sep}}^{n} = \sqrt{\sum_i \sum_{j>i} (r_i - r_j)^2}$, which is a measure of the total distance between vacancy pairs in each cluster. For structures (a), (b), and (c) shown in the insets of Fig. 2.9, r_{sep}^{35} =118, 152, and 167 Å, respectively, highlighting the significant expansion associated with both the MD and MD-regressed LKMC clusters relative to the lattice-cluster KMC structure.

The enhanced cluster diffusivity and morphological changes discussed above are attributable directly to the incorporation of an effective free energy into the LKMC

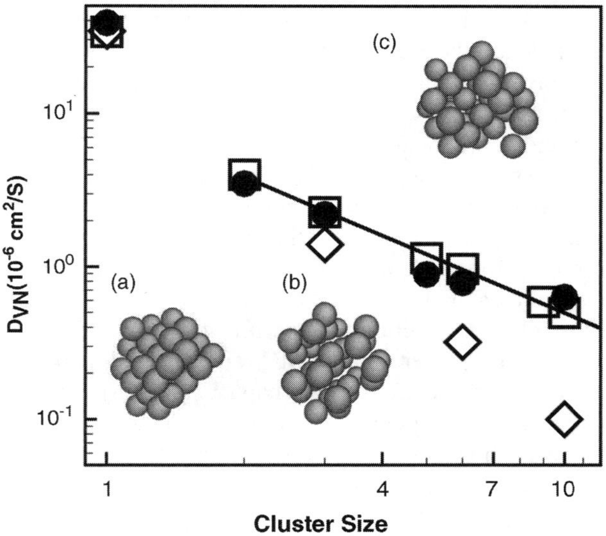

Figure 2.9 Comparison of cluster diffusivities predicted by MD-regressed LKMC model (solid circles), LKMC model parameterized with on-lattice structures (diamonds) and MD (squares). Insets: 35-vacancy cluster configurations: (a) lattice-cluster LKMC structure; (b) MD-regressed LKMC structure; and (c) MD structure.

bonding energies. The free energy includes enthalpic and entropic contributions, the latter consisting of both vibrational and configurational components. When combined with a mechanistically reasonable model, the regression process allows for the implicit capture of what is strictly a continuous space phenomenon. In other words, although the actual configurations mapped in Fig. 2.7 cannot be represented explicitly in a LKMC simulation, the collective entropic contribution of these structures to the overall cluster properties can be captured in an effective manner using the model-to-model regression approach.

We conclude this section with a final example that demonstrates the uniqueness and robustness of the LKMC modeling described here. Here we perform additional regressions to the MD evolution data in Fig. 2.8 in which the LKMC bond-screening physics introduced in Section 2.3.1 were omitted from the rate calculation for each lattice hopping event. Significantly, with this model it was not possible to obtain a qualitatively good fit to the MD data for any values of the effective bond energies. This result demonstrates that the regressions are not overspecified (i.e., even with nine fitting parameters, it is still not possible to obtain good agreement unless the model physics are fundamentally correct). Moreover, this test shows clearly how the model-to-model regression approach can be applied to develop better mechanistic accuracy in coarse-grained models. In the absence of interaction screening, the model predicts cluster binding energies that increase too rapidly with cluster size (i.e., linearly), which inhibits evolution at later times. This problem cannot be "corrected" by finding new bond energies; it is a fundamental mechanistic flaw. As noted in Section 2.3, the particular screening model used here may or may not be unique, but it is sufficient to explain quantitatively all the MD data presented in this chapter.

Some limitations of the model-to-model regression approach should be mentioned. First and most important, the approach lacks transferability because the (configurational and vibrational) entropic contribution included in the effective bonding free energies is a strong function of temperature. Unless this functionality can be modeled analytically, multiple regressions at different temperatures are required to develop an LKMC simulation that can be applied to variable-temperature situations. Second, although the LKMC model generated here is much more computationally efficient than direct MD simulation, it is still not sufficiently coarse-grained to allow connection to experimentally realistic situations, particularly at high temperature, where the LKMC rates are high and the time steps are relatively small. In the final section of this chapter, we describe briefly recent efforts aimed at developing a computational framework for coarse-graining LKMC simulations to bridge the gap between atomistic phenomena and continuum scale processing.

2.4 SPATIAL COARSE GRAINING OF LKMC SIMULATIONS

The basic idea behind coarse graining is neither new nor unique to the applications of interest in this chapter. Examples can be found for almost every type of simulation technique and for a broad range of materials systems. The principal aim of spatially coarse-grained LKMC (CGLKMC) simulations, in particular, is to group multiple atomic lattice sites together into coarse-grained cells, which then constitute the basis for a new coarse lattice. The coarse-grained system evolves temporally by a sequence of coarse-grained events that are related to some average property of the microscopic events in the atomic LKMC system [40]. Such coarse graining can greatly increase the efficiency of an LKMC simulation in two ways. First, each event on the coarse lattice represents a larger displacement in the phase space of the simulation and therefore a longer time step. Second, and perhaps more important, the coarse-graining operation reduces the dimensionality of the simulation phase space, leading to fewer events that must be considered at a given time.

In this section we discuss recent developments toward the systematic coarse graining of LKMC simulations in which strongly interacting (and therefore aggregating) particles are present [41]. The discussion is developed using simple cubic grids in two dimensions, but the majority of the concepts are readily extended to more complex lattice symmetries in three dimensions. Throughout, the term *microscopic* is used to refer to processes taking place on the original lattice (which is comprised of single-atom sites), and the terms *coarse* or *mesoscopic* refer to processes occurring on the coarse-grained lattice. In the following discussion, each coarse cell contains q fine-grid (atomic) sites and is completely characterized by an occupancy number, $0 \leq \eta \leq q$ (Fig. 2.10). A key step in formulating a CGLKMC simulation is the closure rule, which dictates how the single-atom events are averaged within a coarse-grained cell to produce a set of coarse events.

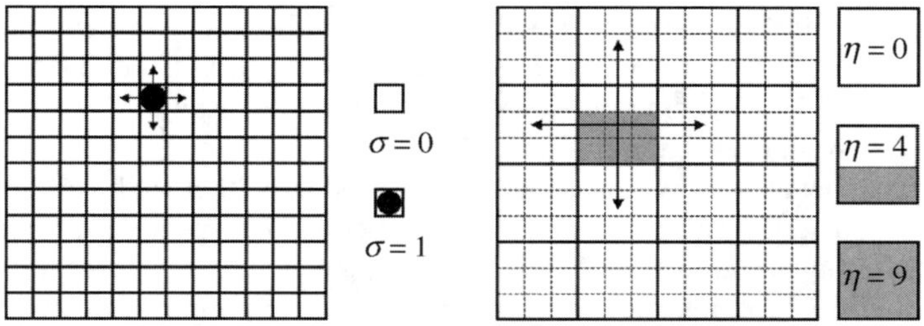

(a) (b)

Figure 2.10 Schematic representations of (a) fine-grid ("microscopic") lattice, and (b) corresponding coarse-grained ("mesoscopic") lattice ($q = 9$). (From [41], with permission.)

The major assumptions required for constructing a CGLKMC approximation have been addressed extensively by Vlachos and co-workers, and are addressed here only briefly; see refs. [40,42] for further details. The most important of these is the fact that the microscopic processes within a coarse cell are fast compared to the overall system dynamics (i.e., that local quasi-equilibrium is maintained within a coarse cell at all times). This quasi-equilibrium is a function only of the local mesoscopic system properties, such as local mass, temperature, and potential energy, and does not depend explicitly on time. The degrees of freedom corresponding to these fast microscopic processes are therefore "averaged out" of the overall system dynamics. In essence, a time-scale separation is achieved, similar in nature to a pseudo-steady-state approximation made in continuous systems.

2.4.1 General Expressions for Hopping Rates on the Coarse-Grained Lattice

The essence of the coarse-graining procedure is to compute the overall jump rate for a single particle from a cell k to an adjacent destination cell l_d on the coarse lattice, subject to the quasi-equilibrium assumption discussed above. The total rate for the coarse event is given by [41]

$$\bar{r}(k \rightarrow l_d) = \sum_{x \in D_k} \frac{1}{q} \left[\sum_{y \in D_l} (1-\sigma_y)\sigma_x r_x \right] = \sum_{x \in D_k} \langle (1-\sigma_y)\sigma_x r_x \rangle_y = q \langle (1-\sigma_y)\sigma_x r_x \rangle_{xy}, \tag{2.7}$$

where σ_x is the occupancy state (0 or 1) at site x within the fine-grid lattice, $\langle \cdot \rangle_y$ represents an average over the y-sites in cell l_d, and $\langle \cdot \rangle_{xy} \equiv \langle \cdot \rangle$ is an average over all possible x and y sites on their respective coarse cells. The microscopic rates, r_x, which represent hopping events on the fine grid, are specified by the physics of the particular problem being considered [see Eq. (2.1)]. In the current discussion, a simple Arrhenius model is applied, so that

$$r_x = \nu_0 \exp\left(-\frac{E_x}{k_B T} \right), \tag{2.8}$$

where E_x is an energy barrier that depends only on the initial configuration of the system (i.e., before the hop). In the present discussion it is assumed that the hopping barrier for particle i, residing on site x, is simply the total binding energy for that particle in its initial configuration (i.e., $E_x = E_i^b$).

Assuming that all sites within a coarse cell are equally likely to be occupied, $\langle \sigma_x \rangle$ and $\langle \sigma_y \rangle$ in Eq. (2.7) can be replaced by η_k/q and η_{ld}/q, respectively, where η_k and η_{ld} are the occupancy numbers of cells k and l_d, respectively. With this simplification, and inserting Eq. (2.8) into Eq. (2.7), the average coarse hopping rate from cell k to cell l_d is given by

$$\bar{r}(k \rightarrow l_d) = \frac{\eta_k}{q}(q - \eta_{ld})v_c\left\langle \exp\left(-\frac{E_x}{k_B T}\right)\right\rangle, \tag{2.9}$$

where the average is evaluated over particles on coarse cell k, and the coarse attempt frequency is given by $v_c = v_0/q$ because the hop length on the coarse lattice is equal to the distance between the centers of the coarse cells.

The method used to compute the average of the microscopic rates in Eq. (2.9) defines the coarse-graining approach. One common approach is to assume that the interparticle interactions are weak, and therefore that

$$\left\langle \exp(-\tilde{E}_i^b)\right\rangle \approx \left\langle 1 - \tilde{E}_i^b \right\rangle = 1 - \left\langle \tilde{E}_i^b \right\rangle \approx \exp\left(-\left\langle \tilde{E}_i^b \right\rangle\right), \tag{2.10}$$

where $\tilde{E}_i^b = E_i^b/k_B T$ is the dimensionless binding energy for particle i and $\left\langle \tilde{E}_i^b \right\rangle$ is the average over the coarse cell of the dimensionless binding energy for one particle. The weak interaction assumption can be coupled with another important assumption, the local mean field (LMF) assumption [40], in which it is assumed that particles are distributed homogeneously within a coarse cell. Most significantly, the combination of these two assumptions allows for the analytic evaluation of the coarse rates in Eq. (2.9). However, as demonstrated in this section, these assumptions are rather limiting in the case of aggregating particles, which are generally strongly interacting.

2.4.2 Binding Energy Decomposition for Efficient Coarse-Grained LKMC Simulation

The total binding energy that any single particle experiences generally extends over several microscopic sites and possibly multiple coarse cells. Assuming that particle interactions are pairwise additive, calculation of the rate average in Eq. (2.9) can be greatly simplified by decomposing the total particle binding energy into contributions from the same cell particles (intracell interactions) and particles on different cells (intercell interaction):

$$E_i^b = E_i^{\text{intra}} + \sum_{j=1}^{N} E_i^{\text{inter},l_j}, \tag{2.11}$$

where the index l_j refers to coarse cells at the jth neighbor shell of reference coarse cell k. The total number of neighbor shells, N, that need to be considered depends on the range of particle interactions as well as the degree of coarse graining.

Although the decomposition in Eq. (2.11) is conceptually useful, both the intra- and intercell interaction contributions are still governed by microscopic particle distributions that are, unfortunately, coupled across the entire interaction distance. This coupling can be removed through additional assumptions. For example, while the intracell particle distribution is still, in principle, a function of particle distributions in all neighboring coarse cells, particles in the same cell are (on average) much closer to each other than to those in neighboring cells. As a result, the intracell binding energies between particles in reference cell k can be assumed to be only weakly affected by the distributions in the surrounding cells, and the cell can be considered to be isolated so that $E_i^{\text{intra}} \sim E_i^{\text{intra}}(k)$. Similar approximations can be made for the intercell contributions at each neighbor shell; that is, the binding energy between an average particle (i) on cell k and particles on cell $l_{j\alpha}$ is, to first order, independent of the particle distributions on other cells. That is,

$$E_i^{\text{inter},l_j} = \sum_{\alpha} E_i^{\text{inter},l_j}(k, l_{j\alpha}), \tag{2.12}$$

where the index j represents a particular neighbor shell and α represents a particular coarse cell in neighbor shell j. The validity of the decoupling assumptions described here is addressed later in this section.

2.4.3 Numerical Computation of the Coarse Lattice Interaction Potential: Intracell

The average coarse binding energies in Eqs. (2.11) and (2.12), which represent a coarse lattice potential, must be calculated numerically for all but the simplest cases. We apply the Wang–Landau Monte Carlo (WLMC) [43] method to build rate averages. Although other sampling methods, such as metropolis Monte Carlo, can be used, the WLMC method is very efficient for strongly interacting systems because it is not hindered by moves that require large energy changes.

The procedure used to perform the averaging is similar for both the intra- and intercell cases, and here we illustrate the basic steps. Assuming, as in Section 2.4.2, that the average intracell binding energy can be computed on the basis of an isolated coarse cell, η particles are initialized in a single coarse cell with a coarse-graining level q, as shown in Fig. 2.11a. A total system energy, E_s, is defined for a given microscopic configuration of N_p particles such that

$$E_s = E_0 - 0.5 \sum_{i=1}^{N_p} E_i^b \equiv E_0 - E_{\text{tot}}^b, \tag{2.13}$$

where E_0 is an arbitrary reference energy and E_{tot}^b therefore is the total binding energy within the coarse cell at a given microscopic configuration. Two functions, the configurational density of states, $g(E_s)$, and the visit histogram, $h(E_s)$, are initialized

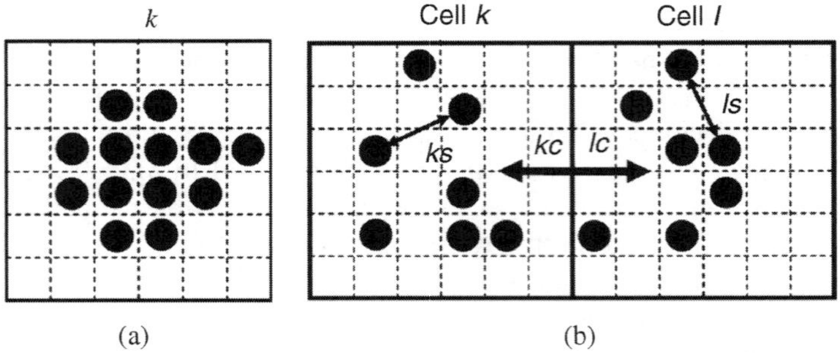

Figure 2.11 (a) Intracell WLMC domain; (b) intercell WLMC domain for first-nearest-neighbor situation. The labels "s" and "c" represent same-cell and cross-cell interactions, respectively. Coarse-graining level is $q = 36$ in both cases. (From [41], with permission.)

to unity and zero, respectively. Both the $g(E_s)$ and $h(E_s)$ distributions, which are continuous variables, are discretized into 0.01-eV energy bins.

Wang–Landau Monte Carlo moves are performed by moving a randomly selected particle to a vacant location picked at random from all sites within the coarse cell. The WLMC acceptance/rejection criterion for accepting a move from system energy level E_s^1 to E_s^2 is given by

$$p(E_s^1 \rightarrow E_s^2) = \min\left[\frac{g(E_s^1)}{g(E_s^2)}, 1\right]. \tag{2.14}$$

Within the WLMC algorithm, each time an energy level E_s is visited by the simulation, $g(E_s)$ is multiplied by a factor $f > 1$ so that $g_{\text{new}}(E_s) = g_{\text{old}}(E_s) \times f$, where the initial value of f is $\exp(1)$. Concurrently, the visit histogram is updated by $h_{\text{new}}(E_s) = h_{\text{old}}(E_s) + 1$. The simulation proceeds until a minimum flatness (deviation from the mean) criterion is achieved in the function $h(E_s)$, which is taken to be 85%. Once the flatness criterion is achieved, the value of f is reduced according to the rule $f_{i+1} = \sqrt{f_i}$, and $h(E_s)$ is reset to zero for the next stage. The criterion used to end the simulations is chosen to be $f = 1.000001$.

The procedure above leads to a converged value for the density of states, from which the probability distribution function at any temperature can be computed [i.e., $p(E_s) = g(E_s)\exp(-E_s/k_BT)$]. The average of any property, A, can be computed using this distribution by

$$\langle A \rangle = \int p(E_s) A(E_s)\, dE_s, \tag{2.15}$$

where $A(E_s)$ is the value of the property for a configuration with system energy level E_s. The average rate for a hop in the coarse cell is then given by

$$\left\langle \exp\left(-\frac{E_1^b}{k_BT}\right)\right\rangle = \left\langle \left\langle \exp\left(-\frac{E_1^b}{k_BT}\right)\right\rangle\Big|_{E_s}\right\rangle = \int p(E_s) \left\langle \exp\left(-\frac{E_1^b}{k_BT}\right)\right\rangle\Big|_{E_s} dE_s, \tag{2.16}$$

where the inner average is taken over all particles at a given configuration with system energy E_s. The most convenient inputs to a CGLKMC simulation are averaged binding energies, not transition rates, and an effective average binding energy for a single particle is defined by

$$\bar{E}_1^b = -k_B T \ln\left(\left\langle \exp\left(-\frac{E_1^b}{k_B T}\right)\right\rangle\right). \tag{2.17}$$

The averaging scheme above, which we term the *average transition rate* (ATR) *framework*, can be performed directly on the binding energies, which is appropriate only in the weak binding limit, as discussed in Section 2.4.2. The direct averaging of binding energies is referred to as the *average binding energy* (ABE) *framework*. Note that the ATR is simply an exponentially weighted version of the ABE.

Examples of the intracell average binding energies obtained using the procedure above are shown in Fig. 2.12 as a function of the cell occupancy fraction. A constant interaction potential between the particles was employed up to the sixth-nearest-neighbor (6NN) shell with the interaction strength given by $\beta J \equiv J/k_B T = 1.5$. A total of four curves are shown in Fig. 2.12. The first two (upper curves, labeled ABE) assume that the interaction is weak and perform the averaging in Eq. (2.16) directly on the binding energies, while the second two (lower curves, denoted ATR) are based on the use of Eqs. (2.16) and (2.17) without any assumptions. In both cases the curves denoted by symbols represent averages computed based on trajectories subject to the interaction potential, while curves without symbols represent averages computed on

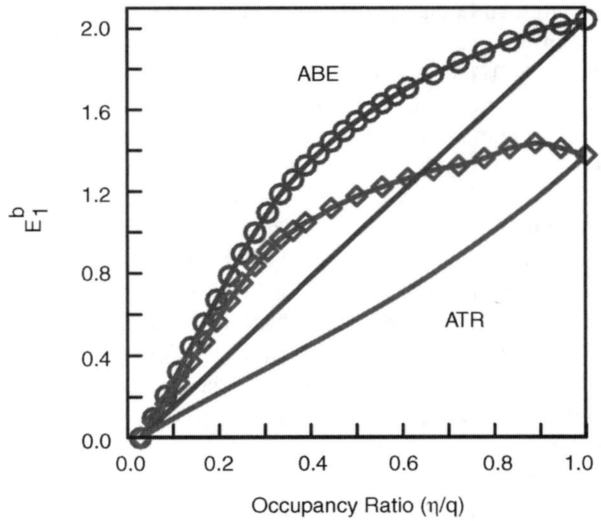

Figure 2.12 Intracell average binding energy as a function of cell occupancy ratio. Solid lines, LMF approximation; symbols and lines, interacting particles model. Upper curves, direct averaging of binding energies (ABE); lower curves, averaging of transition rates (ATR). System information: $q = 36$, $\beta J = 1.5$, $L = 6$NN. (From [41], with permission.)

the basis of interactionless particle trajectories. In other words, the average binding energies represented by the line-only curves assume that the particles interact only by a site-exclusion interaction; this is simply the local mean field (LMF) assumption. How the particle trajectories are connected to the particle interactions is referred to as the *closure rule*.

In both the ATR and ABE cases, the average binding energy predicted by trajectories that are based on the particle interaction potential is always higher than the corresponding LMF value. This feature is a consequence of the fact that interacting particles tend to cluster together, which leads to a higher effective binding energy per particle for all occupancy ratios. In the LMF case, the particles are homogeneously distributed throughout the coarse cell at all times. Of course, both models converge to the same value at $\eta = 0$ and $\eta = q$, the latter case being a result of geometric confinement. The deviation between the LMF and interacting particle cases can be interpreted as a measure of system nonideality, which of course arises from the interactions between particles. It is also important to note that the ABE approximation leads to significant deviations from the ATR case (whether or not the LMF approximation is applied). The effective binding energies obtained in the ATR framework are lower than those obtained using the ABE framework at each occupancy level. This is because strongly bound particles in the ATR scheme contribute less to the overall average, due to the exponential weighting.

2.4.4 Numerical Computation of the Coarse Lattice Interaction Potential: Intercell

A similar procedure can be used to develop averaged intercell interactions. In this case, two interacting coarse cells are employed, as shown in Fig. 2.11b for the case of first-nearest-neighbor (1NN) coarse cells. There are two additional complications in the case of intercell interactions. First, it is now necessary to generate a two-dimensional binding energy surface that is a function of the occupancy fraction in each of the two cells. Second, the interactions that govern the particle trajectories during the WLMC averaging simulations must be specified carefully according to the assumptions highlighted in Section 2.4.2. The various possibilities are illustrated in Fig. 2.11b, which shows schematically how particles in two adjacent coarse cells, k and l, can interact with each other during the WLMC sampling. In the LMF closure rule, no interactions are considered in the WLMC trajectories, and the particles in both cells are distributed homogeneously. Note that in all cases, particles are not allowed to move between cells (i.e., η_k and η_l are fixed during a given WLMC averaging simulation). At the other extreme is the fully coupled closure rule, where particles in both cells interact with each other as well as with particles in the other cell. This closure rule is denoted by the notation *kb–lb*, where *b* stands for "both" self (*s*) and cross (*c*) interactions, and *k* and *l* denote the cell indices. In the same notation scheme, the LMF closure rule is denoted by *kr–lr*, where *r* denotes "random," or interactionless trajectories within a cell.

Intermediate closure rules are also mathematically possible. For example, the rule denoted by *ks–ls* describes a system in which the particles in each cell interact with

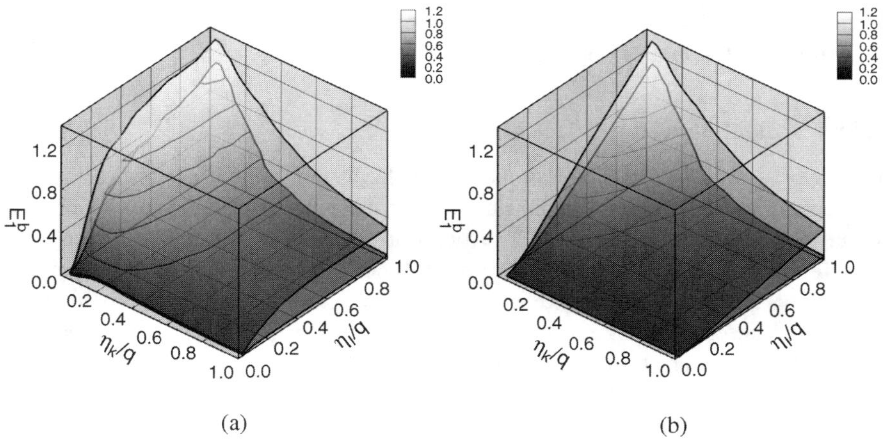

Figure 2.13 Intercell binding energy (due to first-nearest-neighbor coarse cells) for the (a) *kb–lb*, and (b) *kb–lr* closure rules, respectively. In both cases, the upper surface represents averaging within the ABE framework, while the lower surface represents the ATR framework. System characteristics: $q = 36$, $\beta J = 1.5$, $L = 6$NN. (From [41], with permission.)

each other but not with particles in the other cell, while the *kc–lc* rule represents a situation in which particles interact only with particles that reside on the other cell. Clearly, some of the closure rule possibilities are difficult to justify physically, and only a subset of the mathematically possible rules are considered. Shown in Fig. 2.13 are the average binding energies computed using the *kb–lb* and *kb–lr* closure rules. In the latter case, particles in the k cell interact with all particles, while particles in the l cell are assumed to be distributed homogeneously. Physical justification for the *kb–lr* closure rule comes from assuming that secondary coupling interactions arising from other cells surrounding cell l act as strongly on the particles as do those in cell k, leading to a roughly homogeneous distribution therein. In general, the presence of such secondary coupling implies that the *kb–lb* intercell interactions are likely to be overestimated. This issue is addressed further in the ensuing discussion. Binding energy surfaces are shown for both ATR and ABE frameworks. Note that as in the intracell case, the ATR averaging leads to lower binding energies than those predicted by ABE.

Overall, the two closure rules lead to qualitatively similar binding energy surfaces. The largest intercell binding energy between the average particle in cell k and those in cell l occurs when cell k is mostly empty and cell l is mostly populated. This is intuitively reasonable because under these conditions, the particles in cell k can interact with the maximum number of particles in cell l. Finally, intercell binding energies for other interaction shells in the coarse lattice can be computed in the same way as for the 1NN case discussed above. Subject to the nature of the interaction potential being used, the intercell binding energies generally decrease quickly with each interaction shell, and for the 6NN interaction potential used in the examples above, only the 2NN coarse cell interactions need to be considered. The reduced interaction distance on the coarse lattice provides an important contribution to the overall computational savings achieved by coarse graining.

2.4.5 Validation of the CGLKMC

In this final section, the predictions of CGLKMC are compared with a fine-grid reference simulation for both closure rules presented in Section 2.4.4. The test simulation setup is as follows: 900 particles are initially distributed uniformly on a two-dimensional square lattice containing 129,600 sites, which are coarse-grained at a level of $q = 36$ (36 fine-grid sites per coarse-grained cell). A 6NN fine-grid interaction potential with constant interaction strength, $\beta J = 1.5$, is applied. This potential is sufficient to lead to strong aggregation of the particles and provides a stringent test of the CGLKMC framework presented in this section. During each simulation, the cluster size distribution is monitored periodically. A *cluster* is defined as a group of particles that are connected at the second-nearest-neighbor distance on the coarse-grid lattice (i.e., the range of the potential). Two moments of the cluster size distribution are used to monitor system evolution. These are the average cluster size, M_2/M_1, and the total cluster number, M_0, where all moments of the size distribution are defined as $M_n = \sum_s s^n X_s$, where X_s is the number of clusters of size s and n is the moment order. Configurations generated by the fine-grid simulations are mapped onto the coarse lattice to enable a quantitatively consistent comparison between the two simulations.

First, it is found that no closure rule is able to reproduce the fine-grid evolution adequately when the ABE averaging framework is applied. This result is not unexpected given that the ABE framework is valid only in the limit of weak interactions. In the following discussion, therefore, only average binding energies computed with the ATR framework are considered. Shown in Fig. 2.14 are the size distributions generated by the fine-grid reference simulation and the CGLKMC predictions for both the *kb–lb* and *kb–lr* closure rules. The general features of the aggregation process include a rapid nucleation and growth phase driven primarily by monomer diffusion, followed by a slower ripening phase.

Both CGLKMC simulations are in excellent agreement with the fine-grid results at short times, but the *kb–lb* model becomes kinetically arrested after $t \sim 0.01$ s, only to accelerate about two time decades later. This behavior is probably due to over-prediction of the coupling between cells k and l. In other words, by assuming that all other cells are empty, the interactions between particles in any two occupied cells are likely to increase. By contrast, excellent results are generated with the *kb–lr* model. As discussed previously, the *kb–lr* closure rule attempts to capture particle coupling between cells k and l, while assuming that the cell l population is uniformly (or randomly) distributed because of (equal) interactions with other surrounding cells. This is only an approximation, but it appears to be a reasonable one because, on average, all cells are equally likely to be populated. All other possible closure rule combinations lead to substantially weaker agreement with the fine-grid results than that obtained with *kb–lr*. The only exception is the *kc–lr* rule, which provides almost identical results because particles in the k cell are drawn to each other by the

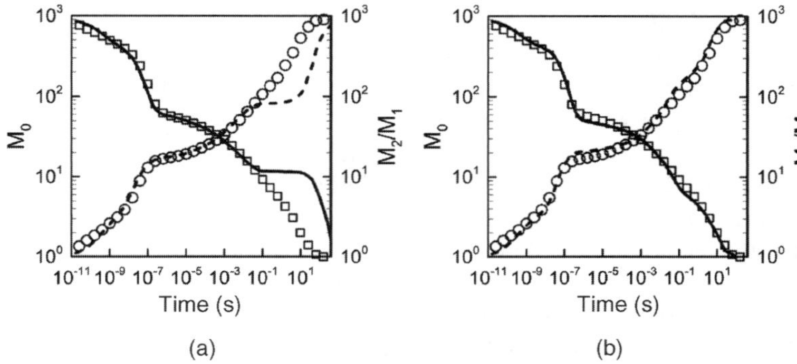

Figure 2.14 Cluster size distribution comparison between fine-grid LKMC and CGLKMC with (a) *kb–lb*, and (b) *kb–lr* closure rules ($q = 36$). Symbols, fine-grid LKMC: circles, average cluster size (M_2/M_1); squares, total number of clusters (M_0). Lines, CGLKMC ($q = 36$): dashed line, average cluster size (M_2/M_1); solid line, total number of clusters (M_0). (From [41], with permission.)

interaction with particles in the *l* cell even if they are not interacting with each other. Further examples demonstrating the robustness of the coarse-graining approach presented in this section are provided in ref. [41].

2.5 CONCLUSIONS AND FUTURE DIRECTIONS

In this chapter we presented a sequential multiscale modeling effort aimed at generating physically realistic lattice kinetic Monte Carlo (LKMC) simulations of atomic aggregation in crystalline systems. Using the well-understood example of vacancy aggregation in crystalline silicon, it was demonstrated that quantitatively capturing even relatively simple phenomena with coarse-grained models such as LKMC is not a trivial endeavor. Notably, the entropic effects that govern vacancy cluster thermodynamics and transport in silicon at very high temperature cannot be predicted on the basis of a ground-state analysis. Consequently, any coarse-grained models built and parameterized with this information cannot be expected to be predictive.

The LKMC simulation framework applied in this work retains full atomic resolution while coarse graining out the vibrational dynamics of lattice atoms, which are the principal limitation in molecular dynamics simulations. Moreover, the LKMC simulation approach removes the need to consider lattice atoms explicitly and, instead, focuses the computational effort on the species of interest (i.e., vacancies). These large reductions in problem dimensionality require that the LKCM physics be specified by a lattice potential model containing several parameters, only some of which can be computed directly with independent molecular simulations. A key element of the strategy presented in this chapter, therefore, was to employ large-scale molecular dynamics simulations to generate a high-resolution database with which coarse-grained models can be rigorously parameterized and validated.

The time- and length-scale limitations of the molecular dynamics method restrict the application of the sequential approach presented here to physical phenomena that can be captured at the nanoscale. Careful choice of simulation conditions, such as high temperature and concentrations, however, can greatly extend the effective time scale of such simulations. Moreover, numerous efforts aimed at extending the scope of brute-force molecular dynamics offer promising avenues for extending the length- and time-scale horizons of the sequential approach described in this chapter. One such example is the feature-activated molecular dynamics (FAMD) approach presented in Section 2.2, which efficiently focuses the computational expense of MD integration onto the localized domains containing the species of interest. Combinations of such approaches with efforts to extend the accessible time scale of MD will lead to powerful atomistic simulation tools.

In the final section of this chapter, recent efforts aimed at further coarse-graining of the LKMC model physics were presented. An atomic-scale lattice kinetic Monte Carlo simulation, while vastly more efficient than standard molecular dynamics, is still limited to microscale phenomena and therefore cannot be applied directly to the process scale. Whereas the coarse-graining efforts are currently limited to simple lattice geometries and interaction models, future work will be focused on extending the general technique to address more complex situations. An adaptively coarse-grained LKMC simulation, which can reduce its resolution as the microstructure in a system evolves and coarsens, offers an extremely powerful method to systematically link atomic-scale information with process scale constraints.

REFERENCES

1. Sinno T, Dornberger E, Brown RA, von Ammon W, Dupret F. Modeling and simulation for defect engineering of CZ-grown silicon crystals. *Mater. Sci. Eng. Rep.* **R28**:149, 2000.
2. Huang H, Gilmer GH, de la Rubia TD. An atomistic simulator for thin film deposition in three dimensions. *J. Appl. Phys.* **84**:3636, 1998.
3. Chen LQ. Phase-field models for microstructure evolution. *Annu. Rev. Mater. Res.* **32**:113, 2002.
4. Nakai K, Ikari A, Huber A, Kaeckl W, von Ammon W,Void formation mechanism in floating zone silicon crystal. Presented at the 4th International Symposium on the Advanced Science Technology of Silicon Materials, JSPS, Hawaii, FL, Nov. 2004.
5. Vanhellemont J, Dornberger E, Esfandyari J, Kissinger G, Trauwaert MA, Bender H, Graf D, Lambert U, von Ammon W. Defects in as-grown silicon and their evolution during heat treatments. *Mater. Sci. Forum* **19**,1997.
6. Stillinger FH, Weber TA. Computer simulation of local order in condensed phases of silicon. *Phys. Rev. B* **31**:5262, 1985.
7. Tersoff J. Empirical interatomic potential for silicon with improved elastic properties. *Phys. Rev. B* **38**:9902, 1988.
8. Bazant MZ, Kaxiras E, Justo JF. Environment-dependent interatomic potential for bulk silicon. *Phys. Rev. B* **56**:8542, 1997.

9. Tadmor EB, Ortiz M, Phillips R. Quasicontinuum analysis of defects in solids. *Philos. Mag. A* **73**:1529, 1996.
10. Broughton JQ, Abraham FF, Bernstein N, Kaxiras E. Concurrent coupling of length scales: methodology and application. *Phys. Rev. B* **60**:2391, 1999.
11. Rudd RE, Broughton JQ. Concurrent coupling of length scales in solid-state systems. *Phys. Status Solidi B* **217**:251, 2000.
12. Wagner GJ, Liu WK. Coupling of atomistic and continuum simulations using a bridging scale decomposition. *J. Comput. Phys.* **190**:249, 2003.
13. Jacobsen J, Cooper BH, Sethna JP. Simulations of energetic beam depositions: from picoseconds to seconds. *Phys. Rev. B* **58**:15847, 1998.
14. Frewen TA, Kapur SS, Haeckl W, von Ammon W, Sinno T. A microscopically accurate continuum model for void formation during semiconductor silicon processing. *J. Cryst. Growth* **279**:258, 2005.
15. Prasad M, Sinno T. An internally consistent approach for modeling solid-state aggregation: II Mean-field representations of atomistic processes. *Phys. Rev. B* **68**:45207, 2003.
16. Vashishta P, Kalia RK, Nakano A. Multimillion atom molecular dynamics simulations of nanostructures on parallel computers. *J. Nanoparticle Res.* **5**:119, 2003.
17. Prasad M, Sinno T. An internally consistent approach for modeling solid-state aggregation: I Atomistic calculations of vacancy clustering in silicon. *Phys. Rev. B* **68**:45206, 2003.
18. Stillinger FH. Rigorous basis of Frenkel-band theory of association equilibrium. *J. Chem. Phys.* **38**:1486, 1963.
19. Prasad M, Sinno T. Feature-activated molecular dynamics: an efficient approach for atomistic simulation of solid-state aggregation phenomena. *J. Chem. Phys.* **121**:8699, 2004.
20. Prasad M, Sinno T. Feature activated molecular dynamics: parallelization and application to systems with globally varying mechanical fields. *J. Comput.-Aided Mol. Des.* **12**:17, 2005.
21. Chu JW, Voth GA. Allostery of actin filaments: Molecular dynamics simulations and coarse-grained analysis. *Proc. Natl. Acad. Sci. USA* **102**:13111, 2005.
22. Voter AF, Montalenti F, Germann TC. Extending the time scale in atomistic simulation of materials. *Annu. Rev. Mater. Res.* **32**:321, 2002.
23. Voter AF. Classically exact overlayer dynamics: Diffusion of rhodium clusters on Rh(100). *Phys. Rev. B* **34**:6819, 1986.
24. Fichthorn KA, Weinberg WH. Theoretical foundations of dynamic Monte-Carlo simulations. *J. Chem. Phys.* **95**:1090, 1991.
25. Dai J, Kanter J, Kapur SS, Seider WD, Sinno T. On-lattice kinetic Monte Carlo simulations of point defect aggregation in entropically influenced crystalline systems. *Phys. Rev. B* **72**:134102, 2005.
26. Law ME, Gilmer GH, Jaraiz M. Simulation of defects and diffusion phenomena in silicon. *MRS Bull.* **25**:45, 2000.
27. Henkelman G, Jonsson H. Long time scale kinetic Monte Carlo simulations without lattice approximation and predefined event table. *J. Chem. Phys.* **115**:9657, 2001.
28. El-Mellouhi F, Mousseau N, Lewis LJ. Kinetic activation-relaxation technique: an off-lattice self-learning kinetic Monte Carlo algorithm. *Phys. Rev. B* **78**:153202, 2008.

29. Zwanzig R. Nonequilibrium Statistical Mechanics, Oxford University Press, Oxford, UK, 2001.
30. Fichthorn KA, Scheffler M. Island nucleation in thin-film epitaxy: a first-principles investigation. *Phys. Rev. Lett.* **84**:5371, 2000.
31. La Magna A, Coffa S, Colombo L. Role of extended vacancy–vacancy interaction on the ripening of voids in silicon. *Phys. Rev. Lett.* **82**:1720, 1999.
32. Dai J, Seider WD, Sinno T. Lattice kinetic Monte Carlo simulations of defect evolution in crystals at elevated temperature. *Mol. Simul.* **32**:305, 2006.
33. Rauls E, Frauenheim Th. Entropy of point defects calculated within periodic boundary conditions. *Phys. Rev. B* **69**:155213, 2004.
34. Mishin Y, Sorensen MR, Voter AF. Calculation of point defect entropy in metals. *Philos. Mag. A* **81**:2591, 2001.
35. Stillinger FH. A topographic view of supercooled liquids and glass formation. *Science* **267**:1935, 1995.
36. Sciortino F, Kob W, Tartalia P. Inherent structure entropy of supercooled liquids. *Phys. Rev. Lett.* **83**:3214, 1999.
37. Kapur SS, Prasad M, Crocker JC, Sinno T. Role of configurational entropy in the thermodynamics of clusters of point defects in crystalline solids. *Phys. Rev. B* **72**: 014119, 2005.
38. Sinno T. A bottom-up multiscale view of point defect aggregation in silicon. *J. Cryst. Growth* **303**:5, 2007.
39. Frewen TA, Sinno T, Haeckl W, von Ammon W. A systems-based approach for generating quantitative models of microstructural evolution in silicon materials processing. *Comput. Chem. Eng.* **29**:713, 2005.
40. Katsoulakis MA, Majda AJ, Vlachos DG. Coarse-grained stochastic processes and Monte Carlo simulations in lattice system. *J. Comput. Phys.* **186**:250, 2003.
41. Dai J, Seider WD, Sinno T. Coarse-grained lattice kinetic Monte Carlo simulations of systems with strongly interacting particles. *J. Chem. Phys.* **128**:194705, 2008.
42. Chatterjee A, Vlachos DG. Multiscale spatial Monte Carlo simulations: multigriding, computational singular perturbation, and hierarchical stochastic closures. *J. Chem. Phys.* **124**:064110, 2006.
43. Wang F, Landau DP. Efficient, multiple–range random walk algorithm to calculate the density of states. *Phys. Rev. Lett.* **86**:2050, 2001.

3

MULTISCALE MODELING OF RARE EVENTS IN SELF-ASSEMBLED SYSTEMS

DMITRY I. KOPELEVICH
Department of Chemical Engineering, University of Florida, Gainesville, Florida

3.1 INTRODUCTION

Surfactants (or amphiphiles) are molecules that contain both hydrophobic and hydrophilic segments. In an aqueous solution, surfactants self-assemble into aggregates in order to shield the hydrophobic tails from unfavorable interactions with water. Surfactants may assemble into a wide variety of structures, depending on multiple factors, including temperature and sizes of surfactant tail and head groups. Several examples of self-assembled structures are shown in Fig. 3.1. The variety of self-assembled structures increases even more in ternary oil–water–surfactant mixtures. For example, spherical micelles (Fig. 3.1a) may become swollen due to the addition of oil to their hydrophobic core and transform into microemulsion droplets (i.e., oil droplets covered by a surfactant monolayer; Fig. 3.1c).

Equilibrium properties of self-assembled aggregates are currently reasonably well understood, owing, in particular, to development of thermodynamic [1,2] and elastic [3–5] models as well as field-theoretic computational tools [6–8]. Less is known about the dynamics of formation and disintegration of self-assembled structures, transitions between structures, and transport of solute molecules through them. These dynamics play a key role in various technological applications, including oil recovery, detergency, synthesis of nanomaterials, development of drug-delivery

Multiscale Modeling of Particle Interactions: Applications in Biology and Nanotechnology,
Edited by Michael R. King and David J. Gee

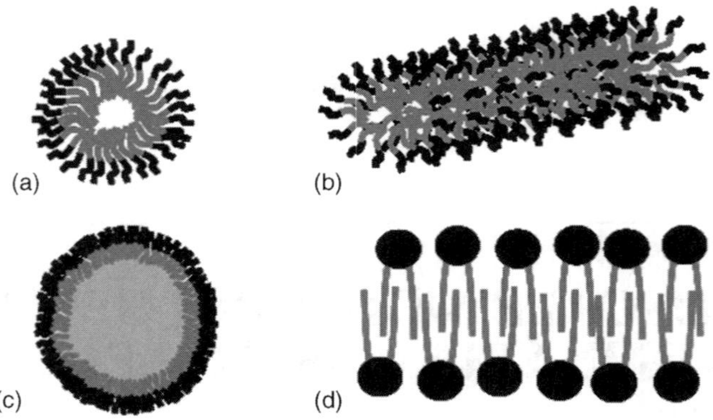

Figure 3.1 Examples of self-assembled structures: (a) spherical micelle; (b) wormlike micelle; (c) oil-in-water microemulsion droplet; (d) lipid bilayer. Hydrophilic head groups are shown in black, hydrophobic tail groups are shown in dark gray, and oil is shown in light gray. Water is not shown for clarity.

vehicles, and defect-free nanostructured templates [9,10]. For example, droplets of water-in-oil microemulsion can be used as reactors for synthesis of nanoparticles [11,12]. The reactants are encapsulated within a droplet and the reaction rate and size of the resulting nanoparticles are controlled by coalescence, fusion, and possibly, fission of the self-assembled structures containing the reactants. The fusion and fission of droplets takes place by rearrangement of the surfactant molecules adsorbed on the droplet surface, and understanding the mechanism of this rearrangement may lead to improved control over the size of synthesized nanoparticles.

Solutions of some self-assembled structures (e.g., wormlike micelles) are capable of a reversible change of their rheological properties in response to an external stimulus. Such fluids with controlled rheology may enable flow control in microfluidic devices and enhanced oil recovery. A solution of wormlike micelles is reminiscent of a polymer solution but with an important difference: The wormlike micelles can form and break bonds reversibly [13] or transform into smaller self-assembled structures (e.g., spherical micelles or vesicles). Similarly to the fusion and fission of microemulsion droplets, the formation and destruction of the bonds between wormlike micelles, as well as the transition between wormlike micelles and other structures, takes place by rearrangement of surfactant molecules. These transitions can be induced, for example, by a shear flow or a temperature change, and may result in a dramatic change in the fluid properties. For example, it has recently been shown [14] that a temperature-induced transition between wormlike micelles and vesicles leads to a 1000-fold decrease in viscosity.

Transitions between self-assembled structures of amphiphilic molecules also play a crucial role in various biological processes. For example, a bilayer composed of amphiphilic lipid molecules (Fig. 3.1d) is an essential constituent of a biomembrane, with other membrane components (e.g., proteins) embedded in the bilayer. Many processes in a living cell, such as endocytosis, exocytosis, and subcellular trafficking, involve fission and fusion of lipid vesicles with cell and organelle membranes [15,16]. Similarly to the examples discussed above, the fission and fusion require transient reorganization of the bilayer configuration into highly curved nonbilayer intermediate structures.

In addition to self-assembly and structural transitions of amphiphilic aggregates, mass transfer through self-assembled interfaces plays a key role in various applications and biological processes. For example, solute transport across surfactant monolayers covering microemulsion droplets finds applications in separations, drug delivery, and detoxification [17,18]. Solute transport into and across cellular membranes is involved in many biological processes, such as anesthesia and toxicity. Although most transport processes of molecules and ions across cell membranes are regulated by proteins, relatively small solutes (e.g., water, oxygen, urea) are known to cross the membrane without a regulatory mechanism [19–21].

Processes within self-assembled systems often involve nontrivial interplay between different time and length scales. Events on the molecular scale often affect the macroscopic scale (e.g., energetics of fission of wormlike micelles controls rheological properties of their solution [22]), and vice versa (e.g., macroscopic shear induces rearrangement of surfactant molecules, leading to transition from vesicles to micelles [23]). Detailed understanding of these processes therefore requires development of a model that links the relevant scales.

Simulations focusing on specific scales have made significant contributions to our understanding of the dynamics of amphiphilic systems. These scales can be divided into (1) *molecular* (*microscopic*), which can be studied by molecular dynamics simulations [24–26]; (2) *mesoscopic*, for which various models involving coarse-grained particle dynamics have been developed, including Brownian dynamics [27,28] and dissipative particle dynamics [29,30]; and (3) *macroscopic*, which can be modeled by continuum models, such as dynamic field-theoretic methods [31,32]. However, connection between these scales is currently incomplete.

In this chapter we present multiscale models for two representative processes: (1) formation and disintegration of a spherical micelle, and (2) transport of a small solute across a surfactant monolayer covering a microemulsion droplet. All self-assembled structures considered are composed of nonionic ethoxylated surfactants $C_iH_{2i+1}{-}[OCH_2CH_2]_j{-}OH$ (abbreviated C_iEO_j). It is demonstrated that these seemingly simple processes involve complex cooperative dynamics between different degrees of freedom of the molecular systems.

Existing models for kinetics of formation and disintegration of spherical micelles [33,34] rely on several assumptions. The crucial assumption is whether the self-assembly takes place through a stepwise addition of single monomers (i.e., individual surfactant molecules) or by coalescence of several small surfactant clusters. Besides purely academic interest, knowledge of the preferred mechanism of self-assembly in solution may, for example, provide means to improve the quality of nanostructured templates produced by deposition of surfactants on a solid surface [10]. Development of a complete model to predict the dominant micellar formation/disintegration mechanism is outside the scope of this chapter, and we focus on a specific aspect of this model. Assuming the stepwise monomer addition–removal mechanism, we obtain addition and removal rates and assess the role of the micellar microstructure in these processes.

The second process considered in this chapter, solute release from a microemulsion droplet, is somewhat similar to removal of a surfactant molecule from a micelle. Perhaps the most significant difference is a much smaller curvature of microemulsion droplets in comparison with spherical micelles. In our analysis we completely neglect the droplet curvature and approximate the solute release from a droplet by its transport across a flat surfactant-covered oil–water interface. It is demonstrated that in addition to providing a static energy barrier, a surfactant monolayer influences the solute transport through dynamic coupling of the solute with capillary waves at the interface.

3.2 METHODS

The molecular scale of the considered systems is modeled by a coarse-grained (CG) molecular dynamics (MD) model, i.e., a model that maps groups of several atoms onto a single coarse-grained bead. Larger scales are modeled by stochastic differential equations (SDEs), which are solved either analytically or numerically. The necessary background information is reviewed in Sections 3.2.1 and 3.2.2. One method commonly used to establish connections between MD simulations and the SDE model—that of constrained MD simulations—is reviewed in Section 3.2.3.

3.2.1 Molecular Dynamics Simulations

MD simulations enable analysis of a physical system with atomistic precision and access time and length scales down to $O(1\text{ ps})$ and $O(1\text{ Å})$, respectively. In their simplest form, these simulations consist of solving Newton's equations of motion,

$$m_i \ddot{r}_{i,\alpha} = -\frac{\partial U(\mathbf{r}_1, \mathbf{r}_2, \ldots, \mathbf{r}_\mathrm{N})}{\partial r_{i,\alpha}}, \qquad i = 1, \ldots, N; \quad \alpha = x, y, \text{ or } z, \tag{3.1}$$

for each atom of the system considered. In Eq. (3.1), m_i and $\mathbf{r}_i \equiv (r_{i,x}, r_{i,y}, r_{i,z})$ are the mass and coordinates of the ith atom and $U(\mathbf{r}_1, \mathbf{r}_2, \ldots, \mathbf{r}_\mathrm{N})$ is the potential energy of the system. The potential energy is usually obtained either from *ab initio* quantum mechanics calculations or by fitting MD simulation results to experimental data. Numerical solution of Eq. (3.1) is usually performed using one of the variations of the Verlet algorithm [35]. This algorithm belongs to a class of symplectic integrators (i.e., it conserves the phase space volume of the system) [36]. This is a very attractive feature for a numerical algorithm used in MD simulations since it implies that the total energy of the system is almost conserved in the simulations.

Simulation of Eq. (3.1) yields the system trajectory in the NVE (microcanonical) ensemble. The acronym NVE implies that the simulations conserve the total number

N of the particles in the system, the system volume V, and the total energy E. Since experiments are typically performed at constant pressure P and temperature T, it would be more natural to perform simulations in the NPT (constant N, P, and T) ensemble. This can be accomplished by utilizing one of the pressure and temperature coupling schemes [37]. In this chapter we use the Berendsen temperature and pressure coupling schemes [38] to keep the constant pressure of 1 bar and the constant temperature of 300 K.

Some atomistic details obtained by MD simulations, such as vibration of a C–H bond in a methylene group of an alkane chain, are irrelevant in analysis of a complex system. Therefore, to improve computational efficiency, it is desirable to omit insignificant details and focus on simulations of relevant degrees of freedom. This is accomplished by coarse-grained molecular dynamics (CGMD) models. In these models, groups of atoms (such as several methylene groups of an alkane chain) are approximated as a single coarse-grained bead. Resolution of CGMD models is intermediate between the atomistic MD simulations and mesoscopic simulation methods (e.g., Brownian dynamics and dissipative particle dynamics). Although atomistic details are not resolved explicitly in CGMD, their effects on the relevant degrees of freedom are implicitly embedded in the force field. Coarse-grained force fields are often obtained by matching certain characteristics (e.g., bead–bead correlation functions or forces acting between beads) with predictions of an atomistic model [39,40].

Several CGMD models have been introduced and applied to simulations of various complex molecular systems [40–43]. In this chapter we use the model proposed by Marrink *et al.* [42]. This model has been shown to yield good agreement with experiments and atomistic simulations for a wide range of molecular systems and, in particular, can accurately reproduce densities and mutual solubilities of water and alkanes [42], as well as the interfacial tension at an oil–water interface [44]. In the current chapter, all molecules are modeled using two types of beads: hydrophobic tail bead (denoted as T) and hydrophilic head bead (denoted as H). This notation is different from that used in ref. [42], where four bead types are introduced in order to describe a wide range of amphiphilic molecules. The beads denoted H and T in this chapter correspond to beads C (apolar) and P (polar) in ref. [42]. We use this different notation, as it is more common in the literature on surfactant systems.

Both H and T beads are spherical and have the same mass, $m = 72$ amu. Interaction between two nonbonded beads is modeled by the Lennard-Jones (LJ) potential with the same effective diameter, $\sigma = 0.47$ nm, for all beads. The character of interactions is modeled through the values of the LJ energy parameter ε. The interaction between two H beads is highly attractive ($\varepsilon_{HH} = 5$ kJ/mol), between two T beads is slightly attractive ($\varepsilon_{TT} = 3.4$ kJ/mol), and between H and T beads is almost purely repulsive ($\varepsilon_{HT} = 1.8$ kJ/mol). All LJ potentials are smoothly shifted to zero from 0.9 nm to the cutoff distance $1.2\,\text{nm} \approx 2.5\sigma$. The LJ interaction between chemically connected beads is excluded and the interactions between these beads are modeled by harmonic potentials for the bond length and angle vibrations. The force constant for the bond length potential is $K_{bond} = 1250\,\text{kJ/mol·nm}^2$ and the corresponding equilibrium bond length is $R_{bond} = \sigma = 0.47$ nm for all bonds. The harmonic potential for the bond-angle vibrations is of cosine type with the force constant $K_{angle} = 25$ kJ/mol·rad^2 and the equilibrium angle $\theta_0 = 180°$ for all bond angles.

A single tail bead T represents four methylene or methyl groups of a hydrocarbon chain, and a single head bead H represents four water molecules [42]. The original work of Marrink *et al.* [42] does not contain a model for head groups of ethoxylated surfactants C_iEO_j. In our recent work [44,45] we proposed to use a single head bead H to approximate two ethoxy groups so that T_nH_m approximates an ethoxylated surfactant $C_{4n}EO_{2m}$. We note that unlike the models for water molecules and hydrocarbon chains, this model of the ethoxy group has not been optimized to match atomistic MD simulations or experimental data. Moreover, the terminal −OH group of ethoxylated surfactants is not modeled explicitly. Since the goal of the current work is to connect GCMD simulations with SDEs, we have refrained from refining the model for the ethoxy groups. Comparison with experimental data for the critical micelle concentration [44] and the monolayer density and interfacial tension at surfactant-covered interfaces [45] indicates that this model correctly reproduces qualitative features of systems containing ethoxylated surfactants. Quantitative agreement between CGMD simulations and the experimental data is also adequate, considering the approximations involved in the model.

In this chapter we consider the formation of spherical micelles composed of H_4T_4 surfactants and transport of a small hydrophobic solute (modeled by a single bead T) across water–hexadecane (H/T_4) interfaces covered by H_3T_3, H_5T_5, and H_7T_7 surfactant monolayers. All MD simulations are performed using a GROMACS simulations package [46]. Plots of instantaneous molecular configurations are generated using a VMD package [47].

3.2.2 Modeling of Rare Events: Stochastic Differential Equations

Modeling even relatively simple processes in self-assembled systems, such as formation and disintegration of spherical micelles, is currently out of reach of direct MD simulations. Although several atomistic MD simulations of micelle formation have been performed [24,25], the self-assembly dynamics observed in these simulations is likely to deviate significantly from that realized in an experimental system. MD simulations of surfactants initially randomly distributed in solution assemble into micelles within tens of nanoseconds, which is orders of magnitude faster than the self-assembly time scale inferred from experiments [9]. The fast self-assembly time scale predicted by MD simulations is caused by the relatively small size of a simulation box [45,48]. Indeed, the dimensions of the box are limited by computational capabilities and are typically on the order of 10 nm. In laboratory experiments of micellar formation, the surfactant concentration is typically on the order of the critical micelle concentration (CMC), which is $O(10^{-6}$ mol/L$) \approx O(10^{-6}$ molecule/nm$^3)$ for ethoxylated surfactants [49]. Since a typical micelle contains $O(100)$ surfactant molecules, performing a simulation of micellar formation at concentration close to CMC would require a system volume of $O(10^8$ nm$^3)$, which is out of reach of MD simulations, even if a coarse-grained model is used.

An obvious consequence of performing MD simulations at high surfactant concentration is an underestimate of the time scale of diffusion of individual surfactant molecules prior to their aggregation, leading to an underestimate of the self-assembly time. Perhaps a less obvious consequence is that a self-assembly mechanism observed in MD simulations may also be incorrect. When surfactant molecules are confined to a small volume, they are likely to aggregate into several small clusters, which proceed to fuse into larger clusters or micelles, as observed, for example, in MD simulations of Marrink *et al.* [25]. On the other hand, in a dilute surfactant solution collisions between different clusters may be less likely than collisions of clusters with monomers, since the cluster concentration is likely to be smaller than the monomer concentration. In this case, the self-assembly process will be dominated by the addition of single monomers to clusters rather than cluster fusion. To determine whether self-assembly takes place through cluster fusion, monomer addition to clusters, or some combination of these two mechanisms, it is necessary to obtain rates of various aggregation steps, as well as cluster and monomer diffusivities. We show in Section 3.3.1 that even a single monomer addition to a surfactant cluster may take $O(100\,\text{ns})$, which is comparable to, or even exceeds, the time scale of formation of an entire micelle predicted by direct MD simulations. Hence, it is necessary to develop a multiscale approach to estimate correctly rates of individual steps of the micellar formation.

Micellar disintegration is even slower than its formation. For example, removal of a single monomer from a micelle involves exposure of the surfactant tail to the aqueous environment. This energetically unfavorable configuration creates a high free-energy barrier for the monomer removal. Therefore, a monomer spends a long time inside a micelle until thermal interactions with the monomer surroundings provide it with sufficient energy to overcome the barrier and leave the micelle. Such a process is called an *activated process* or a *rare event*. Activated processes often play a crucial role in the dynamics of self-assembled systems. For example, release of a hydrophobic solute from an oil-in-water microemulsion droplet is extremely slow since the oily interior of the droplet provides the most favorable environment to the solute, and leaving this environment involves overcoming a high free-energy barrier.

System configurations corresponding to energy barriers are referred to as *transition states*. Prediction of rates of rare events requires detailed knowledge of system behavior in configurations close to those of transition states. Direct MD simulations of rare events are extremely time consuming since most of the computational time is spent sampling less relevant low-energy configurations, and transition states are sampled during an exponentially small fraction of time.

One of the earliest theoretical methods of predicting rates of activated events is the transition-state theory (TST) [50]. In this approach it is assumed that a small number of key degrees of freedom (referred to as *reaction degrees of* freedom or *reaction coordinates*) can adequately describe a transition of the system from one state to another. For brevity, we sometimes refer to such transitions as reactions. For example, a natural choice of a reaction coordinate for a monomer addition or removal to or from a micelle is the distance between centers of mass of the monomer and the micelle. In the case of solute transport across a surfactant monolayer, a natural choice for the reaction coordinate is the distance between the solute center of mass and the monolayer dividing surface. All remaining degrees of freedom are considered to constitute a *thermal bath* whose role is to provide both source and sink for thermal energy of the reaction degrees of freedom. This ensures that the reaction degrees of freedom remain in thermal equilibrium with their surroundings while the system is in a low-energy state (e.g., a solute is inside a microemulsion droplet). Once the reaction degrees of freedom gain sufficient energy to overcome the barrier, it is assumed that their interaction with the thermal bath is negligible while the system crosses the energy barrier.

The latter assumption is valid if the time scale of interactions between the thermal bath and the reaction degrees of freedom is much slower than the time scale of crossing the barrier. Although this is a reasonable assumption for gas-phase reactions, it fails for most transport processes in soft condensed matter. The coupling between reaction degrees of freedom and the thermal bath during the process of climbing and overcoming the free-energy barrier is taken into account by an extension of the TST proposed by Kramers [51]. Kramers' theory and most of its generalizations use SDEs, such as the Langevin equations discussed below, and model thermal bath by stochastic components of these equations.

For example, a system with a single reaction coordinate ξ can be described by the following *Langevin equation*:

$$m\ddot{\xi}(t) + \gamma(\xi)\dot{\xi}(t) + \frac{dG(\xi)}{d\xi} = \Gamma(t;\xi). \tag{3.2}$$

Here m is the reduced mass corresponding to the reaction coordinate ξ and $G(\xi)$ is the potential of mean force acting on the reaction degrees of freedom. It follows from statistical mechanics arguments [57] that $G(\xi)$ is equal to the free energy of the system at a given value of ξ. We sometimes refer to the mean force as the *deterministic force* to distinguish it from the *random force* $\Gamma(t;\xi)$ on the right-hand side of Eq. (3.2). $\Gamma(t;\xi)$ together with the friction term $\gamma(\xi)\dot{\xi}(t)$ captures the interaction of the reaction degree of freedom with the thermal bath. The stochastic force $\Gamma(t;\xi)$ is a Gaussian variable with zero mean and is often assumed to be Markovian (i.e., to have zero correlation time). This assumption is valid if there is a significant time-scale separation between the reaction degrees of freedom and the degrees of freedom constituting the thermal bath. Moreover, $\Gamma(t;\xi)$ is related to the friction coefficient $\gamma(\xi)$ by the fluctuation–dissipation theorem,

$$\langle\Gamma(t;\xi)\Gamma(t+\tau;\xi)\rangle = 2k_B T\gamma(\xi)\delta(\tau). \tag{3.3}$$

where k_B is the Boltzmann constant and T is the temperature of the system. The fluctuation–dissipation theorem implies that the energy exchange of the reaction degree of freedom with the thermal bath is such that, on average, the energy loss due to friction is compensated by the energy gain due to the stochastic force $\Gamma(t;\xi)$. Of course, the instantaneous energy of the reaction degree of freedom is not conserved in a stochastic system.

In most systems involving soft condensed matter, inertia of the reaction degree of freedom is negligible compared with the friction and stochastic forces acting on it. In this high-friction limit, the Langevin equation, Eq. (3.2), becomes a first-order SDE,

$$\gamma(\xi)\dot{\xi}(t) + \frac{dG(\xi)}{d\xi} = \Gamma(t;\xi). \tag{3.4}$$

This equation can be solved readily for arbitrary friction coefficient $\gamma(\xi)$ and free energy $G(\xi)$ to obtain the average time $T_{a\to b}[\gamma, G]$ of transport along the reaction coordinate from $\xi = a$ to $\xi = b$ [52]:

$$T_{a\to b}[\gamma, G] = \frac{1}{k_B T}\int_a^b \gamma(y) e^{G(y)/k_B T} dy \int_a^y e^{-G(z)/k_B T} dz. \tag{3.5}$$

If $\xi = a$ and $\xi = b$ correspond to system configurations on opposite sides of the energy barrier, Eq. (3.5) predicts the average reaction time. Moreover, if the energy barrier ΔG is sufficiently high, $\Delta G \gg k_B T$, and $G(\xi)$ can be approximated by parabolas near both the maximum $\xi_{\max}$ and the minimum $\xi_{\min}$ of G, we obtain the celebrated Kramers' result [51] for the average reaction time in the high-friction limit:

$$T_{a\to b}[\gamma, G] = \frac{2\pi\gamma(\xi_{\max})}{\omega_{\min}\omega_{\max}} e^{\Delta G/k_B T}. \tag{3.6}$$

Here $\omega_{\max}$ and $\omega_{\min}$ are the frequencies at the top of the energy barrier and the bottom of the energy well near $\xi = a$, respectively.

Equation (3.6) can easily be generalized to systems with multiple reaction coordinates, $\xi = (\xi_1, \xi_2, \ldots, \xi_n)$. However, if the energy barrier is not sufficiently high, Eq. (3.6) and its generalizations are not applicable and it is necessary to use Eq. (3.5) or its generalizations to obtain the average reaction time. Analytical solution of multidimensional SDEs is a formidable task for systems with complex free-energy landscapes $G(\xi)$ and, in order to predict reaction rates in these systems, we must turn to numerical simulations of SDEs.

At first sight it may appear that a numerical solution of a SDE requires sampling of the Markovian random force $\Gamma(t)$ at every moment of time t. However, it is sufficient to generate a random variable closely related to $\Gamma(t)$ only at discrete moments of time. To illustrate this, consider movement of a Brownian particle immersed in a homogeneous fluid. In this case, reaction coordinates ξ correspond to Cartesian coordinates of the Brownian particle, the deterministic force acting on the particle vanishes, $\nabla G(\xi) = 0$, and interactions between the particle and solvent molecules are modeled by the thermal bath with the constant friction coefficient γ. A step of the numerical method can be obtained by integrating Eq. (3.4) from time t_n to time $t_{n+1} \equiv t_n + \Delta t$. For one-dimensional Brownian motion this yields

$$\xi(t_{n+1}) = \xi(t_n) + R_n, \tag{3.7}$$

where

$$R_n \equiv \frac{1}{\gamma}\int\limits_{t_n}^{t_{n+1}} \Gamma(t)dt, \tag{3.8}$$

is a random variable. From the properties of $\Gamma(t)$, it follows that R_n is a Gaussian variable with zero mean and

$$\langle R_n R_m \rangle = \frac{1}{\gamma^2}\int\limits_{t_n}^{t_{n+1}}\int\limits_{t_m}^{t_{m+1}} \langle \Gamma(t)\Gamma(s)\rangle dt\, ds = \delta_{mn}\frac{2k_B T\Delta t}{\gamma}. \tag{3.9}$$

The last equality in Eq. (3.9) follows from the fluctuation–dissipation theorem, Eq. (3.3).

In the case of $\gamma = \text{const}$ and $G'(\xi) = 0$, iterations of Eq. (3.7) yield an exact solution of Eq. (3.4) in a sense that the values of $\xi(t_1)$, $\xi(t_2)$,. . ., $\xi(t_n)$,. . . produced by these iterations correspond to a realization of the stochastic process generated by Eq. (3.4). Numerical solution of SDE in a more general case requires numerical approximations to integrals involving position-dependent $\gamma(\xi)$ and $G(\xi)$, similarly to numerical solutions of deterministic differential equations.

SDEs with multiple degrees of freedom are frequently used in modeling of colloidal and polymer solutions. Numerical solutions of SDEs in this context are usually referred to as *Brownian dynamics* (BD) *simulations*. In Section 3.3.2.2 we discuss BD simulations of micelles and surfactant monomers. In these simulations, micelles are modeled by spheres and their simulations are performed using the method of Ermak and McCammon [53], while the monomers are approximated by rigid rods and their simulations are performed using the method of Cobb and Butler [54].

3.2.3 Establishing Connection between MD and SDE Models

Application of a stochastic model to a particular system requires consistency of the model parameters [e.g., γ and G of Eq. (3.4)] with molecular details of the system. In principle, these parameters can be obtained using the Mori–Zwanzig projection [55], which yields an SDE from a molecular model by projecting fast degrees of freedom onto the friction and random force of the stochastic model. However, with the exception of a few simple cases, calculation of these projections is a formidable task. Therefore, it is necessary to use a computational approach to connect stochastic and molecular models.

In this chapter we use constrained MD simulations [20,56] as a basis to establish the MD–SDE connection. In Section 3.3 we show that results of this method often need to be amended. However, constrained simulations provide a useful starting point for the development of multiscale models. As the name suggests, the constrained MD simulations are performed with constrained reaction coordinate ξ. This constraint is implemented by an external force $F^c(\xi;\mathrm{t})$ applied to the reaction degree of freedom ξ.

$F^c(\xi; t)$ balances the deterministic and random forces at each moment of time t. Since the mean of the random force is zero, F^c should, on average, balance the deterministic force,

$$G'(\xi) = \langle F^c(\xi) \rangle. \tag{3.10}$$

A rigorous proof of Eq. (3.10) [57,58] demonstrates that it holds when the reaction coordinate ξ corresponds to the distance between two points whose coordinates are defined as linear combinations of Cartesian coordinates of particles. The systems considered in this chapter satisfy this requirement and we use Eq. (3.10) to obtain the potential of mean force without applying corrections that would have been necessary in a more general case [57,58].

Instantaneous deviations of the constraint force F^c from its average should balance the random force $\Gamma(t; \xi)$; that is,

$$\Gamma(t; \xi) = F^c(t; \xi) - \langle F^c(\xi) \rangle. \tag{3.11}$$

The friction coefficient $\gamma(\xi)$ can be obtained from the autocorrelation function of $\Gamma(t; \xi)$ using the fluctuation–dissipation theorem. Recall that in the formulation of this theorem and in the corresponding Langevin equation [i.e., Eq. (3.2)], we assumed that $\Gamma(t; \xi)$ has zero correlation time. However, stochastic forces obtained from MD simulations, as well as stochastic forces in real physical systems, have nonzero (albeit typically very small) correlation time. This fact is taken into account by the *generalized Langevin equation* (GLE),

$$m\ddot{\xi} + \int_{-\infty}^{t} \tilde{\gamma}(t-\tau; \xi)\dot{\xi}(\tau)d\tau + \frac{dG(\xi)}{d\xi} = \Gamma(t; \xi). \tag{3.12}$$

In this equation, both the friction (second term of the equation) and the random force $\Gamma(t; \xi)$ have nonzero memory. As in Eq. (3.2), $\Gamma(t; \xi)$ obeys the Gaussian distribution and has zero mean. The fluctuation–dissipation theorem for GLE takes the non-Markovian nature of the force into account [56,59],

$$\langle \Gamma(t; \xi)\Gamma(t + \tau; \xi) \rangle = k_B T \tilde{\gamma}(\tau; \xi), \tag{3.13}$$

and therefore differs from the fluctuation-dissipation theorem for Eq. (3.2).

The fluctuation–dissipation relationship, Eq. (3.13), allows us to compute the memory friction kernel $\tilde{\gamma}(\tau; \xi)$ from the autocorrelation function (ACF) of $\Gamma(t; \xi)$. The latter can readily be obtained from constrained MD simulations and Eq. (3.11). If the correlation time of the stochastic force is negligible in comparison with the time scale of the reaction coordinate, the memory friction kernel can be approximated as

$$\tilde{\gamma}(\tau, \xi) = 2\gamma(\xi)\delta(\tau), \tag{3.14}$$

thus transforming Eqs. (3.12)–(3.13) into Eqs. (3.2)–(3.3). Moreover, Eqs. (3.13) and (3.14) allow us to obtain the friction coefficient $\gamma(\xi)$ from constrained MD simulations,

$$\gamma(\xi) = \frac{1}{k_B T} \int_0^\infty \langle \Gamma(\xi; t)\Gamma(\xi; t+\tau) \rangle d\tau. \tag{3.15}$$

Initial conditions for the constrained simulations discussed in Section 3.3 are obtained from an equilibrated system by applying an artificial force to the reaction degree of freedom. This force pulls ξ over a range of values and creates a series of system configurations corresponding to different values of ξ. For each of these configurations, we constrain ξ using the SHAKE algorithm [60] and equilibrate the system obtained for at least 10 ns. This is followed by a production run with duration ranging from 190 to 490 ns, depending on the system.

3.3 RESULTS

3.3.1 Equilibrium Systems

3.3.1.1 Spherical Micelles Equilibrium configurations of spherical micelles composed of H_4T_4 surfactants are obtained by simulations of self-assembly [45]. Randomly dispersed surfactant molecules assemble in micellar aggregates within tens of nanoseconds. As discussed in Section 3.2.2, the self-assembly takes place on this short time scale because the surfactant concentration significantly exceeds CMC. A snapshot of a typical equilibrated system is shown in Fig. 3.2a. Multiple self-assembly simulations are performed for systems with various surfactant concentrations. This yields multiple samples of surfactant clusters with aggregation number N

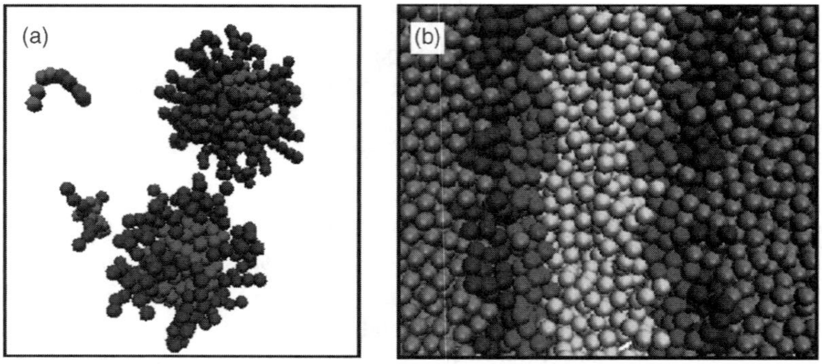

Figure 3.2 (a) Aqueous solution of H_4T_4 micelles and monomers; (b) H_3T_3 monolayers at hexadecane–water interfaces. Surfactant head and tail beads are shown by blue and red spheres, respectively. Solvent molecules are represented by cyan (water) and yellow (oil) beads. In plot (a), water molecules are omitted for clarity. (*See insert for color representation of figure.*)

ranging from 2 to 88. Structures of these clusters are stored in a database and are used to initialize constrained simulations. In what follows we denote a surfactant aggregate of size N by A_N.

To generate the initial configuration for constrained simulations of aggregate A_N, a cluster of size N is selected randomly from the database and is placed in a cubic cell of size $10 \times 10 \times 10\,\text{nm}^3$. The simulation cell is then solvated by 6000 water beads and equilibrated for 75 ns. After the equilibration, a monomer within the cluster is selected randomly and pulled by a constant force applied at its center of mass [45].

3.3.1.2 Surfactant Monolayers We investigate solute transport across hexadecane–water interfaces covered by H_3T_3, H_5T_5, or H_7T_7 monolayers. The simulation cell sizes are $10 \times 10 \times 10\,\text{nm}^3$ for an H_3T_3 monolayer, $12 \times 12 \times 12\,\text{nm}^3$ for an H_5T_5 monolayer, and $10 \times 10 \times 26\,\text{nm}^3$ for an H_7T_7 monolayer. The surfactant coverage of all monolayers considered is 2 molecules/nm^2.

H_3T_3 and H_5T_5 monolayers are prepared by simulations of self-assembly of ternary oil–water–surfactant mixtures with equal or nearly equal mass fractions of oil and water [44]. Random molecular dispersions assemble into a system containing two interfaces within 100 ns. An example of such a self-assembled system is shown in Fig. 3.2b. Interfaces covered by H_7T_7 surfactants are prepared using a different approach, since self-assembly simulations are inefficient for these long surfactants. In this case, hexadecane, water, and surfactant molecules are placed initially into three different rectangular boxes. After minimizing the energy of each of these systems, they are stacked next to each other in the following sequence: water–surfactant–oil–surfactant–water. The obtained system is then equilibrated for 100 ns [44].

Due to periodic boundary conditions imposed in MD simulations, all considered systems contain two identical oil–water interfaces. In our analysis we focus on one of these interfaces.

3.3.2 Kinetics of Formation and Disintegration of Spherical Micelles

3.3.2.1 Degrees of Freedom Relevant to Monomer Addition As discussed in Section 3.2.2, a natural choice of the reaction coordinate ξ for addition or removal of a surfactant monomer to or from a surfactant cluster is the distance between the cluster and monomer centers of mass. The ξ-dependence of the potential of mean force $G(\xi)$ and the friction coefficient $\gamma(\xi)$ are computed using constrained MD simulations discussed in Section 3.2.3. The $G(\xi)$ and $\gamma(\xi)$ obtained are qualitatively similar for all surfactant clusters A_N, $N = 2, \ldots, 88$, considered [45]. Typical examples of $G(\xi)$, $\gamma(\xi)$, and radial density profiles of A_N are shown in Fig. 3.3. The free energy approaches a minimum at some value of the reaction coordinate $\xi = \xi_0$. This value corresponds to the equilibrium location of the monomer in the cluster. As the distance ξ between the monomer and the cluster centers of mass increases above ξ_0, the exposure of the monomer tail to the unfavorable polar environment grows, leading to an increase in the free energy $G(\xi)$. Interaction between the cluster and the monomer vanishes once separation between them becomes sufficiently large. The value of ξ corresponding to this separation is denoted by ξ_{free} [i.e., $G'(\xi) = 0$ for $\xi > \xi_{\text{free}}$]. The obtained

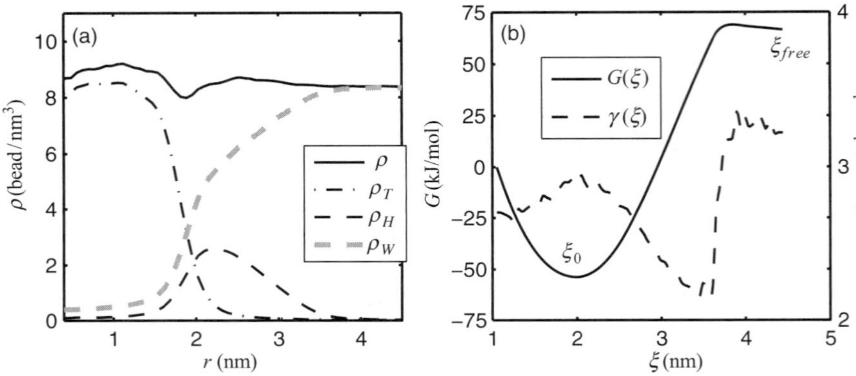

Figure 3.3 (a) Dependence of the total density (ρ) and the density of surfactant tails (ρ_T), head groups (ρ_H), and water (ρ_W) for micelle A_{64} as a function of distance r from the center of the micelle; (b) dependence of the free energy G and the friction coefficient γ on the distance ξ between centers of mass of the monomer and the micelle A_{64}. (From [45], with permission. Copyright © 2008, American Institute of Physics.)

free-energy profiles $G(\xi)$ predict significant energy barriers for monomer removal and negligible barriers for monomer addition, in agreement with free-energy profiles obtained for similar coarse-grained models [27,61].

A negligibly small energy barrier for monomer addition implies that the monomer will enter a cluster as soon as it approaches one. Indeed, applying Eq. (3.5) to compute the average time $T_{\xi_{\text{free}} \to \xi_0}[\gamma, G]$ of monomer entry into a cluster after the first monomer–cluster contact, we obtain addition times $T_{\xi_{\text{free}} \to \xi_0} \approx 200$ ps for all considered clusters A_N, $N = 2, \ldots, 88$ [45]. Therefore, monomer addition appears to be controlled completely by diffusion, so that the fusion between a monomer and a cluster takes place as soon as they approach each other. In fact, this is often assumed in studies of micellar kinetics and is a basis for application of the Smoluchowski theory of diffusion-controlled colloidal aggregation [62] to micellar formation [25,61,63].

However, MD simulations of a monomer addition to a micelle indicate that calculations based on the Langevin equation (3.4) underestimate the average addition time by orders of magnitude. A typical example is MD simulations of a micelle A_{87} and a monomer in a $10 \times 10 \times 10\,\text{nm}^3$ box filled with water. The initial distance between the centers of mass of the micelle and the monomer exceeds ξ_{free}, so they are independent of each other at the start of the simulation. The small cell size implies that the diffusion time will be relatively small (on the order of a few nanoseconds). It is observed that for some initial conditions, the monomer does not enter the micelle for $O(100\,\text{ns})$, which is significantly larger than the estimate of the mean addition time, $T_{\xi_{\text{free}} \to \xi_0} \approx 220$ ps, based on the one-dimensional Langevin equation (3.4).

A detailed examination of the simulations reveals that the monomer makes multiple attempts to merge with the micelle. However, the vast majority of these attempts are unsuccessful and the monomer enters a micelle only if the following two conditions are satisfied: (1) the hydrophobic tail of the monomer is pointing toward the micelle, and (2) the monomer approaches the micelle near an exposed area of its

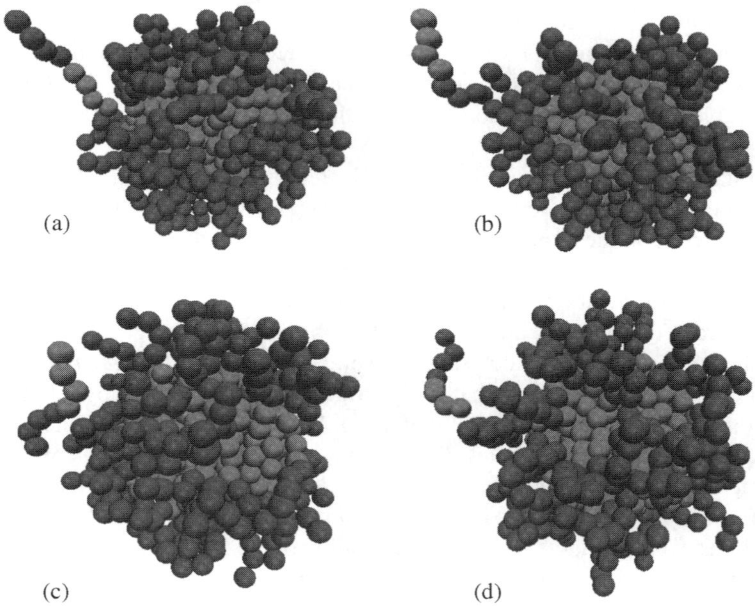

Figure 3.4 MD simulations of attempts of monomer entry into micelle A_{87}. Hydrophilic head beads and hydrophobic tail beads are shown by blue and red spheres, respectively. (a) Both the monomer and the micelle are in a favorable configuration leading to a successful entry; (b, c) the monomer, with an unfavorable orientation, approaches an exposed hydrophobic patch on the surface of the micelle; (d) the monomer, with a favorable orientation, approaches the hydrophilic micellar corona (i.e., the micelle is in an unfavorable configuration). The attempts at monomer entry fail in cases (b) to (d). (*See insert for color representation of figure.*)

hydrophobic core [45]. An example of a favorable configuration leading to the monomer entry into the micelle is shown in Fig. 3.4a. Figures 3.4b–d provide examples of unfavorable configurations of the monomer or the micelle, leading to unsuccessful monomer attempts to enter into the micelle. Figure 3.4b shows a monomer approaching an exposed hydrophobic patch on the micellar surface. However, the hydrophilic head group of the monomer is pointing toward the micelle, resulting in the repulsion of the monomer. Similarly, Fig. 3.4c shows repulsion of the monomer approaching the micelle with a favorable microstructure but with an unfavorable orientation (i.e., it deviates significantly from the normal to the micellar surface). Finally, if the monomer has a favorable orientation but reaches the hydrophilic region of the micellar surface, it is repelled, as shown in Fig. 3.4d.

These repulsive steric interactions are not visible in the one-dimensional free-energy profiles, such as that shown in Fig. 3.3b. To capture these interactions, it is necessary to take at least two additional degrees of freedom into consideration: the monomer orientation, ξ_2, and the cluster microstructure, ξ_3. These additional degrees of freedom cannot be neglected since their time scale is comparable to that of the original translational degree of freedom, $\xi \equiv \xi_1$, during the monomer addition. Indeed, analysis of the ACF of orientation of a freely floating monomer shows that the time scale of the monomer orientation ξ_2 is on the order of 1 ns. This time scale is comparable with that of the monomer addition to a micelle under favorable conditions, O(0.1 ns).

Similar conclusions are reached for the cluster microstructure ξ_3. The aspect of the cluster microstructure relevant to the monomer addition is distribution of hydrophobic patches on the cluster surface (i.e., areas of the hydrophobic core of a cluster exposed to the solvent). At each point in time, each point on the cluster surface corresponds to either the hydrophilic corona or a hydrophobic patch. The hydrophobic patches are very dynamic, due to the high mobility of surfactant molecules. This is demonstrated in Fig. 3.5, which shows several examples of *dynamic patch distribution* (i.e., the fraction of time that each surface point is occupied by a hydrophobic patch during a relatively short period of time). The duration τ_p of time interval used in calculations of the dynamic patch distributions is 200 ps. This τ_{p} value corresponds to the time scale of addition of a monomer to a cluster if both the monomer orientation and the cluster surface are in a favorable configuration. For most of the patches, their central areas remain stable for the entire duration of the 200-ps time interval, whereas their border areas are exposed for only a fraction of this time. Moreover, the lifetime of some small patches is less than 200 ps. Nevertheless, it is clear that some of the patches

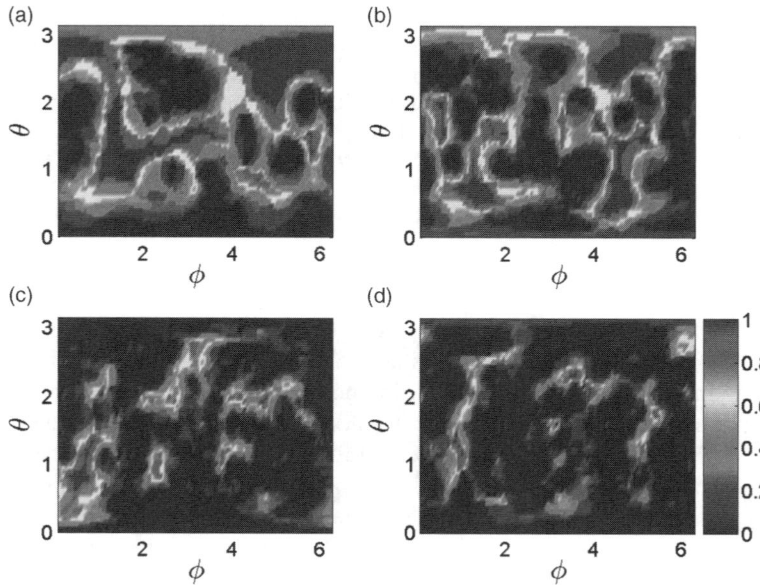

Figure 3.5 Fraction of time that cluster surfaces are occupied by hydrophobic patches during a period of 200 ps. Cluster aggregation numbers are (a) $N = 16$, (b) $N = 32$, (c) $N = 64$, and (d) $N = 88$. The surface of each cluster is parameterized by spherical angles θ and ϕ, with the origin of the system of coordinates located at the cluster center of mass. (*See insert for color representation of figure.*)

remain stable within the 200-ps period, which implies that they will remain stable during a monomer addition. Repeating the analysis of the dynamic patch distributions for time intervals of various durations τ_p, we observe that depending on the cluster size, average lifetimes of hydrophobic patches range from $O(1\,\text{ns})$ to $O(10\,\text{ns})$.

The average monomer removal times predicted by the one-dimensional Langevin equation are on the order of 1 to 10 ms for clusters with aggregation numbers $N \geq 10$ [45]. This is several orders of magnitude slower than the characteristic time scales of the monomer rotation and the hydrophobic patch dynamics. Therefore, the additional degrees of freedom ξ_2 and ξ_3 are not relevant during monomer removal and are slaved to the original reaction coordinate ξ_1 during this process.

On the other hand, ξ_2 and ξ_3 play a significant role in the monomer addition; that is, a one-dimensional free-energy curve $G(\xi_1)$ should be replaced by a hypersurface $G(\xi_1, \xi_2, \xi_3)$ in order to analyze this process. This hypersurface will prescribe a high free-energy barrier for monomer entry into the micelle if the monomer approaches the micelle with an unfavorable orientation or if the patch distribution on the micellar surface is unfavorable. The one-dimensional free-energy profile shown in Fig. 3.3b is computed under the assumption that both ξ_2 and ξ_3 are slaved to ξ_1. Indeed, once MD simulations are started for a given ξ_1, the degrees of freedom ξ_2 and ξ_3 adjust toward their most favorable configuration within the equilibration time of the simulations. In effect, the one-dimensional free-energy profile $G(\xi_1)$ corresponds to the free energy on the *minimum energy path* (MEP) for the removal and addition of a monomer. However, during the addition of a monomer, the dynamics of ξ_1 is too fast for ξ_2 and ξ_3 to approach the MEP.

3.3.2.2 Multidimensional Model for Monomer Addition Reconstruction of the free-energy surface $G(\xi_1, \xi_2, \xi_3)$ from MD simulations, as well as simulations of the resulting multidimensional SDEs, is an extremely difficult task. Instead, to assess the effect of the additional degrees of freedom, we consider a relatively simple model that takes these degrees of freedom into account. In this model, a monomer is approximated by a rigid rod with hydrophobic and hydrophilic ends. A surfactant cluster is approximated as a sphere with each point on the surface assigned to either the hydrophilic corona or a hydrophobic patch to mimic its surface structure. For each cluster A_N, a large set of possible surface structures is generated from equilibrium MD simulations and is stored in a database. To obtain a surface structure, we split the MD trajectory into segments of length $\tau_p = 200$ ps, which corresponds to the time scale of the monomer addition along the MEP. For each of these segments, we obtain the hydrophobic areas that are exposed to water for the entire period τ_p. These areas correspond to hydrophobic patches that are expected to remain stable during monomer entry into the cluster. The spatial distribution of these stable patches is then added to the database and the process is repeated for the next trajectory segment.

We assume the following conditions of the monomer entry into the cluster: (1) the hydrophobic side of the monomer is in contact with a hydrophobic patch on the cluster surface; and (2) the angle between the monomer direction and the outward normal to the cluster at the point of contact does not exceed 40°. The latter condition is based on the distribution of favorable monomer orientations obtained from the constrained simulations. For brevity, in what follows we refer to the monomer entry into the cluster as a *reaction*.

Until the monomer and the cluster come into contact with each other, they are assumed to undergo independent Brownian motions modeled by BD simulations (see Section 3.2.2). The parameters of the BD model (i.e., the monomer and cluster diffusivities) are obtained from equilibrium MD simulations. Once there is contact between the monomer and the cluster, we check the reaction conditions. If they are satisfied, the monomer enters the cluster successfully and the simulation stops. Otherwise, a BD step is performed to repel the monomer from the cluster in a random direction.

The monomer addition rate is obtained using the method of Northrup et al. [64]. This method was developed to obtain the rate of a diffusion-controlled reaction between two particles (which may diffuse to infinite separation) from finite BD trajectories. Initial conditions for the simulations are chosen so that the initial separation, b, between the centers of mass of the monomer and the cluster is sufficiently large for the reactive flux to be independent of the particle orientations and microstructures. Initial orientations of both the monomer and the cluster are chosen at random. Since b is chosen so that the reactive flux $k_D(b)$ at b is purely diffusive, this flux can be predicted by the Smoluchowski model [62] for colloidal aggregation,

$$k_D(b) = 4\pi D_0 b. \tag{3.16}$$

Here D_0 is the diffusion coefficient for relative motion of the cluster and the monomer.

The reaction process can be split into the following steps: (A) independent diffusion of the monomer and the micelle until the distance between their centers of mass reaches b, and (B) either addition of the monomer to the cluster or diffusion of the monomer and cluster to infinite separation. Since steps A and B are independent of each other, the monomer–cluster reaction rate is

$$k = k_D(b)p_\infty(b). \tag{3.17}$$

Here $p_\infty(b)$ is the probability that the cluster and monomer, initially separated by distance b, will react rather than diffuse to infinite separation. To obtain $p_\infty(b)$, one can perform a series of simulations with initial cluster–monomer distance b and count the fraction of simulations that led to a successful reaction rather than diffusion to infinite separation. Obviously, infinite separation cannot be achieved in BD simulations. Instead, BD trajectories are truncated at some finite distance $q > b$ between the cluster and the monomer. These truncated trajectories allow one to obtain the probability $p_q(b)$ that the cluster and monomer, initially separated by distance b, will eventually react rather than become separated by distance q. Once $p_q(b)$ is obtained, $p_\infty(b)$ is computed using the following relationship [64]:

$$p_\infty(b) = \frac{p_q(b)}{1-[1-p_q(b)][k_D(b)/k_D(q)]}. \tag{3.18}$$

The values of b and q used in BD simulations of the monomer addition are 8 and 30 nm, respectively. This value for b is sufficiently large to ensure homogeneous reactive flux at the starting point of the BD simulations, and the value for q is sufficiently large to ensure that a significant fraction of BD trajectories is successful.

Since the lifetime of a surface patch typically exceeds the time scale of the monomer addition under favorable conditions, we assume that the surface patch distribution remains stable during the monomer approach and attempted entry. If an attempted monomer entry is not successful, there is a large probability that it will be followed by another attempt at a nearby point. The time between these two attempts is short and it is expected that the configuration of the cluster surface will remain relatively unchanged. Hence, in the current model we keep the surface structure constant during the sequential collisions with the monomer until the monomer either leaves a neighborhood of the cluster or enters the cluster. The surface structure is replaced every time the distance between the centers of mass of the cluster and monomer reaches q. A new surface patch distribution is randomly selected from the database prepared from results of the MD simulations prior to the BD simulations.

BD simulation were performed for clusters A_N with aggregation number N ranging from 9 to 88. The obtained addition rates are shown in Fig. 3.6a. For comparison, this plot also shows predictions of the Smoluchowski model for rates of the purely diffusion-controlled aggregation,

$$k_{N,N+1} = 4\pi(R_N + R_1)D_0, \tag{3.19}$$

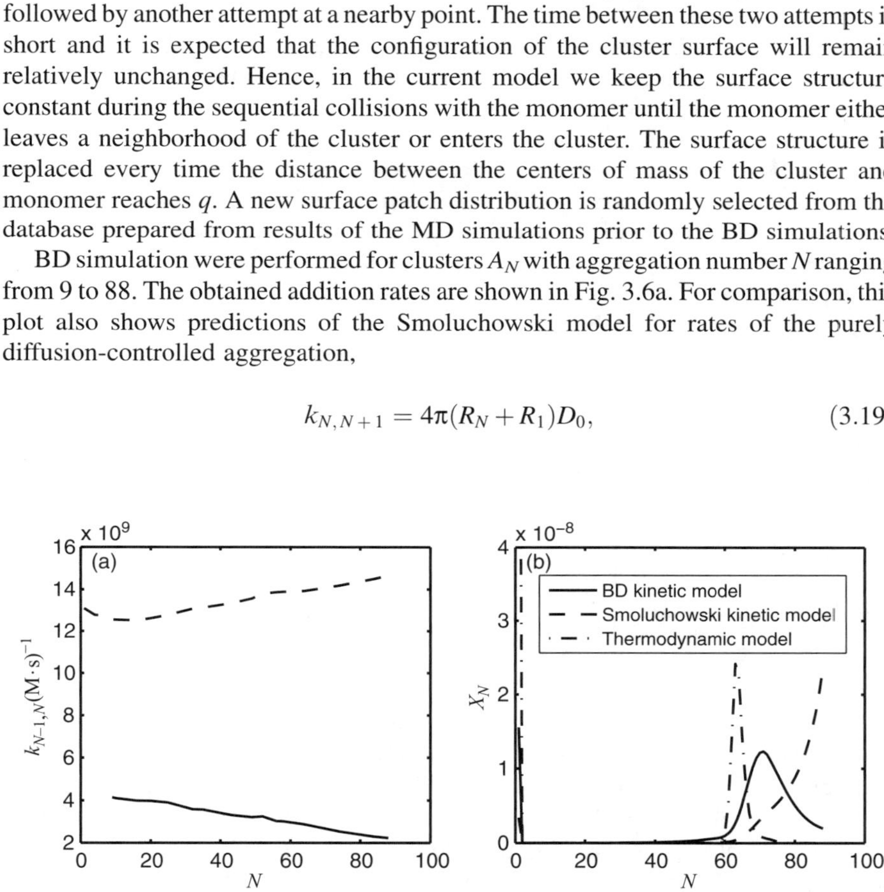

Figure 3.6 (a) Rate constants $k_{N,N+1}$ of addition of monomers to clusters A_N obtained from the BD simulations (solid line) and the Smoluchowski model Eq. (3.19) (dashed line); (b) concentration $X_N = NC_N$ of surfactant molecules contained in clusters A_N computed at total surfactant concentration of 2×10^{-7} M. Predictions of the kinetic equations (3.21) and (3.22) with the monomer addition rates determined from the BD simulations and the Smoluchowski model are shown by the solid and dashed lines, respectively. Predictions of the thermodynamic theory are shown by the dashed–dotted line. (From [45], with permission. Copyright © 2008, American Institute of Physics.)

where R_N is the radius of cluster A_N and $R_1 \approx 0.8$ nm is the largest radius of gyration of the monomer. The Smoluchowski model corresponds to the reaction rate obtained when the degrees of freedom ξ_2 and ξ_3 are neglected. This model significantly overestimates the addition rates. More importantly, the additional degrees of freedom lead to a qualitative difference in the dependence of the addition rates on the cluster aggregation number N. The addition rates obtained from BD model decrease with increasing aggregation number N, whereas the Smoluchowski model predicts that the addition rates increase with increasing N for $N \geq 16$.

3.3.2.3 Model Validation To validate the model developed for the addition and removal of the monomers to surfactant clusters, we compare the equilibrium cluster size distribution predicted by this model with the distribution obtained from an independent thermodynamic calculation [1,45]. The equilibrium distribution corresponds to a steady-state solution of kinetic equations for micellar formation and disintegration. We assume that self-assembly takes place through stepwise addition or removal of surfactant monomers to or from clusters [33] and neglect the contributions of cluster fusion and fission. In this case, the kinetic equations for micellar formation/disintegration are

$$A_1 + A_{N-1} \underset{k_{N,N-1}}{\overset{k_{N-1,N}}{\rightleftharpoons}} A_N, \quad N = 2, 3, \ldots, \tag{3.20}$$

where $k_{N\text{-}1,N}$ is the rate constant for addition of a monomer to cluster $A_{N\text{-}1}$, and $k_{N,N\text{-}1}$ is the rate constant for removal of a monomer from cluster A_N. The molar concentration C_N of aggregates A_N obeys the following set of master equations:

$$\frac{dC_1}{dt} = \sum_{N=1}^{\infty} (k_{N+1,N} C_{N+1} - k_{N,N+1} C_N C_1), \tag{3.21}$$

$$\frac{dC_N}{dt} = k_{N-1,N} C_{N-1} C_1 + k_{N+1,N} C_{N+1} - k_{N,N+1} C_N C_1 - k_{N,N-1} C_N, \qquad N > 1. \tag{3.22}$$

BD simulations to obtain the monomer addition rates $k_{N,N+1}$ were performed for clusters with aggregation number $N \geq 9$. For smaller N, the spherical approximation of the cluster surface becomes inaccurate. Moreover, the surface becomes less defined for these small clusters and hence the location of the hydrophobic patches cannot be delineated adequately by mapping them to the cluster surface as was done for the larger clusters. Therefore, the rates of monomer addition to clusters A_N with $N < 9$ are approximated by extrapolation of the rates obtained by BD simulations for the larger clusters. The cluster size distribution at the total surfactant concentration of 2×10^{-7} M obtained from the solution of the master equations (3.21)–(3.22) is compared with the prediction of the thermodynamic theory in Fig. 3.6b. The cluster size distribution predicted by the kinetic equations is somewhat wider and the aggregation number of the most probable micelle, $N = 70$, slightly exceeds the prediction of the thermodynamic theory, $N = 64$. The value obtained for CMC is 1.06×10^{-8} M, which is an order of magnitude smaller than the prediction of the thermodynamic theory, 1.0×10^{-7} M. Reasons for this discrepancy and possible improvement of the stochastic model are discussed in Section 3.5.

Despite this discrepancy, taking the additional degrees of freedom ξ_2 and ξ_3 into account leads to a qualitative improvement of the obtained cluster size distribution. For comparison, in Fig. 3.6b we also show the cluster size distribution obtained using the monomer addition rates predicted by the Smoluchowski model. This distribution increases continuously with the cluster aggregation number N. This qualitative difference arises because the Smoluchowski addition rates grow with N when $N \geq 16$, whereas the steric constraints due to ξ_2 and ξ_3 lead to the decrease of the addition rates with increase of N (see Fig. 3.6a).

3.3.3 Diffusion Across Surfactant-Covered Oil–Water Interfaces

3.3.3.1 Monolayer Structure and Static Barriers to Solute Transport Rates of molecular transport across surfactant-covered oil–water interfaces are closely correlated with structures of these interfaces. Therefore, our development of a stochastic model for molecular transport is preceded by investigations of relevant structural properties of surfactant-covered interfaces. A typical density profile of a surfactant monolayer at the hexadecane–water interface is shown in Fig. 3.7a. The system of coordinates is oriented so that the z axis is normal to the interface and points to the water phase, and the x–y plane corresponds to the dividing surface of the interface.

The total density of the surfactant-covered interface approaches a minimum at the center of the interfacial region, due to dewetting caused by hydrophobic repulsion of hydrophobic and hydrophilic groups. Similar dewetting is observed at oil–water interfaces not covered by surfactants [44,65]. This density minimum is surrounded by two maxima caused primarily by a local increase in the surfactant density. The inhomogeneity of the surfactant density is due to stretching of the surfactant molecules near the monolayer dividing surface followed by their compression away from the surface [44]. The bonds closest to the dividing surface tend to orient themselves in the normal direction to reduce the interaction between the hydrophilic and hydrophobic beads. As the distance between a bond and the dividing surface increases, contact between the beads connected by this bond and the beads on the opposite side of the dividing surface becomes negligible. This significantly relaxes the enthalpic constraints on the bond orientation and the entropic contribution to the free energy becomes more important. The entropic force causes the surfactants to coil up even at the expense of a local density increase. For longer surfactants, the entropic force is larger and is capable of overcoming larger enthalpic repulsions between the beads, causing higher local surfactant density. These arguments are confirmed by the surfactant density profiles ρ_{surf} for the H_3T_3, H_5T_5, and H_7T_7 monolayers shown in Fig. 3.7b.

The total density ρ within the monolayer is increased further due to penetration of solvent molecules into the monolayer. The solvent density profiles ρ_{solv} for the three monolayers considered are shown in Fig. 3.7c. The density of water in the head-group region is higher than the density of hexadecane in the tail-group region, since the smaller water molecules are able to penetrate the space between surfactant molecules

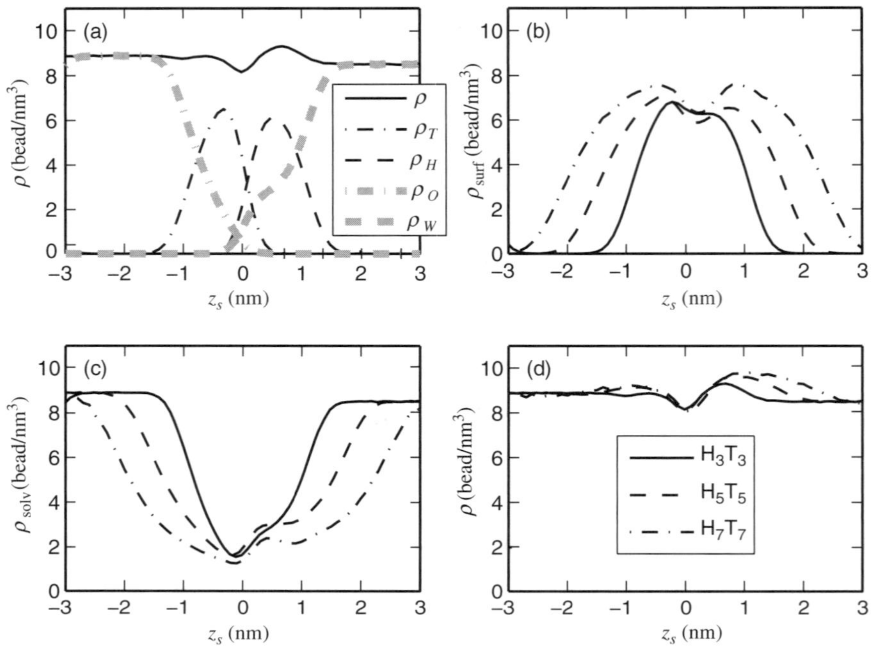

Figure 3.7 (a) Density profile of a hexadecane–water interface covered by an H_3T_3 monolayer; (b)–(d) comparison of (b) surfactant, (c) solvent, and (d) total density profiles of H_3T_3 (solid line), H_5T_5 (dashed line), and H_7T_7 (dashed–dotted line) monolayers. (From [44], with permission. Copyright © 2008, American Institute of Physics.)

more effectively than the bulkier hexadecane molecules. This leads to a larger total density maximum in the head-group region than in the tail-group region. The solvent density inside a monolayer decreases with increase in the surfactant length since the larger density of longer surfactants provides a larger barrier for the solvent penetration into the monolayer. However, increase of the surfactant density ρ_{surf} with increase of the surfactant length is more substantial than decrease of the solvent density ρ_{solv}. Therefore, the total monolayer density, $\rho = \rho_{\text{surf}} + \rho_{\text{solv}}$, increases with the surfactant length, as can be seen in Fig. 3.7d.

In this section we consider transport of a small hydrophobic solute modeled by a single hydrophobic bead T. As discussed in Section 3.2.2, the reaction coordinate ξ for the solute transport is the distance between the solute center of mass and the monolayer dividing surface. Therefore, ξ coincides with the z-coordinate of the solute and, in what follows, we refer to this coordinate as z_s.

Free-energy profiles for the hydrophobic solute in H_3T_3, H_5T_5, and H_7T_7 monolayers are shown in Fig. 3.8a. As expected, the solute free energy is significantly higher on the water-rich side of the interface than on the oil-rich side. In addition to this overall energy increase, the solute traveling from the oil-rich side to the water-rich side encounters two local energy maxima and a minimum. Good correlation between

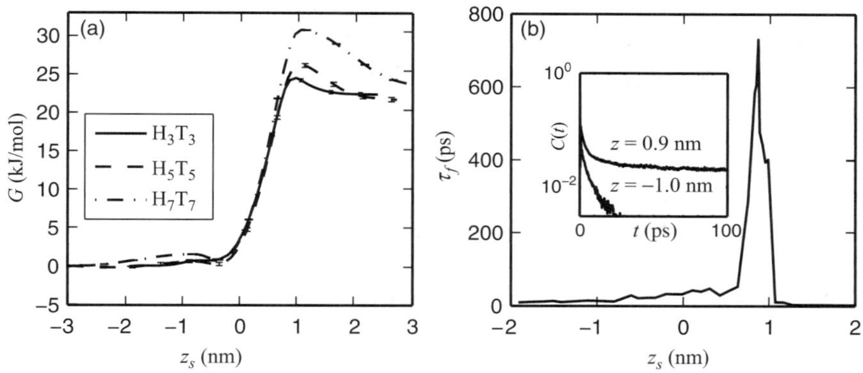

Figure 3.8 (a) Free energy G of a small spherical hydrophobic solute at the hexadecane–water interface covered by the H_3T_3, H_5T_5, and H_7T_7 surfactants. (From [44], with permission. Copyright © 2008, American Institute of Physics.) (b) The correlation time $\tau_f(z_s)$ of the random force $\Gamma(t, z_s)$ acting on the solute constrained at a distance z_s away from the dividing surface of the H_3T_3 monolayer. The inset shows examples of ACF of the random force $\Gamma(t, z_s)$ for solute positions corresponding to the long ($z_s = 0.9$ nm) and short ($z_s = -1.0$ nm) fluctuation decay times. ACF are normalized so that $C(t = 0; z_s) = 1$.

the density and energy extrema (Figs. 3.7d and 3.8a, respectively) indicates that the energy extrema are caused, in large part, by excluded volume interactions of the solute with the solvents and surfactants. There is a slight displacement of the free-energy extrema relative to the corresponding density extrema. This displacement occurs because, in addition to the excluded volume effects captured by the density profiles, free energy is affected by hydrophobic–hydrophilic interactions between the solute and its surroundings [44].

The free-energy maximum located in the head-group region of a surfactant monolayer is significantly higher than that in the tail-group region. This is explained by (1) a lower height of the density maximum in the tail region (Fig. 3.7d), and (2) more favorable interactions of the hydrophobic solute with its surroundings in the tail region. Therefore, the most significant energy barrier for the solute transport is located in the head-group region of the monolayer. The height of this barrier is higher in monolayers consisting of longer surfactants, which is consistent with the trends of the corresponding density maximum.

Effect of the curvature of a microemulsion droplet on the barrier for solute transport through the droplet interface can be assessed from a comparison of the flat interfaces, considered here, with the opposite extreme: highly curved micellar surfaces. A micelle can be transformed into a microemulsion droplet by adding oil to its hydrophobic core, which will increase its radius and hence decrease the interfacial curvature. A smaller interfacial curvature corresponds to a smaller angle between neighboring surfactant molecules (i.e., a tighter packing of the surfactant head groups). This suggests that a flatter interface leads to a higher barrier for the solute transport. Indeed, Figs. 3.3a and 3.7d indicate that the local density maximum in the head-group region of a micelle is much smaller than the corresponding maximum in a flat monolayer. This is consistent with a negligible local free-energy maximum for removal of a relatively bulky surfactant molecule from a micelle (Fig. 3.3b). In comparison, there may be a substantial local energy maximum for transport of a much smaller solute across a flat monolayer (Fig. 3.8a).

3.3.3.2 Coupled Dynamics of Interface Fluctuations and Solute Transport In addition to the free-energy profiles obtained in Section 3.3.3.1, development of a stochastic model for the solute transport requires knowledge of properties of the random force acting on the solute. In this section we demonstrate that this force may deviate significantly from the Markovian assumption (3.3). This deviation is caused by relatively slow capillary waves at the interface, which implies that the interface fluctuations are dynamically coupled with the solute transport. The mechanism of this solute–interface coupling is sufficiently general and is expected to influence solute transport across various flexible or fluid membranes.

In this section we focus on transport of a small hydrophobic solute across the H_3T_3 surfactant monolayer. Correlation times $\tau_f(z_s)$ of the random forces $\Gamma(z_s; t)$ acting on the solute constrained at various distances z_s away from the dividing surface of the H_3T_3 monolayer are shown in Fig. 3.8b. The time scale of the random force fluctuations is extremely sensitive to the solute location, with $\tau_f(z_s)$ varying by two orders of magnitude in a region less than 0.5 nm wide. From comparison of Fig. 3.8a and b, it is evident that the slow force fluctuations take place near the free-energy barrier for the solute transport, which suggests that the slow fluctuations play a substantial role in the solute transport.

The fluctuations of the force acting on the solute are related directly to fluctuations of the solvent and surfactant density around the solute. Fast density fluctuations correspond to diffusive motion of individual particles, whereas slow density fluctuations correspond to capillary waves at the interface. Interestingly, the slow force fluctuations are observed on only one side of the interface, while the interface fluctuations are expected to cause slow density fluctuations on both sides of the dividing surface.

To understand this asymmetry, it is necessary to consider the monolayer fluctuations in more detail. In this analysis we use a moving system of coordinates with the origin corresponding to the solute projection onto the x–y plane. The *instantaneous dividing surface* $z = h(x, y; t)$ at time t is defined as a surface passing through N_s *monolayer pivot points* $(x_j(t), y_j(t), z_j(t))$. Here N_s is the number of surfactant molecules in the monolayer, and the jth pivot point at time t, $(x_j(t), y_j(t), z_j(t))$, is defined as the midpoint of the bond connecting the tail and head groups of the jth surfactant molecule, $j = 1, \ldots, N_s$.

Instantaneous dividing surfaces averaged over MD simulations with the solute constrained at $z = z_s$ provide a starting point for understanding the asymmetry of the force fluctuations. We denote the time-averaged surfaces as $z = h^{(0)}(\mathbf{r}; z_s)$, where $\mathbf{r}$ denotes the distance between a point on the xy plane and the solute projection onto this plane. Surfaces $z = h^{(0)}(\mathbf{r}; z_s)$ corresponding to four representative solute positions,

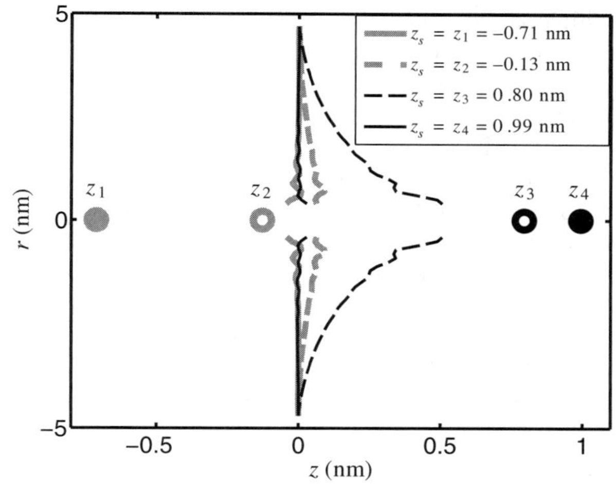

Figure 3.9 Average dividing surfaces $h^{(0)}(\mathbf{r}; z_s)$ corresponding to the solute constrained at various positions z_s. To emphasize the rotational symmetry of the dividing surfaces, the symmetric reflections of these surfaces with respect to the z-axis are also plotted. Dividing surfaces exhibiting significant deviations from the unperturbed planar interface are shown by dashed lines and the corresponding solute locations are shown by hollow circles. Solute locations that do not lead to significant interface perturbations are indicated by filled circles, and the corresponding dividing surfaces are shown by solid lines. The gray (black) lines and circles correspond to solute located on the oil (water)-rich sides of the interface.

$z_s = z_i$, $i = 1,\ldots,4$; $z_1 < z_2 < z_3 < z_4$, are shown in Fig. 3.9. The average dividing surface remains unperturbed when the solute is located in the oil-rich phase, even if the solute is relatively close to the xy plane, as illustrated by $h^{(0)}(\mathbf{r}; z_s = z_1)$. As the solute approaches the xy plane, it starts pushing the interface toward the water-rich phase [see $h^{(0)}(\mathbf{r}; z_s = z_2)$]. The interface deforms to maximize wetting of the solute by hydrophobic groups and to minimize the solute interactions with hydrophilic groups. As the solute crosses the dividing surface and moves into the water-rich phase, the protrusion into this phase grows, due to attraction between the solute and hydrophobic groups located on the opposite side of the interface. The protrusion height reaches its maximum when $z_s \approx z_3$.

Further solute displacement into the water-rich phase requires larger interface deformation to ensure that the hydrophobic groups are within the interaction range of the solute. Eventually, an increase in the interfacial energy due to such deformations becomes too large to be compensated by the favorable interactions of the solute with the hydrophobic groups. Therefore, once the solute passes through the point corresponding to the maximum protrusion height, the protrusion disappears fairly quickly and the interface returns to its unperturbed state when $z_s \approx z_4$.

The strong dependence of the average interface shape on the solute position is a key factor leading to the slow fluctuations of the random force $\Gamma(z_s; t)$. To see this, consider a solute–interface system with the solute constrained at some $z_s = z_0$. In this system, the dividing surface fluctuates around its average $h^{(0)}(\mathbf{r}; z_0)$. Remove the constraint at time t_0 and consider the system configuration at time $t_1 = t_0 + \Delta t$, where Δt is on the order of the time scale of the fast fluctuations of $\Gamma(z_s; t)$. Assume that a small solute displacement in the z direction which happened between t_0 and t_1 corresponds to a substantial change in the average interface shape. In this case, the instantaneous interface shape at time t_1 will be different from the average shape $h^{(0)}(\mathbf{r}; z_s(t_1))$, since the time scale of the interface relaxation is much larger than the time scale of the fast fluctuations of $\Gamma(z_s; t)$. The unfavorable interface configuration caused by its delayed response will lead to a strong restoring force pulling the solute back toward $z_s = z_0$. This force corresponds to the slow component of the force $\Gamma(z_s; t)$, and its time scale corresponds to the interface relaxation time. This relaxation time is on the order of hundreds of picoseconds [66], which is consistent with the time scale of the slow component of $\Gamma(z_s; t)$ in the region of the strong solute–interface coupling (Fig. 3.8b).

The restoring force is especially strong in regions of large sensitivity of the interface deformation to the solute position. In such regions, a small solute displacement causes a large change in the average protrusion magnitude and hence a strong restoring force. These arguments suggest that the slow component of the random force $\Gamma(z_s; t)$ is the largest in the region $z_3 \leq z_s \leq z_4$ shown in Fig. 3.9, which is consistent with the region of the large correlation time of $\Gamma(z_s; t)$ obtained directly from MD simulations (Fig. 3.8b).

To fully understand the solute–interface coupling, it is necessary to develop a stochastic model in which both the solute coordinate and the interface configuration are treated as reaction coordinates. A phenomenological model for the solute–interface dynamics was proposed by Daikhin *et al.* [67] for an ionic solute at an interface between two electrolyte liquids. In ref. [66] we used constrained MD simulations to validate and extend this model for a wider class of interfaces, including nonionic fluid–fluid and surfactant-covered interfaces. Moreover, the latter work established a method to obtain all stochastic model parameters directly from MD simulations, thus enabling quantitative predictions of solute transport rates.

The main assumption of this stochastic model is that the interface and solute dynamics can be described adequately by (1) the potential of mean force, which depends on the solute coordinate z_s and the interface configuration, and (2) random forces acting on the solute and the interface. Since the capillary waves at the interface are linear, it is convenient to use Fourier series to describe the configuration $h(x,y; t)$ of the instantaneous dividing surface,

$$h(x, y; t) = \sum_{|\mathbf{k}| \leq k_{\mathrm{cut}}} \hat{h}_{\mathbf{k}}(t) e^{i(k_x x + k_y y)}, \tag{3.23}$$

where $\mathbf{k} = (k_x, k_y)$ is a wavevector and $k_{\mathrm{cut}} = 2\pi\,\mathrm{nm}^{-1}$ is the cutoff wavevector magnitude, which corresponds approximately to two diameters of the coarse-grained beads.

As shown in ref. [66], the potential $G(z_s, \{\hat{h}_{\mathbf{k}}\})$ of mean force acting between the solute located at position z_s and the interface in configuration defined by Fourier coefficients $\{\hat{h}_k\}$ is

$$G(z_s; \{\hat{h}_\mathbf{k}\}) = G_0(z_s) + \frac{1}{2}\sum_{\mathbf{k}\neq 0} \alpha_\mathbf{k}(z_s)|\hat{h}_\mathbf{k} - \hat{h}_\mathbf{k}^{(0)}(z_s)|^2. \tag{3.24}$$

Here $\hat{h}_\mathbf{k}^{(0)}(z_s)$ are the Fourier coefficients of the averaged dividing surface $h^{(0)}(\mathbf{r}; z_s)$, and the coefficients $\alpha_\mathbf{k}(z_s) > 0$ specify the fluctuation magnitude of the interface mode $\hat{h}_\mathbf{k}$ when the solute is constrained at $z = z_s$. The function $G_0(z_s)$ specifies the potential of the mean force acting on the solute when the interface shape coincides with $h^{(0)}(\mathbf{r}; z_s)$. The solute–interface coupling is described by the sum on the right-hand side of Eq. (3.24). This sum is finite due to natural cutoff values for magnitudes of the wavevector $\mathbf{k}$. The cutoff for large $|k|$ (i.e., short waves) is due to finite molecular size. For small $|k|$ (i.e., long waves), the cutoff corresponds to the wavelength at which the gravity and/or nonlinear effects become sufficiently large to limit the magnitude of the wave fluctuations.

From Eq. (3.24) it follows that the system dynamics is described by the following system of Langevin equations:

$$\gamma_s(z_s)\frac{dz_s}{dt} = -G'_0(z_s) - \frac{1}{2}\frac{d}{dz_s}\left[\sum_{\mathbf{k}\neq 0}\alpha_\mathbf{k}(z_s)\left|\hat{h}_\mathbf{k} - \hat{h}_\mathbf{k}^{(0)}(z_s)\right|^2\right] + \Gamma_s(t; z_s), \tag{3.25}$$

$$\gamma_\mathbf{k}(z_s)\frac{d\hat{h}_\mathbf{k}}{dt} = -\alpha_\mathbf{k}(z_s)\left[\hat{h}_\mathbf{k} - \hat{h}_\mathbf{k}^{(0)}(z_s)\right] + \Gamma_\mathbf{k}(t; z_s), \quad \mathbf{k} \neq 0 \tag{3.26}$$

In these equations, $\gamma_s(z_s) > 0$ and $\gamma_\mathbf{k}(z_s) > 0$ are friction coefficients and $\Gamma_s(t; z_s)$ and $\Gamma_\mathbf{k}(t; z_s)$ are Markovian random forces acting on the solute and interface modes, respectively. These friction and stochastic forces arise due to thermal interactions between individual molecules. All stochastic forces are independent of each other and satisfy the fluctuation–dissipation theorem,

$$\langle \Gamma_i(t; z_s)\Gamma_j^*(t+\tau; z_s)\rangle = 2k_BT\gamma_i(z_s(t))\delta_{ij}\delta(\tau), \tag{3.27}$$

where the indices i and j take on all possible values referring to the solute (s) or a wavevector ($\mathbf{k} \neq 0$) of an interface mode.

The non-Markovian forces due to interaction between the collective degrees of freedom of the monolayer (i.e., the monolayer Fourier modes) and the solute are captured by the second term on the right-hand side of Eq. (3.25) and the first term on the right-hand side of Eq. (3.26). The latter equation indicates that in addition to the interface deformation, solute may influence the magnitude, $k_BT/\alpha_\mathbf{k}(z_s)$, and the correlation time, $\tau_\mathbf{k}(z_s) = \gamma_\mathbf{k}(z_s)/\alpha_\mathbf{k}(z_s)$, of the interface fluctuations.

The effect of the capillary interface fluctuations on the solute dynamics is significant if $\alpha_\mathbf{k}(z_s)$ or $\hat{h}_\mathbf{k}^{(0)}(z_s)$ exhibit strong dependence on z_s. In this case, the time scale of the force fluctuations is determined by the time scales $\tau_\mathbf{k}(z_s)$ of the corresponding interface modes. Whereas the dependence of $\alpha_\mathbf{k}(z_s)$ on the solute position is relatively weak [66], $\hat{h}_\mathbf{k}^{(0)}(z_s)$ exhibits strong dependence on z_s, as illustrated in Fig. 3.10a. The gradient of $\hat{h}_\mathbf{k}^{(0)}(z_s)$ is much larger on the hydrophilic

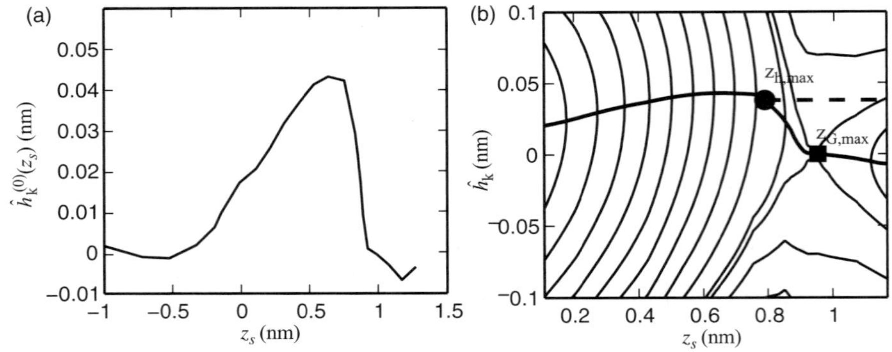

Figure 3.10 (a) Dependence of $\hat{h}_{\mathbf{k}}^{(0)}$ on z_s for a wavevector k with magnitude 0.64 nm^{-1}; (b) projection of the free energy $G(z_s; \{\hat{h}_{\mathbf{k}}\})$ on the plane z_s–$\hat{h}_{\mathbf{k}}$ with $|k| = 0.64$ nm^{-1}. The minimal energy path is shown by a thick solid line and the path assumed in the calculation of the mean transport time $T[\gamma_s, \tilde{G}]$ is shown by the dashed line. The point $z_{h,\max}$ corresponding to the maximum of $\hat{h}_{\mathbf{k}}^{(0)}(z_s)$ is shown by a circle and the point $z_{G,\max}$ corresponding to the maximum of $G_0(z_s)$ is shown by a square.

side of the interface, which is consistent with the physical picture of the protrusion formation due to attraction between the solute and the hydrophobic beads. Equation (3.25) then implies that the contribution of the interface fluctuations to the solute dynamics is significant only on the water side of the interface, which is consistent with the observation that the random force fluctuations are slow only on one side of the interface (Fig. 3.8b). Moreover, the gradient of $\hat{h}_{\mathbf{k}}^{(0)}(z_s)$ is the largest for z_s corresponding to the peaks of the force correlation time $\tau_f(z_s)$.

To conclude this section, we note that the presence of the solute in the region of a high gradient of $\hat{h}_{\mathbf{k}}^{(0)}(z_s)$ may lead to a significant increase of the relaxation time $\tau_{\mathbf{k}}$ of the interface fluctuations. For example, the correlation time $\tau_{\mathbf{k}}$ corresponding to a wavevector of magnitude $|k| = 0.64$ nm^{-1} increases from 285 ps in the absence of the solute, to 517 ps when the solute is located in the region of high gradient of $\hat{h}_{\mathbf{k}}^{(0)}(z_s)$ [66].

3.3.3.3 Solute Transport Rate Equations (3.25) and (3.26) imply that the coupled solute–interface dynamics corresponds to random motion on the multidimensional free-energy landscape $G(z_s; \{\hat{h}_{\mathbf{k}}\})$. From Eq. (3.24) it follows that the minimum energy path (MEP) on the surface $G(z_s; \{\hat{h}_{\mathbf{k}}\})$ is a path such that $\hat{h}_{\mathbf{k}} = \hat{h}_{\mathbf{k}}^{(0)}(z_s)$ for all $\mathbf{k}$ at every solute position z_s. The system free energy on this path, $G_{\mathrm{MEP}}(z_s; \{\hat{h}_{\mathbf{k}}\}) = G_0(z_s)$, does not depend on the interface configuration and it may appear that the solute–interface coupling does not affect the solute transport. However, the solute–interface coupling influences the MEP geometry, which in turn affects the solute transport.

This situation is illustrated in Fig. 3.10b, which shows projections of $G(z_s; \{\hat{h}_k\})$ and MEP on a $z_s - \hat{h}_{\mathbf{k}}$ plane in the region of strong solute–interface coupling. Once the solute passes point $z_{h,\max}$ corresponding to the maximum of $\hat{h}_{\mathbf{k}}^{(0)}(z_s)$, MEP makes a sharp turn toward $\hat{h}_{\mathbf{k}}^{(0)} = 0$. The origin of the sharp decrease of $\hat{h}_{\mathbf{k}}^{(0)}(z_s)$ is disappearance of the interface protrusion once it becomes energetically unfavorable. This is expected to be a common feature of transport across flexible interfaces. Moreover, this turn is likely always to directly precede point $z_{G,\max}$ corresponding to the transition state, because once the protrusion disappears, the solute energy is determined only by its interactions with the surrounding unfavorable phase.

The prediction of transport rate in a system with multiple reaction coordinates requires generalization of Eq. (3.5) to a stochastic system with multiple degrees of freedom. This generalization can be developed, for example, if the system motion, on average, takes place along the MEP. In this case, the integrations are performed along the MEP and the system dynamics in directions transversal to the MEP is captured by factors corresponding to frequencies of oscillations in these directions. The underlying assumption of this generalization is that the time scale of approach to the MEP is much faster than the time scale of motion along the MEP.

This assumption fails for the coupled solute–interface dynamics when the solute is located between points $z_{h,\max}$ and $z_{G,\max}$. Due to the proximity of these points, the difference between the free energies along the MEP passing through these points is $G_0(z_{G,\max}) - G_0(z_{h,\max}) = 2.5$ kJ/mol. This is close to the average contribution of each interface mode to the free energy, $k_B T = 2.5$ kJ/mol at $T = 300$ K. In other words, the driving force for approach toward MEP is comparable with the driving force for motion along the MEP between $z_{h,\max}$ and $z_{G,\max}$. In addition, the friction coefficients of the interface modes, $\gamma_{\mathbf{k}} = O(10^{14}$ kg/mol·s), exceed the solute friction coefficient, $\gamma_s = O(10^{12}$ kg/mol·s), by two orders of magnitude [66]. These two observations imply that near $z_{G,\max}$ the change of the interface configuration is much slower than the solute transport. Therefore, the solute will pass through point $z_{G,\max}$ before the disappearance of the protrusion formed when the solute was at $z_{h,\max}$.

To estimate the effect of the slow interface relaxation on the solute transport rate, we assume that the interface configuration remains unchanged for $z_s \geq z_{h,\max}$. The change of $\hat{h}_{\mathbf{k}}^{(0)}(z_s)$ with increase of z_s is much more gradual when $z_s \leq z_{h,\max}$ (Fig. 3.10a), which allows us to assume that the system moves along MEP when $z_s \leq z_{h,\max}$. The free energy along the assumed path is

$$\tilde{G}(z_s) = \begin{cases} G_0(z_s), z_s \leq z_{h,\max}, \\ G_0(z_s) + \dfrac{1}{2}\displaystyle\sum_{\mathbf{k}\neq 0} \alpha_{\mathbf{k}}(z_s)\left|\hat{h}_{\mathbf{k}}^{(0)}(z_{h,\max}) - \hat{h}_{\mathbf{k}}^{(0)}(z_s)\right|^2, \qquad z_s \geq z_{h,\max}. \end{cases} \tag{3.28}$$

Weak dependence of $\alpha_{\mathbf{k}}(z_s)$ on z_s allows us to neglect changes in frequencies ω_j of the system oscillations in the directions transversal to the MEP when $z_s \leq z_{h,\max}$. The changes in ω_j can also be neglected for $z_s \geq z_{h,\max}$ since the interface is assumed to be frozen on the time scale of the solute motion in this region. These assumptions allow us to estimate the mean time of the solute transport across the interface using Eq. (3.5) with $\gamma = \gamma_s$ and $G = \tilde{G}$. The solute friction coefficient $\gamma_s(z_s)$ is obtained from Eq. (3.15) using ACF of the Markovian force $\Gamma_s(t; z_s)$ acting on the solute. The latter is obtained by removing the slowly decaying component from ACF of the random force $\Gamma_{MD}(t; z_s)$ measured in the constrained MD simulations. The mean transport time obtained is $T[\gamma_s, \tilde{G}] = 4.13 \times 10^4$ s.

For comparison, a naive use of MD results based on the assumption that the system follows the MEP and the friction coefficient γ_{MD} corresponds to ACF of $\Gamma_{MD}(t; z_s)$ leads to an overestimate of the mean transport time, $T[\gamma_{MD}, G_0] = 1.45 \times 10^5$ s. This is caused by an overestimate of the friction coefficient due to incorrect treatment of the interface fluctuation modes. On the other hand, if we use the corrected friction coefficient γ_s while still assuming that the motion takes place along the MEP, the mean transport time is underestimated, $T[\gamma_s, G_0] = 2.45 \times 10^4$ s. These observations suggest that the effect of the dynamic solute–interface coupling is somewhat similar to friction acting on the solute. This effective friction is neglected in calculation of $T[\gamma_s, G_0]$ and is overestimated in calculation of $T[\gamma_{MD}, G_0]$, since in the latter case the slow and fast contributions to the stochastic force are not separated correctly.

3.4 DISCUSSION

In this chapter we considered the development of models connecting coarse-grained MD simulations with stochastic models for the following rare events in self-assembled systems:(A) removal of a surfactant monomer from a surfactant cluster, (B) addition of a monomer to a cluster, and (C) solute transport across a surfactant monolayer. Each of these processes can be thought of as a transport process along a translational degree of freedom ξ. For processes A and B, this translational degree of freedom is the distance ξ_1 between the monomer and cluster centers of mass. For process C, this degree of freedom is the distance z_s between the solute center of mass and the dividing surface of the monolayer. Whereas process A is adequately described by a stochastic model involving only the translational degree of freedom (with all other degrees of freedom contributing only to thermal bath), processes B and C involve cooperative dynamics of several degrees of freedom, which should be included explicitly in the stochastic model.

The relative importance of different degrees of freedom can be assessed by comparing their time scales in the neighborhood of a transition state. A one-dimensional stochastic model provides an adequate model of a rare event if the system dynamics near the transition state can be reduced to stochastic dynamics of a point moving along the minimum energy path (MEP). This assumption holds if the time scale of approach toward the MEP can be neglected in comparision with motion along the MEP. This assumption breaks down if time scales of motion in directions transversal to the MEP are comparable to the time scale of motion along the MEP near the transition state. This situation is observed in process C, where the time scale of the interface deformation is comparable to that of the solute motion across the interface near the free-energy barrier (Fig. 3.10b). In process B, the motion toward the MEP may be retarded significantly if at least one of the additional degrees of freedom (monomer orientation ξ_2 or the cluster surface microstructure ξ_3) is in an unfavorable configuration. In this case the approach towards the MEP will involve overcoming a high free-energy barrier caused by an unfavorable configuration. This motion will, in fact, be much slower than motion along MEP once it is reached!

Although the approach toward the MEP is dominated by different factors in processes A, B, and C, the major factors governing the free energy along the MEP—density of solute and surfactants and hydrophobic/hydrophilic interactions—are the same for all processes considered. In fact, even the approach toward the MEP in these processes appears to involve all of the additional degrees of freedom discussed in this chapter, albeit with varying importance.

For example, micellar surface fluctuations appear to be involved in processes A and B, similarly to the monolayer fluctuations in process C. In MD simulations of a cluster–monomer system with constrained translational degree of freedom ξ_1, we observe formation of protrusions at the micellar surface. Recall that protrusion formation plays a key role in the solute–interface coupling (see Fig. 3.9 and its discussion). This suggests the existence of coupling between the monomer and the cluster surface fluctuations, which should manifest itself in a slowly fluctuating component of the stochastic force $\Gamma(t;\ \xi)$ acting on the solute. Indeed, a slowly decaying component is observed for ACF of $\Gamma(t;\ \xi)$ at some values of ξ. However, this component is relatively small in the micellar systems considered and is therefore neglected in our analysis of processes A and B. This smallness of the slow random force component is caused by the relatively small amplitude of the micellar surface fluctuations and the relatively few available modes of vibration in comparison with the flat interface.

Similarly to the micellar microstructure in process B, the monolayer microstructure is likely to be involved in process C. The aspect of the monolayer microstructure most relevant to the solute transport is the distribution of pores accessible to the solute inside the monolayer. This conjecture is confirmed by our MD simulations of transport of a small hydrophobic solute from water to oil across the H_3T_3 monolayer. This monolayer imposes a relatively low barrier for solute transport, and once the solute enters a monolayer, it is transported to the oil phase within 100 to 200 ps. However, the solute may spend significant amounts of time in the water phase while attempting unsuccessfully to enter the monolayer, similarly to a surfactant monomer attempting to enter a surfactant cluster. The solute entry into the monomer is successful and leads to water-to-oil transport only when the solute enters the monolayer through a sufficiently large pore. We were able to neglect the monolayer microstructure in our analysis of process C because in Section 3.3.3 we focused on the oil-to-water transport of a hydrophobic solute, which is much slower than the fluctuations of the monolayer microstructure.

3.5 FUTURE DIRECTIONS

3.5.1 Improvement of the Model for Micellar Formation and Disintegration

As we discussed in Section 3.3.2, a stochastic model for addition of a surfactant monomer to a surfactant cluster should consider interaction of the translational degree of freedom ξ_1 with the monomer orientation ξ_2 and the cluster microstructure ξ_3. The model developed for the coupled dynamics of ξ_1, ξ_2, and ξ_3 yields a qualitative improvement over the one-dimensional model for ξ_1 only (Fig. 3.6). However, approximations introduced into the multidimensional stochastic model lead to quantitative differences between predictions of this model and the results of thermodynamic calculations.

The current model assumes that the range of the monomer orientations leading to a successful entry into the cluster is the same for all hydrophobic patches on the cluster surface. However, our observations of unconstrained MD simulations of the monomer entry indicate that the range of the favorable monomer orientations depends on the patch area and on whether the monomer contacts the patch in its central or border region. In addition, a monomer whose orientation deviates significantly from the normal to the cluster surface is less likely to enter the cluster than is a monomer oriented along the normal. A more accurate representation of the reaction conditions would assign probabilities of a successful monomer entry into the cluster based on the monomer orientation, patch size, and the location of the patch–monomer contact. Moreover, in the current model we assume that the patch shape and location do not change during several consecutive contacts with the monomer. A more accurate reaction model would include changes in the patch shape and location due to fluctuations of the surfactants within the cluster, and cluster interaction with the monomer during an entry attempt.

The reaction conditions for the current model were determined from (1) constrained simulations (to obtain distribution of favorable monomer orientations), and (2) equilibrium MD simulations of individual clusters (to obtain surface patch structures). To implement most of the corrections outlined above it is necessary to decipher short-scale dynamics of interactions between the monomer and the cluster. The methods used in the current work are not applicable to investigations of these nonequilibrium processes. One possible approach is to follow the method of refs. [68,69] and obtain the reaction conditions based on a series of short-scale nonequilibrium simulations initialized at various initial surfactant and cluster configurations. Multiple repetition of such simulations with the same initial condition for the reaction coordinates (ξ_1, ξ_2, ξ_3) but different initial thermal bath configurations (e. g., locations of the solvent molecules) will yield probability P(ξ_1, ξ_2, ξ_3) of the monomer entry into the cluster for given values of the reaction coordinates. Such probabilities obtained for a range of the reaction coordinates can then be incorporated into the stochastic model to improve the reaction conditions. Similarly, one can use these short-scale simulations to assess the effects of the monomer–cluster interactions on patch shape and location.

Another possible source of the discrepancy between stochastic and thermodynamic models is the assumption that the stepwise monomer addition or removal is the only mechanism of micelle formation and disintegration. To complete this model, it is necessary to take the fusion and fission of surfactant clusters into account. As in the case of the monomer addition to clusters, the dynamics of the hydrophobic patches on the cluster surfaces plays a crucial role in the cluster fusion. Our MD simulations of cluster fusion demonstrate that prior to fusion the molecules within the clusters rearrange themselves so that the hydrophobic cores of the clusters are exposed to each other. Therefore, further understanding of the cluster surface dynamics should enable development of a model for cluster fusion.

3.5.2 Extension to Other Self-Assembled Systems

In this chapter we discussed models for micelles and monolayers composed of nonionic surfactants. Most systems of technological and biological importance contain ionic amphiphiles, which requires extension of the developed models to ionic self-assembled structures. This extension is also of theoretical interest since it is expected that long-range electrostatic interactions will strengthen coupling between different degrees of freedom. An additional collective degree of freedom in ionic systems—distribution of counterions in solution—is also expected to contribute to the self-assembly and transport processes and may lead to new phenomena.

The mechanisms of rare events considered in this chapter are rather general and are expected to govern similar processes in other self-assembled systems. For example, fusion of surfactant clusters is similar to a large number of other processes occurring in self-assembled systems, such as transitions from spherical to wormlike micelles and fusion of lipid vesicles. Understanding the dynamics of microstructure of self-assembled aggregates gained in the analysis of micellar formation is therefore expected to aid in the development of models for these more complex systems.

As we discussed in Section 3.3.3, the main condition for the coupling between the solute and interface dynamics is attraction between the solute and one of the phases separated by the interface. This condition is satisfied in a large number of interfacial systems, and therefore the solute–interface coupling is expected to affect transport across most fluid and flexible membranes. We have already seen in Section 3.4 that a monomer partially inserted into a micelle leads to protrusion formations, implying that the micellar fluctuations are coupled (in this case, weakly) with the monomer dynamics.

The solute–interface coupling is also likely to contribute to interactions of lipid membranes with solute particles. Several studies of transport across lipid membranes (e.g., [20,70]) report a significant local increase in the friction coefficient. A probable cause for this increase is a strong solute–membrane coupling modeled by a random force in a one-dimensional Langevin equation [66]. In addition to affecting the solute transport through a bilayer, the dynamic membrane–solute coupling may alter the physical properties of the lipid membranes. Interaction of a cellular membrane with particles embedded into it is believed to play a crucial role in various biological processes, ranging from the toxicity of antimicrobial peptides [71] and certain nanoparticles [72] to anesthesia [73]. The dynamic membrane–solute coupling may therefore contribute to these processes.

Acknowledgments

This research was supported by the National Science Foundation through a Career Award (grant CBET-0644089) and grant CTS-0500090. Computational resources were provided in part by the University of Florida High-Performance Computing Center.

REFERENCES

1. Israelachvili J. *Intermolecular and Surface Forces*, 2nd ed., Wiley, New York, 1989.
2. Nagarajan R, Ruckenstein E. Theory of surfactant self-assembly: a predictive molecular thermodynamic approach. *Langmuir* **7**:2934–2969, 1991.
3. Helfrich W. Elastic properties of lipid bylayers: theory and possible experiments. *Z. Naturforsch.* **28**:693, 1973.
4. Hamm M, Kozlov MM. Tilt model of inverted amphiphilic mesophases. *Eur. Phys. J. B* **6**:519–528, 1998.
5. Hamm M, Kozlov MM. Elastic energy of tilt and bending of fluid membranes. *Eur. Phys. J. E* **3**:323–335, 2000.
6. Gompper G, Klein S. Ginzburg–Landau theory of aqueous surfactant solutions. *J. Phys. II France* **2**:1725, 1992.
7. Fredrickson GH, Ganesan V, Drolet F. Field-theoretic computer simulation methods for polymers and complex fluids. *Macromolecules* **35**:16–39, 2002.
8. Christopher PS, Oxtoby DW. Density functional model of surfactant mesostructures. *J. Chem. Phys.* **117**:9502, 2002.
9. Patist A, Kanicky JR, Shukla PK, Shah DO. Importance of micellar kinetics in relation to technological processes. *J. Colloid Interface Sci.* **245**:1–15, 2002.
10. Paruchuri VK, Nalaskowski J, Shah DO, Miller JD. The effect of cosurfactants on sodium dodecyl sulfate micellar structures at a graphite surface. *Colloide Surf. A* **272**:157–163, 2006.
11. Adair JH, Li T, Kido T, Havey K, Moon J, Mecholsky J, Morrone A, Talham DR, Ludwig MH, Wang L. Recent developments in the preparation and properties of nanometer-size spherical and platelet-shaped particles and composite particles. *Mater. Sci. Eng.* **R23**:139–242, 1998.
12. Karanikolos GN, Alexandridis P, Itskos G, Petrou A, Mountziaris TJ. Synthesis and size control of luminescent ZnSe nanocrystals by a microemulsion-gas contacting technique. *Langmuir* **20**:550–553, 2004.
13. Walker LM. Rheology and structure of worm-like micelles. *Curr. Opin. Colloid. Interface. Sci.* **6**:451–456, 2001.
14. Davies TS, Ketner AM, Raghavan SR. Self-assembly of surfactant vesicles that transform into viscoelastic wormlike micelles upon heating. *J. Am. Chem. Soc.* **128**:6669–6675, 2006.
15. Burger KNJ. Greasing membrane fusion and fission machineries. *Traffic* **1**:605–613, 2000.
16. Chernomordik LV, Kozlov MM. Protein–lipid interplay in fusion and fission of biological membranes. *Annu. Rev. Biochem.* **72**:175–207, 2003.
17. Moulik SP, Paul BK. Structure, dynamics and transport properties of microemulsions. *Adv. Colloid Interface Sci.* **78**:99–195, 1998.
18. Bagwe RP, Kanicky JR, Palla BJ, Patanjali PK, Shah DO. Improved drug delivery using microemulsions: rationale, recent progress, and new horizons. *Crit. Rev. Ther. Drug Carrier Sys.* **18**:77–140, 2001.

19. Finkelstein A. Water and nonelectrolyte permeability of lipid bilayer membranes. *J. Gen. Physiol.* **68**:127–135, 1976.
20. Marrink SJ, Berendsen HJC. Simulation of water transport through a lipid membrane. *J. Phys. Chem.* **98**:4155–168, 1994.
21. Paula S, Volkov AG, Deamer DW. Permeation of halide anions through phospholipid bilayers occurs by the solubility-diffusion mechanism. *Biophys. J.* **74**:319–327, 1998.
22. Kwon SY, Kim MW. Topological transition in aqueous nonionic micellar solutions. *Phys. Rev. Lett.* **89**:258302, 2002.
23. Mendes E, Oda R, Manohar C, Narayanan J. A small-angle neutron scattering study of a shear-induced vesicle to micelle transition in surfactant mixtures. *J. Phys. Chem. B* **102**:338–343, 1998.
24. Maillet J-B, Lachet V, Coveney PV. Large scale molecular dynamics simulation of self-assembly processes in short and long chain cationic surfactants. *Phys. Chem. Chem. Phys.* **1**:5277–5290, 1999.
25. Marrink SJ, Tieleman DP, Mark AE. Molecular dynamics simulations of the kinetics of spontaneous micelle formation. *J. Phys. Chem. B* **104**:12165–12173, 2000.
26. Garde S, Yang L, Dordick JS, Paulaitis ME. Molecular dynamics simulation of C_8E_5 micelle in explicit water: Structure and hydrophobic solvation thermodynamics. *Mol. Phys.* **14**:2299–2306, 2002.
27. von Gottberg FK, Smith KA, Hatton TA. Dynamics of self-assembled surfactant systems. *J. Chem. Phys.* **108**:2232–2244, 1998.
28. Noguchi H, Takasu M. Fusion pathways of vesicles: a Brownian dynamics simulation. *J. Chem. Phys.* **115**:9547, 2001.
29. Jury S, Bladon P, Cates M, Krishna S, Hagen M, Ruddock N, Warren P. Simulations of amphiphilic mesophases using dissipative particle dynamics. *Phys. Chem. Chem. Phys.* **1**:2051–2056, 1999.
30. Kranenburg M, Venturoli M, Smit B. Molecular simulations of mesoscopic bilayer phases. *Phys. Rev. E* **67**:060901(R), 2003.
31. Qi S, Wang Z-G. Kinetic pathways of order–disorder and order–order transitions in weakly segregated microstructured systems. *Phys. Rev. Lett.* **76**:1679–1682, 1996.
32. Ganesan V, Pryamitsyn V. Dynamical mean-field theory for inhomogeneous polymeric systems. *J. Chem. Phys.* **118**:4345–4348, 2003.
33. Aniansson EAG, Wall SN. On the kinetics of step-wise micelle association. *J. Phys. Chem.* **78**:1024, 1974.
34. Neu JC, Cañizo JA, Bonilla LL. Three eras of micellization. *Phys. Rev. E* **66**:061406, 2002.
35. Verlet L. Computer experiments on classical fluids: I Thermodynamical properties of Lennard-Jones molecules. *Phys. Rev.* **159**:98–103, 1967.
36. Berendsen HJC. *Simulating the Physical World: Hierarchical Modeling from Quantum Mechanics to Fluid Dynamics*, Cambridge University Press, Cambridge, UK, 2007.
37. Nosé S. Constant temperature molecular dynamics methods. *Prog. Theor. Phys. Suppl.* **103**:1, 1991.
38. Berendsen HJC, Postma JPM, van Gunsteren WF, DiNola A, Haak JR. Molecular dynamics with coupling to an external bath. *J. Chem. Phys.* **81**:3684–3690, 1984.
39. Reith D, Pütz M, Müller-Plathe F. Deriving effective mesoscale potentials from atomistic simulations. *J. Comput. Chem.* **24**:1624–1636, 2003.

40. Izvekov S, Voth GA. A multiscale coarse-graining method for biomolecular systems. *J. Phys. Chem. B* **109**:2469–2473, 2005.

41. Srinivas G, Klein ML. Molecular dynamics simulations of self-assembly and nanotube formation by amphiphilic molecules in aqueous solution: a coarse-grain approach. *Nanotechnology* **18**:205703, 2007.

42. Marrink SJ, de Vries AH, Mark AE. Coarse grained model for semiquantitative lipid simulations. *J. Phys. Chem. B* **108**:750–760, 2004.

43. Marrink SJ, Risselada HJ, Yefimov S, Tieleman DP, de Vries AH. The MARTINI force field: coarse grained model for biomolecular simulations. *J. Phys. Chem. B* **111**:7812–7824, 2007.

44. Gupta A, Chauhan A, Kopelevich DI. Molecular modeling of surfactant covered oil–water interfaces: dynamics, microstructure, and barrier for mass transport. *J. Chem. Phys.* **128**:234709, 2008.

45. Mohan G, Kopelevich DI. A multi-scale model for kinetics of formation and disintegration of spherical micelles. *J. Chem. Phys.* **128**:044905, 2008.

46. van der Spoel D, Lindahl E, Hess B, Groenhof G, Mark AE, Berendsen HJC. GROMACS: Fast, flexible, and free. *J. Comput. Chem.* **26**:1701–1718, 2005.

47. Humphrey W, Dalke A, Schulten K. VMD: Visual molecular dynamics. *J. Mol. Graph.* **14**:33–38, 1996.

48. Pool R, Bolhuis PG. Accurate free energies of micelle formation. *J. Phys. Chem. B* **109**:6650–6657, 2005.

49. Mukerjee P. Mysels KJ, Critical Micelle Concentrations of Aqueous Surfactant Systems. *Nat. Stand. Ref. Data Ser., Nat. Bur. Stand. (Washington, DC)* **36**,1971.

50. Eyring H. The activated complex in chemical reactions. *J. Chem. Phys.* **3**:107, 1935.

51. Kramers HA. Brownian motion in a field of force and the diffusion model of chemical reactions. *Physica* **7**:284, 1940.

52. Gardiner GW. *Handbook of Stochastic Methods for Physics, Chemistry and the Natural Sciences*, Springer-Verlag, Berlin, 1983.

53. Ermak DL, McCammon JA. Brownian dynamics with hydrodynamic interactions. *J. Chem. Phys.* **69**:1352–1360, 1978.

54. Cobb PD, Butler JE. Brownian dynamic simulations of concentrated suspensions of rigid fibers: relationship between short-time diffusivities and the long-time rotational diffusion. *J. Chem. Phys.* **12**:054908, 2005.

55. Zwanzig R. *Nonequilibrium Statistical Mechanics*, Oxford University Press, New York, 2001.

56. Roux B, Karplus M. Ion transport in a gramicidin-like channel: dynamics and mobility. *J. Phys. Chem.* **95**:4856–4868, 1991.

57. Sprik M, Ciccotti G. Free energy from constrained molecular dynamics. *J. Chem. Phys.* **109**:7737–7744, 1998.

58. den Otter WK, Briels WJ. The calculation of free-energy differences by constrained molecular-dynamics simulations. *J. Chem. Phys.* **109**:4139–4146, 1998.

59. Haynes GR, Voth GA, Pollak E. A theory for the activated barrier crossing rate constant in systems influenced by space and time dependent friction. *J. Chem. Phys.* **101**:7811–7822, 1994.

60. Ryckaert JP, Ciccotti G, Berendsen HJC. Numerical integration of the Cartesian equations of motion of a system with constraints: molecular dynamics of *n*-alkanes. *J. Comput. Phys.* **23**:327–341, 1977.

61. Pool R, Bolhuis PG. Prediction of an autocatalytic replication mechanism for micelle formation. *Phys. Rev. Lett.* **97**:18302, 2006.

62. Smoluchowski MV. Engineering biological structures of prescribed shape using self-assembling multicellular systems. *Z. Phys. Chem.* **92**:129–168, 1917.

63. Evans DF, Wennerström H. *The Colloidal Domain: Where Physics, Chemistry, Biology and Technology Meet*, 2nd ed., Wiley–VCH, New York, 1999.

64. Northrup SH, Allison SA, McCammon JA. Brownian dynamics simulation of diffusion-influenced bimolecular reactions. *J. Chem. Phys.* **80**:1517–1524, 1984.

65. Patel HA, Nauman EB, Garde S. Molecular structure and hydrophobic solvation thermodynamics at an octane–water interface. *J. Chem. Phys.* **119**:9199–9206, 2003.

66. Gupta A, Chauhan A, Kopelevich DI. Molecular transport across fluid interfaces: coupling between solute dynamics and interface fluctuations. *Phys. Rev. E* **78**:041605, 2008.

67. Daikhin LI, Kornyshev AA, Kuznetsov AM, Urbakh M. ITES fluctuations induced by easily transferable ions. *Chem. Phys.* **319**:253–260, 2005.

68. Hummer G, Kevrekidis IG. Coarse molecular dynamics of a peptide fragment: free energy, kinetics, and long-time dynamics computations. *J. Chem. Phys.* **118**:10762, 2003.

69. Kopelevich DI, Panagiotopoulos AZ, Kevrekidis IG. Coarse kinetic approach to rare events: application to micelle formation. *J. Chem. Phys.* **122**:044908, 2005.

70. Kopelevich DI, Bonzongo JC, Tasseff RA, Gao J, Ban Y-M, Bitton G. Potential toxicity of fullerenes and molecular modeling of their transport across lipid membranes, In: *Nanoscience and Nanotechnology: Environmental and Health Impacts*, ed. VH Grassian, Wiley, Hoboken, NJ, pp. 235–260, 2008.

71. Epand RM, Vogel HJ. Diversity of antimicrobial peptides and their mechanisms of action. *Biochim. Biophys. Acta* **1462**:11–28, 1999.

72. Leroueil PR, Hong S, Mecke A, Baker JR, Orr BG, Holl MMB. Nanoparticle interaction with biological membranes: Does nanotechnology present a Janus face? *Acc Chem. Res.* **40**:335–342, 2007.

73. Eckenhoff RG. Promiscuous ligands and attractive cavities: How do the inhaled anesthetics work? *Mol. Intervent.* **1**:258–268, 2001.

SUGGESTED ADDITIONAL READINGS

Introduction to stochastic modeling and methods of solution of SDE:

Gardiner GW. *Handbook of Stochastic Methods for Physics, Chemistry and the Natural Sciences*, Springer-Verlag, Berlin, 1983.

Introduction to self-assembled and interfacial systems:

Israelachvili J. *Intermolecular and Surface Forces*, 2nd ed., Wiley, New York, 1989.

Safran SA. *Statistical Mechanics of Surfaces, Interfaces, and Membranes*, Westview Press, Boulder, CO, 2003.

A possible approach to development of a model for microstructure dynamics of self-assembled aggregates is based on percolation theory [44]; good introductory text on percolation theory:

Stauffer D, Aharony A. *Introduction to Percolation Theory*, Revised 2nd ed., Taylor & Francis, London, 1994.

4

CONTINUUM DESCRIPTION OF ATOMIC SHEETS

CONSTANTINE POZRIKIDIS

Department of Chemical Engineering, University of Massachusetts, Amherst, Massachusetts

4.1 INTRODUCTION

Viewed from a distance, a particulate atomic lattice forming a planar or curved surface resembles a material sheet. A planar graphene sheet consisting of a hexagonal lattice of carbon atoms appears like a thin material layer, and a cylindrical carbon nanotube appears like a thin-walled conduit. Because in applications of interest in nanotechnology the typical macroscopic size of atomic lattices is comparable to the interatomic distance of its constituents, the particulate nature of the sheet may not be ignored. In contrast, the typical size of a three-dimensional lattice comprising a solid crystalline material is typically much smaller than both its macroscopic size and the spatial scale of an imposed deformation, and the Cauchy–Born rule bridging microscopic to macroscopic deformation can safely be applied (e.g., [1,2]). The rule essentially relates a set of base atomic lattice vectors defining a simple, dual, or multiple lattice before deformation to corresponding lattice vectors after deformation in terms of the macroscopic deformation gradient. The rule does not provide information on the atomic arrangement inside a unit cell.

Efforts to develop continuum theories for describing the mechanical behavior of atomic sheets are faced with conceptual difficulties regarding the definition and computation of macroscopic kinematic and dynamic properties from atomic positions, and vice versa. Even in the idealized case of a homogeneous deformation in the

Multiscale Modeling of Particle Interactions: Applications in Biology and Nanotechnology, Edited by Michael R. King and David J. Gee

planar or cylindrical geometry, hidden degrees of freedom are encountered, determining the separation of the constituent triangular lattices comprising the hexagonal lattice of the graphene sheet. To compute equilibrium configurations, energy minimization must be carried out with respect to this inner displacement, amounting to inner relaxation [3]. This degree of freedom does not have a counterpart in the standard theory of elastic continua. Additional conceptual and practical subtleties arise in the case of inhomogeneous deformation.

Molecular dynamics and molecular mechanics simulations allow us to extract useful macroscopic mechanical, transport, and electronic measures by considering the dynamics and statics of the miscrostructure under specific conditions. However, our ultimate goal is to surpass the case-by-case approach by developing a general theoretical framework that allows us to describe kinematics and dynamics rationally, and eventually to make reliable engineering predictions. Because of the high regularity of the atomic lattice, statistical averaging is not appropriate and formulations bridging particulate to continuum mechanics are required. Certain fundamental ideas toward this goal and recent progress are discussed in this chapter.

In Section 4.2, the atomic structure of graphene sheets and cylindrical nanotubes is discussed. In Section 4.3, a method of constructing perfectly cylindrical, pristine nanotubes is outlined based on homogeneous deformation and inner displacement of the unrolled graphene sheet, endowing us with five scalar degrees of freedom. In Section 4.4, fundamental concepts from the membrane theory of thin shells are reviewed, and a preliminary correspondence is made between continuum mechanics and the mechanics of particulate sheets. In Section 4.5, a method for quantifying the kinematics of particulate sheets is proposed, and sample results are presented for graphene sheets and nanotubes. In Sections 4.6 and 4.7, the deformation and energetics of a model particulate sheet is considered where the particles are arranged at the vertices of a hexagonal lattice and are connected to three neighbors by elastic springs. An overview and closing remarks are presented in Section 4.8.

4.2 PLANAR GRAPHENE SHEETS AND NANOTUBES

A planar layer of carbon atoms forms a periodic structure called a *graphene sheet* (e.g., [4]). Pencil lead consists of a stack of overlaying graphene sheets that easily separate when sheared in pencil writing. A perfect graphene sheet in the xy plane consists of a doubly periodic hexagonal lattice defined by two base vectors,

$$\mathbf{a}_1 = \rho\,(\sqrt{3}, 0), \qquad \mathbf{a}_2 = \rho\,\frac{\sqrt{3}}{2}\,(1, \sqrt{3}), \tag{4.1}$$

where ρ is the radius of each hexagonal cell, as depicted in Fig. 4.1. Note that the lengths of these vectors are equal, $|\mathbf{a}_1| = |\mathbf{a}_2| \equiv \alpha = \sqrt{3}\,\rho$. The centers of the hexagonal cells are located at the vertices of a planar triangular Bravais lattice whose vertices are positioned at

$$\mathbf{x} = \mathbf{x}_0 + i\mathbf{a}_1 + j\mathbf{a}_2, \tag{4.2}$$

where $\mathbf{x}_0$ is an arbitrary center and i and j are two positive or negative integers.

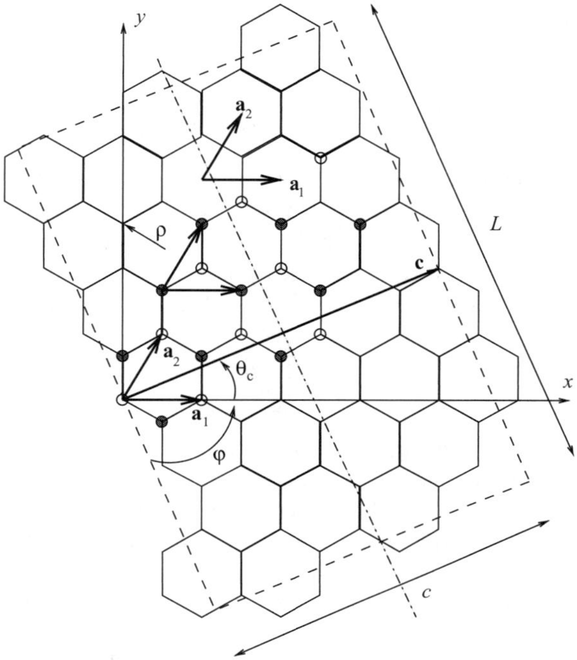

Figure 4.1 Graphene sheet consisting of hexagonal cells. The carbon atoms are arranged at the vertices of two interwoven planar Bravais lattices. Atoms on lattice A are marked by open circles, and atoms on lattice B are marked by filled circles. The chiral vector depicted, **c**, corresponds to $n_1 = 3$ and $n_2 = 2$. The dashed lines enclose one repeated unit with axial periodicity L and circumferential arc length c. To form a nanotube, the repeated unit is rolled around the dotted–dashed line, which then becomes the centerline of the nanotube.

The carbon atoms themselves are arranged at the vertices of a hexagonal lattice consisting of two interwoven Bravais lattices, denoted as A and B. In Fig. 4.1, atoms on lattice A are shown as open circles, and atoms on lattice B are shown as filled circles. Lattice B is displaced with respect to lattice A in the y direction by the shift $\mathbf{d} = \rho \mathbf{e}_y$, where $\mathbf{e}_y$ is the unit vector along the y axis.

To construct a nanotube, we specify a positive integer, n_1, and a second zero or nonnegative integer, $n_2 \leq n_1$, and introduce the chiral vector

$$\mathbf{c} = n_1 \mathbf{a}_1 + n_2 \mathbf{a}_2 = c_x \mathbf{e}_x + c_y \mathbf{e}_y, \tag{4.3}$$

where $\mathbf{e}_x$ is the unit vector along the x axis. The chiral vector corresponding to $n_1 = 3$ and $n_2 = 2$ is shown in Fig. 4.1. Substituting the expressions for the base vectors given in Eq. (4.1), we find that the length of the chiral vector is

$$c \equiv |\mathbf{c}| = \alpha \sqrt{n_1^2 + n_1 n_2 + n_2^2} \tag{4.4}$$

and the chiral angle is

$$\theta_c \equiv \arctan \frac{c_y}{c_x} = \arctan \frac{\sqrt{3}\, n_2}{2\, n_1 + n_2}. \tag{4.5}$$

When $n_2 = 0$, we find that $\theta_c = 0$, and when $n_1 = n_2$, we find that $\theta_c = \pi/6$. As a preliminary, we introduce the greatest common divisor (gcd) of n_1 and n_2, denoted as $n \equiv \mathrm{gcd}(n_1, n_2)$, with the understanding that $\mathrm{gcd}(n_1, 0) = n_1$.

The graphene sheet is now rolled around an axis that is normal to the chiral vector and thus rotated with respect to the x axis in a clockwise direction by the angle

$$\phi = \frac{\pi}{2} - \theta_c, \tag{4.6}$$

as shown in Fig. 4.1. To generate one repeated unit of the nanotube, we identify one repeated graphene unit represented by the rectangular area enclosed by the dashed lines in Fig. 4.1 and roll it around the axis that is parallel to the downward-sloping rectangular sides, drawn with a dotted–dashed line in Fig. 4.1. When $n_2 = 0$, we obtain an armchair pattern where some hexagon sides are oriented along the tube axis. When $n_2 = n_1$, we obtain a zigzag pattern where some hexagon sides are oriented in the circumferential direction.

The circumferential arc length of the tube is equal to the length of the chiral vector, c, and the tube radius is $a = c/2\pi$. The axial period of the repeated unit is

$$L = \frac{\sqrt{3}\, c}{Rn}, \tag{4.7}$$

where $R = 3$ if the difference $n_1 - n_2$ is a multiple of $3n$, $\mathrm{mod}(n_1 - n_2, 3n) = 0$, and $R = 1$ otherwise (e.g., [5]). On geometrical grounds, we find that the number of hexagonal faces comprising a repeated unit is

$$N_F = \frac{2c^2}{Rn\alpha^2}. \tag{4.8}$$

The tube shape exhibits an n-fold rotational symmetry around the tube axis at any cross section normal to the x axis at the two end caps of each period.

To generate a nanotube in practice it is convenient to rotate the graphene sheet in the counterclockwise direction by the angle ϕ so that the tube axis points along the horizontal x' axis and the chiral vector points along the vertical y' axis, as shown in Fig. 4.2a. In the rotated frame, the base vectors become

$$\mathbf{a}'_i = \begin{bmatrix} \cos\phi & -\sin\phi \\ \sin\phi & \cos\phi \end{bmatrix} \cdot \mathbf{a}_i, \tag{4.9}$$

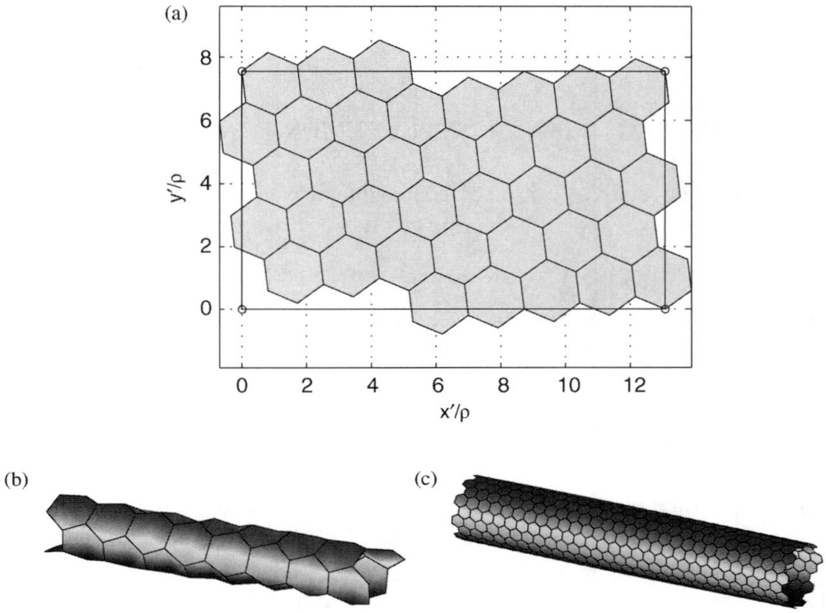

Figure 4.2 (a) Repeated unit of the graphene for $n_1 = 3$ and $n_2 = 2$, rotated by the angle ϕ so that the tube axis becomes horizontal, resulting in the tube shape shown in (b); (c) repeated unit of a tube for $n_1 = 10$ and $n_2 = 9$.

for $i = 1, 2$. Explicitly,

$$\begin{aligned} \mathbf{a}'_1 &= \rho\sqrt{3}\,(\cos\phi,\ \sin\phi), \\ \mathbf{a}'_2 &= \rho\,\frac{\sqrt{3}}{2}\,(\cos\phi - \sqrt{3}\sin\phi,\ \sin\phi + \sqrt{3}\cos\phi). \end{aligned} \tag{4.10}$$

The rotated chiral vector is $\mathbf{c}' = n_1\mathbf{a}'_1 + n_2\mathbf{a}'_2$. Because of the identity

$$(2n_1 + n_2)\cos\phi - n_2\sqrt{3}\sin\phi = 0, \tag{4.11}$$

originating from Eq. (4.5), the x' component of $\mathbf{c}'$ is zero, confirming that $\mathbf{c}'$ is aligned with the y' axis. The length of the chiral vector remains unchanged, $|\mathbf{c}'| \equiv c' = c$.

The Cartesian coordinates of the tube indicated by a "hat" arise from the transformation

$$\hat{x} = x', \qquad \hat{y} = a\cos\frac{y'}{a}, \qquad \hat{z} = a\sin\frac{y'}{a} \tag{4.12}$$

as depicted in Fig. 4.2b. A periodic unit of a tube with $n_1 = 10$ and $n_2 = 9$ is shown in Fig. 4.2c.

4.3 HOMOGENEOUS INNER DISPLACEMENT AND DEFORMATION

Prior to rolling, the relative position of the two interwoven atomic lattices in the planar graphene sheet can be altered by perturbing the shift from $\rho\mathbf{e}_y$ to

$$\mathbf{d} = \rho(e_y + \eta), \tag{4.13}$$

where the vector η is the dimensionless *inner displacement*. In the rotated coordinates in the $x'y'$ plane, the lattice shift becomes

$$\mathbf{d}' = \rho \begin{bmatrix} \eta_x \cos\phi - (1+\eta_y)\sin\phi \\ \eta_x \sin\phi + (1+\eta_y)\cos\phi \end{bmatrix}. \tag{4.14}$$

Three additional homogeneous transformations can be applied with no effect on the hexagonal pattern:

1. The tube surface can be twisted uniformly so that the vertices of the precursor graphene sheet depicted in Fig. 4.2a are displaced vertically along the y' axis by the distance tcx'/L, as though they underwent a *shearing* deformation, where t is a dimensionless twist coefficient. Because of the aforementioned n-fold symmetry of the tube cross section at the two ends of a repeated unit, when t is zero, an integer, or a multiple $1/n$, twisting has no effect on the axial period, L.
2. The tube can be stretched along its length by the dimensionless stretch, s, so that the x' or $\hat{x}$ coordinate of each vertex is multiplied by s, and the axial period of each repeated unit becomes sL.
3. The tube can be expanded in the radial direction normal to the $\hat{x}$ axis by the dimensionless stretch r, so that the y' coordinate and the radial $\hat{\sigma}$ coordinate of each vertex are both multiplied by r, and the tube radius becomes ra.

One repeated unit of the precursor graphene sheet for $n_1 = 10$, $n_2 = 9$, $\eta = \mathbf{0}$, $t = 1.0$, $s = 2.0$, and $r = 1.0$ is shown in Fig. 4.3a. Tube shapes after twisting and elongation are shown in Fig. 4.3b–d. Because shearing deformation of the graphene along the x' axis is possible only for a sequence of discrete rates that displace the top of a repeated unit with respect to the bottom by multiples of the axial wavelength, it is not an acceptable mode of deformation.

After shearing along the y' axis, stretching along the x' axis, and expansion along the y' axis, the base vectors in the rotated graphene sheet become

$$\tilde{\mathbf{a}}_i' = \begin{bmatrix} 1 & 0 \\ 0 & r \end{bmatrix} \cdot \begin{bmatrix} s & 0 \\ 0 & 1 \end{bmatrix} \cdot \begin{bmatrix} 1 & 0 \\ tc/L & 1 \end{bmatrix} \cdot \begin{bmatrix} \cos\phi & -\sin\phi \\ \sin\phi & \cos\phi \end{bmatrix} \cdot \mathbf{a}_i. \tag{4.15}$$

(a)

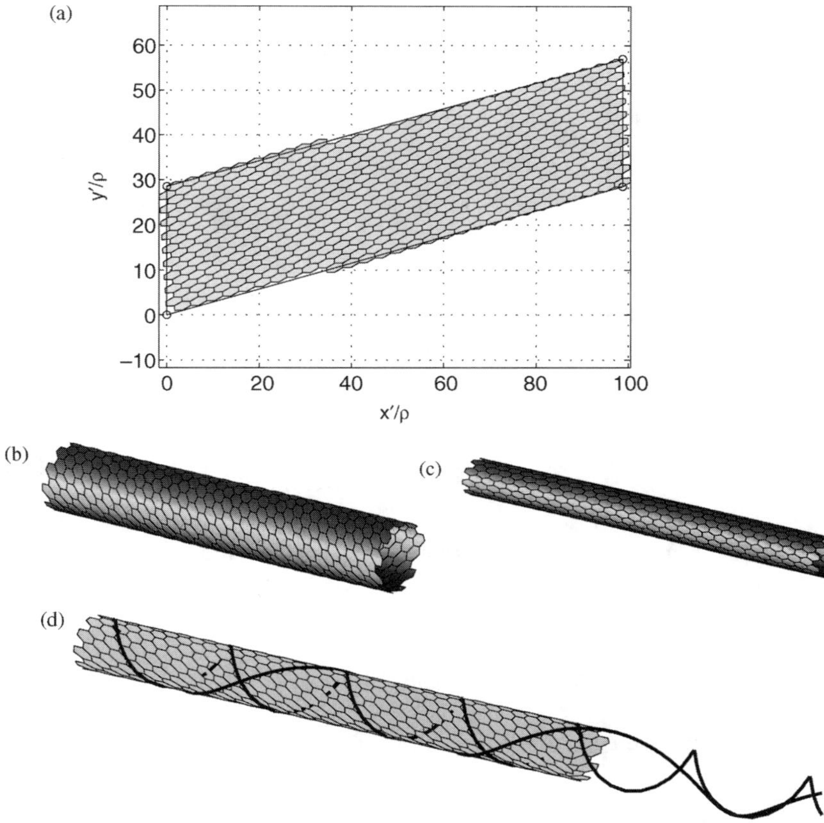

(b) (c) (d)

Figure 4.3 (a) Repeated unit of a deformed graphene sheet for $n_1 = 10$, $n_2 = 9$, inner displacement $\eta = \mathbf{0}$, twist $t = 1.0$, stretch $s = 2.0$, and expansion $s = 1.0$. Repeated unit of a tube for $n_1 = 10, n_2 = 9$, and $\eta = \mathbf{0}$, (b) twisted by the factor $t = 1$, (c) stretched by a factor of 2, $s = 2$, and (d) twisted by $t = 1$ and stretched by $s = 2$. Part (d) also shows the spiral lines corresponding to multiples of each base vector.

Multiplying the three matrices on the right-hand side, we find that

$$\tilde{\mathbf{a}}'_i = \begin{bmatrix} s & 0 \\ rtc/L & r \end{bmatrix} \cdot \begin{bmatrix} \cos\phi & -\sin\phi \\ \sin\phi & \cos\phi \end{bmatrix} \cdot \mathbf{a}_i. \tag{4.16}$$

The first matrix on the right-hand side of Eq. (4.16) implements lattice deformation due to a restricted deformation gradient. Because the graphene sheet must be rolled up seamlessly into the cylindrical shape, the deformation excludes shearing along the x' axis. Explicitly, the base vectors are given by

$$\tilde{\mathbf{a}}_1' = \rho\,\sqrt{3}\left[s\cos\phi,\quad r\left(\sin\phi + \frac{tc}{L}\cos\phi\right)\right],$$
$$\tilde{\mathbf{a}}_2' = \rho\,\frac{\sqrt{3}}{2}\left[s\,(\cos\phi - \sqrt{3}\sin\phi), r\left(\sin\phi + \sqrt{3}\cos\phi + \frac{tc}{L}(\cos\phi - \sqrt{3}\sin\phi)\right)\right]. \tag{4.17}$$

Using identity (4.11), we find that the x' component of the transformed chiral vector, $\tilde{\mathbf{c}} = n_1\tilde{\mathbf{a}}_1 + n_2\tilde{\mathbf{a}}_2$, is zero, while the y' component is equal to rc. The lattice shift becomes

$$\tilde{\mathbf{d}}' = \rho\begin{bmatrix} s & 0 \\ rtc/L & r \end{bmatrix}\cdot\begin{bmatrix} \cos\phi & -\sin\phi \\ \sin\phi & \cos\phi \end{bmatrix}\cdot(\mathbf{e}_y + \eta). \tag{4.18}$$

Next, we define the vector $\tilde{\mathbf{a}}_3 = \tilde{\mathbf{a}}_1 - \tilde{\mathbf{a}}_2$ and note that the three vectors $\tilde{\mathbf{a}}_1$, $\tilde{\mathbf{a}}_2$, and $\tilde{\mathbf{a}}_3$ connect the three closest neighbors of any vertex on the graphene sheet. Squaring the definition of $\tilde{\mathbf{a}}_3$, we find that

$$|\tilde{\mathbf{a}}_3|^2 = |\tilde{\mathbf{a}}_1|^2 + |\tilde{\mathbf{a}}_2|^2 - 2\,\tilde{\mathbf{a}}_1\cdot\tilde{\mathbf{a}}_2. \tag{4.19}$$

Using this relation, we compute the length of the chiral vector as

$$|\tilde{\mathbf{c}}| = \left(n_1^2|\tilde{\mathbf{a}}_1|^2 + n_2^2|\tilde{\mathbf{a}}_2|^2 + n_1n_2(|\tilde{\mathbf{a}}_1|^2 + |\tilde{\mathbf{a}}_2|^2 - |\tilde{\mathbf{a}}_3|^2)\right)^{1/2}. \tag{4.20}$$

In summary, we have described a geometrical method for constructing a cylindrical nanotube with a given chirality in terms of five geometrical parameters, including the twist, radial expansion, axial stretch, and two components of the inner displacement of the interwoven Bravais lattices defining the unrolled, planar, hexagonal graphene sheet.

4.4 MEMBRANE MECHANICS

We begin making a correspondence between continuum mechanics and the mechanics of atomic sheets by reviewing fundamental concepts from the continuum membrane theory of thin shells. Applications of membrane theory can be found in the study of structural domes, elastic sheets, and membranes of biological shells. The cornerstone of continuum mechanics is the deformation gradient, $\mathbf{F}$, relating an infinitesimal material vector before and after deformation by the linear relation $\mathrm{d}\mathbf{x} = \mathbf{F}\cdot\mathrm{d}\mathbf{X}$, where $\mathbf{X}$ is the position of a material point particle in an elastic medium

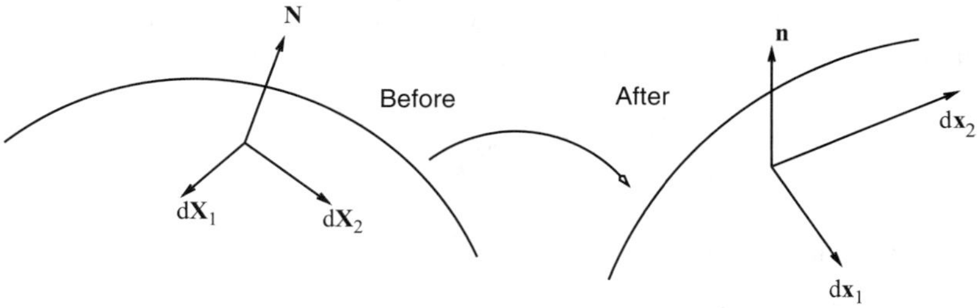

Figure 4.4 Zero-thickness elastic sheet showing two tangential material vectors before and after deformation.

before deformation, and $\mathbf{x}$ is the position of the same point particle after deformation. Both $d\mathbf{X}$ and $d\mathbf{x}$ are material vectors with a distinct and permanent identity.

For a zero-thickness elastic sheet such as the biological or polymeric membrane illustrated in Fig. 4.4, we introduce the surface deformation gradient, $\mathbf{F}^S$, defined from the expressions

$$d\mathbf{x} = \mathbf{F}^S \cdot d\mathbf{X}, \qquad \mathbf{F}^S \cdot \mathbf{N} = 0, \tag{4.21}$$

where $\mathbf{N}$ is the unit normal vector before deformation. The second equation shows that $\mathbf{N}$ is an eigenvector corresponding to the null eigenvalue. Because the vector $d\mathbf{x}$ is tangential in the deformed configuration, $\mathbf{n} \cdot \mathbf{F}^S = \mathbf{0}$, where $\mathbf{n}$ is the normal vector after deformation. Thus, the left eigenvector of the surface deformation gradient corresponding to the null eigenvalue is the normal vector in the deformed configuration. In the absence of deformation, $\mathbf{F}^S = \mathbf{P}$, where

$$\mathbf{P} = \mathbf{I} - \mathbf{N}\mathbf{N} \tag{4.22}$$

is the tangential projection tensor in the undeformed configuration, and $\mathbf{I}$ is the 3×3 identity matrix. Since $\mathbf{N}$ is a unit vector, the trace of $\mathbf{P}$ is equal to 2.

In practice, the nine components of the surface deformation gradient at a point can be computed from the images of two infinitesimal or small material vectors whose cross product defines the normal vector, and the second condition in Eq. (4.21). Enforcing these equations, we obtain nine scalar equations for an equal number of unknowns. For example, a three-dimensional membrane can be discretized into planar triangular elements defined by vertex nodes, and the deformation gradient at each vertex of a triangle can be computed in terms of the deformation of the edges of the triangle, regarded as small material vectors. The deformation gradient computed in this fashion over each triangle sharing a node will differ by an amount that decreases as the triangle size becomes smaller due to the discretization error associated with the triangulation. A similar computation can be performed for curved triangles defined by vertex, edge, and possibly interior nodes.

To relate membrane deformation to surface tensions developing in the plane of the membrane, we introduce the right Cauchy–Green surface deformation tensor,

$$\mathbf{U}^{S^2} \equiv \mathbf{F}^{S^T} \cdot \mathbf{F}^S, \tag{4.23}$$

and the left Cauchy–Green surface deformation tensor

$$\mathbf{V}^{S^2} \equiv \mathbf{F}^S \cdot \mathbf{F}^{S^T}, \tag{4.24}$$

where the superscript T denotes the matrix transpose. Both $\mathbf{U}^{S^2}$ and $\mathbf{V}^{S^2}$ are symmetric, $\mathbf{U}^{S^{2T}} = \mathbf{U}^{S^2}$ and $\mathbf{V}^{S^{2T}} = \mathbf{V}^{S^2}$. In the absence of deformation, $\mathbf{U}^S = \mathbf{P}$ and $\mathbf{V}^S = \mathbf{P}$. Moreover, we introduce the right Green–Lagrange surface strain tensor

$$\mathbf{E}^S = \frac{1}{2}\,(\mathbf{U}^{S^2} - \mathbf{P}). \tag{4.25}$$

Since $\mathbf{E}^S \cdot \mathbf{N} = \mathbf{0}$, the undeformed normal vector, $\mathbf{N}$, is an eigenvector $\mathbf{E}^S$ corresponding to the null eigenvalue. In the absence of deformation, $\mathbf{F}^S = \mathbf{0}$.

The left Cauchy–Green surface deformation tensor, $\mathbf{V}^{S^2}$, has two positive eigenvalues λ_1^2 and λ_2^2, with two associated orthogonal tangential unit eigenvectors $\mathbf{v}_1$ and $\mathbf{v}_2$, and a third zero eigenvalue. The positive values λ_1 and λ_2 are the principal surface stretches or extension ratios. The principal directions of the Eulerian (Cauchy) surface tension tensor, τ, coincide with the principal directions of $\mathbf{V}^{S^2}$. We refer to τ simply as the surface tension tensor.

The second surface Piola–Kirchhoff tensor, $\mathbf{S}^S$, derives from a surface strain-energy function, W_S, by the relations

$$S_{ij}^S = \frac{\partial W_S}{\partial E_{ij}^S}. \tag{4.26}$$

A small increment of the surface strain tensor causes a corresponding small increment in the stress tensor,

$$\delta S_{ij}^S = C_{ijkl}^S\, \delta E_{kl}^S, \tag{4.27}$$

where C_{ijkl}^S is the incremental surface modulus tensor. In terms of the second surface Piola–Kirchhoff tensor, the surface tension tensor is

$$\tau = \frac{1}{J_S}\, \mathbf{F}^S \cdot \mathbf{S}^S \cdot \mathbf{F}^{S^T}, \tag{4.28}$$

where $J_S = \lambda_1 \lambda_2$ is the dilatation of an infinitesimal membrane patch.

Now referring to local Cartesian coordinates with two axes parallel to the local principal directions of the Cauchy tension tensor, we use Eqs. (4.26) and (4.28) and find that the principal surface tensions are given by

$$\tau_1^P = \frac{1}{\lambda_2}\,\frac{\partial W_S}{\partial \lambda_1}, \qquad \tau_2^P = \frac{1}{\lambda_1}\,\frac{\partial W_S}{\partial \lambda_2}. \tag{4.29}$$

Introducing the unit eigenvectors, $\mathbf{v}_1$ and $\mathbf{v}_2$, we write

$$\tau = \frac{1}{\lambda_2}\,\frac{\partial W_S}{\partial \lambda_1}\,\mathbf{v}_1\mathbf{v}_1 + \frac{1}{\lambda_1}\,\frac{\partial W_S}{\partial \lambda_2}\,\mathbf{v}_2\mathbf{v}_2. \tag{4.30}$$

The surface strain-energy function, W_S, depends on the surface deformation gradient by way of surface strain invariants. One possible set of invariants used to model the membrane of red blood cells is

$$I_1 \equiv 2\,\mathrm{Trace}(\mathbf{E}^S) = \lambda_1^2 + \lambda_2^2 - 2, \qquad I_2 \equiv J_S^2 - 1 = \lambda_1^2\lambda_2^2 - 1. \tag{4.31}$$

Substituting in Eq. (4.30) the expressions

$$\frac{\partial W_S}{\partial \lambda_1} = \frac{\partial W_S}{\partial I_1}\frac{\partial I_1}{\partial \lambda_1} + \frac{\partial W_S}{\partial I_2}\frac{\partial I_2}{\partial \lambda_1} = 2\,\lambda_1\,\frac{\partial W_S}{\partial I_1} + 2\,\lambda_1\,\lambda_2^2\,\frac{\partial W_S}{\partial I_2} \tag{4.32}$$

and

$$\frac{\partial W_S}{\partial \lambda_2} = \frac{\partial W_S}{\partial I_1}\frac{\partial I_1}{\partial \lambda_2} + \frac{\partial W_S}{\partial I_2}\frac{\partial I_2}{\partial \lambda_2} = 2\,\lambda_2\,\frac{\partial W_S}{\partial I_1} + 2\,\lambda_1^2\,\lambda_2\,\frac{\partial W_S}{\partial I_2}, \tag{4.33}$$

we find that

$$\tau = 2\frac{\lambda_1}{\lambda_2}\left(\frac{\partial W_S}{\partial I_1} + \lambda_2^2\frac{\partial W_S}{\partial I_2}\right)\mathbf{v}_1\mathbf{v}_1 + 2\frac{\lambda_2}{\lambda_1}\left(\frac{\partial W_S}{\partial I_1} + \lambda_1^2\frac{\partial W_S}{\partial I_2}\right)\mathbf{v}_2\mathbf{v}_2. \tag{4.34}$$

Rearranging, we obtain

$$\tau = 2\lambda_1\lambda_2\frac{\partial W_S}{\partial I_2}(\mathbf{v}_1\mathbf{v}_1 + \mathbf{v}_2\mathbf{v}_2) + \frac{2}{\lambda_1\lambda_2}\,\frac{\partial W_S}{\partial I_1}(\lambda_1^2\,\mathbf{v}_1\mathbf{v}_1 + \lambda_2^2\,\mathbf{v}_2\mathbf{v}_2) \tag{4.35}$$

or

$$\tau = 2\,\lambda_1\lambda_2\frac{\partial W_S}{\partial I_2}\,\mathbf{p} + \frac{2}{\lambda_1\lambda_2}\,\frac{\partial W_S}{\partial I_1}\,\mathbf{V}^2, \tag{4.36}$$

where

$$\mathbf{p} = \mathbf{I} - \mathbf{nn} \tag{4.37}$$

is the tangential projection tensor in the deformed configuration. Note that if $\partial W_S/\partial I_1 = 0$, the tensions are isotropic. Equation (4.36) allows us to compute the surface tensions from the surface deformation gradient, subject to a given surface strain energy function.

Performing a differential force balance on the membrane in the absence of an external surface load applied on either side of the membrane, we derive the equilibrium equation

$$\nabla_S \cdot \tau = \mathbf{0}, \tag{4.38}$$

where $\nabla_S \equiv \mathbf{p} \cdot \nabla$ is the tangential gradient. An inhomogeneous term must be included in the presence of a surface load.

4.4.1 Cylindrical Membrane

As an application, we consider a cylindrical membrane with circular cross section and only allow modes of deformation that preserve the circular cylindrical shape. In cylindrical polar coordinates, (x, σ, φ), where the x axis is coaxial with the tube centerline, the coordinates of a point particle in the reference state are (X, Σ, Φ), and the corresponding unit vectors are $\mathbf{e}_X, \mathbf{e}_\Sigma, \mathbf{e}_\Phi$. The coordinates of the same point particle in the deformed state are (x, σ, φ), and the corresponding unit vectors are $\mathbf{e}_x, \mathbf{e}_\sigma, \mathbf{e}_\varphi$, where $\mathbf{e}_x = \mathbf{e}_X$. The unit normal vectors in the reference and deformed state are, respectively, $\mathbf{N} = \mathbf{e}_\Sigma$ and $\mathbf{n} = \mathbf{e}_\sigma$. A differential tangential material vector before and after deformation can be described as

$$d\mathbf{X} = \mathbf{e}_X\, dX + \mathbf{e}_\Phi \Sigma\, d\Phi, \qquad d\mathbf{x} = \mathbf{e}_x\, dx + \mathbf{e}_\varphi \sigma\, d\varphi. \tag{4.39}$$

In the case of a homogeneous deformation, the point particle coordinates are mapped as

$$x = sX, \quad \sigma = r\Sigma, \quad \varphi = c + \Phi + \chi X, \tag{4.40}$$

where s is the axial stretch, r is the radial expansion, χ is the dimensionless twist per axial length, and c is an inconsequential constant expressing global rotation. Thus,

$$d\mathbf{x} = \mathbf{e}_x s\, dX + \mathbf{e}_\varphi r\Sigma\, (d\Phi + \chi\, dX). \tag{4.41}$$

In the undeformed configuration, $s = 1$, $r = 1$, and $\chi = 0$. Using Eq. (4.39), we find that

$$d\mathbf{x} = [s\mathbf{e}_x\mathbf{e}_X + \mathbf{e}_\varphi(r\mathbf{e}_\Phi + \chi\sigma\mathbf{e}_X)] \cdot d\mathbf{X}. \tag{4.42}$$

The expression inside the square brackets is the surface deformation gradient, given by

$$\mathbf{F}^S = s\mathbf{e}_x\mathbf{e}_X + r\mathbf{e}_\varphi\mathbf{e}_\Phi + \chi\sigma\mathbf{e}_\varphi\mathbf{e}_X. \tag{4.43}$$

It is reassuring to confirm that $\mathbf{F}^S \cdot \mathbf{N} = \mathbf{F}^S \cdot \mathbf{e}_\Sigma = \mathbf{0}$ and $\mathbf{n} \cdot \mathbf{F}^S = \mathbf{e}_\sigma \cdot \mathbf{F}^S = \mathbf{0}$.

The right Cauchy–Green surface deformation tensor is given by

$$\begin{aligned}\mathbf{U}^{S^2} &\equiv \mathbf{F}^{S^T} \cdot \mathbf{F}^S = (s\mathbf{e}_X\mathbf{e}_x + r\mathbf{e}_\Phi\mathbf{e}_\varphi + \chi\sigma\mathbf{e}_X\mathbf{e}_\varphi) \cdot (s\mathbf{e}_x\mathbf{e}_X + r\mathbf{e}_\varphi\mathbf{e}_\Phi + \chi\sigma\mathbf{e}_\varphi\mathbf{e}_X)\\ &= (s^2 + \chi^2\sigma^2)\mathbf{e}_X\mathbf{e}_X + r\chi\sigma(\mathbf{e}_\Phi\mathbf{e}_X + \mathbf{e}_X\mathbf{e}_\Phi) + r^2\mathbf{e}_\Phi\mathbf{e}_\Phi,\end{aligned} \tag{4.44}$$

the left Cauchy–Green surface deformation tensor is given by

$$\begin{aligned}\mathbf{V}^{S^2} &\equiv \mathbf{F}^S \cdot \mathbf{F}^{S^T} = (s\mathbf{e}_x\mathbf{e}_X + r\mathbf{e}_\varphi\mathbf{e}_\Phi + \chi\sigma\mathbf{e}_\varphi\mathbf{e}_X) \cdot (s\mathbf{e}_X\mathbf{e}_x + r\mathbf{e}_\Phi\mathbf{e}_\varphi + \chi\sigma\mathbf{e}_X\mathbf{e}_\varphi)\\ &= s^2\mathbf{e}_x\mathbf{e}_x + s\chi\sigma(\mathbf{e}_\varphi\mathbf{e}_x + \mathbf{e}_x\mathbf{e}_\varphi) + (r^2 + \chi^2\sigma^2)\mathbf{e}_\varphi\mathbf{e}_\varphi,\end{aligned} \tag{4.45}$$

and the left Green–Lagrangian surface strain tensor is given by

$$\mathbf{E}^S = \frac{1}{2}\left[(s^2 + \chi^2\sigma^2 - 1)\mathbf{e}_X\mathbf{e}_X + r\chi\sigma(\mathbf{e}_\Phi\mathbf{e}_X + \mathbf{e}_X\mathbf{e}_\Phi) + (r^2 - 1)\mathbf{e}_\Phi\mathbf{e}_\Phi\right]. \tag{4.46}$$

The eigenvalues of $\mathbf{V}^{S^2}$ depend on s, χ, and r, but are independent of position over the tube surface. Accordingly, the derivatives with respect to the invariants I_1 and I_2 can be expressed with respect to derivatives with respect to s, r, and χ, and subsequently evaluated by analytical or numerical methods.

Three modular problems naturally arise. In the first problem, the tube is stretched along its axis in the absence of a circumferential force, while the interior and exterior pressures are kept constant. In the second problem, the tube is twisted around its axis in the absence of an axial force, while the interior and exterior pressures are kept constant. In the third problem, the tube expands under the influence of a pressure difference between the inside and outside, in the absence of an axial and circumferential force. If the tube is made of an orthotropic material with principal axes in the axial and circumferential directions, stretching does not induce twist.

These three modes of deformation can also be applied to a cylindrical nanotube with arbitrary chirality. To compute undeformed or deformed tube shapes, the atomic energy is minimized with respect to s, r, χ, and the two scalar components of the homogeneous inner displacement. Unconstrained minimization produces equilibrium configurations dependent on the tube chirality. Constrained minimization produces response curves that can be used to extract elastic moduli and critical conditions for disintegration in axial extension, radial expansion, and twist [6–8].

The direct correspondence between a cylindrical elastic tube and a nanotube is possible only because of the uniformity of the surface deformation gradient and inner displacement field.

4.4.2 Small Deformation

Consider a slightly deformed membrane regarded as a two-dimensional curved continuum, and introduce the displacement field $\mathbf{v}$ defined by $\mathbf{x} = \mathbf{X} + \mathbf{v}$. The transpose of the surface deformation gradient can be expressed in the form

$$\mathbf{F}^{S^T} = \mathbf{P} + \nabla_S \mathbf{v} = \mathbf{P} \cdot (\mathbf{I} + \nabla \mathbf{v}). \tag{4.47}$$

Linearizing the left Cauchy–Green surface deformation tensor with respect to $\mathbf{v}$, we find that

$$\mathbf{V}^{S^2} \equiv \mathbf{F}^S \cdot \mathbf{F}^{S^T} \simeq \mathbf{P} + 2\,\varepsilon^S, \tag{4.48}$$

where

$$\varepsilon^S = \frac{1}{2}\left[\nabla_S \mathbf{v} + (\nabla_S \mathbf{v})^T\right] \tag{4.49}$$

is the symmetric surface strain tensor. We are thus led to considering the following functional form for the surface strain energy function,

$$W_S = G(\varepsilon^S). \tag{4.50}$$

The work increment associated with an arbitrary deformation of a control surface D is

$$\delta E = \iint_D \delta W_S \, dS = \iint_D \tau_{ij}\, \delta\varepsilon_{ij}^S \, dS, \tag{4.51}$$

where τ_{ij} is the surface tension tensor conjugate to the surface strain tensor and dS is a differential surface area. Setting the variation to zero in the absence of a surface load and integrating by parts, we derive the equilibrium equation (4.38).

4.4.3 Flexural Stiffness

An extended formulation is necessary for a thin shell supporting tangential tensions, transverse tensions, and bending moments endowing the shell with flexural stiffness. The mathematical formulation requires constitutive equations in terms of the undeformed and deformed surfaces curvatures, and culminates in generalized equilibrium equations amenable to numerical methods [9].

4.5 ATOMIC SHEETS

We seek to develop the counterpart of the membrane theory outlined in Section 4.4 for planar and curved particulate sheets. In the case of a planar atomic sheet described by one Bravais lattice or two superposed Bravais lattices with the same base vectors, we may use the Cauchy–Born rule to relate the lattice vectors before deformation, $\mathbf{A}_i$, to the lattice vectors after deformation, $\mathbf{a}_i$, in terms of the macroscopic deformation gradient, $\mathbf{F}$,

$$\mathbf{a}_i = \mathbf{F} \cdot \mathbf{A}_i \tag{4.52}$$

for $i = 1, 2$ (e.g., [1]). A tacit assumption is that the deformation gradient is constant over many cells, and the deformation is sufficiently small that material and geometrical nonlinearities do not arise. After deformation, the planar lattice can be rolled up to yield a rectilinear nanotube. The strong assumptions underlying the Cauchy–Born rule motivates us to pursue a generalized framework.

Consider a three-dimensional surface defined by a curved hexagonal atomic lattice, as shown in Fig. 4.5a. A particle has three closest neighbors, and each neighbor has two other closest neighbors. According to the Tersoff–Brenner potential for a hexagonal carbon lattice [10], the energy of the central atom (blue) depends on the relative position of its first and second neighbors. Ten atoms are thus involved in an energy cluster. The central atom is labeled 1, and the peripheral atoms are labeled 2 through 9 in Fig. 4.5a. The atomic energy is a highly nonlinear function of the atomic positions.

A surface can be envisioned passing through the 10 particles, as shown in Fig. 4.5b, and the normal direction at the location of the central particle can be defined by sensible interpolation. For example, the unit normal vector can be set equal to the average of the three unit normal vectors on each flat triangle defined by the central atom and one pair of its closest neighbors. In an improved representation, a local Cartesian system $x'y'z'$ can be introduced with the z' axis aligned with the normal vector, and the surface can be described with a

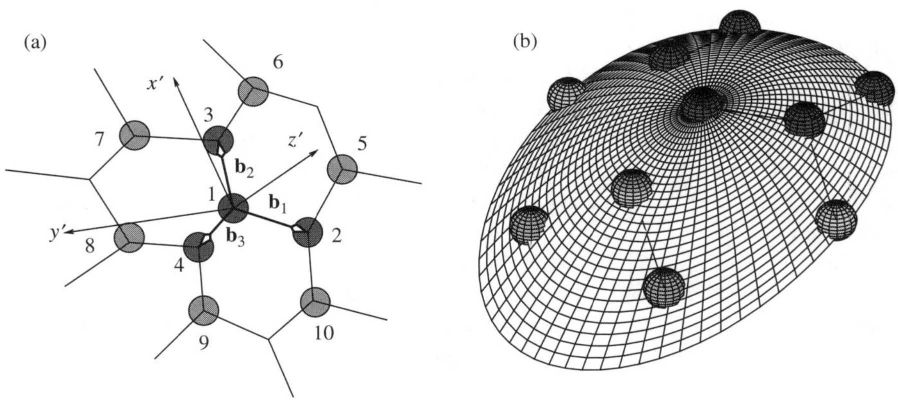

Figure 4.5 (a) Planar or curved hexagonal atomic lattice representing a graphene sheet; (b) a cubic surface can be defined passing through the 10 particles. (*See insert for color representation of figure.*)

complete cubic function as

$$z = f(x, y, z) = a_0 + a_1 x' + a_2 y' + a_3 x'^2 + a_4 x' y' + a_5 y'^2 \\ + a_6 x'^3 + a_7 x'^2 y' + a_8 x' y'^2 + a_9 y'^3. \tag{4.53}$$

The number of coefficients, a_0 to a_9, is precisely equal to the number of atoms, and the cubic is uniquely defined. In fact, because the atomic arrangement is analogous to that of the Pascal triangle, the molecular surface can be described by a higher-order polynomial, provided that the directional curvatures of the surface at the central atom are sufficiently small compared to the inverse of the interatomic distances.

If the particulate sheet is deformed, the three bond vectors originating from the central atom, $\mathbf{b}_i$, shown in Fig. 4.5a, will distort. By analogy with a zero-thickness elastic sheet, we may write

$$\mathbf{b}_i = \mathbf{F} \cdot \mathbf{B}_i \quad \text{for} \quad i = 1, 2, 3, \qquad \mathbf{F} \cdot \mathbf{N} = \mathbf{0}, \tag{4.54}$$

providing us with 12 scalar equations for the nine components of the *particulate surface deformation gradient,* **F**. Consequently, the particulate surface deformation gradient defined with respect to the bonds is not single valued for arbitrary deformations, and can be computed either for one chosen pair of bonds and the normal vector abandoning the third bond, or for the triplet of bonds abandoning the normal vector. The third approach fails in the case of a planar sheet subject to an arbitrary in-plane atomic displacement.

Authors have bypassed this essential difficulty by implicitly or explicitly focusing their attention on spatially homogeneous deformations where the atomic displacement is a linear function of position, and the strain tensor is uniform over the particulate medium. By definition, then, the surface deformation gradient is single valued and well defined. Unfortunately, continuum theories built on these premises are restricted in scope and lack the necessary generality. Other authors have regarded the atomic displacements, $\mathbf{u}_i$, as discrete realizations of a continuous and differentiable function of surface coordinates and applied Taylor series expansions to write $\mathbf{u}(\mathbf{x}) = \mathbf{u}_i + (\mathbf{x} - \mathbf{x}_i) \cdot (\nabla \mathbf{u})_i$, but did not discuss the existence of the deformation gradient, $(\nabla \mathbf{u})_i$. Some authors have extended the Taylor series to second order to obtain the high-order Cauchy–Born rule, as discussed in Section 4.6. The uniqueness of the second-order deformation gradient $(\nabla\nabla \mathbf{u})_i$ is even more doubtful for general deformations.

The nonuniqueness of the particulate surface deformation gradient can be resolved by introducing an additional vectorial variable defined at the position of each particle, termed the *local inner displacement* [11]. Once this is done, the atomic energy can be expressed in terms of surface fields that have counterparts in the theory of continua.

With reference to the central atom shown in Fig. 4.5a, we achieve closure by introducing the local inner displacement, η, defined by the equations

$$\mathbf{b}_i = \mathbf{F} \cdot \mathbf{B}_i + \eta \quad \text{for } i = 1, 2, 3, \qquad \mathbf{F} \cdot \mathbf{N} = \mathbf{0}. \tag{4.55}$$

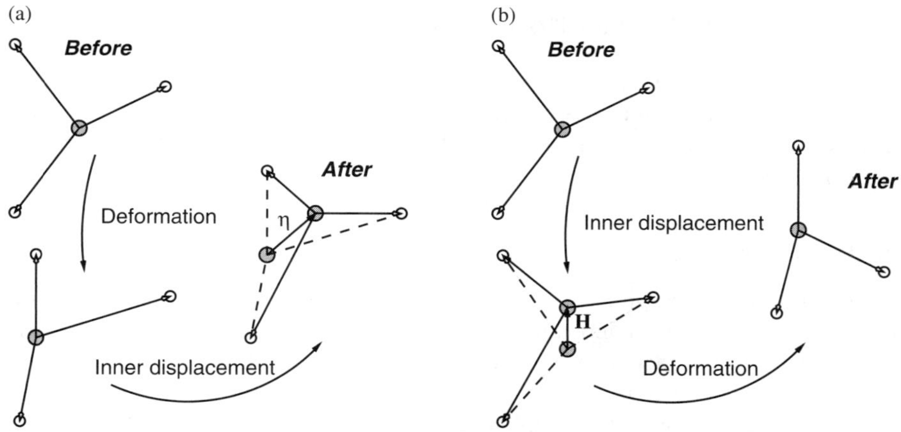

Figure 4.6 The central atom undergoes inner displacement (a) after or (b) before a locally homogeneous deformation in order to render the deformation gradient single valued.

Physically, introducing the inner displacement amounts to shifting the central atom *after* a locally homogeneous deformation, as illustrated in Fig. 4.6a. From a practical standpoint, introducing the inner displacement allows us to balance the number of equations (12) with the number of unknowns (nine for the deformation gradient and three for the inner displacement), and thereby render the surface deformation gradient single valued at every atom. When $\eta = \mathbf{0}$, the particulate sheet locally behaves like a continuum.

The second equation in Eq. (4.55) reveals that the surface deformation gradient is singular. By definition, the right eigenvector corresponding to the zero eigenvalue is the normal vector in the undeformed configuration. Setting the corresponding left eigenvector equal to the normal vector in the deformed configuration, $\mathbf{n} \cdot \mathbf{F} = \mathbf{0}$, we find that $\mathbf{n} \cdot (\mathbf{b}_i - \eta) = \mathbf{0}$, which ensures that the inner displacement resides in an average tangential space defined by the three bonds.

In an essentially equivalent approach, we introduce the inner displacement *before* deformation, $\mathbf{H}$, defined from the equations

$$\mathbf{b}_i = \mathbf{F} \cdot (\mathbf{B}_i + \mathbf{H}) \quad \text{for } i = 1, 2, 3, \qquad \mathbf{F}^S \cdot \mathbf{N} = \mathbf{0}, \tag{4.56}$$

where

$$\mathbf{F} \cdot \mathbf{H} = \eta. \tag{4.57}$$

Physically, this amounts to adjusting the central atom *before* mapping, so that the surface deformation gradient becomes single valued, as illustrated in Fig. 4.6b. The second equation in Eq. (4.55) ensures that only the tangential component of this adjustment is relevant. In practical computation, adjustment after deformation, expressed by Eq. (4.55), is preferred over adjustment before deformation, expressed by Eq. (4.56).

The preceding expressions apply for a central atom that belongs to Bravais lattice A. For a central atom that belongs to Bravais lattice B, the direction of the inner displacement is reversed to ensure that the results are independent of the lattice labeling, A or B.

4.5.1 Planar Sheet

For deformation of the graphene in the xy plane, the last row and last column of the particulate surface deformation gradient can be filled with zeros, and the z component of the inner displacement can be set to zero at the outset. The formulation yields a system of six equations for the four nonzero components of the surface deformation gradient and the two nonzero components of the inner displacement in the xy plane. The number of unknowns exactly balances the number of equations derived from the deformation of the three planar bonds.

As an illustration we consider the doubly periodic in-plane deformation of the planar graphene in the $(10, 5)$ chiral mode, as shown in Fig. 4.7a and b. In this example, the atomic displacements describe a homogeneous shearing deformation along the y' axis with shear strain $s = 0.05$, superimposed on the periodic inhomogeneous deformation

$$v_x = \delta_x \rho \sin\frac{2\pi x'}{L}\sin\frac{y'}{a}, \qquad v_y = \delta_y \rho \sin\frac{2\pi x'}{L}\sin\frac{y'}{a}, \tag{4.58}$$

where the y' axis is aligned with the chiral vector, $\delta_x = 0.5$ and $\delta_y = 0.5$. The Tersoff–Brenner potential induced is displayed in Fig. 4.7c, with the light color denoting high values and the dark color denoting low values of the potential. Figure 4.7d and e show the bond network before and after deformation, with arrows indicating the exaggerated inner displacement. Figure 4.7f shows a plot of the magnitude of the inner displacement, and Fig. 4.7g shows corresponding plots of the four nonzero components of the planar deformation gradient.

Now we roll the graphene to form a $(10, 5)$ chiral nanotube and compute the three-dimensional surface deformation gradient and accompanying inner displacement by solving a system of 12 linear equations for each atom, as discussed previously in this section. The unit normal vector before deformation, **N**, is directed normal to the tube axis. Figure 4.8a shows a plot of the magnitude of the inner displacement, Fig. 4.8b and c show the distribution of the first and second invariants of the surface Cauchy–Green tensor, and Fig. 4.8d shows the distribution of the atomic potential. This example visually confirms the physical consistency of the various quantities displayed.

In summary, we have described the kinematics of a planar or curved atomic lattice in terms of a computable single-valued surface deformation gradient and a computable inner displacement field. The inner displacement has been used previously to describe the kinematics of homogeneous deformation where the surface deformation

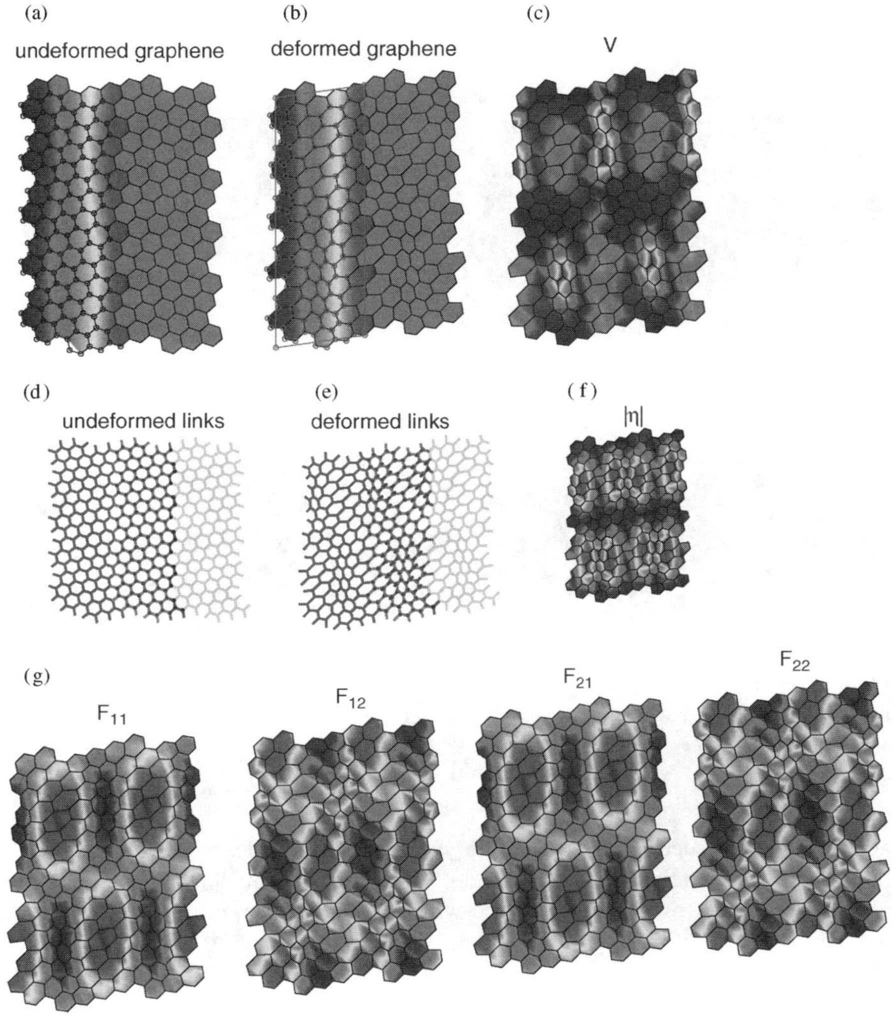

Figure 4.7 (a–c) Doubly periodic deformation of the planar graphene sheet in the $(10, 5)$ chiral mode, and induced Tersoff–Brenner potential; (d,e) bond network before and after deformation (the arrows indicate the exaggerated inner displacement, η); (f) plot of the magnitude of the inner displacement, $|\eta|$; (g) four components of the associated deformation gradient. (*See insert for color representation of figure.*)

gradient is the same at all atoms and the inner displacement alternates in sign across the two constituent triangular lattices [6–8]. Examples can be found in the stretching, expansion, and twist of cylindrical nanotubes.

The three-dimensional curvature of the particulate sheet is implicit in the surface deformation gradient and local inner displacement. If both are given, the local geometry of a particle cluster can be readily reconstructed. A different

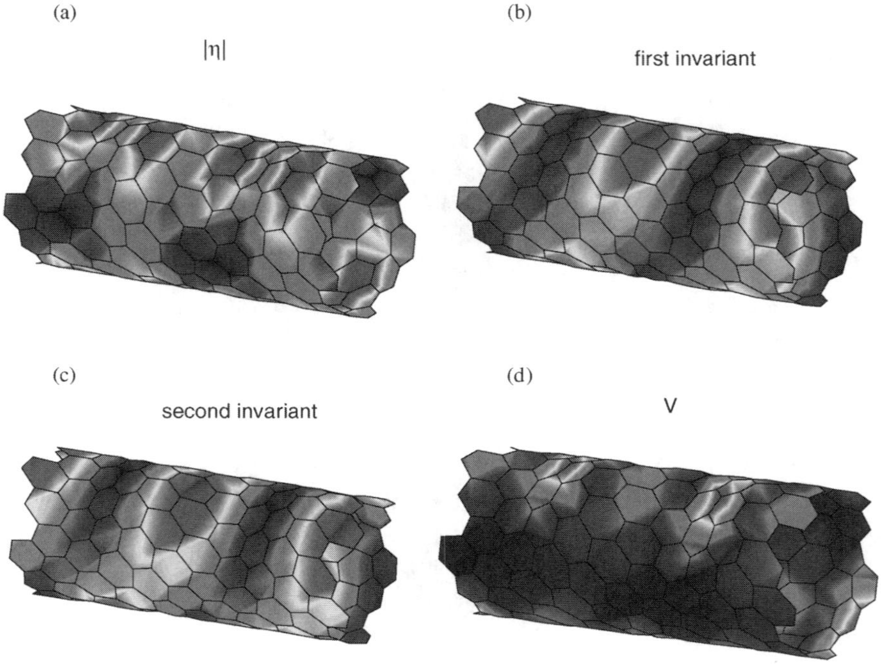

Figure 4.8 The graphene has been rolled up into a nanotube. (a) Plot of the magnitude of the inner displacement; (b,c) first and second invariants of the surface Cauchy–Green tensor; (d) atomic potential. (*See insert for color representation of figure.*)

approach was pursued by Arroyo and Belytschko [12], who described the kinematics in terms of a tangential deformation field followed by warping with an exponential mapping in order to construct a shape with a specified curvature at every atomic position.

4.6 HEXAGONAL LATTICES WITH ELASTIC LINKS

It is illuminating to consider the deformation of a finite patch of a planar hexagonal lattice in the xy plane, where each particle is connected to its three immediate neighbors with identical elastic springs, as shown in Fig. 4.9. The equilibrium length of the springs, d_0, is set equal to the length of the springs in the initial (undeformed) configuration, R. To construct a patch, we retain only a specific collection of hexagonal cells determining the patch shape. The patch has interior and boundary nodes. An interior node participates in three hexagonal faces, whereas a boundary node participates in one or two hexagonal faces. An interior node has three neighbors, whereas a boundary node may have two or three neighbors; in either case, the neighbors can be interior or boundary nodes.

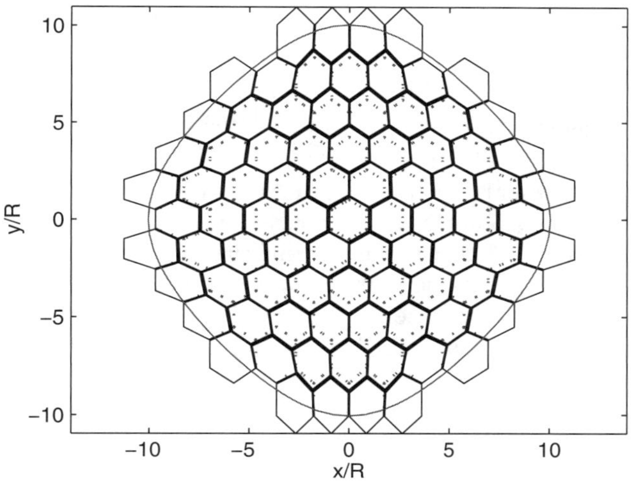

Figure 4.9 Cubic extensional deformation of a circular patch. The dotted lines show the links after deformation and before equilibration, and the solid lines show the links after equilibration.

The mean elastic energy of the patch arises by summing the energy of the individual links,

$$\mathcal{V} = \frac{k_s}{2N_s} \sum_i N_s (d_i - d_0)^2, \tag{4.59}$$

where k_s is the spring constant, the sum runs over all N_s springs, and d_i is the spring length. In practice, the total energy can be computed by running over the N_I interior nodes and summing the energy of three springs connecting these nodes to their neighbors,

$$\mathcal{V} = \frac{k_s}{2N_s} \sum_{i=1}^{N_I} \left[\sum_{j=1}^{3} c_{ij} (d_{ij} - d_0)^2 \right], \tag{4.60}$$

where $c_{ij} = 0.5$ if a neighbor of the ith node is an interior node, and $c_{ij} = 1$ otherwise. The total number of springs is

$$N_s = \sum_{i=1}^{N_i} \left(\sum_{j=1}^{3} c_{ij} \right). \tag{4.61}$$

In the numerical experiments, the lattice is deformed, the boundary nodes are held fixed, and the equilibrium positions of the interior nodes are computed by minimizing the mean elastic energy with respect to the x and y coordinates of the interior nodes. The procedure crudely amounts to solving a boundary value problem with essential (Dirichlet) boundary conditions, where the boundary displacement is specified and the inner displacement field is computed on the condition of nodal equilibrium.

In the numerical computations, the atomic positions after deformation and before equilibration are given by the incomplete linear-cubic field

$$\begin{bmatrix} x' \\ y' \end{bmatrix} = \begin{bmatrix} A_{11} & A_{12} \\ A_{21} & A_{22} \end{bmatrix} \begin{bmatrix} x \\ y \end{bmatrix} + \frac{1}{d_0} \begin{bmatrix} a_3 x^3 \\ b_3 y^3 \end{bmatrix}, \tag{4.62}$$

where $\mathbf{A}$ is a dimensionless linear strain tensor and a_3 and b_3 are dimensionless parameters introduced to implement nonlinear boundary deformation.

When $a_3 = 0$ and $b_3 = 0$, we obtain a homogeneous boundary deformation. If, in addition, the matrix $\mathbf{A}$ is diagonal with identical diagonal elements equal to $1 + \varepsilon$, we obtain isotropic expansion with perfect hexagonal cells at equilibration. In that case, the mean elastic energy is a quadratic function of ε, and the modulus of isotropic expansion of the patch is equal to the spring constant multiplied by the square of the undeformed spring length, $E_{\text{iso}} \equiv \partial^2 \mathcal{V} / \partial \varepsilon^2 = k_s d_0^2$. For any other type of deformation, the orientation of the deformation field relative to the hexagonal lattice is significant.

4.6.1 Extensional Deformation of a Circular Patch

To construct the circular patch shown in Fig. 4.9, we retain only hexagonal cells whose center resides inside a circle of radius $8R$ centered at the origin. The graph in this figure illustrates the equilibrium structure for a cubic boundary deformation field where $\mathbf{A} = \mathbf{I}$, $a_3 = b_3 = 0.004$, and $\mathbf{I}$ is the identity matrix. A deformed circle is also shown to establish a frame of reference.

Figure 4.10 shows plots of the discrete deformation gradient field and inner displacement field. The light struts correspond to the state of the lattice before equilibration, and the bold struts correspond to the state of the lattice after equilibration. The information displayed in this figure is duplicated in the three-dimensional plots shown in Fig. 4.11. Because of the cubic boundary displacement, the diagonal components of $\mathbf{F}$ reach minimum values at the center of the circular patch, and increase along the x or y axis. Significant nondiagonal components are observed due to the particulate nature of the material. The inner displacement is very small at the center of the patch and increases toward the boundaries.

These results illustrate explicitly the establishment of an inhomogeneous and smoothly varying discrete deformation field and the onset of a smoothly varying inner displacement field.

4.6.2 Extensional Deformation of a Square Patch

It is instructive to consider the linear deformation of a square patch, setting $a_3 = 0$ and $b_3 = 0$. First, we consider an extensional area-preserving deformation where the boundary nodes are displaced according to the nominal deformation gradient

$$\mathbf{A} = \begin{bmatrix} 1+\varepsilon & 0 \\ 0 & 1-\varepsilon \end{bmatrix}, \tag{4.63}$$

where ε is a dimensionless constant.

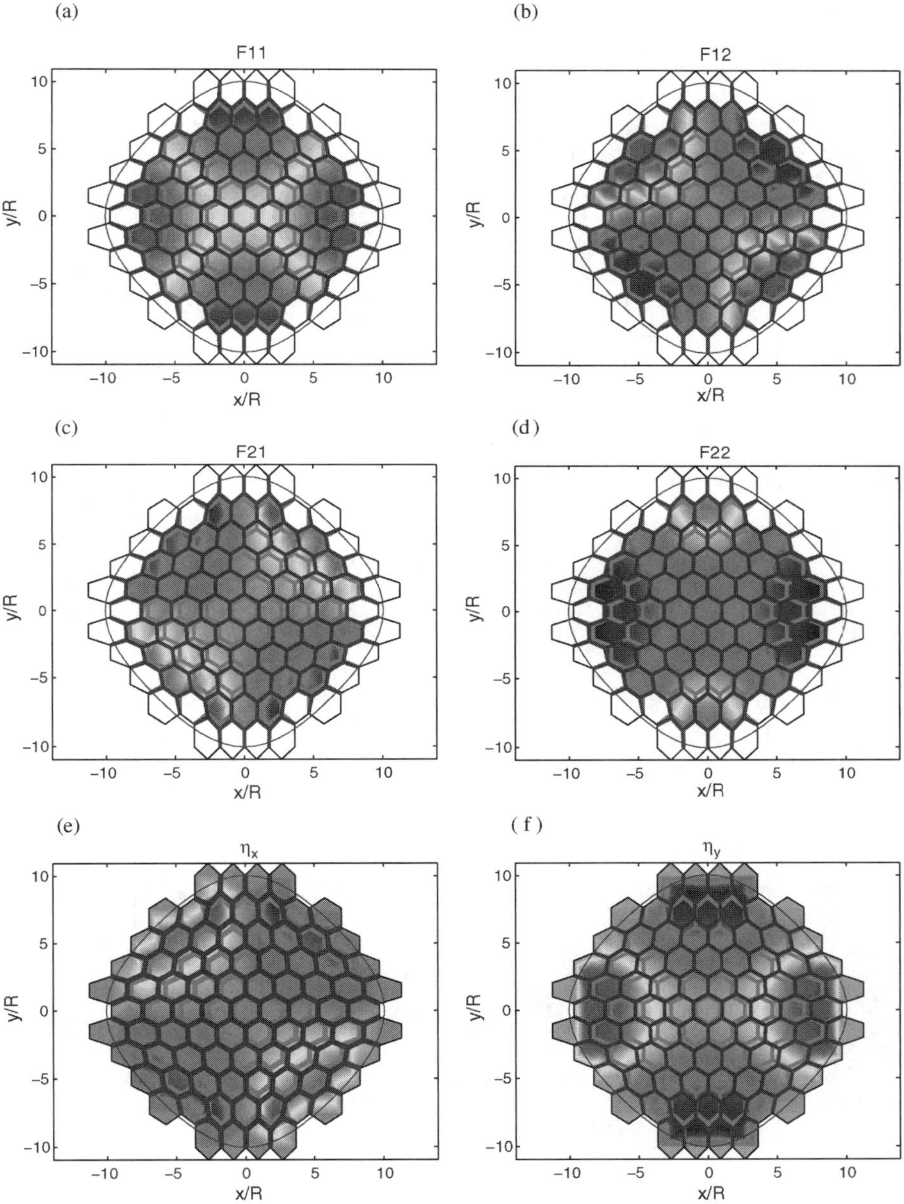

Figure 4.10 Deformation of a circular patch of a hexagonal lattice whose atoms are connected with elastic springs, subject to a specified cubic boundary displacement: (a–d) four components of the discrete deformation gradient field; (e,f) two components of the inner displacement field. (*See insert for color representation of figure.*)

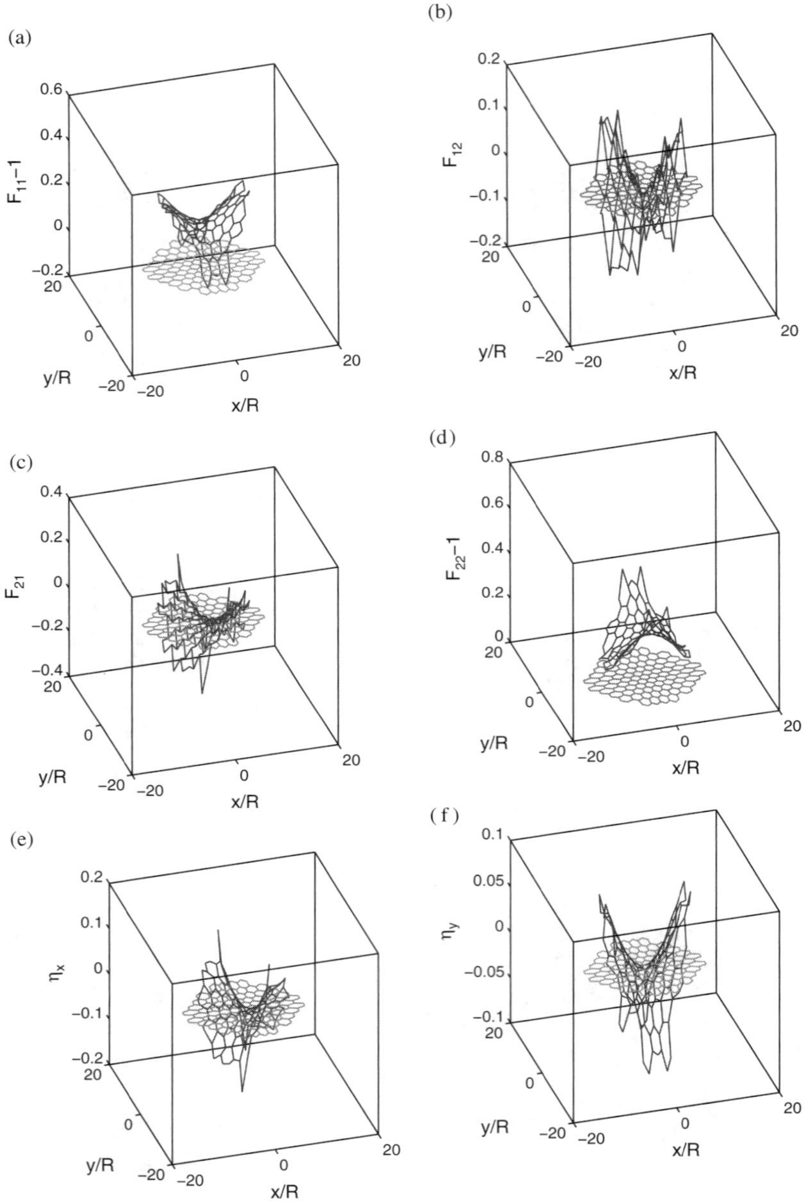

Figure 4.11 Three-dimensional perspectives of the plots shown in Fig. 4.10.

Figure 4.12a–d show equilibrium shapes for $\varepsilon = 0.2$ and several patch side lengths, $2R$, $4R$, $8R$, and $16R$. The dotted lines show the links after deformation and before equilibration. The solid lines show the links after equilibration. As the size of the patch increases, the difference between the dotted and solid links

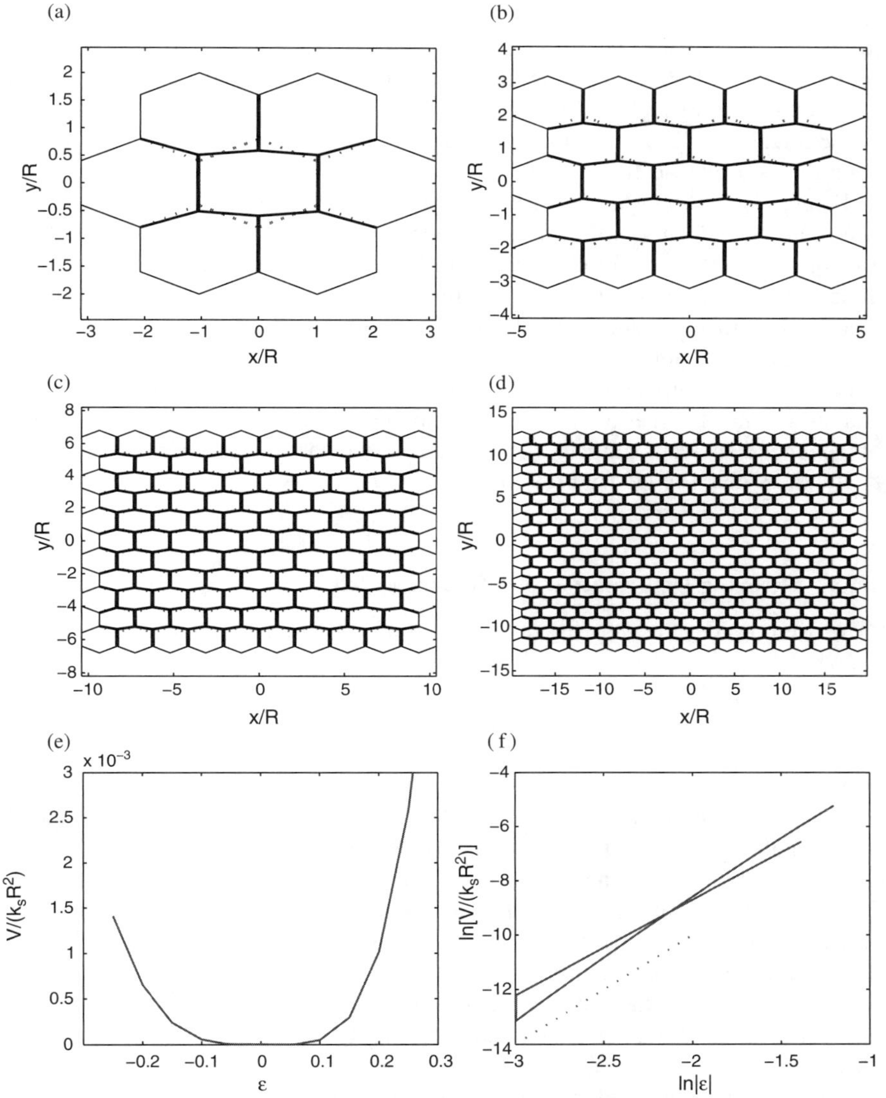

Figure 4.12 Extensional, area-preserving deformation of a square patch with side length (a) $2R$, (b) $4R$, (c) $8R$, and (d) $16R$. (e) Mean elastic energy plotted against the stretch, ε. (f) A log-log plot reveals that the potential is a quartic function of ε; one curve corresponds to positive values of ε, and the second curve corresponds to negative values of ε. The slope of the dotted line in (f) is equal to 4.

persists, yielding a nonzero resultant inner displacement field aligned predominantly with the y axis. These calculations demonstrate that a homogeneous boundary deformation without inner displacement may induce a finite discrete inner displacement field independent of the patch size. The scaled mean elastic energy of the patch, V/k_sR^2, is 0.0013, 0.0093, 0.0010, 0.0010, and 0.0011, respectively, for patch length side $2R$, $4R$, $8R$, $16R$, and $24R$. We observe rapid convergence with respect to the patch size. Figure 4.12e and f are graphs of the mean elastic energy plotted against the deformation parameter ε on a linear and log-log scale, for a patch with side length equal to $8R$. The data reveal that the potential is a quartic function of ε.

Figure 4.13 shows the distribution of the four components of the discrete deformation gradient and inner displacement fields over a patch with side length $8R$, for $\varepsilon = 0.2$. The information displayed in this figure is duplicated in the three-dimensional plots shown in Fig. 4.14. In the interior of the patch, the deformation gradient is consistent with that inducing the boundary deformation. Fluctuations are observed near the left and right edges due to the discrete nature of the particulate medium. The x component of the inner displacement is near zero in the patch interior, and shows small fluctuations near the left and right edges. In contrast, the y component of the inner displacement is significant and nearly flat in the patch interior, and shows fluctuations near the left and right edges.

4.6.3 Shearing Deformation of a Square Patch

Next, we consider the shearing deformation of a square patch where

$$\mathbf{A} = \begin{bmatrix} 1 & \varepsilon \\ 0 & 1 \end{bmatrix} \tag{4.64}$$

and ε is a dimensionless shear parameter. Figure 4.15a–d show equilibrium shapes for $\varepsilon = 0.2$ and square patch side length $2R$, $4R$, $8R$, and $16R$. The dotted lines show the links after the homogeneous shearing deformation and before equilibration. Figure 4.15e and f are graphs of the mean elastic energy plotted against the shear parameter ε on a linear and log-log scale, for a patch with side length equal to $8R$. The data reveal that the potential is a quartic function of ε.

Figure 4.16 shows the distribution of the four components of the deformation, and inner displacement over the patch with side length $8R$, for $\varepsilon = 0.2$. The information displayed in this figure is duplicated in the three-dimensional plots in Fig. 4.17. As in the case of extensional deformation, the discrete deformation gradient is consistent with that inducing the boundary deformation in the interior of the patch. Small fluctuations are observed near the left and right edges. In contrast, the x component of the inner displacement is nearly flat in the patch interior and shows fluctuations near the left and right edges. The y component of the inner displacement is near zero in the patch interior and shows small fluctuations near the left and right edges. These calculations confirm that a homogeneous shearing boundary deformation without inner displacement may induce a finite inner displacement field independent of the patch size. Similar results are obtained when shearing occurs along the y axis.

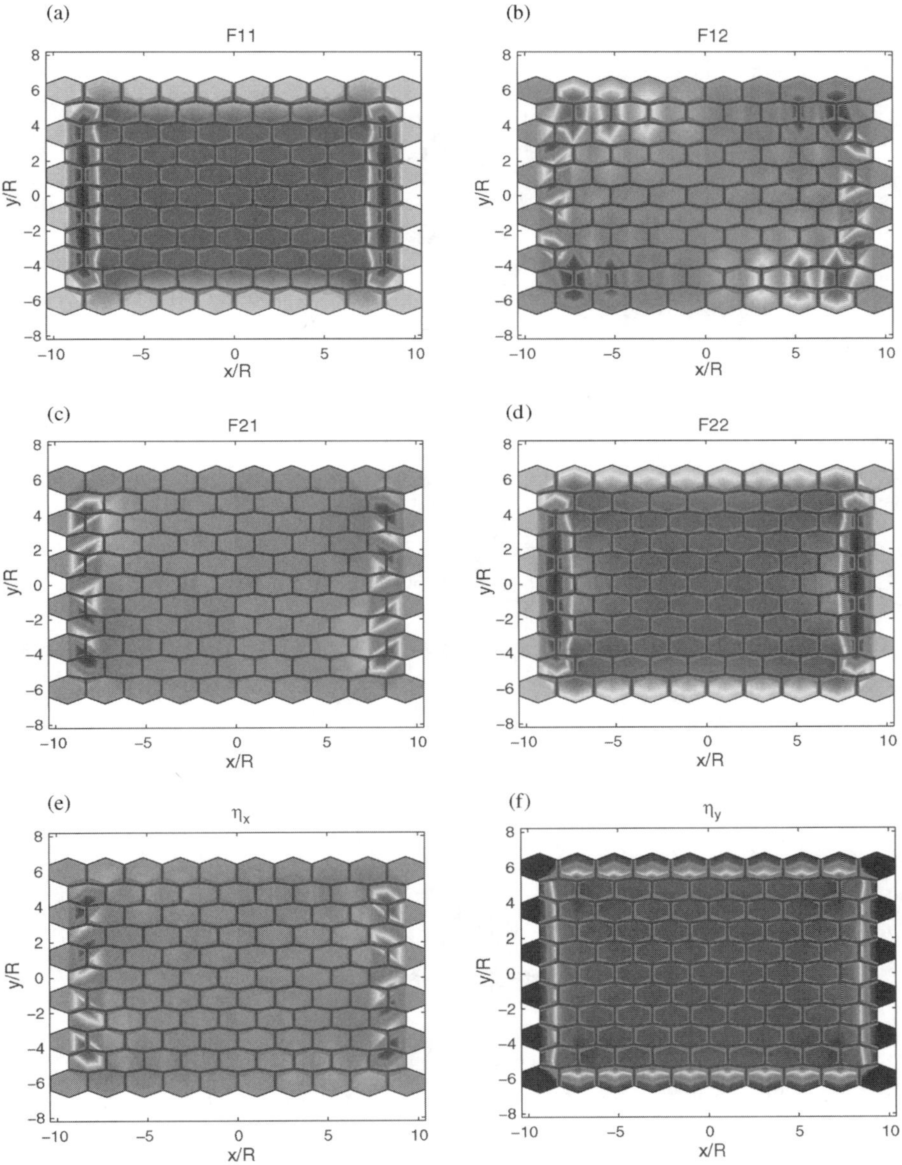

Figure 4.13 (a–d) Distribution of the four components of the discrete deformation gradient; (e,f) the inner displacement over a patch with side length $8R$ in extensional area-preserving boundary deformation with $\varepsilon = 0.2$. (*See insert for color representation of figure.*)

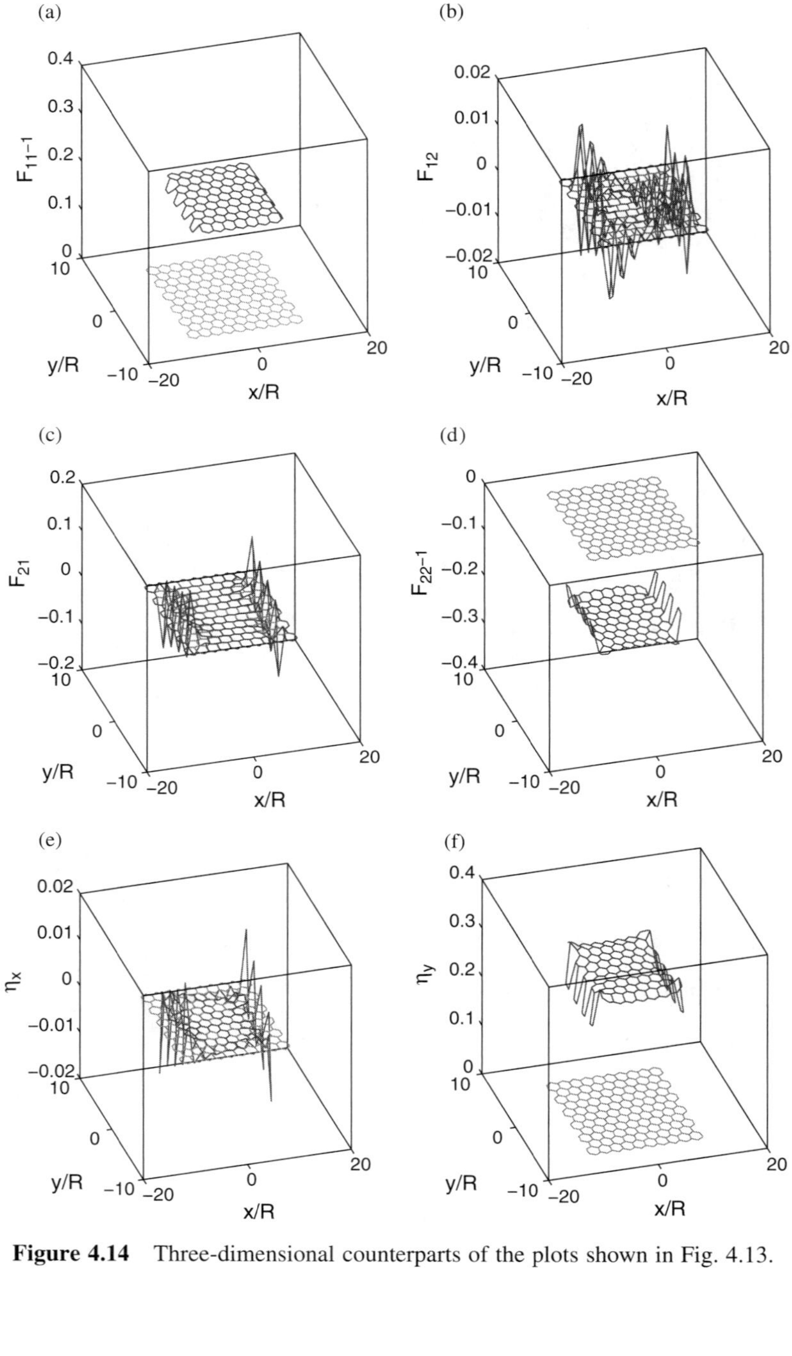

Figure 4.14 Three-dimensional counterparts of the plots shown in Fig. 4.13.

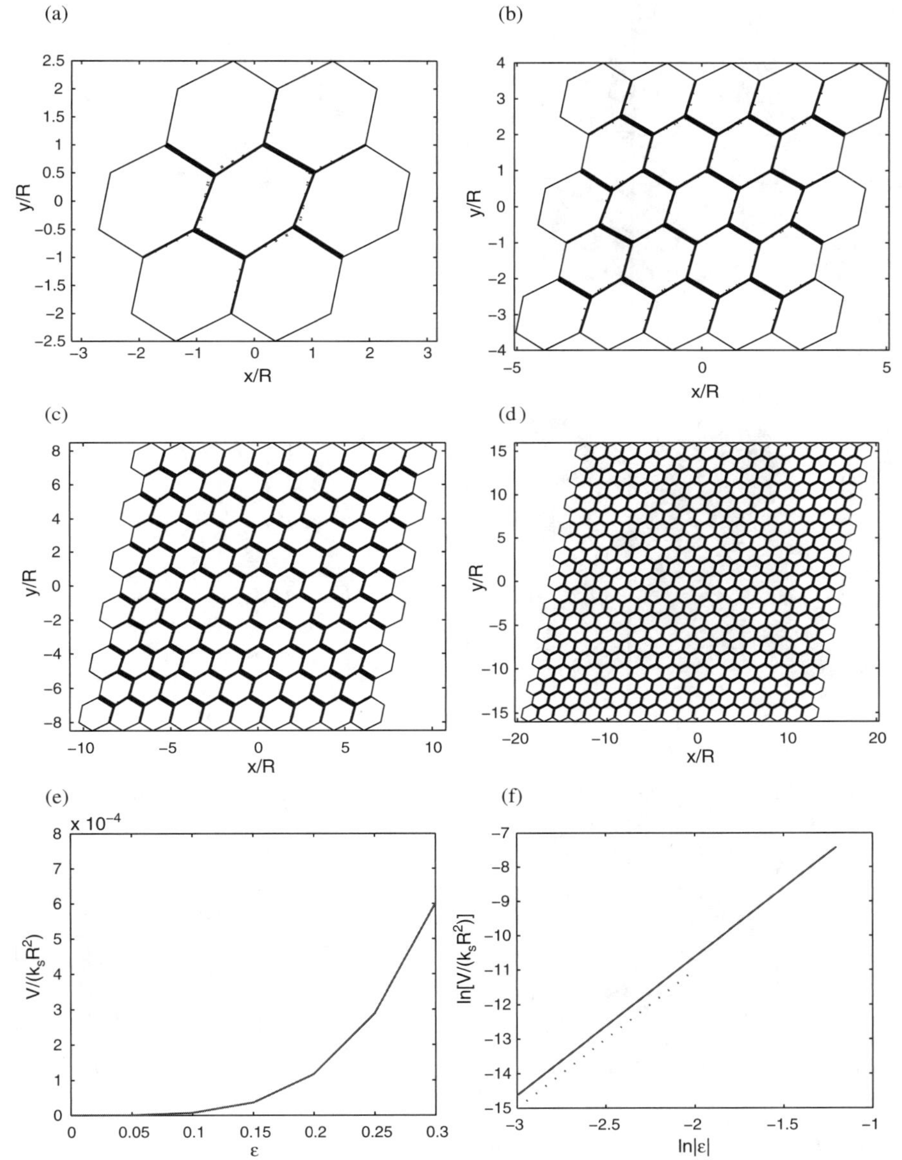

Figure 4.15 Shearing deformation of a square patch with side length (a) $2R$, (b) $4R$, (c) $8R$, and (d) $16R$, all for $\varepsilon = 0.2$. (e) Mean elastic energy of a square patch in shearing deformation, plotted against the stretch, ε. (f) A log-log plot reveals that the potential is a quartic function of ε. The slope of the dotted line in (f) is equal to 4.

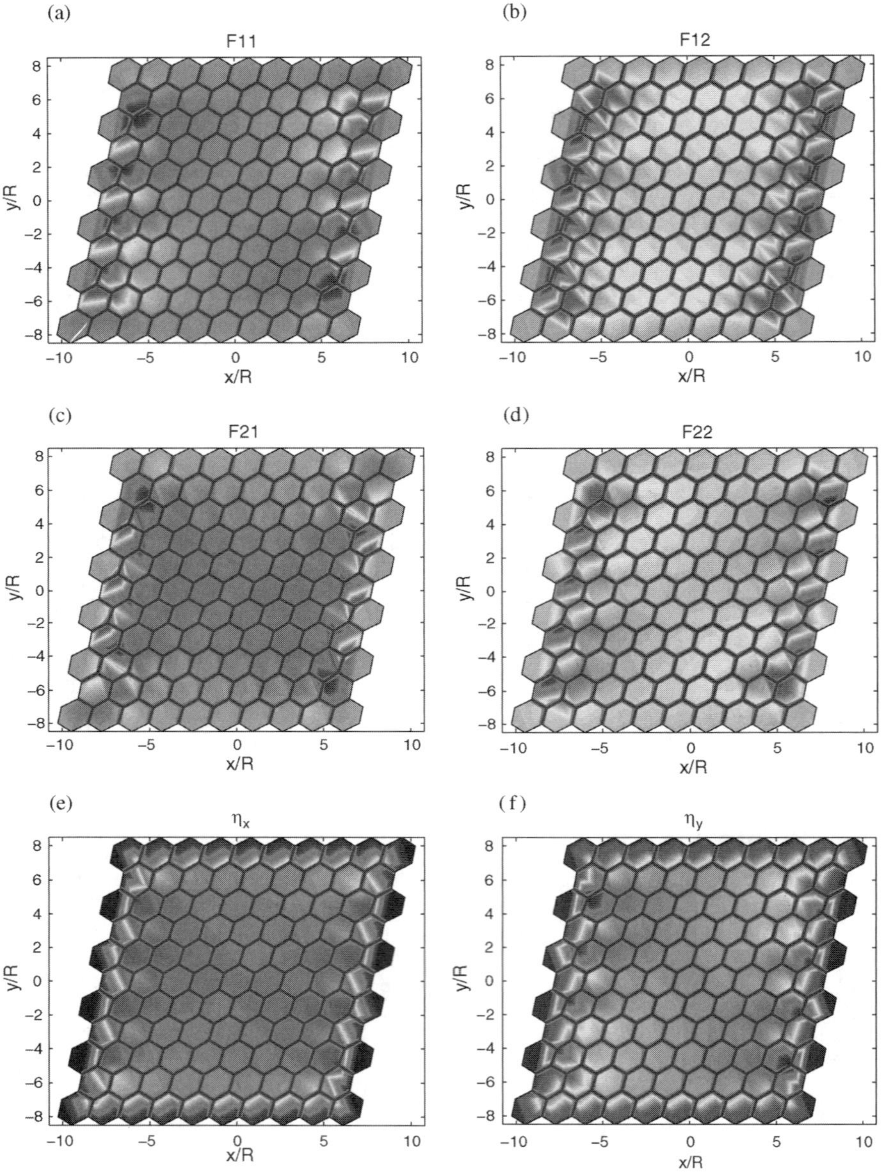

Figure 4.16 (a–d) Distribution of the four components of the discrete deformation gradient; (e,f) inner displacement over a patch with side length $8R$ in extensional area-preserving boundary deformation with $\varepsilon = 0.2$. (*See insert for color representation of figure.*)

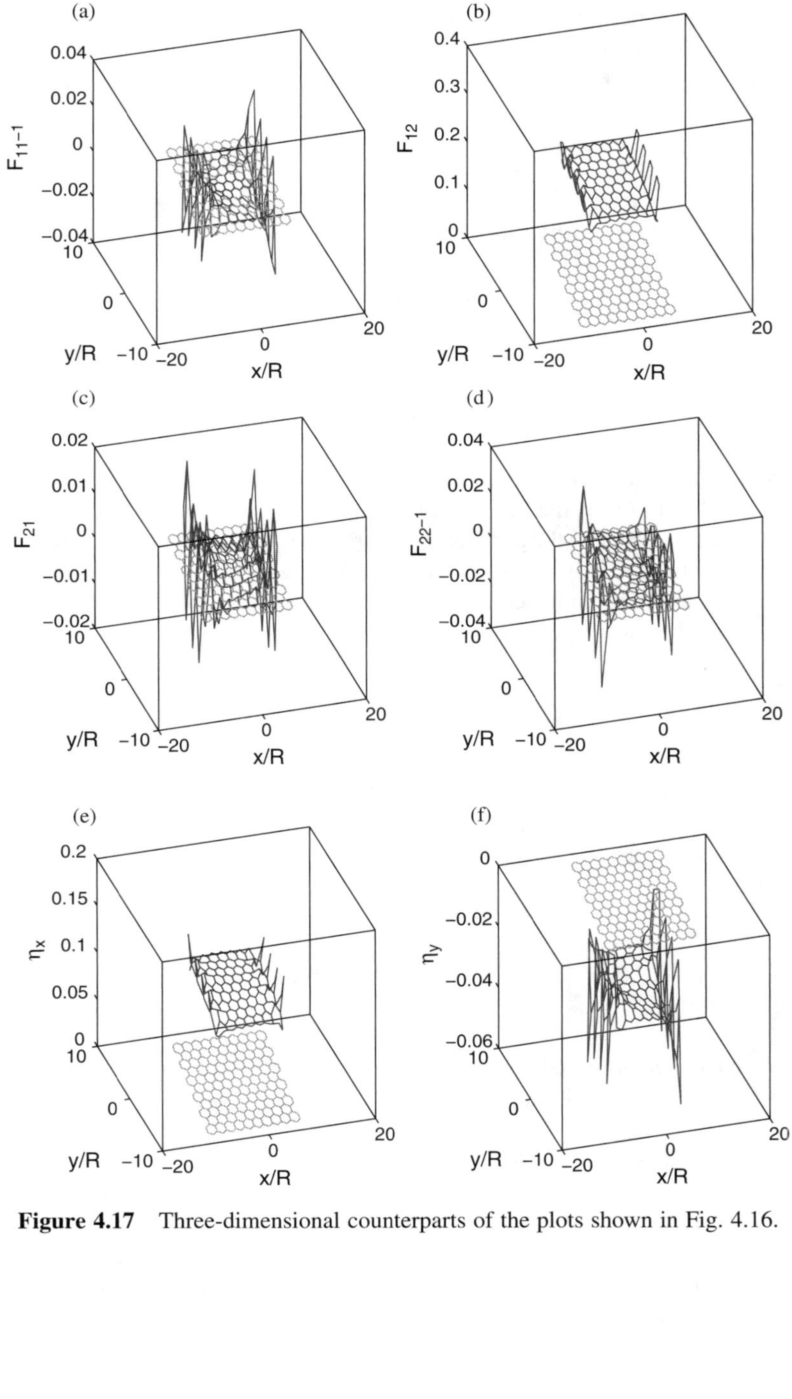

Figure 4.17 Three-dimensional counterparts of the plots shown in Fig. 4.16.

4.7 ENERGETICS OF HEXAGONAL LATTICES WITH ELASTIC LINKS

Since the particulate deformation gradient and inner displacement uniquely determine the position of the neighbors of each particle, the potential of the ith particle in the planar elastic lattice discussed in Section 4.6 can be expressed by the functional form

$$\Phi_i = \mathcal{F}(\mathbf{F}_i, \eta_i). \tag{4.65}$$

Invariance to rotation of the four-particle cluster requires that

$$\Phi_i = \mathcal{F}(\mathbf{V}_i^2, \mathbf{H}_i), \tag{4.66}$$

where

$$\mathbf{V}_i^2 \equiv \mathbf{F}_i \cdot \mathbf{F}_i^T \tag{4.67}$$

is the particulate left Cauchy–Green surface deformation tensor. Note that the second argument in Eq. (4.66) is the inner displacement *before* deformation. Because of the inherent anisotropy due to the particulate nature of the atomic sheet, the first argument on the right-hand side may not be replaced by the two scalar invariants, I_1 and I_2, as is customary in the continuum mechanics of surfaces. We can sensibly spread the atomic potential over the surface represented by the atomic lattice to obtain a surface density function,

$$\phi = \mathcal{G}(\mathbf{V}^2, \mathbf{H}). \tag{4.68}$$

In numerical practice, spreading can be done using radial basis functions or other interpolation methods for scattered data.

Next, we consider the potential of a collection of N atoms residing away from boundaries,

$$\Phi_{\text{total}} = \sum_{j=1}^{N} \Phi_j(\mathbf{V}_j^2, \mathbf{H}_j). \tag{4.69}$$

Using the principle of virtual displacements, we find that the force load of the ith atom in a specified geometrical configuration is

$$\mathbf{F}_i = \frac{\partial \Phi_{\text{total}}}{\partial \mathbf{x}_i} = \sum_{j=1}^{N} \frac{\partial \Phi_j}{\partial \mathbf{x}_i}, \tag{4.70}$$

where the derivative with respect to $\mathbf{x}_i$ is computed holding the position of all other particles fixed. Because of the simplified two-body interaction, only four terms in the sum survive, corresponding to $j = i$ and three other values of j labeling the closest neighbors of the ith atom. The derivatives of Φ_j with respect to $\mathbf{x}_i$ can be recast in terms of derivatives with respect to $\mathbf{V}_j^2$ and $\mathbf{H}_j$ using the chain rule,

$$\frac{\partial \Phi_j}{\partial \mathbf{x}_i} = \frac{\partial \Phi_j}{\partial \mathbf{V}_j^2}\frac{\partial \mathbf{V}_j^2}{\partial \mathbf{x}_i} + \frac{\partial \Phi_j}{\partial \mathbf{H}_j}\frac{\partial \mathbf{H}_j}{\partial \mathbf{x}_i}, \tag{4.71}$$

where $\partial \mathbf{H}_i / \partial \mathbf{x}_i = \mathbf{I}$. This expression provides us with a venue for computing the pointwise atomic load from derivatives with respect to the $\mathbf{V}_j^2$ and $\mathbf{H}_j$.

We can apply the classical membrane formulation to a particulate sheet for the purpose of identifying effective tension fields and deriving discrete equilibrium equations. The counterpart of expression (4.47) at an atom labeled 1 and three neighbors labeled 2, 3, and 4 on a slightly deformed atomic lattice is

$$\mathbf{F}_1^T = \mathbf{P} + \sum_{j=2,3,4} \mathbf{T}^{(j)} \cdot (\mathbf{v}_j - \mathbf{v}_1), \tag{4.72}$$

where $\mathbf{v}_i$ are the atomic displacements and $\mathbf{T}^{(j)}$ are three-index transfer matrices dependent on the local atomic topology and satisfying $\mathbf{N}_1 \cdot \mathbf{T}^{(j)} = \mathbf{0}$. When $\mathbf{v}_j = \mathbf{0}$ for $j = 2, 3, 4$, $\mathbf{F}_1^S = \mathbf{P}$ for any central displacement, $\mathbf{v}_1$, causing a local inner displacement alone, and thus

$$\sum_{j=2,3,4} \mathbf{T}^{(j)} = \mathbf{0}. \tag{4.73}$$

The discrete symmetric surface strain tensor at the location of the central atom is

$$\varepsilon_1 = \frac{1}{2} \sum_{j=2,3,4} (\mathbf{T}_j + \mathbf{T}^{(j)^T}) \cdot (\mathbf{v}_j - \mathbf{v}_1), \tag{4.74}$$

where the transpose applies to the first two indices.

The inner displacement before deformation can be expressed in the corresponding form

$$\mathbf{H}_1 = \sum_{j=2,3,4} \mathbf{S}^{(j)} \cdot (\mathbf{v}_j - \mathbf{v}_1), \tag{4.75}$$

where $\mathbf{S}^{(j)}$ are new transfer matrices. When $\mathbf{v}_j = \mathbf{0}$ for $j = 2, 3, 4$, $\mathbf{H}_1 = \mathbf{v}_1$; thus,

$$\sum_{j=2,3,4} \mathbf{S}^{(j)} + \mathbf{I} = \mathbf{0}. \tag{4.76}$$

Since a locally isotropic deformation does not induce an inner displacement,

$$\sum_{j=2,3,4} \mathbf{S}^{(j)} \cdot \mathbf{A} \cdot (\mathbf{X}_j - \mathbf{X}_1) = \mathbf{0}, \tag{4.77}$$

where $\mathbf{A}$ is an arbitrary matrix.

The work increment for a collection of N atoms can be expressed in the form

$$\delta\Phi = \sum_{i=1}^{N} \delta\Phi_i \equiv \sum_{i=1}^{N} (\tau_i : \delta\varepsilon_i^S + \chi_i \cdot \delta\mathbf{H}_i), \tag{4.78}$$

where τ_i is a discrete surface tension field and χ_i is a discrete surface force field. Thus,

$$\tau_i = \frac{\partial\Phi}{\partial\varepsilon_i^S}, \qquad \chi_i = \frac{\partial\Phi}{\partial\mathbf{H}_i}. \tag{4.79}$$

These derivatives can be expressed in terms of derivatives with respect to the individual atomic displacements using the chain rule, and then evaluated by numerical methods. Discrete equilibrium equations may be derived by substituting Eqs. (4.74) and (4.75) into Eq. (4.78) and then involving the principle of virtual displacements. Although the procedure is undoubtedly cumbersome, it does provide us with a rational framework for developing equilibrium equations on the particulate sheet and thereby establishing an analogy with continuum media for small deformations.

The discrete surface field χ_i is conjugate to the discrete inner displacement field. In the theory of thin shells, the surface energy depends on the surface strain and curvature tensor whose conjugate fields are the tangential tension tensor and tangential bending moment tensor. By analogy, then, the surface field χ_i embodies the effect of bending moments associated with changes in the curvature of the particulate sheet.

4.8 STRAIN GRADIENT FORMULATION

We now turn our attention to the energetics of a planar or curved graphene sheet. Since the Tersoff–Brenner potential of each carbon atom in the hexagonal lattice depends on the position of the neighbors as well as on the position of the neighbors of the neighbors, the mechanical response of the particulate sheet is classified as nonlocal. We note that the particulate deformation gradient and inner displacement of an atom and its three neighbors uniquely determine the position of the first- and second-generation neighbors, and express the potential at the central atom numbered 1 in the functional form

$$\Phi_1 = \mathcal{F}(\mathbf{F}_1, \mathbf{F}_2, \mathbf{F}_3, \mathbf{F}_4, \eta_1, \eta_2, \eta_3, \eta_4), \tag{4.80}$$

involving $4 \times (9 + 3) = 4 \times 12 = 48$ scalar unknowns computed using 36 scalar equations.

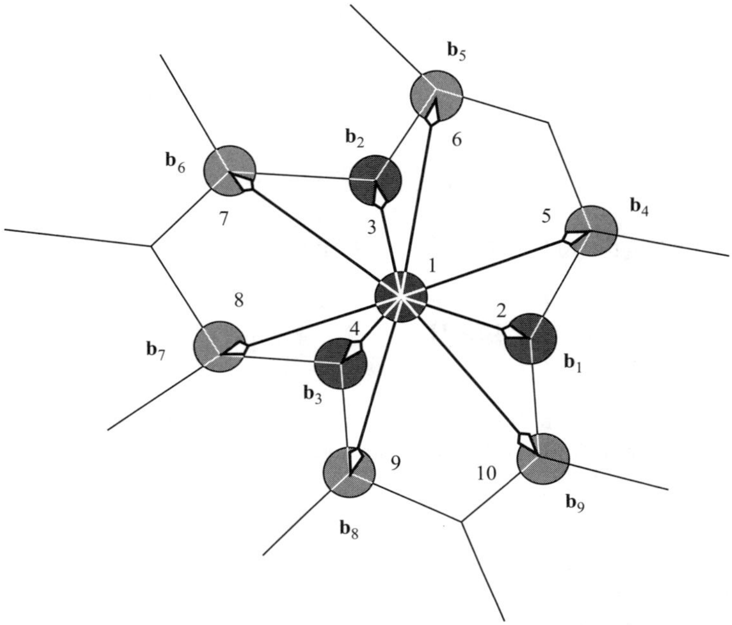

Figure 4.18 Nine bonds originate from the central atom up to the first and second neighbors. (*See insert for color representation of figure.*)

Alternatively, the energy functional is expressed in terms of the first surface deformation gradient, $\hat{F}_{ij}$, second deformation gradient, G_{ijk}, and local inner displacement, η_i. Denoting the nine vectors subtended between a chosen atom and its first and second neighbors by $\mathbf{B}_i$ before deformation and by $\mathbf{b}_i$ after deformation, as shown in Fig. 4.18, we write

$$\mathbf{b}_i = \hat{\mathbf{F}} \cdot \mathbf{B}_i + \eta + \frac{1}{2}\mathbf{G} : \mathbf{B}_i\mathbf{B}_i \tag{4.81}$$

for $i = 1, 2, \ldots, 9$, and also require that

$$\hat{\mathbf{F}} \cdot \mathbf{N} = \mathbf{0}. \tag{4.82}$$

Because only the symmetric part of $\mathbf{G}$ with respect to the second and third indices defined in terms of $3 \times 6 = 18$ scalars is relevant, we may set

$$G_{ijk} = G_{ikj}. \tag{4.83}$$

We have nine unknowns in $\hat{\mathbf{F}}$ (full matrix), three unknowns in η, and 18 unknowns in $\mathbf{G}$, a total of 30 unknowns. The union balances exactly the number of equations corresponding to nine bonds and the normal vector before or after deformation. Accordingly, the local inner displacement, first deformation gradient, and second deformation gradient are uniquely defined. The Tersoff–Brenner potential of the jth atom takes the form

$$V_j = \Phi(\hat{\mathbf{F}}_j, \mathbf{G}_j,\ \eta_j), \tag{4.84}$$

where $\mathbf{G}_i$ incorporates the combined effect of local geometrical nonlinearity and curvature.

An alternative of expression (4.81) corresponding to Eq. (4.56) is

$$\mathbf{b}_i = \tilde{\mathbf{F}} \cdot (\mathbf{B}_i + \mathbf{H}) + \frac{1}{2}\tilde{\mathbf{G}} : (\mathbf{B}_i + \mathbf{H})\,(\mathbf{B}_i + \mathbf{H}) \tag{4.85}$$

for $i = 1, 2, \ldots, 9$, where $\tilde{\mathbf{F}}$ and $\tilde{\mathbf{G}}$ are alternative first and second deformation gradients. The Tersoff–Brenner potential of the jth atom takes the form

$$V_j = \Phi(\tilde{\mathbf{F}}_j, \tilde{\mathbf{G}}_j, \mathbf{H}_j). \tag{4.86}$$

The Tersoff–Brenner potential of a collection of N atoms on a planar or curved sheet may now be expressed in the form

$$\Phi_{\text{total}} = \sum_{j=1}^{N} \Phi_j(\tilde{\mathbf{F}}_j, \mathbf{G}_j, \mathbf{H}_j). \tag{4.87}$$

Using the principle of virtual displacements, we find the ith atomic force load,

$$\mathbf{F}_i = \frac{\partial \Phi_{\text{total}}}{\partial \mathbf{x}_i} = \sum_{j=1}^{N} \frac{\partial \Phi_j}{\partial \mathbf{x}_i}. \tag{4.88}$$

We can repeat the small deformation theory outlined in Section 4.7 to derive expressions for localized generalized tensions and accompanying equilibrium equations. The details are tedious, yet straightforward.

This formulation bears a striking similarity to the strain gradient theory of three-dimensional elastic media developed by Toupin [13] and Mindlin [14,15], and advanced by Fleck and Hutchinson [16]. In the strain gradient formulation, a small material vector before deformation is related to the same material vector after deformation by the counterpart of Eq. (4.81) in the absence of inner displacement,

$$d\,\mathbf{x} = d\,\mathbf{X} \cdot (\mathbf{I} + \nabla \mathbf{v}) + \frac{1}{2} d\,\mathbf{X}\, d\,\mathbf{X} : \nabla\nabla \mathbf{v}, \tag{4.89}$$

where $\mathbf{v}$ is a continuous displacement field and $\hat{\mathbf{F}}^T = \mathbf{I} + \nabla \mathbf{v}$. In the linear strain gradient theory, the strain energy function, w, depends on the symmetric strain tensor $\varepsilon = \frac{1}{2}[\nabla \mathbf{v} + (\nabla \mathbf{v})^T]$, and on the symmetric second deformation gradient, $\zeta \equiv \nabla\nabla \mathbf{v}$.

The work increment of a material in a volume Ω is

$$\delta E = \iiint_{\Omega} \delta w \, d\, V = \iiint_{\Omega} (\sigma_{ij}\, \delta\varepsilon_{ij} + \tau_{ijk}\, \delta\zeta_{ijk})\, d\, V, \tag{4.90}$$

where σ_{ij} and τ_{ijk} are generalized stresses. Setting the variation to zero and integrating by parts, we derive the equilibrium equation

$$\nabla \cdot \sigma - \nabla\nabla : \tau = 0. \tag{4.91}$$

Because this is a high-order differential equation, six boundary conditions on the displacement field or traction field are required.

The use of the second deformation gradient to describe the inhomogeneous deformation of atomic microstructures arranged on simple (Bravais) lattices was advocated by Sunyk and Steinmann [17]. Wang et al. [18,19] regarded the atomic displacement, $\mathbf{u}_i$, of a planar or curved graphene sheet as discrete realizations of a continuous and differential function of surface coordinates, $\mathbf{u}$, and applied Taylor series expansions to write the counterpart of Eq. (4.89),

$$\mathbf{u}(\mathbf{x}) = \mathbf{u}_i + (\mathbf{u} - \mathbf{u}_i) \cdot (\nabla \mathbf{u})_i + \frac{1}{2} (\mathbf{u} - \mathbf{u}_i)\, (\mathbf{u} - \mathbf{u}_i) : \nabla\nabla \mathbf{v}, \tag{4.92}$$

termed the high-order Cauchy–Born rule. Our discussion indicates that this expansion is valid only for a restricted class of deformations.

4.9 SUMMARY

We have discussed a kinematic framework for describing the arbitrary deformation of planar and curved hexagonal atomic lattices composing graphene sheets and nanotubes. Kinematics is described in terms of a discrete surface deformation gradient and associated inner displacement, both evaluated at the position of each individual atom. The inner displacement accounts for the separation of the two Bravais lattices composing the hexagonal lattice of the graphene. In an improved formulation, kinematics is described in terms of a discrete first and second surface deformation gradient and associated inner displacement. Knowledge of these kinematic variables at the position of an atom allows us to compute the atomic energy and atomic load, and thereby derive equilibrium equations. The analysis offers a rational method for developing discrete equilibrium equations analogous to those developed for thin membranes.

The apparent similarity between the hexagonal lattice of the graphene and the hexagonal lattice of a two-dimensional cellular solid suggests that micropolar medium theory offers a possible framework for describing the deformation of the graphene [11]. The most interesting aspect of the micropolar-medium theory is the dependence of the energy density function on the microrotation angle, ϕ, in addition to the deformation gradient. In fact, the main intention of the micropolar-medium theory is to introduce an additional degree of freedom for the purpose of capturing nonlocal material response. In the case of a particulate sheet, additional degrees of freedom are provided by the inner displacement.

REFERENCES

1. Born M, Huang K. *Dynamical Theory of Crystal Lattices*, Oxford University Press, New York, 1954.
2. Ericksen JL. On the Cauchy–Born rule. *Math. Mech. Solids* **13**:199–220, 2008.
3. Tadmor EB, Smith GS, Bernstein N, Kaxiras E. Mixed finite element and atomistic formulation for complex crystals. *Phys. Rev. B* **59**:235–245, 1999.
4. Dresselhaus MS, Dresselhaus G, Saito R. Physics of CNTs. *Carbon* **33**:883–891, 1995.
5. Damnjanović M, Milošević I, Vuković T, Sredanović R. Symmetry and lattices of single-wall nanotubes. *J. Phys. A* **32**:4097–4104, 1999.
6. Jiang H, Zhang P, Liu B, Huang Y, Geubelle PH, Gao H, Hwang KC. The effect of nanotube radius on the constitutive model for carbon nanotubes. *Comp. Mater. Sci.* **28**:429–442, 2003.
7. Chandraseker K, Mukherjee S. Coupling of extension and twist in single-walled carbon nanotubes. *J. Appl. Mech.* **73**:315–326, 2006.
8. Treister Y, Pozrikidis C. Numerical study of equilibrium shapes and deformation of single-wall carbon nanotubes. *Comp. Mater. Sci.* **41**:383–408, 2008.
9. Pozrikidis C. Effect of membrane bending stiffness on the deformation of capsules in simple shear flow. *J. Fluid Mech.* **440**:269–291, 2001.
10. Brenner DW. Empirical potential for hydrocarbons for use in simulating the chemical vapor deposition of diamond films. *Phys. Rev. B* **42**:9458–9471, 1990.
11. Pozrikidis C. Mechanics of hexagonal atomic lattices. *Int. J. Solids Struct.* **45**:732–745, 2008.
12. Arroyo M, Belytschko T. Finite crystal elasticity of carbon nanotubes based on the exponential Cauchy–Born rule. *Phys. Rev. B* **69**:115415, 2004.
13. Toupin RA. Elastic materials and couple stresses. *Arch. Rational Mech. Anal.* **11**:385–414, 1962.
14. Mindlin RD. Micro-structure in linear elasticity. *Arch. Rational Mech. Anal.* **16**:51–78, 1964.
15. Mindlin RD. Second gradient of strain and surface tension in linear elasticity. *Int. J. Solids Struct.* **1**:417–438, 1965.
16. Fleck NA, Hutchinson JW. Strain gradient plasticity. *Adv. Appl. Mech.* **33**:295–361, 1997.
17. Sunyk R, Steinmann P. On higher gradients in continuum-atomistic modelling. *Int. J. Solids Struct.* **1**:417–438, 2003.
18. Guo X, Wang JB, Zhang HW. Mechanical properties of single-walled carbon nanotubes based on higher order Cauchy–Born rule. *Int. J. Solids Struct.* **43**:1276–1290, 2006.
19. Wang JB, Guo X, Zhang HW, Wang L, Liao JB. Energy and mechanical properties of single-walled carbon nanotubes predicted using the higher order Cauchy–Born rule. *Phys. Rev.* **73**:115428, 2006.

5

COULOMBIC DRAGGING AND MECHANICAL PROPELLING OF MOLECULES IN NANOFLUIDIC SYSTEMS

PETR KRÁL AND BOYANG WANG

Department of Chemistry, University of Illinois at Chicago, Chicago, Illinois

5.1 INTRODUCTION

We are entering the nanotechnology era, during which chemical, physical, and biological systems and phenomena will be prepared and controlled at the nanoscale. The enormous potential hidden in the nanoscale was predicted in 1960 by the great science wizard Richard Feynman [1], and over the decades his words have largely been confirmed. In the future, nanoscale systems with biological and inorganic origins may coevolve with the goal to develop hybrid "living nanostructures." In this chapter we discuss transport phenomena at the interface of nanoscale systems, with potential applications in physics, chemistry, biology, medicine, and other fields.

In recent years, huge progress has been made in nanofluidics, where transportation of fluids through many types of micro- and nanochannels has been realized [2,3]. Channels and other fluid elements can now be integrated into lab-on-a-chip (LOC) or micro total analysis systems (μ-TASs), which can be prepared using many experimental techniques [4]. In these systems, fluid transport is realized by channel

Multiscale Modeling of Particle Interactions: Applications in Biology and Nanotechnology,
Edited by Michael R. King and David J. Gee

and digital microfluidics [5,6]. In the first case, fluid passes through channels [7] made of silica, glass [8], polymers [9,10], or DNA molecules [11,12]. A quasi-one-dimensional flow can also be realized on material surfaces without the physical presence of channel walls [13,14]. Fluids can be pumped electrically [15,16], optically [17], biomechanically [18], and by other means. In digital microfluidics, droplets of various sizes are transported on material surfaces with embedded electrodes. Controlling the shape and motion of droplets is of practical importance in many technologies [19–22]. For example, using tiny droplets one could study up to 10^{10} reactions in 1 mL of emulsion [23]. Today's LOC systems closely resemble electronic circuitry systems [24], with fluidic diodes [25], transistors [26], and other control elements known from electronics. Nanofluidic devices built from these elements [27] can be used for controlled transportation and separation of DNA [28–30], proteins [31,32], colloids [33], and cells [34].

Many emerging questions in nanofluidics are related to the sizes of the channels, in some cases just a few nanometers in diameter, and materials used for their construction, the possibilities of transportation and control of individual molecules in vitro and in vivo, integrating biological systems with nanosystems, and so on. Biological systems have developed sophisticated transport methods that can be modified and used in nanofluidics. Cells can transport molecules and micelles [35] along tubules by kinesin [36] and use other molecular motors to realize numerous types of motions [37]. Some of these motors have been extracted from cells and used in vitro [38,39]. New molecular motors have also been designed [40–43], which could manipulate single molecules, viruses, and bacteria, as an alternative to using glass microfibers, optical or magnetic tweezers [44–46], scanning tunneling (STM) [47], and atomic force microscopy (AFM) [48–51]. Of particular importance is the delivery of genes, various biomolecules, or drugs through cell membranes without relying on natural protein pathways [52]. This task could be realized by attaching molecules to polymers [53], nanofibers [54], carbon nanotubes [55], and other nanostructures [56] that can penetrate the cell membranes. The lab-on-a-chip concept [57] could, in principle, be modified to study single molecules [58–61], and extended to the lab-on-a-cell concept, where one could transport individual biomolecules inside cells [62,63].

Carbon (CNT) and boron–nitride (BNT) nanotubes [64–73] have many unique properties that can be useful in nanofluidics. They are chemically stable, possess conjugated bonds that can conduct electrons, and are mechanically very strong. They can be modified by physisorption [74,75] or by covalent bonding of molecular ligands [76], and have many potential applications, such as hybrid materials [77], chemical sensors [78], and drug-delivery tools [79,80]. Whereas CNTs can be both conducting and semiconducting, BNTs have large energy bandgaps and cannot screen electric fields very well [81]. Carbon compounds such as diamond [82,83] or graphene-based nanostructures [84,85] are also ideal biosubstrates. Since they are water resistant, they could be modified to be hydrophobic or hydrophilic and allow for the activation of attached molecules [86]. These properties make these materials suitable for the fabrication of channels for nanofluidic applications.

Here we describe two transport methods that could be used to manipulate individual molecules: molecular assemblies, and nanostructures in air or solution. In the first method, molecules can be dragged on the external surfaces of carbon and boron–nitride nanotube arrays [87] with ionic solutions passing through them. In the second method, molecules could be pumped by molecular propellers [88]. We introduce the methods by modeling them with classical nonequilibrium molecular dynamics simulations (typical movies of the simulations are attached to the published papers).

5.2 COULOMBIC DRAGGING OF MOLECULES ON EXTERNAL SURFACES OF NANOTUBES

In the recent past, intensive experimental activity has been devoted to our theoretical predictions of molecular drag phenomena around the surfaces of CNTs. We had shown [89] that atoms and molecules intercalated inside CNTs or adsorbed at their surfaces could be pumped by electrons passing through them (see Fig. 5.1a). Subsequent experimental observations [90,91] have confirmed that these effects could efficiently transport materials at the nanoscale. We have also shown that nanotube electrons, highly exposed to external stimuli, could be driven by fluids passing around them [92] (see Fig. 5.1b). This effect has also been studied in several modifications, where static and oscillatory flows of liquids were measured [93,94]. The flow of gases was also detected [95]. These phenomena have rich potential (patented) applications. Further possibilities of molecular transportation in CNTs emerged with the molecular dynamics simulations of Hummer et al. [96], showing that water can enter and pass through narrow CNTs in the form of chains, due to large van der Waals coupling to the conjugated carbon bonds. The experiments have demonstrated that unusually low friction conditions allow fast passage of gases [97,98] and liquids [99–102] through CNT membranes, and easy mutual slipping of multiwall nanotubes [103]. The atomic smoothness and hydrophobicity of the CNT wall leads to transport rates that are several orders of magnitude higher

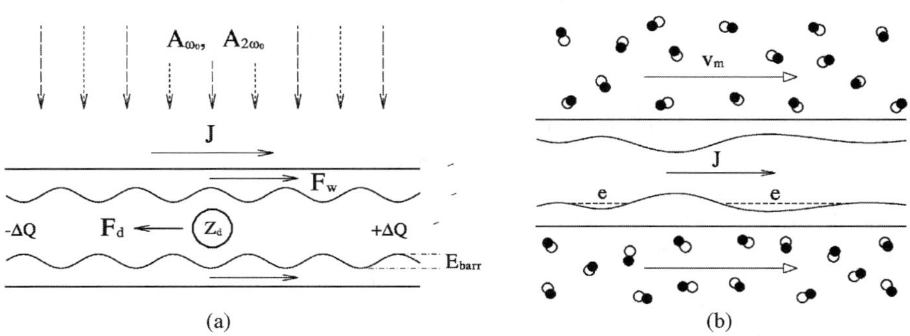

Figure 5.1 (a) Dragging of atoms and molecules intercalated inside carbon nanotubes by electric current induced by light; (b) electron drag in nanotube by polar liquid flowing around it.

than those expected according to classical fluid transport theory (Knudsen theory for rarefied gases and Poiseuille laminar flow for continuum fluids). The fluid behaves similarly in other narrow channels, but in CNTs the speed of passage is much faster and could be described by corrected classical transport equations [104]. The passage of water and ionic solutions through nanochannels [105,106], CNTs' interior and exterior, could also be controlled by chemistry [107,108] and electrostatic gating [109].

5.2.1 Dragging of Single Molecules by Separately Flowing Polar Liquids

Recently we have demonstrated that dragging of polar molecules and ions adsorbed on the nanotube surfaces can be realized not only by their coulombic scattering with passing electrons [89–91] but also by their scattering with polar solutions and individual ions transported inside the nanotubes [87,110]. A simple realization of the phenomenon is illustrated in Fig. 5.2a, where a zwitterion molecule $NH^+{}_3{-}(CH{=}CH)_4{-}CH_4{-}(CH{=}CH)_6{-}CO_2{}^-$ is adsorbed on the surface of the (14,0) single-walled carbon nanotube (SWNT) due to moderate van der Waals binding [87]. A (14,0) SWNT is a zigzag type of carbon nanotube with 14 benzene units wrapped around its circumference [65]. Driving of the zwitterion is realized by coulombic scattering with water passing under pressure through the CNT [99]. Here,

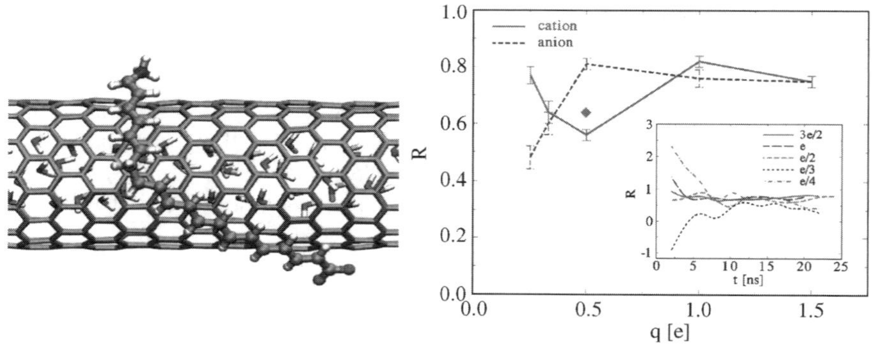

(a) (b)

Figure 5.2 (a) The $NH_3{}^+(CH{=}CH)_3CH_2(CH{=}CH)_6CO_2{}^-$ zwitterion molecule adsorbed on the surface of the (14,0) CNT is driven by the flow of water passing inside. The water flow going to the left is realized at $T = 300$ K. (b) The ratio R of the distances traveled by the adsorbed ions $CH_2{=}CH{-}CH{=}CH{-}NH_3{}^+$ and $CH_2{=}CH{-}CH{=}CH{-}CH_2O^-$, with different effective total charges q, and water inside the (14, 0) CNT. The small rectangle at $q \approx 0.5$e applies to the $NH_3{}^+(CH{=}CH)_3CH_2(CH{=}CH)_6CO_2{}^-$ zwitterion molecule with correct (unscaled) charges. Inset: The ratio between the average speed of the $CH_2{=}CH{-}CH{=}CH{-}CH_2O^-$ anion and the flowing water plotted with the time of driving (averaging). For shorter averaging times t, the ratio R deviates significantly from the converged values, especially in ions with small charges that diffuse faster on the fluctuating potential $\varphi(r)$ of the moving water column. (*See insert for color representation of figure.*)

screening of the fluctuating electric fields is neglected since CNTs are not considered to be polarizable.

The coulombic potential energy E_C between the molecule adsorbed on the nanotube and all the water molecules flowing inside the tube is

$$E_C = \sum_i q_i \phi(r_i), \quad \phi(r) \approx \frac{1}{4\pi\varepsilon} \sum_j \frac{Q_j}{|r - r_j|}. \tag{5.1}$$

Here, q_i and Q_j are the charges of the ith atom in the adsorbed molecule and jth atom in the water chain (O or H), respectively and r_i and r_j are the Cartesian coordinates of atom i above the CNT surface and atom j of water inside the CNT, respectively. The charges are effectively reduced by the size of the dielectric constant ε of the (semiconducting) nanotube and water. The ab initio quantum mechanical calculations [110] show that thin semiconducting single-walled carbon nanotubes [111,112], such as (4,3) CNTs, and all BNTs [81] do not screen the electric fields to any significant extent. These systems have large bandgaps and relatively small dielectric constants [113], so that the field can emanate through their walls. In the present simulations, we use generic zigzag SWNTs with (n,0) ($n = 11$, 14, 17) and neglect screening (discussed later in more detail).

We model the transport phenomena with molecular dynamics simulation realized with the NAMD package [114–116]. The parameters of atoms in aliphatic groups are obtained from ab initio quantum chemistry calculations [117,118], using the Gaussian03 package [119] and added to the CHARMM27 force field. This system is equilibrated as an NVT ensemble (fixed number of particles, volume, and temperature). The long-range electrostatic forces are computed by the particle-mesh Ewald method [120], and the time step is 1 fs. We also use Langevin dynamics [121] (at $T = 300$ K) with a small damping coefficient 0.01 ps^{-1} to avoid unphysical dissipation of the momenta [122], and leave the tube free to vibrate, except for a few fixed atoms that hold it in place. In the simulations, the water flow is induced by applying a force of 0.01 $kcal/mol^1 \cdot nm^1$, oriented along the tube, on the oxygen atom of each water molecule with certain z coordinates. This generates a pressure of $P \approx 70$ atm [123], causing the water molecules to flow with an average velocity of $v_W \approx 10$ to 20 m/s. In the simulations, we cannot use very low pressures, due to long simulation times. The speed of water and the dragged molecules scale linearly with the pressure, so low pressures can be used in the experiments.

In Fig. 5.2b, we illustrate the drag phenomenon on the ratio R between the total distances traversed by the attached molecules and molecular ions on the surface of the CNT and the water flowing inside the tube. The last is calculated via the number of water molecules passing through the tube entrance in one direction minus those passing in the opposite direction, divided by their linear density. The ratios R are obtained from dragging the systems for $t \approx 100$ ns for different effective total charges q of the $CH_2{=}CH{-}CH{=}CH{-}NH_3{}^+$ and $CH_2{=}CH{-}CH{=}CH{-}CH_3{-}O^-$ ions, to reflect the effect of the dielectric constant ε in Eq. (5.1). The small rectangle at $q \approx 0.5$e applies to the $NH_3{}^+(CH{=}CH)_3CH_2(CH{=}CH)_6CO_2{}^-$ zwitterion molecule with unscaled charges. In the inset, the time dependence of the ratio R is also shown for the $CH_2{=}CH{-}CH{=}CH{-}CH_3{-}O^-$ anion. For shorter averaging time t, the ratio R deviates significantly from the converged values, especially for ions with small charges that diffuse faster on the fluctuating potential $\varphi(r)$.

In this phenomenon, the adsorbed molecule diffuses almost freely on the nanotube surface. But if the adsorbed molecule is sufficiently polar or even charged, its random motion is determined by the fluctuating and slowly "shifting" coulombic potential $\varphi(r)$, generated by the flowing polar liquid [89,92] rather than by the lattice-modulated van der Waals interactions with the CNT. Therefore, the center of mass of the diffusion trajectory slowly shifts with the fluctuating relief of $\varphi(r)$. The more polar the molecule is, the more locked (localized) it is in the potential wells of the fluctuating potential $\varphi(r)$ generated around the string of flowing polar molecules [124]. As a result, the molecular motion starts to gain a more deterministic character, resembling the motion on an "elastic transportation belt." This drag phenomenon is distantly related to electroosmotic transport, with many practical applications [125–129].

The methodology outlined offers unique possibilities for the controlled manipulation of molecules in systems where other methods might be less practical. Here, we present it in several more configurations, where the pumped medium is formed by assemblies of polar molecules and ions that are positioned in air or solvated in various solutions. The dragging medium is formed by individual ions driven by electric fields applied along the tube axis or by ionic solutions pushed by pressure. The experiments might also be realized on arrays of nanotubes, as explained later.

5.2.2 Dragging of Molecular Assemblies by Separately Field-Driven Ions

Here, we discuss dragging nanodroplets in air on the surfaces of CNTs and BNTs realized by individual ions driven inside the nanotubes by electric fields. Although inserting individual naked ions inside nanotubes might be problematic, practically the same driving can be achieved if we pump ionic solutions through the nanotubes [99–102,130–132]. This is because the electric field outside the nanotubes is only slightly weaker in the presence of water inside the tube.

In Fig. 5.3a, we illustrate driving of $N_W = 400$ water molecules, positioned on the surface of semiconducting (10,0) single-walled CNT, by a single Na^+ cation (or Cl^- anion), solvated at a distance in the water droplet. The 96-Å-long tube is held in a box of approximately $1000\,nm^3$, fitting the coexistence of gas and liquid phases at $T = 300\,K$. If the system is left to relax for several nanoseconds, a dynamical equilibrium is established, where a water nanodroplet condenses on the nanotube around the ion, similar to the situation with other fibers [133]. These very small droplets evaporate rapidly and their water molecules might join larger droplets present in the system, with smaller surface tensions. But this effect should be relatively slow when the droplets are larger ($N_W > 5000$ to 10,000) or for droplets covered with lipids (micelles) [35], as discussed later.

Let us return briefly to the problem of screening. We have calculated ab initio the electrostatic potential φ generated above a (4,3) CNT (bandgap of 1.28 eV) [112] by

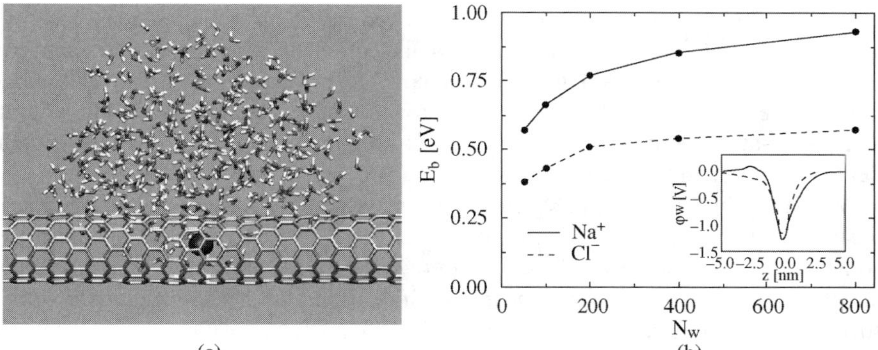

Figure 5.3 (a) A nanodroplet of $N_W = 400$ water molecules dragged on the surface of (10,0) carbon nanotube by a single Na^+ ion intercalated inside the tube, driven by the electric field of $E_0 = 0.1$ V/nm applied along the tube z axis; (b) binding energy E_b between the Na^+ and Cl^- ions and water nanodroplets containing N_W molecules (neglecting gas-phase molecules). Inset: Electrostatic potential φ_W along the tube axis and the Na^+ ion inside.

a Li^+ ion located in its center, and φ generated above a (5,5) BNT (bandgap of 5.5 eV) [81] by a Na^+ ion. In the calculations, the 5-nm-long tubes are kept neutral and frozen, since their structure is rather rigid. The potentials φ are obtained from NBO (natural bond orbital) atomic charges using the B3LYP density functional and the 3–21g basis set in Gaussian03 [119]. The potential φ of Li^+ is decreased by about 25% due to screening, while that of Na^+ is decreased by about 10%. The same results are obtained in the presence of an electric field of $E = 0.1$ V/nm applied along the tube axis. Moreover, if the ions are shifted along the nanotube axis by a small distance d, the total energy changes by about Ed. Note that the presence of large screening in metallic SWNTs does not imply that molecular dragging by individual ions or ionic solutions is negligible. This is because two ions separated by the wall of a metallic SWNT can still "see each other" via their screening charges that interact in the monolayer. Our ab initio calculations done in the Hartree–Fock approximation on long metallic single-walled carbon nanotubes have shown that the coupling of two such ions is reduced by only about 70%, due to this imperfect screening. These results show that the screening of the ionic field is small in selected CNTs and BNTs, and the dragging might work even in metallic nanotubes (not by electric field, though). With this in mind, we model the ion-droplet dynamics in some typical nanotubes and, for simplicity, consider them to be nonpolarizable.

To estimate the possibility of dragging the droplets by individual ions, we have calculated the ion-droplet binding energy, E_b, in the absence of electric fields in nonpolarizable CNTs. In Fig. 5.3b, we show E_b obtained at $T = 300$ K and averaged over 10,000 frames, separated by 500-fs intervals. The binding energy is several times smaller than the hydration energy of the Na^+ and Cl^- ions, $E_{solv} = 7.92$ and 6.91 eV [134], respectively, and it does not depend much on the size of larger droplets with $N_W = 100$ to 800 waters. These energies, which are calculated for the dielectric constant $\varepsilon = 1$, should be decreased proportionally when the polarization nanotubes is included, but the binding should be preserved in larger droplets. In the inset we also show two images of the one-dimensional electrostatic potential generated by the water droplet ($N_W = 400$) along the SWNT axis. The polarized water molecules create a steep potential well of depth about 1 eV around the ion that moves together with it. The potential gradient of 0.5 to 1 V/nm blocks the ion from leaving the well even at fields of $E \approx 0.1$ to 0.2 V/nm.

We can estimate the coupling energy E_b analytically by assuming that the ion is located a distance of $d \approx 0.35$ nm above a flat surface of water with the dielectric constant $\varepsilon_W \approx 80$. This gives $E_b \approx -e^2/4d\ (\varepsilon_W - 1)/(\varepsilon_W + 1) \approx -1$ eV, in good agreement with Fig. 5.3b. The fact that in the simulations the Na^+ ion binds twice as strongly to the nanodroplet as does the Cl^- ion is caused by the character of the water polarization: The Na^+ ion attracts the O atom from water molecules, which is twice as charged as the H atoms, and the Cl^- ion attracts just one of the two H atoms. The large difference between E_b and the bulk solvation energies is caused by the fact that ions solvated in bulk water are surrounded by two to three times more water molecules that are about twice as close to them. These binding energies are decreased proportionally if the nanotube polarization is included, as discussed above.

We continue by discussing dragging of the droplets with the ions in the presence of electric fields aligned along the nanotubes. The field acts on the entire system except on the nanotube, which is treated as nonpolarizable. In Fig. 5.4a, we show the velocity, v_W, of the nanodroplets with N_W water molecules. The data are calculated from 50-ns simulations (ca.50 rounds along the CNT) at $E_0 = 0.1$ V/nm and $T = 300$ K; this statistical averaging is fully sufficient [87]. The velocity obtained is proportional to the electric field and depends strongly on the droplet size. It has practically the same value whether Na^+ or Cl^- ions are used for dragging. To test the scalability of this dragging phenomenon, we model a droplet with $N_W = 10{,}000$, located between two parallel (10,0) CNTs separated by a 5-nm distance. If one Na^+ ion is placed in each nanotube, and both ions are driven by a field of $E_0 = 0.1$ V/nm, the droplet moves together with the ion pair at a high velocity of $v_W = 6.6$ nm/ns, due to a small contact with the CNTs.

In all these cases, the character of the nanodroplet motion on the CNTs might be closer to sliding [135] than rolling [136], due to partial wetting of the CNT surface with a large van der Waals binding. In macroscopic systems, the droplet velocity for both mechanisms of motion is controlled by the momentum or energy dissipation of water layers sliding inside the droplet [135]. These mechanisms give the qualitative dependence of the droplet velocity $v_W \sim eE/r\eta$, where the droplet radius is $r \sim N_W^{1/3}$ and the water viscosity is $\eta \sim 1/T$ [137]. The results in Fig. 5.4a roughly confirm this dependence even for the motion of nanoscale droplets, but the driving velocity scales more steeply with the number of water molecules. In the inset, we also show for $N_W = 400$ that the temperature dependence is almost linear, $v_W \sim T$, as expected. To clarify more clearly the motion of the nanodroplet, we test its dragging by a Na^+ ion that is solvated in it directly. The velocity obtained for the droplet is about 20% larger than when the ion is inside the tube. This is probably caused by the tendency of the droplet to roll, since the dragging force acts close to its center rather than on its periphery at the nanotube surface.

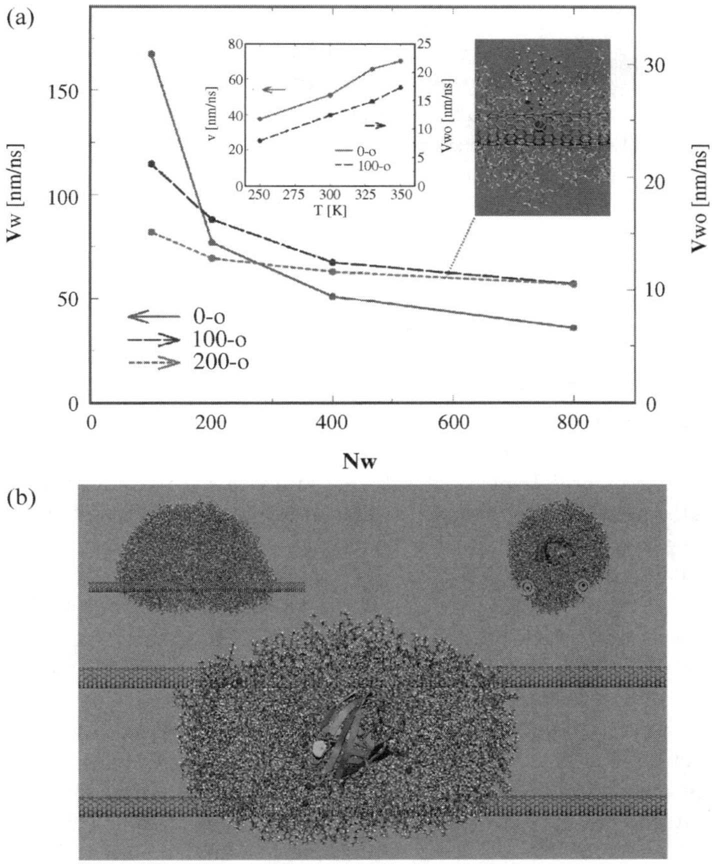

Figure 5.4 (a) Dependence of the nanodroplet velocity on the number of its water molecules N_W. Dragging of the droplet in the presence of $N_O = 100$ and 200 octane molecules is considered as well. Right inset: Visualization of the nanodroplet with $N_W = 50$ in $N_O = 200$ oil molecules. Left inset: Temperature dependence of the droplet velocity for $N_W = 400$. (b) Three views on a micelle formed by phospholipids and water molecules, which contains a GFP protein and five Na^+ ions neutralizing it [water is partly removed to see the protein (bottom and right up)]. It is driven by two Na^+ ions intercalated inside two CNTs. (*See insert for color representation of figure.*)

We also check if boron–nitride nanotubes could potentially be used for ion-droplet dragging. BNTs screen electric fields very little, but it is not clear whether their large surface polarity may prevent this dragging. To test this possibility, we briefly simulate dragging of droplets with $N_W = 100$ and 400 waters on the (5,5) BNT. The atomic charges of the B and N atoms are calculated ab initio [119] and approximated by the values of 1.0e and −1.0e, respectively [104,138]. The BNT is highly polar but not very polarizable (neglected). In the simulations, one Na^+ ion is driven inside the tube by an electric field $E = 0.1$ V/nm, and $T = 300$ K. The molecules of the droplets with

$N_W = 100$ and 400 waters form on the (5,5) BNT an open and closed ring, respectively, formed by one or two water layers. These droplets move with an average velocity of $v_W = 0.28$ and 1.29 nm/ns, respectively. The droplet with $N_W = 400$ waters moves about 40 times slower on the BNT than on the CNTs, due to strong coupling to the highly polar BNT surface. We also obtain $v_W \sim 1/N_W$, showing a stronger dependence on N_W than on the CNT surfaces. This can be understood by the fact that the driving of the flattened nanodroplets is close to pure sliding, whereas the friction is proportional to the contact area, or N_W.

5.2.2.1 Driving in the Presence of Oil The character of the nanodroplet motion could be altered dramatically if a monolayer of oil is adsorbed on the nanotube surface. In Fig. 5.4 (left), we show that the presence of $N_O = 100$ and 200 octane molecules decreases the droplet velocity v_{WO} by an order of magnitude, due to friction between water and oil. Smaller droplets, $N_W < 100$, are attached to the ion by a narrow "neck" passing through the oil layer (see inset in Fig. 5.4). Larger droplets, $100 < N_W < 200$, are more or less spherical, submerged significantly inside the oil, and share a very small surface area with the CNT. In analogy to a water droplet inside bulk oil, their driving could be described by Stokes' law, which is largely valid at the nanoscale [152,153]. Here it gives $F = -6\pi r \eta v_{WO}$, where $F = eE_0 = 16$ pN is the drag force acting on the droplet, r is the droplet radius, and $\eta \approx 0.54\,\text{mPa}\cdot\text{s}$ is the viscosity of octane at $T = 300$ K. For $N_W = 100$ we find that $r \approx 3.5$ Å, so $v_{WO} = 4.5$ m/s. This value is three to four times smaller than that obtained in the simulations, due to incomplete coverage of the nanodroplet by oil.

From the point of potential applications, it is interesting to find out if the dragged droplets can serve as reservoirs able to carry molecular cargo. In Fig. 5.4b we show dragging of a reverse micelle [35] on two (10,0) CNTs with one Na^+ ion per tube. The micelle is formed by 10,152 water molecules and 472 lipid molecules. The amphiphilic lipid molecule is 1,2-dipentanoyl-3-phosphatidylcholine (PPOPC), which is similar to POPC, a common biological membrane phospholipid [139]. PPOPC has twice shorter hydrophobic chains than POPC, to preserve good contact between the micelle and the CNTs [110]. The interior of the micelle contains a GFP protein [140,141] and five Na^+ ions that neutralize it [142]. The shortened lipids should sufficiently cover the CNTs and prevent potential unfolding of the proteins on their surfaces. When the ions are driven by an electric field of $E_0 = 0.03$ V/nm, the micelle moves relatively slowly, with $v < 0.2$ nm/ns. In the presence of lipids the friction is proportional to the number of CNTs and so is the driving force on the ions, but the stabilization of the droplet on the CNT pair (array) is significantly better.

If the nanotube is fully submerged in water instead of oil, the dynamics of the field-driven ion becomes very different [143]. We simulate this situation in a (10,0) CNT, placed in the driving field of $E = 0.1$ V/nm. We have obtained at $T = 240$, 270, and 300 K the following velocities $v_{Na} = 464.7$, 622.5, 733.7 m/s and $v_{Cl} = 284.5$, 401.6, 547.9 m/s, respectively. These velocities are four to five times larger than those of the ions dragging the water droplet. This is because waters around the submerged nanotube rearrange fast, locally, when they react to the field-driven ion. At higher temperatures, the ions move faster, since their binding to the water molecules is less stable [124]. The velocities of the Cl^- ion are smaller than those of the Na^+ ion, because Cl^- easily attracts the light H atoms that are not bound (frustrated) in water molecules at the nanotube surface.

5.2.2.2 Transient Ion Capture by Nanodroplets Finally, we discuss the dynamical stability of the coupled ion-droplet pair. In larger electric fields or when the nanodroplet is small, the ion might get released, and in a model with periodic boundary conditions, it might be recaptured later by the nanodroplet. In Fig. 5.5 we plot for $E = 0.02$ V/nm the trajectory of the ion that left the droplet of $N_W = 20$ and was recaptured in the next run around it. The ion's trajectory is shown by the dark line, and the time-frame separation on the vertical axis is 100 fs. The electric field along the CNT axis created by the ion-polarized droplet is plotted by regions of different color densities. The positive and negative field regions that lock the ion are obtained from the derivative taken at the sides of the potential well (see the inset of Fig. 5.3b).

The ion released from the droplet goes once around the tube and reapproaches the droplet with a velocity of $v_{\text{ini}} \approx 1400$ m/s at $t \approx 5$ ps (bottom). After it gets closer to the droplet, the water molecules fast become polarized (Fig. 5.5, inset, at $z = -1.3$ nm,

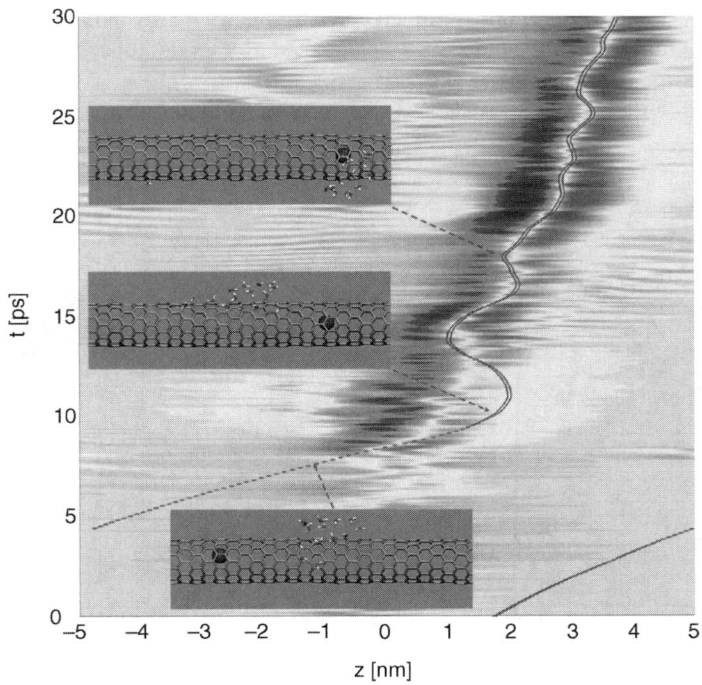

Figure 5.5 Trajectory of the Na^+ ion that is recaptured by a water nanodroplet with $N_W = 20$. The axial position and the time of ion motion are shown on the horizontal and vertical axes, respectively. The electric field generated by the water molecules along the CNT axis is plotted by contours. (*See insert for color representation of figure.*)

$t = 7$ ps). The ion first passes around the droplet, only to be attracted back by several chained molecules protruding from the droplet (Fig. 5.5, inset, at $z = 1.7$ nm, $t = 10$ ps). This is possible since a chain of five hydrogen-bonded and aligned water molecules generate at a distance of 1 nm a field of ≈ 0.17 V/nm, which is opposite, and almost an order of magnitude larger than, the external field. Deceleration of the ion by this large induced field causes the coupled system to gain high coulombic potential energy. Thus, the ion position oscillates three or four times, before it is fully seized back by the droplet (Fig. 5.5, inset, at $z = 2$ nm, $t = 18$ ps). The two start to move together at a much smaller velocity $v_{\text{end}} \approx 130$ m/s, while the waters are already interconnected. If the droplet does not catch the ion within several of its runs around the periodic system, the ion might heat and evaporate the droplet temporally. The transient oscillations observed in this ion catching closely resemble quasiparticle formation in condensed matter systems.

5.2.3 Dragging of Solvated DNA by Separately Flowing Ionic Solutions

Here we investigate dragging of DNA on the surfaces of nanotubes immersed in physiological solutions, where the dragging is realized by pressurizing another ionic solution through the nanotubes. The system is shown in Fig. 5.6 (top), where we display a 50-base-long single-stranded DNA molecule that binds by π–π interactions [144] to the surface of the (25,0) CNT, used as a model tube. The neutral ionic solution inside the nanotube consists of 1850 waters, as well as 15 Na^+ and 15 Cl^- ions. Outside the nanotube, we put 49 Na^+ cations to neutralize the DNA and add

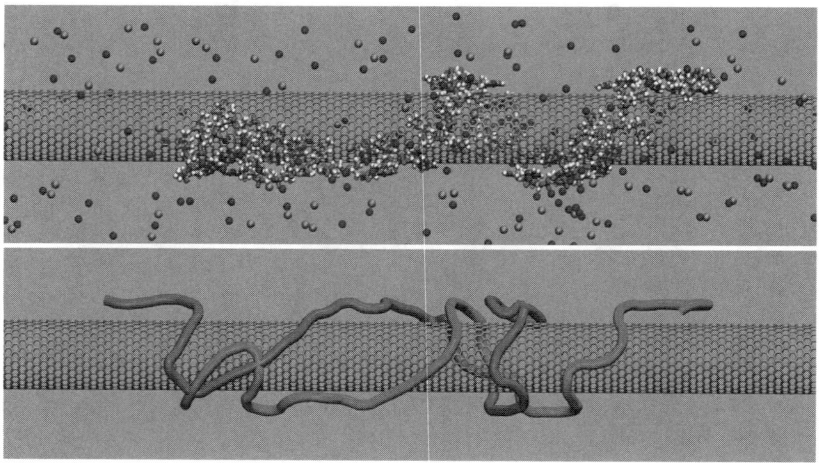

Figure 5.6 (top) A 50-base-long single-stranded DNA molecule, with the sequence CCTTCAGTGG CCGGTCATTG ATGAAGCCCT GAGGAACAAG GACACTCCGG, driven on the surface of a (25,0) CNT by the ionic solution flowing inside; (bottom) motion to the left of the DNA during a 7.4-ns simulation is clearly visible. (*See insert for color representation of figure.*)

33,600 waters, as well as 100 Na^+ and 100 Cl^- ions, to model the physiological salt concentration (1%) inside the human body. Since the nanotube is neutral, we do not have to investigate explicitly effects associated with electric double layers [145]. Such effects should be described automatically in all-atom simulations.

During the simulations, the solution inside the nanotubes is pumped at $T = 350$ K and $P = 70$ atm (significantly smaller pressures would be used in experiments). The CNT is also not polarizable, as in the previous cases. The DNA on the nanotube surface moves with an average speed of $v \approx 0.3$ nm/ns [see Fig. 5.6 (bottom)]. We have noticed a large dependence of dragging on temperature. Recent experiments [146] have shown that friction between nanotubes and polymers attached to them is highly temperature dependent and drops significantly for $T > 350$ K. The simulations confirm that DNA dragging might be realized, but at room temperature with low pressure in the dragging ionic solution and realistic screening of the CNT, the speed of the DNA can be several orders of magnitude slower than that found in our simulations. It is quite expensive to simulate the dragging of DNA by full atomistic simulations, and we also cannot easily coarse-grain this nonequilibrium system and preserve realistic dragging conditions. Therefore, the simulation cannot, at this moment, give reliable values of the dragging velocities. Nevertheless, the theoretical results are compelling enough to conclude that DNA drag is experimentally possible. This methodology could be used in the delivery and manipulation of the genetic material and drugs inside cells [53–56].

5.2.3.1 Experimental Realization of the Molecular Dragging on Nanotube Arrays

Gases [97,98] and liquids [99–102] were transported successfully through carbon nanotube membranes. The drag described above could be realized efficiently in LOC systems shown schematically in Fig. 5.7. It is an array of CNTs or BNTs connected by two (metallic) bonding frames that hold the nanotubes in parallel positions. The tubes

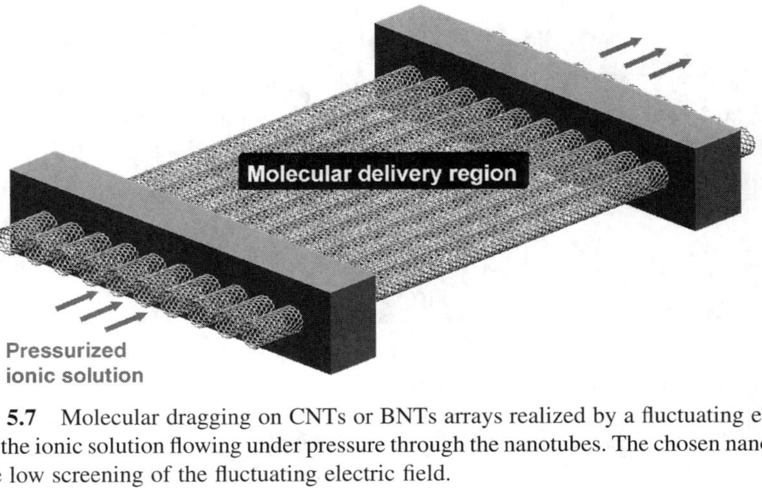

Figure 5.7 Molecular dragging on CNTs or BNTs arrays realized by a fluctuating electric field of the ionic solution flowing under pressure through the nanotubes. The chosen nanotubes provide low screening of the fluctuating electric field.

are opened at both ends by the application of reactive ion etching [98]. The ionic solution could be passed by the pressure of several atmospheres through the nanotubes [132]. The two frames isolate the pressurized side regions from the molecular delivery region. The central region can be used for molecular dragging in air and physiological solutions by the passing ionic solution. These devices should allow manipulation of individual molecules [61], molecular assemblies, and nanostructures in the complex environments of cells. The methodology is expected to lead to important applications in nanofluidics, where it can complement current LOC techniques and be used in molecular delivery, separation, and desalination, with an unprecedented level of control.

5.3 MOLECULAR PUMPING AND PROPELLING AT THE NANOSCALE

The second molecular transport method that we discuss is focused on molecular propellers that can pump liquids at the nanoscale and propel nanosystems [88]. In Fig. 5.8 we display the designed and modeled bulk (a) and surface (b) propellers in water solvent that are formed by functionalized carbon nanotubes [147]. The bulk propeller can pump liquid along the tube z axis by two blades formed by pyrene molecules, attached to opposite sides of the (8,0) CNT and tilted with respect to its axis, due to two chiral centers. The surface propeller pumps water orthogonal to the tube axis by four larger blades (see the inset of Fig. 10b) aligned straight along the axis.

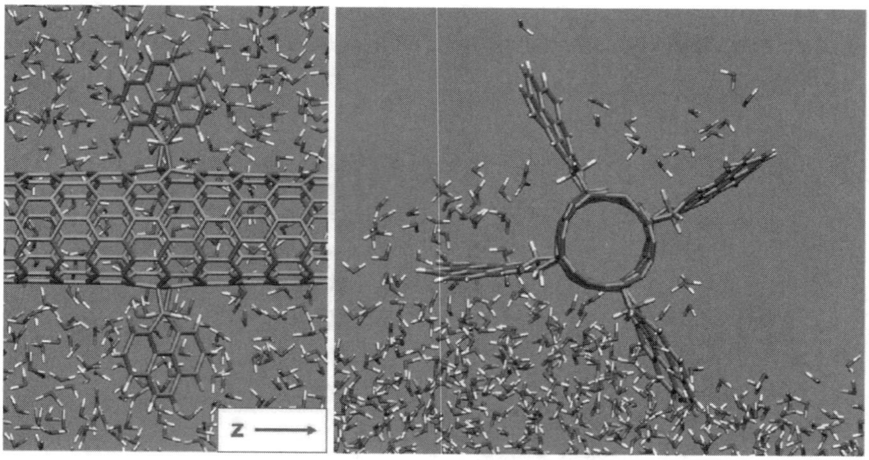

(a) (b)

Figure 5.8 The model bulk (a) and surface (b) molecular propellers pump water along the tube (z) axis and orthogonal to it, respectively. Both systems are based on the (8,0) CNTs and have covalently attached aromatic (hydrophobic) blades. Water is partly removed from the front to uncover propeller detail. (*See insert for color representation of figure.*)

With the advances in synthetic chemistry, these systems could be realized by cyclic addition reactions [148]. Similar molecular propellers with large blades attached to a benzene ring have recently been prepared by cyclodehydrogenation [149].

The blades in the molecular propellers can be designed to optimally fit the goals that we wish to pursue in a given chemical or biological environment. We have shown that their pumping efficiency can be very sensitive to the hydrophobic and hydrophilic character of liquids. In a hydrophobic propeller, the charges of the H atoms at the tips of the pyrene blades were set to 0.12e, and those of the nearest aromatic carbon atoms to −0.12e. In a hydrophilic propeller, these charges were chosen to be 0.30e and −0.30e, respectively [150]. A torque **T** is applied to the two ends of the CNT propeller, with the tube fixed against side motion. The propellers were embedded inside different solvents and the molecular dynamics simulations were performed in the NPT ensemble (fixed number of particles, pressure and temperature, variable volume), where the pressure is kept constant ($P = 1$ atm) using the Langevin piston method [121] and periodic boundary conditions are applied.

Figure 5.9a shows the temperature dependence of the rotation rates of the model bulk propeller (Fig. 5.8a). The rotation rates are plotted in the unit of rounds per nanosecond. It has the hydrophobic and hydrophilic design and the results are obtained in hydrophobic dichloromethane (DCM) and hydrophilic (water) solvents. The data are averaged over approximately 50 rotations with an applied torque of $\mathbf{T} = 0.2\,\text{nN}\cdot\text{nm}$. As the system is heated above the (normal) freezing points of the solvents, $T_{\text{DCM},f} = 175$ K and $T_{\text{water},f} = 273$ K, the rotation rates grow, due to lower solvent viscosities. Interestingly, the hydrophilic propeller rotates more slowly, since its polar blades interact more strongly with both solvents. Substantially slower rotation occurs in water that forms hydrogen bonds with its polar blades [151]. These results agree with Stokes' law, which is largely valid at the nanoscale even in the presence of slipping boundary conditions [152,153]. We can write it as $\mathbf{T} \approx K\mu\omega$, where **K** is a constant determined by the shape of the propeller, μ is the dynamic

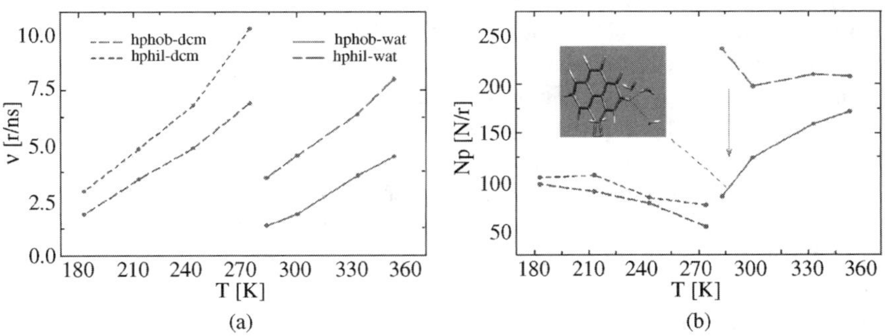

Figure 5.9 (a) Rotation rates (round/ns) of the bulk hydrophobic ("pho") and hydrophilic ("phi") propellers in water and DCM solvents as a function of temperature; (b) pumping rates (molecules/round) of these propellers. Inset: Formation of hydrogen bonds between the hydrophilic blades and water can reduce the pumping rate dramatically.

viscosity, and ω is the angular velocity of the rotation. Since the viscosity is $\mu \sim 1/T$ [137], the angular velocity $\omega = 2\pi\nu$ is approximately proportional to temperature for a constant torque **T**, in agreement with Fig. 5.9a.

Figure 5.9b shows the temperature dependence of the pumping rates of the propeller. The pumping rates are plotted in units of number of pumped molecules per round. Solvent molecules are pumped by collisions with the propeller's blades. If they were noninteracting point particles (ideal gas), all the molecules present in the volume that the two blades can reach should be pumped out in a half rotation. In the DCM solvent, coulombic and van der Waals interactions between the molecules and between them and the blades are weak, so the solvent molecules resemble an ideal gas that "slips" on the blades in many independent scattering events [154]. A similar situation occurs in the hydrophilic propeller, as seen in Fig. 5.9b. In the polar water solvent, molecules form clusters, transiently held together by hydrogen bonds [151]. In a hydrophilic propeller, water forms relatively stable hydration shells around the blades [155] that reduce the effective space available for direct contact of the pumped molecules with the blades. This causes a drastic reduction in the pumping rates, as shown by the vertical arrow in Fig. 5.9b. At high temperatures, the hydrogen bonds break down and the pumping rates increase. We can also find that the water molecules form layers [156] around the propeller that are largely preserved even during the rotation [157,158].

Practical propellers that can be used in different types of solutions should have well-tuned blade sizes and shapes. One could imagine using the propellers not only for pumping of molecules, but also for testing the speed of fluids moving around them. Our simulations show that these propellers preserve, in this passive (sensing) mode, the same features (dependence on chemistry) as in the active (pumping) mode.

To see the effect of different sizes of solvent molecules, we have tested pumping of n-alkanes, n-C_nH_{2n+2} ($n = 5$ to 17), which are in a liquid phase at $T = 300$ K and $P = 1$ atm [159]. These alkanes are of the same size or larger than the pyrene blades of the approximate size 0.8×1.1 nm. The sizes of the alkanes and pyrene blade are sketched in Fig. 5.10a. We show their pumping rates Np (number of molecules per round) with a hydrophobic propeller and the applied torque $\mathbf{T} = 0.2$ nN · nm, depending on the alkane length n_C. The pumping rate of alkanes increases with the length of alkane, up to n-octane, for which $n_C = 8$, and decreases for even longer alkanes. In longer alkanes, the number of atoms that are hooked in between the other molecules grows. Their release by the moving blades is hampered by significant alkane–alkane interactions, causing a drop in pumping rates.

From a technological point of view, it is interesting to study pumping of molecules on the surfaces of liquids that can be covered by monolayers of other molecules [160]. Figure 5.10b shows the pumping rates of water with hydrophilic and hydrophobic surface propellers, which have the same atomic charges as in the bulk cases. We present the pumping rate, N_P, which depends on the height, h, of mounting of the surface propeller that is positioned parallel to the water surface and described by the NVT ensemble (9×9 dummy atoms support 3500 waters in a periodic box of $5.5 \times 5 \times 9\,\mathrm{nm}^3$). During the pumping realized with the torque of $\mathbf{T} = 0.16$ nN · nm, the propellers need to push their blades through the water surface. This can be resisted

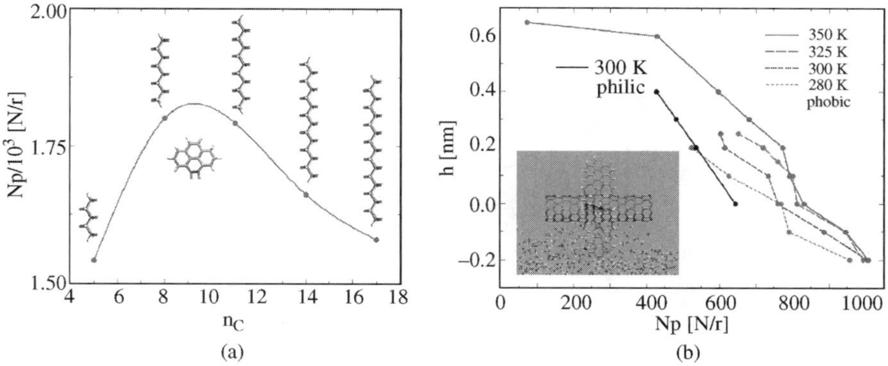

Figure 5.10 (a) Dependence of the pumping rate N_P on the number of carbons, nC, in the lipid molecule n-C_nH_{2n+2} ($n = 5$ to 17), used in the pumped solvent. The lipids are comparable in size to the blade, shown in the inset. (b) Dependence of the pumping rate N_P of water by the hydrophobic and hydrophilic propellers on the height of their mounting. Inset: Water around the hydrophobic propeller, mounted at the height $h = 0.1$ nm and pumping in the back, orthogonal to the tube axis.

by surface tension, depending on the blade's chemical and structural parameters, the height of the propeller's mounting, and its angular velocity. At high temperatures, the propeller can only pump molecules that are thermally lifted up. For the hydrophobic propeller at heights $h = 0.3$ to 0.4 nm and at $T = 350$ K, pumping of water is possible with rates of $N_P < 500$. Here the propeller can rotate fast and irregularly, so it might chop water molecules off the surface and carry them around, as shown in Fig. 5.8b. At lower temperatures, pumping is possible only at lower heights. The pumping rate increases more or less linearly with the depth of the blades. The pumping rates of the hydrophilic propeller at $T = 300$ K are smaller, as in the bulk propeller, and depend less on the height, due to the attraction of water to the blades.

Although the pumping described might sound intuitive and clear from our everyday experience, mechanical pumping of fluids at the nanoscale is far from trivial: nanoscale systems are inevitably described by low Reynolds numbers (ratio of inertial and viscous forces), $R = \rho va/\eta$, where a is the size of the system, v is its velocity, ρ is its density, and η is the viscosity of the fluid in which they move. Therefore, according to the *scallop theorem* [161], any reversible motion would not allow them to pump fluids or propel themselves. In fact, they can do both if their motion is not reversible. Here the irreversibility is caused by the rotational motion, where the system asymmetry, the frequency and amplitude of the motion, and the size and functionalization of the blades determine the character of pumping or propelling.

The described molecular propellers could be used to deliver nanoparticles [164], genetic material [162], and drugs [163] inside cells. For some of these goals they might need to be mounted into nanoscale systems able to perform locomotion [165–167]. Molecular propellers for such biomedical applications can be larger and have diameters of several nanometers to several micrometers. They could also

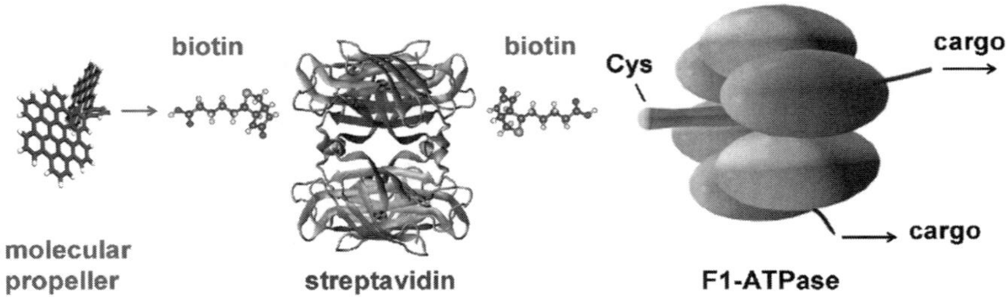

Figure 5.11 Assembly of the propelling unit, as described in the text. (*See insert for color representation of figure.*)

rotate several orders of magnitude more slowly when driven by biological and other less energetic means.

5.3.1 Design and Realization of Molecular Propelling Devices

Molecular propellers could be used in practical devices that might operate on the cellular level, as shown in Fig. 5.11. The system is similar in design to the unit realized experimentally by Soong et al. [39] and later by Dangerfield (CDL-RVNT Singapore). In the current system, the molecular propeller (sufficiently extended in size to allow directional propulsion of the system) is linked by biotin to the *crankshaft* of the nanomotor (the gamma subunit of the F1-ATPase enzyme) by means of streptavidin. The activity of this system is controlled by the concentration of the "motor fuel" (ATP).

Large molecular propellers could be prepared as shown by Simpson et al. [149], where such propellers were synthesized by cyclodehydrogenation of polyphenylene dendrimers. The synthesized propellers had planar and three-dimensional helical forms with C_n ($n = 2$ to 4) symmetries and sizes of 5 to 10 nm. These propellers could be extended further by polymerization to fit the described purposes. Alternatively, the blades could be made from graphene sheets [84,85], which can be cut, chemically modified [141], and attached to the molecular motors by the biotin–streptavidin complex [168].

One could also operate the molecular propellers in nonbiological environments and drive them by different means. Then the CNT shaft in the molecular propellers could potentially be mounted into CNT bearings [103] and attached to various types of molecular motors, driven by optical [169–173], electrical [174], or chemical means [175]. A prototype of a molecular motor driven by electron tunneling is shown in Fig. 5.12. Molecular dynamics simulations show that this motor can be highly efficient [176]. After further optimization, it could drive the molecular propellers described above.

5.3.2 Optimization Techniques in the Design of Nanoscale Machines

We also describe briefly how one could optimize these motile systems chemically: in particular, functionalize their blades and apply them in different situations.

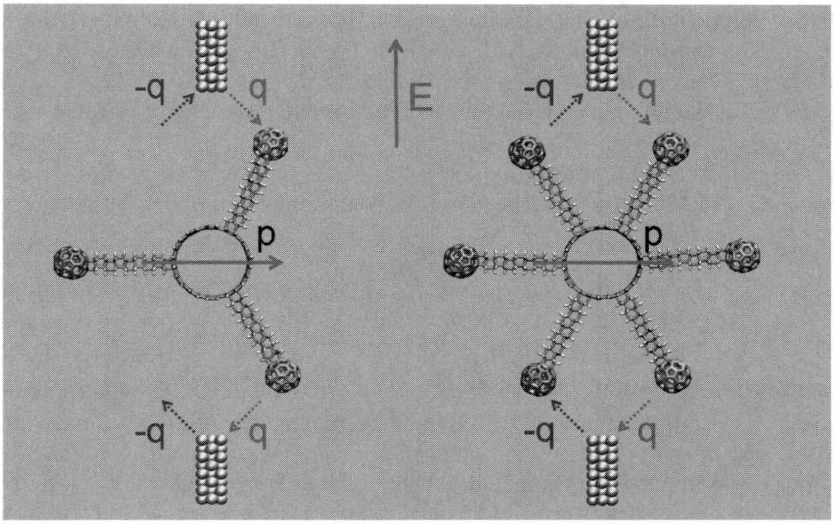

Figure 5.12 Tunneling-driven nanoscale motors with three (left) and six (right) fullerene blades. In an external homogeneous electric field E oriented along the vertical z direction, the electron tunneling from the neutral electrodes to the blades maintains an electric dipole p on the rotor, which is on average orthogonal to the field direction. This dipole is rotated unidirectionally by the electric field. (*See insert for color representation of figure.*)

For example, propellers so designed could pump fluids selectively or be able to sieve some molecules in the fluid. Here we explain the methodology for chemical optimization on the design of highly selective molecular nests on modified material surfaces by doping [141]. The same approach could be used in the optimal design of the functional blades. In this approach, the material surfaces are modified by genetic algorithms [177] that are used in the design of proteins and drugs [178,179], nanostructures [180], and in the optical control of chemical reactions [181,182] and nanoscale processes [183].

One could build the selective molecular nests by doping the surfaces using carefully designed patterns of suitable atoms, which become partially charged upon doping. Nesting of the molecules on the modified surfaces is controlled by their coulombic and van der Waals coupling to the atoms, which must be arranged in such a way that the local electrostatic potential formed above the doped surface selectively binds the molecule chosen. We illustrate this methodology on a graphene layer [84,85], substitutionally doped with B and N atoms [184], with the goal of minimizing the potential energy of the docked molecules.

First, we find the local charges of each dopant and its neighbors. The ab initio calculations, done with the B3LYP exchange-correlation functional in Gaussian03 [119], show that when a single C atom in the graphene layer is substituted by an isolated B or N atom, it carries the charge $Q_B \approx -1.2$e or $Q_N \approx 1.0$e, respectively. We assume that this charge is neutralized within the first three neighboring C atoms, each carrying the charge $Q_C = -Q_{B(N)}/3$. When several dopants have a single C neighbor between them, we charge the C atom by summing contributions from those dopants. The same dopants are not allowed to be first neighbors. However, when different types of dopants (B and N) are first neighbors, we find that their charges are $Q_B \approx -1.0$e and $Q_N \approx 1.0$e. To simplify the simulations, we assume that the charges of dopants are always $Q_B \equiv -1.1$e and $Q_N \equiv 1.1$e.

The total Coulombic interaction energy between the adsorbed molecule and the graphene layer doped by an isolated B (or N) atom is given by

$$E_{\mathrm{B(N)}} = Q_{\mathrm{B(N)}} \sum_{i=1}^{N} \frac{Q_i}{4\pi\,\varepsilon_0} \left(\frac{1}{r_{i0}} - \sum_{i=1}^{N} \frac{1}{3r_{ij}} \right). \tag{5.2}$$

Here Q_i is the charge of the ith (out of n) atom of the molecule, r_{i0} is its distance to the doped site, r_{ij} is its distance to the jth (out of three) neighbor of this site, and ε_0 is the vacuum permittivity.

The nest for the attached molecule can be designed by the following optimization strategy. We assume that every C atom of the graphene layer is a potential doping site. It can be doped by B or N, depending on which of them decreases the coulombic interaction energy in Eq. (5.2), where the surrounding atoms are assumed to be charged as if the dopants are isolated. For simplicity, we use the same approach even in the few rare cases where the B and N atoms are first neighbors. The final charges are assigned to the atoms correctly by the rules explained above. By limiting the number of dopants, we could control the strength of the molecule–surface binding and tune the selectivity of the nesting site. To realize this idea, we introduce a negative energy, E_m, called the level of importance, and dope a given site only when the change of the coulombic energy associated with this process is $E_{\mathrm{B(N)}} < E_m$.

In Fig. 5.13 we demonstrate the electrostatic potential distribution obtained for the doping pattern designed for the AYM peptide, with $E_m = -0.02$ eV, on the graphene sheet (right), and the electrostatic potential distribution for the peptide (left). This peptide consists of three amino acids, alanine (A), tyrosine (Y), and methionine (M), and has an acetylated N terminus and an amidized C terminus. It is designed (using PyMOL) with a beta-sheet secondary structure ($\varphi = -135°$, $\psi = 135°$). The figure displays the peptide nested above the center of the square graphene layer, formed by 684 C atoms. The three amino acids are arranged in the sequence alanine, tyrosine, methionine, from left to right.

The starting configuration of the AYM peptide is obtained from MD simulations realized at $T = 50$ K on the undoped static graphene layer. After it is left there to relax for 1 ns, the plane defined by the three alpha C atoms of its backbone becomes nearly parallel to the layer. We use the resulting AYM configuration to design the initial doping pattern and then run the MD simulations at $T = 300$ K for 100 ps and at $T = 50$ K for 1 ns, to further optimize the peptide's position. Then, we recalculate the doping profile for the peptide and repeat the entire cycle once more. Finally, we take a snapshot of a typical peptide's configuration, shown in Fig. 5.13. It shows the "high quality of docking" that can be seen from the complementarity of the electrostatic potentials created by the peptide and the doped graphene sheet. The highly positive

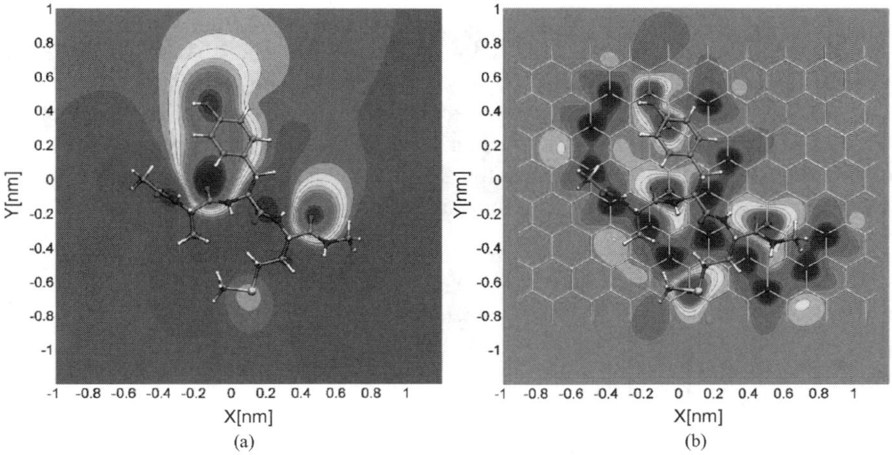

Figure 5.13 (a) Electrostatic potential distribution formed by the AYM peptide docked on the surface of the doped graphene sheet; (b) "complementary" distribution formed by the doped graphene sheet. (*See insert for color representation of figure.*)

and negative potentials above the B and N dopants, respectively, are clearly visible. Practically in all the regions where the peptide has a positive potential, the doped surface creates a negative potential, and vice versa. This complementarity allows efficient immobilization of the peptide in the nesting site.

Figure 5.14 (left) shows the coulombic potential energy E_C of the peptide as it is shifted like a rigid body along the x and y coordinates from the docking site shown in Fig. 5.13. It has a deep minimum of $E_C \approx -1.25$ eV when the peptide is docked, and E_C increases sharply to $E_C \approx 0$ eV as the peptide is moved in either direction. In the inset we also show the full two-dimensional coulombic potential-energy surface of the shifted peptide, where these cuts have been taken. It closely resembles funneling surfaces in protein-folding problems [185]. These results reflect the ability of the proposed methodology to achieve a high-quality molecular docking.

Figure 5.14b shows the AYM peptide's average total binding energy, $\langle E_{\text{tot}} \rangle = \langle E_C \rangle + \langle E_{\text{vdW}} \rangle$, as a function of the level of importance, E_m. The average electrostatic (coulombic) energy, $\langle E_C \rangle$, and the average van der Waals energy, $\langle E_{\text{vdW}} \rangle$, are calculated by averaging the difference (on-surface vs. away from it) over 1000 consecutive frames of the simulation trajectory, with a 1-ps time interval between frames. Binding of the molecule in the nesting site becomes stronger as $E_m \rightarrow 0$. This behavior can be understood from the inset, where we show the dependence of the number of B and N dopants on E_m. The number of dopants increases and becomes large as $E_m \rightarrow 0$, since less energetically significant doping sites become included. This process allows one to optimize the strength of binding by the choice of E_m.

We also estimate the recognition abilities of the nest. To do so, we dock in the nesting site designed for the AYM peptide, other peptides, which differ from it just in the middle Y residue. We replace this residue with different amino acids having neutral side chains, and calculate the binding energy of each peptide. In Fig. 5.14c, we show the total binding energies, $\langle E_{\rm tot}\rangle$, of the 1 to 15 peptides selected: AYM, AWM, AFM, ASM, AQM, ANM, AHM, AMM, ATM, ACM, ALM, AVM, AIM, AGM, and AAM, respectively, in nesting sites designed for the AYM peptide with the levels of importance $E_m = -0.005$, -0.01, -0.02, and -0.04 eV. (The peptides are ordered according to the sum of their total binding energies obtained for $E_m = -0.01$ and -0.02 eV.) We can see that the AYM peptide, and thus the Y residue, is very well recognized by the doping pattern designed for it; for $E_m = -0.01$ and -0.02 eV,

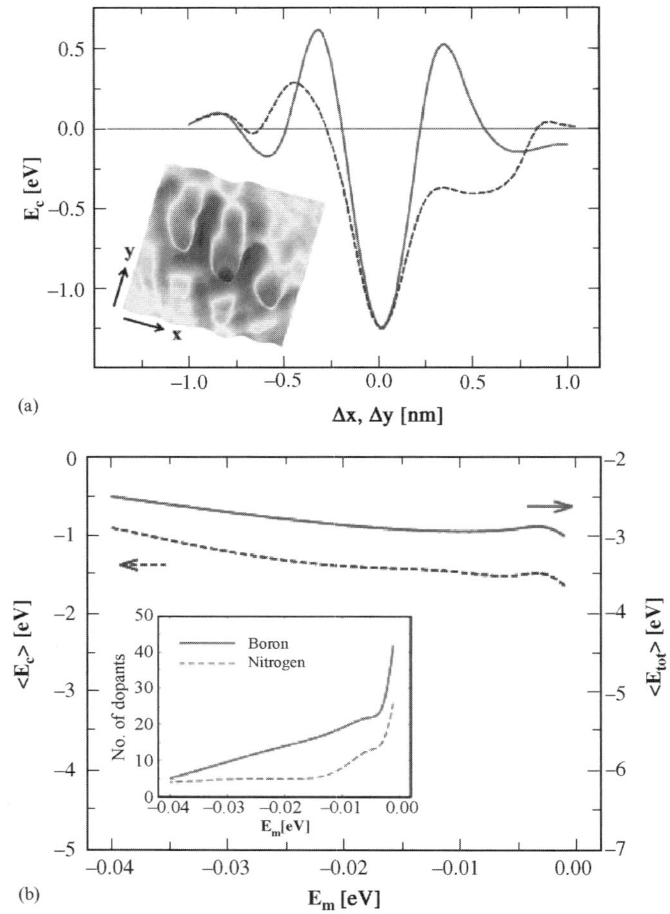

Figure 5.14 (a) Dependence of the coulombic E_C potential energy of the system, when the peptide is shifted in the x (solid) and y (dashed) directions along the surface, away from its docking site. In the inset, the full two-dimensional potential surface of E_C is shown. (b) The average Coulombic (dashed line), $\langle E_C\rangle$, and total (solid line), $\langle E_{\rm tot}\rangle$, binding energy of the AYM peptide to the doped surface. In the inset we show the number of B (solid) and N (dashed) dopants as a function of E_m. (c) The total binding energy of the 15 peptides at four levels of importance, E_m, shown in the upper right corner. [*See insert for color representation of part (a).*]

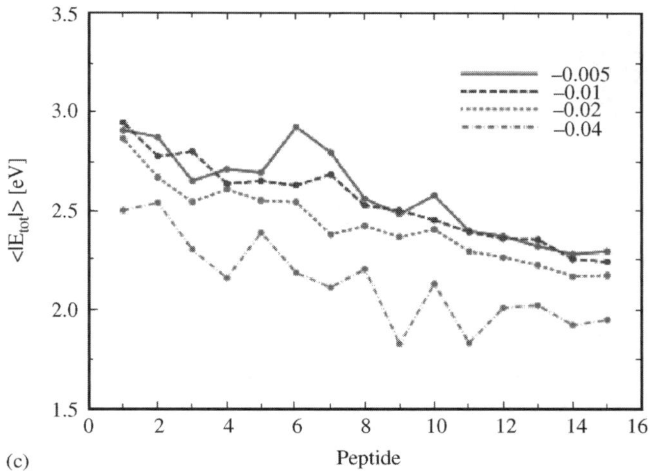

Figure 5.14 *(Continued)*

its $\langle E_{\text{tot}} \rangle$ is 0.15 to 0.65 eV larger than for other peptides. These results show that the nest designed can be highly specific. Since the entire peptide is recognized by its individual amino acids, it might be easier to recognize larger systems.

These nests could operate in polar solvents if the binding sites have larger charges. They could be active on the propeller blades as docking sites, or they can simply help to pump the fluid more efficiently. Here, we test peptide docking in water on modified graphene sheets. To do so we have replaced the B and N dopants by ammonium cation (NH_3^+) and carboxylate anion (CO_2^-), respectively, as charged ligands attached covalently to the surface [148]. The method is tested on a peptide extracted from the small green fluorescence protein (GFP) [140]. The molecule is formed roughly by a "cylinder" made of beta-sheet strands that protect a photoactive group in its interior. On the GFP surface, we find two neighboring strands (105 to 130 residues) that are relatively highly charged. Further details of the graphene–ligand system, the docked peptide, and their simulations are given in ref. [141].

In Fig. 5.15a, we show the initial and final configurations of the two GFP strands above the graphene layer with attached ligands in water (hidden). The charged residues in the initial configuration and the ligands on the graphene layer are displayed in atomistic details. During the simulation, the two strands shift their position a bit, but they stay attached to the ligands. In Fig. 5.15b we calculate by VMD [116] the root-mean-square deviation (RMSD) of all atoms between the initial and the actual configurations of the two-stranded peptide, obtained at each time frame. The RMSD stabilizes after 2.0 ns and fluctuates within 0.5 Å. Therefore, the peptide could be nested selectively above the predesigned ligand pattern on the graphene layer in water, without being dissolved or modified significantly. This example shows that one could design selective nesting sites for biomolecules on material surfaces in water. The same approach could be used in diluted ionic solutions [186]. Practical doping and

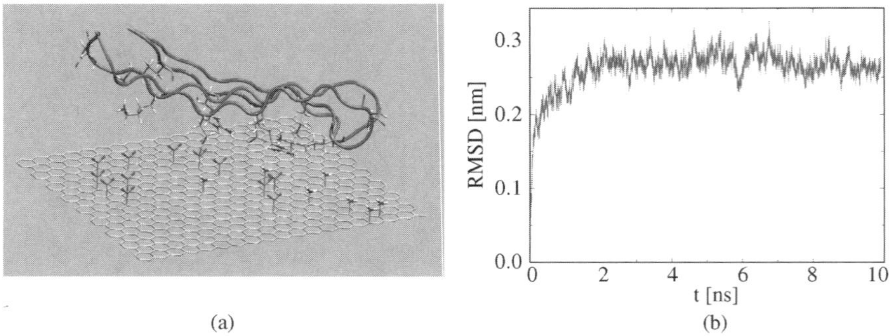

Figure 5.15 (a) Initial (red) and final (green) structures of the 105 to 130 residues from the GFP that are nested on the ligand-doped graphene layer. The peptide sequence is NYK-TRAEVKFEGDTLVNRIELKGIDF, where the 10 bold residues are charged and displayed in the initial structure. (b) RMSD of this peptide obtained during the simulation. (*See insert for color representation of figure.*)

functionalization of the surfaces by atoms and ligands, respectively, could be realized by STM and AFM techniques [50].

The methodology above could also be used to design structures (pores) optimally in the blades of the molecular propellers. In Fig. 5.16c, we show two pores designed in this way in graphene monolayers [187]. The F–N pore (left structure) is terminated by negatively charged nitrogens and fluorines, favoring the passage of cations. The H pore (right structure) is terminated by positively charged hydrogens, favoring the passage of anions. These nanopores show high selectivity for the passage of ions of various charges and sizes [187]. In Fig. 5.16a, we show the trajectories of the Na^+ and Cl^- ions through the F–N pore and H pore, respectively. The optimization

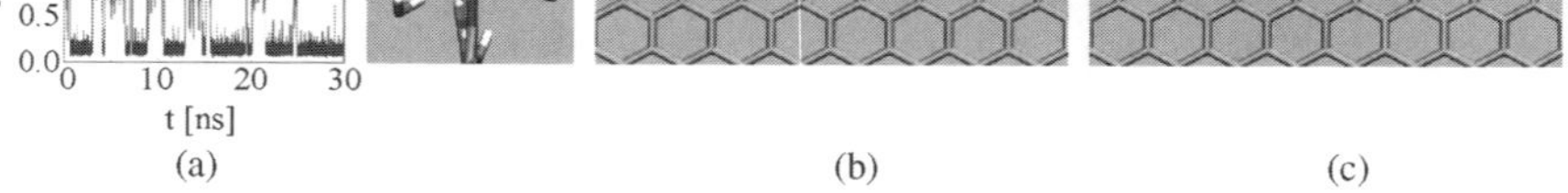

Figure 5.16 (a) Time-dependent distance d between the Na^+ and Cl^- ions and the centers of the F–N pore and H pore, respectively, at the field of $E = 6.25$ mV/nm. The dynamics of passage of these ions through the two pores is very different. (b) While both ions are surrounded by two water half-shells when passing through their pores, only the Cl^- ion has relatively stable binding to the H pore. (c) Detail of the F–N pore and the H pore, respectively, formed in the graphene monolayer. (*See insert for color representation of figure.*)

methodology described could be used to design nanoscale structures that would fit different purposes.

5.4 CONCLUSIONS

We have discussed two complementary methodologies that allow efficient manipulation of molecules at the nanoscale. The first approach is based on dragging of molecules on the surfaces of nanotubes by coulombic scattering with polar or ionic solutions, as well as individual ions, flowing inside the nanotubes. In this approach, one can manipulate individual molecules and their assemblies (droplets) inside gas or liquid phases. This methodology has a unique potential for the manipulation of molecules in biological cells and other highly constricted systems. The second approach might be used in pumping of larger volumes of molecules by mechanical means with molecular propellers. This methodology can be made highly specific to the fluids pumped if we design and optimally functionalize the blades of the molecular propellers used. Potentially the highly specific propellers could sieve for certain molecules solvated in the fluids. This approach could also be used to manipulate cellular components, such as small organelles, to which the molecular propellers could be attached and rotated by various means. These methodologies have the potential to open new avenues in the design and control of hybrid functional systems based on inorganic and organic compounds, polymers, proteins, and nanoparticles.

REFERENCES

1. Feynman RP. There's plenty of room at the bottom: an invitation to enter a new field of physics. *Eng. Sci.* **23**:22, 1960.
2. Adiga SP, Brenner DW. Flow control through polymer-grafted smart nanofluidic channels: molecular dynamics simulations. *Nano Lett.* **5**:2509, 2005.
3. Warrick J, Meyvantsson I, Jub J, Beebe DJ. High-throughput microfluidics: improved sample treatment and washing over standard wells. *Lab Chip* **7**:316, 2007.
4. Mijatovic D, Eijkel JCT, van den Berg A. Technologies for nanofluidic systems: top-down vs. bottom-up – a review. *Lab Chip* **5**:492, 2005.
5. Moon H, Wheeler AR, Garrell RL, Loo JA. An integrated digital microfluidic chip for multiplexed proteomic sample preparation and analysis by MALDI-MS. *Lab Chip* **6**:1213, 2006.
6. Kim W, Guo M, Yang P, Wang D. Microfabricated monolithic multinozzle emitters for nanoelectrospray mass spectrometry. *Anal. Chem.* **79**:3703, 2007.
7. Shui L, Eijkel JCT, van den Berg A. Multiphase flow in micro- and nanochannels. *Sen. Actuat. B* **121**:263, 2007.
8. Zhang X, Haswell SJ. Materials matter in microfluidic devices. *MRS Bull.* **31**:95, 2006.
9. Huh D, Mills KL, Zhu X, Burns MA, Thouless MD, Takayama S. Tunable elastomeric nanochannels for nanofluidic manipulation. *Nat. Mater.* **6**:424, 2007.

10. Phillip WA, Rzayev J, Hillmyer MA, Cussler EL. Gas and water liquid transport through nanoporous block copolymer membranes. *J. Membr. Sci.* **286**:144, 2006.
11. Reisner W, et al. Statics and dynamics of single DNA molecules confined in nanochannels. *Phys. Rev. Lett.* **94**:196101, 2005.
12. Mao Y, Chang S, Yang S, Ouyang Q, Jiang L. Tunable non-equilibrium gating of flexible DNA nanochannels in response to transport flux. *Nat. Nanotechnol.* **2**:366, 2007.
13. Rauscher M, Dietrich S, Koplik J. Shear flow pumping in open micro- and nanofluidic systems. *Phys. Rev. Lett.* **98**:224504, 2007.
14. Sharma R, et al. Nanometer positioning, parallel alignment, and placement of single anisotropic nanoparticles using hydrodynamic forces in cylindrical droplets. *Nano Lett.* **7**:2693, 2007.
15. Kuo T-C, Sloan LA, Sweedler JV, Bohn PW. Manipulating molecular transport through nanoporous membranes by control of electrokinetic flow: effect of surface charge density and Debye length. *Langmuir* **17**:6298, 2001.
16. Hu JS, Chao CYH. A study of the performance of microfabricated electroosmotic pump. *Sen. Actuat. A* **135**:273, 2007.
17. Leach J, Mushfique H, di Leonardo R, Padgetta M, Cooper J. An optically driven pump for microfluidics. *Lab Chip* **6**:735, 2006.
18. Tanaka Y, et al. An actuated pump on-chip powered by cultured cardiomyocytes. *Lab Chip* **6**:362, 2006.
19. Gleiche M, Chi LF, Fuchs H. Nanoscopic channel lattices with controlled anisotropic wetting. *Nature* **403**:173, 2000.
20. Rodolfa KT, Bruckbauer A, Zhou DJ, Schevchuk AI, Korchev YE, Klenerman D. Nanoscale pipetting for controlled chemistry in small arrayed water droplets using a double-barrel pipet. *Nano Lett.* **6**:252, 2006.
21. Fang A, Dujardin E, Ondarcuhu T. Control of droplet size in liquid nanodispensing. *Nano Lett.* **6**:2368, 2006.
22. Linke H, et al. Self-propelled Leidenfrost droplets. *Phys. Rev. Lett.* **96**:154502, 2006.
23. Kelly BT, Baret J-C, Talyab V, Griffiths AD. Miniaturizing chemistry and biology in microdroplets. *Chem. Commun.* 1773, 2007.
24. Gijs MAM. Will fluidic electronics take off? *Nat. Nanotechnol.* **2**:268, 2007.
25. Vlassiouk I, Siwy ZS. Nanofluidic diode. *Nano Lett.* **7**:552, 2007.
26. Karnik R, Fan R, Yue M, Li D, Yang P, Majumdar A. Electrostatic control of ions and molecules in nanofluidic transistors. *Nano Lett.* **5**:**943**:2005.
27. Mielnik MM, Ekatpure RP, Stran LR, Schonfeld F. Sinusoidal crossflow microfiltration device: experimental and computational flowfield analysis. *Lab Chip* **5**:897, 2005.
28. Stein D, van der Heyden FHJ, Koopmans WJA, Dekker C. Pressure-driven transport of confined DNA polymers in fluidic channels. *Proc. Natl. Acad. Sci. USA* **103**:15853, 2006.
29. Turner SWP, Cabodi M, Craighead HG. Confinement-induced entropic recoil of single DNA molecules in a nanofluidic structure. *Phys. Rev. Lett.* **88**:128103, 2002.
30. Fu J, Schoch RB, Stevens AL, Tannenbaum SR, Han J. A patterned anisotropic nanofluidic sieving structure for continuous-flow separation of DNA and proteins. *Nat. Nanotechnol.* **2**:121, 2007.
31. Hosokawa K, Omata M, Sato K, Maeda M. Power-free sequential injection for microchip immunoassay toward point-of-care testing. *Lab Chip* **6**:236, 2006.

32. Yarin AL, Zussman E, Wendorff JH, Greiner A. Material encapsulation and transport in core-shell micro/nanofibers, polymer and carbon nanotubes and micro/nanochannels. *J. Mater. Chem.* **17**:2585, 2007.
33. Saleh OA, Sohn LL. Quantitative sensing of nanoscale colloids using a microchip Coulter counter. *Rev. Sci. Instrum.* **72**:4449, 2001.
34. Even-Ram S, Yamada KM. Cell migration in 3D matrix. *Curr. Opin. Cell Biol.* **17**:524, 2005.
35. Bru R, Sanchez-Ferrer A, Garcia-Carmona F. Kinetic models in reverse micelles. *Biochem. J.* **310**:721, 1995.
36. Brunner C, Wahnes C, Vogel V. Cargo pick-up from engineered loading stations by kinesin driven molecular shuttles. *Lab Chip* **7**:1263, 2007.
37. Schliwa M. Molecular Motors, Wiley-VCH, Hoboken, NJ, 2003.
38. Bull JL, Hunt AJ, Meyhofer E. A theoretical model of a molecular-motor-powered pump. *Biomed. Microdev.* **7**:21, 2005.
39. Soong RK, Bachand GD, Neves HP, Olkhovets AG, Craighead HG, Montemagno CD. Powering an inorganic nanodevice with a biomolecular motor. *Science* **290**:1555, 2000.
40. Kay ER, Leigh DA, Zerbetto F. Synthetic molecular motors and mechanical machines. *Angew. Chem. Int. Ed.* **46**:72, 2007.
41. Astumian RD. Thermodynamics and kinetics of a Brownian motor. *Science* **276**:917, 1997.
42. Astumian RD, Dernyi I. Fluctuation driven transport and models of molecular motors and pumps. *Eur. Biophys. J.* **27**:474, 1998.
43. Kelly TR, Cavero M. Selective monoacylation of a diamine using intramolecular delivery by a DMAP unit. *Org. Lett.* **4**:2653, 2002.
44. Ashkin A, et al. Optical trapping and manipulation of single cells using infrared laser beams. *Nature* **330**:769, 1987.
45. Mehta A, et al. Single-molecule biomechanics with optical methods. *Science* **283**:1689, 1999.
46. Grier DG. A revolution in optical manipulation. *Nature* **424**:810, 2003.
47. Binnig G, Rohrer H. Scanning tunneling microscopy. *IBM J. Res. Dev.* **44**:355, 1986.
48. Falvo MR, Washburn S, Superfine R, Finch M, Brooks FP Jr, Chi V, Taylor RM. Manipulation of individual viruses: friction and mechanical properties. *Biophys. J.* **72**:1396, 1997.
49. Pakes CI, George DP, Ramelow S, Cimmino A, Jamieson DN, Prawer S. Manipulation of single magnetic protein particles using atomic force microscopy. *J. Magn. Magn. Mater.* **272**:e1231 2004.
50. Duwez A-S, et al. Mechanochemistry: targeted delivery of single molecules. *Nat. Nanotechnol.* **1**:122, 2006.
51. Goncalves RP, Scheuring S. Manipulating and imaging individual membrane proteins by AFM. *Surf. Interface Anal.* **38**:1413, 2006.
52. Meyer DI, Krause DIE, Dobberstein B. Secretory protein translocation across membranes: the role of the "docking protein." *Nature* **297**:647, 1982.
53. Pack DW. DNA delivery: timing is everything. *Nat. Mater.* **3**:133, 2004.
54. Knight TE, Melechko AV, Hensley DK, Mann DGJ, Griffin DG, Simpson ML. Tracking gene expression after DNA delivery using spatially indexed nanofiber arrays. *Nano Lett.* **4**:1213, 2004.

55. Chen X, Kis A, Zettl A, Bertozzi CR. A cell nanoinjector based on carbon nanotubes. *Proc. Natl. Acad. Sci. USA* **104**:8218, 2007.
56. Luo D. Nanotechnology and DNA delivery. *MRS Bull.* **30**:2005.
57. Andersson H, van den Berg A. From lab-on-a-chip to lab-in-a-cell. *Proc. SPIE* **5718**:1, 2005.
58. Craighead HG. Future lab-on-a-chip technologies for interrogating individual molecules. *Nature* **442**:387, 2006.
59. Mannion JT, Craighead HG. Nanofluidic structures for single biomolecule fluorescent detection. *Biopolymers* **85**:135, 2006.
60. Kellermayer MSZ. Visualizing and manipulating individual protein molecules. *Physiol. Meas.* **26**:R119 2005.
61. Rhoades E, Gussakovsky E, Haran G. Watching proteins fold one molecule at a time. *Proc. Natl. Acad. Sci. USA* **100**:3197, 2003.
62. Strick T, Allemand J-F, Croquette V, Bensimon D. The manipulation of single biomolecules. *Phys. Today* **54**:46, 2001.
63. Ying L. Single molecule biology: Coming of age. *Mol. BioSyst.* **3**:377, 2007.
64. Iijima S. Helical microtubules of graphitic carbon. *Nature (London)* **354**:56, 1991.
65. Dresselhaus MS, Dresselhaus G, Eklund PC. *Science of Fullerenes and Carbon Nanotubes*, Academic Press, San Diego, CA, 1996.
66. Chico L, Crespi VH, Benedict LX, Louie SG, Cohen ML. Pure carbon nanoscale devices: nanotube heterojunctions. *Phys. Rev. Lett.* **76**:971, 1996.
67. Collins PG, Zettl A, Bando H, Thess A, Smalley RE. Nanotube nanodevice. *Science* **278**:100, 1997.
68. Bachtold A, et al. Scanned probe microscopy of electronic transport in carbon nanotubes. *Phys. Rev. Lett.* **84**:6082, 2000.
69. Park J, et al. Coulomb blockade and the Kondo effect in single-atom transistors. *Nature* **417**:722, 2002.
70. Chen J, et al. Solution properties of single-walled carbon nanotubes. *Science* **282**:95, 1998.
71. Martel R, Schmidt T, Shea HR, Hertel T, Avouris P. Single- and multi-wall carbon nanotube field-effect transistors. *Appl. Phys. Lett.* **73**:2447, 1998.
72. Baughman RH, et al. Carbon nanotube actuators. *Science* **284**:1340, 1999.
73. Gugenberger F, et al. Glass-transition in single-crystal C-60 studied by high-resolution dilatometry. *Phys. Rev. Lett.* **69**:3774, 1992.
74. Zheng M, et al. Structure-based carbon nanotube sorting by sequence-dependent DNA assembly. *Science* **302**:1545, 2003.
75. Wang B, Král P, Thanopulos I. Docking of chiral molecules on twisted and helical nanotubes: nanomechanical control of catalysis. *Nano Lett.* **6**:1918, 2006.
76. Banerjee S, Hemraj-Benny T, Wong SS. Covalent surface chemistry of single-walled carbon nanotubes. *Adv. Mater.* **17**:17, 2005.
77. Barber AH, Cohen SR, Eitan A, Schadler LS, Wagner HD. Fracture transitions at a carbon-nanotube/polymer interface. *Adv. Mater.* **18**:83, 2006.
78. Wong SS, Joselevich E, Woolley AT, Cheung CL, Lieber CM. Covalently functionalized nanotubes as nanometre-sized probes in chemistry and biology. *Nature* **394**:52, 1998.

79. Kam NWS, et al. Nanotube molecular transporters: internalization of carbon nanotube-protein conjugates into mammalian cells. *J. Am. Chem. Soc.* **126**:6850, 2004.
80. Caruthers SD, Wickline SA, Lanza GM. Nanotechnological applications in medicine. *Curr. Opin. Biotechnol.* **18**:26, 2007.
81. Blasé X, Rubio A, Louie SG, Cohen ML. Stability and band gap constancy of boron nitride nanotubes. *Europhys. Lett.* **28**:335, 1994.
82. Yang W, et al. DNA-modified nanocrystalline diamond thin-films as stable, biologically active substrates. *Nat. Mater.* **1**:253, 2002.
83. Wang J, Firestone MA, Auciello O, Carlisle JA. Surface functionalization of ultra-nanocrystalline diamond films by electrochemical reduction of aryldiazonium salts. *Langmuir* **20**:11450, 2004.
84. Novoselov KS, et al. Electric field effect in atomically thin carbon films. *Science* **306**:666, 2004.
85. Stankovich S, et al. Graphene-based composite material. *Nature* **442**:282, 2006.
86. Ferro S, De Battisti A. The 5-V window of polarizability of fluorinated diamond electrodes in aqueous solutions. *Anal. Chem.* **75**:7040, 2003.
87. Wang B, Král P. Dragging of molecules on material surfaces induced by flowing liquids. *J. Am. Chem. Soc.* **128**:15984, 2006.
88. Wang B, Král P. Chemically tunable nanoscale propellers of liquids. *Phys. Rev. Lett.* **98**:266102, 2007.
89. Král P, Tománek D. Laser-driven atomic pump. *Phys. Rev. Lett.* **82**:5373, 1999.
90. Regan BC, Aloni S, Ritchie RO, Dahmen U, Zettl A. Carbon nanotubes as nanoscale mass conveyors. *Nature* **428**:924, 2004 (Regan BC, Shaul A, Zettl AK, U.S. patent 7,341,651).
91. Svensson K, Olin H, Olsson E. Nanopipettes for metal transport. *Phys. Rev. Lett.* **93**:145901, 2004.
92. Král P, Shapiro M. Nanotube electron drag in flowing liquids. *Phys. Rev. Lett.* **86**:134, 2001.
93. Ghosh S, Sood AK, Kumar N.. Carbon nanotube flow sensors. *Science* **299**:1042, 2003 (Sood AK, Ghosh S, U.S. patent 6,718,834 B1).
94. Sood AK, Ghosh S. Direct generation of a voltage and current by gas flow over carbon nanotubes and semiconductors. *Phys. Rev. Lett.* **93**:086601, 2004.
95. Subramaniam C, Pradeep T, Chakrabarti J. Flow-induced transverse electrical potential across an assembly of gold nanoparticles. *Phys. Rev. Lett.* **95**:164501, 2005.
96. Hummer G, Rasaiah JC, Noworyta JP. Water conduction through the hydrophobic channel of a carbon nanotube. *Nature* **414**:188, 2001.
97. Skoulidas AI, Ackerman DM, Johnson JK, Sholl DS. Rapid transport of gases in carbon nanotubes. *Phys. Rev. Lett.* **89**:185901, 2002.
98. Holt JK, et al. Fast mass transport through sub-2-nanometer carbon nanotubes. *Science* **312**:1034, 2006.
99. Majumder M, Chopra N, Andrews R, Hinds BJ. Enhanced flow in carbon nanotubes. *Nature* **483**:44, 2005.
100. Wang Z, Ci L, Chen L, Nayak S, Ajayan PM, Koratkar N. Polarity-dependent electrochemically controlled transport of water through carbon nanotube membranes. *Nano Lett.* **7**:697, 2007.

101. Zhou JJ, Noca F, Gharib M. Flow conveying and diagnosis with carbon nanotube arrays. *Nanotechnology* **17**:4845, 2006.
102. Whitby M, Quirke N. Fluid flow in carbon nanotubes and nanopipes. *Nat. Nanotechnol.* **2**:87, 2007.
103. Cummings J, Zettl A. Low-friction nanoscale linear bearing realized from multiwall carbon nanotubes. *Science* **289**:602, 2000.
104. Qiao R, Aluru NR. Ion concentrations and velocity profiles in nanochannel electroosmotic flows. *J. Chem. Phys.* **118**:4692, 2003.
105. Daiguji H, Yang P, Majumdar A. Ion transport in nanofluidic channels. *Nano Lett.* **4**:137, 2004.
106. Pamme N. Magnetism and microfluidics. *Lab Chip* **6**:24, 2006.
107. Waghe A, Rasaiah JC, Hummer G. Filling and emptying kinetics of carbon nanotubes in water. *J. Chem. Phys.* **117**:10789, 2002.
108. Majumder M, Chopra N, Hinds BJ. Effect of tip functionalization on transport through vertically oriented carbon nanotube membranes. *J. Am. Chem. Soc.* **127**:9062, 2005.
109. Li J, Gong X, Lu H, Li D, Fang H, Zhou R. Electrostatic gating of a nanometer water channel. *Proc. Natl. Acad. Sci. USA* **104**:3687, 2007.
110. Wang B, Král P. Dragging of polarizable nanodroplets by ions solvated at nanometer separations. *Phys. Rev. Lett.* **101**:046103, 2008.
111. Saito R, Fujita M, Dresselhaus G, Dresselhaus MS. Electronic structure of chiral graphene tubules. *Appl. Phys. Lett.* **60**:2204, 1992.
112. Cabria I, Mintmire JW, White CT. Metallic and semiconducting narrow carbon nanotubes. *Phys. Rev. B* **67**:121406(R), 2003.
113. Leonard F, Tersoff J. Dielectric response of semiconducting carbon nanotubes. *Appl. Phys. Lett.* **81**:4835, 2002.
114. Phillips JC, Braun R, Wang W, Gumbart J, Tajkhorshid E, Villa E, Chipot C, Skeel RD, Kale L, Schulten K. Scalable molecular dynamics with NAMD. *J. Comput. Chem.* **26**:1781, 2005.
115. MacKerell AD, et al. All-atom empirical potential for molecular modeling and dynamics studies of proteins. *J. Phys. Chem. B* **102**:3586, 1998.
116. Humphrey W, Dalke A, Schulten K. VMD: Visual molecular dynamics. *J. Mol. Graph.* **14**:33, 1996.
117. Stoyanov SR, Král P. Metallo-pyrrole-capped carbon nanocones. *J. Phys. Chem. B* **110**:21480, 2006.
118. Stoyanov SR, Král P, Wang B. Extended electronic states above metal-doped carbon nanostructures. *Appl. Phys. Lett.* **90**:153110, 2007.
119. Frisch MJ, et al. Gaussian 03, Revision C.02, Gaussian, Inc., Wallingford CT, 2004.
120. Darden T, York D, Pedersen L. Particle mesh Ewald: an $N^*\log(N)$ method for Ewald sums in large systems. *J. Chem. Phys.* **98**:10089, 1993.
121. Feller SE, Zhang YH, Pastor RW, Brooks BR. Constant-pressure molecular-dynamics simulation: the Langevin piston method. *J. Chem. Phys.* **103**:4613, 1995.
122. Servantie J, Gaspard P. Methods of calculation of a friction coefficient: application to nanotubes. *Phys. Rev. Lett.* **91**:185503, 2003.
123. Zhu F, Tajkhorshid E, Schulten K. Pressure-induced water transport in membrane channels studied by molecular dynamics. *Biophys. J.* **83**:154, 2002.

124. Laage D, Hynes JT. A molecular jump mechanism of water reorientation. *Science* **311**:832, 2006.
125. Kuo T, Sloan L, Sweedler J, Bohn P. Manipulating molecular transport through nanoporous membranes by control of electrokinetic flow: effect of surface charge density and Debye length. *Langmuir* **17**:6298, 2001.
126. Kemery P, Steehler J, Bohn P. Electric field mediated transport in nanometer diameter channels. *Langmuir* **14**:2884, 1998.
127. Majumder M, Chopra N, Hinds BJ. Effect of tip functionalization on transport through vertically oriented carbon nanotube membranes. *J. Am. Chem. Soc.* **127**:9062, 2005.
128. Barz DPJ, Ehrhard P. Model and verification of electrokinetic flow and transport in a micro-electrophoresis device. *Lab Chip* **5**:949, 2005.
129. Skoog DA, Leary JJ. Principles of Instrumental Analysis, Saunders College Publishing, Fort Worth, TX, 1992.
130. Dellago C, Naor MM, Hummer G. Proton transport through water-filled carbon nanotubes. *Phys. Rev. Lett.* **90**:105902, 2003.
131. Mann DJ, Halls MD. Water alignment and proton conduction inside carbon nanotubes. *Phys. Rev. Lett.* **90**:195503, 2003.
132. Liu HM, Murad S, Jameson CJ. Ion permeation dynamics in carbon nanotubes. *J. Chem. Phys.* **125**:084713, 2006.
133. McHale G, Newton MI, Carroll BJ. The shape and stability of small liquid drops on fibers. *Oil Gas Sci. Technol. Rev. IFP* **56**:47, 2006.
134. Hitoshi O, Radnai T. Structure and dynamics of hydrated ions. *Chem. Rev.* **93**:1157, 1993.
135. Kim HY, Lee HJ, Kang BH. Sliding of liquid drops down an inclined solid surface. *J. Colloid. Interface. Sci.* **247**:372, 2002.
136. Mahadevan L, Pomeau Y. Rolling droplets. *Phys. Fluids* **11**:2449, 1999.
137. Brancker AV. Viscosity-temperature dependence. *Nature* **166**:905, 1950.
138. Enyashin AN, Seifert G, Ivanovskii AL. Calculation of the electronic and thermal properties of C/BN nanotubular heterostructures. *Inorg. Mater.* **41**:595, 2005.
139. Cevc G. *Phospholipids Handbook*, Marcel Dekker, New York, 1993.
140. van Thor JJ, Georgiev GY, Towrie M, Sage JT. Ultrafast and low barrier motions in the photoreactions of the green fluorescence protein. *J. Biol. Chem.* **280**:33652, 2005.
141. Wang B, Král P. Optimal atomistic modifications of material surfaces: custom-designed nesting sites for biomolecules. *Small* **3**:580, 2007.
142. Baret J-C, Mugele F. Electrical discharge in capillary breakup: controlling the charge of a droplet. *Phys. Rev. Lett.* **96**:016106, 2006.
143. Mills DL. Image force on a moving charge. *Phys. Rev. B* **15**:763, 1977.
144. Zheng M, Jagota A, Semke ED, Diner BA, McLean RS, Lustig SR, Richardson RE, Tassi NG. DNA-assisted dispersion and separation of carbon nanotubes. *Nat. Mater.* 2003 **2**:338.
145. Bogotsky VS. *Fundamentals of Electrochemistry*, Wiley-Interscience, Hoboken, NJ, 2006.
146. Suhr J, Zhang W, Ajayan MP, Koratkar NA. Temperature-activated interfacial friction damping in carbon nanotube polymer composites. *Nano Lett.* **6**:219, 2006.
147. Han J, Globus A, Jaffe R, Deardorff G. Molecular dynamics simulations of carbon nanotube-based gears. *Nanotechnology* **8**:95, 1997.

148. Tasis D, Tagmatarchis N, Bianco A, Prato M. Chemistry of carbon nanotubes. *Chem. Rev.* **106**:1105, 2006.

149. Simpson CD, Mattersteig G, Martin K, Gherghel L, Bauer RE, Rader HJ, Mullen K. Nanosized molecular propellers by cyclodehydrogenation of polyphenylene dendrimers. *J. Am. Chem. Soc.* **126**:3139, 2004.

150. Zhu F, Schulten K. Water and proton conduction through carbon nanotubes as models for biological channels. *Biophys. J.* **85**:236, 2003.

151. Silverstein KAT, Haymet ADJ, Dill KA. The strength of hydrogen bonds in liquid water and around nonpolar solutes. *J. Am. Chem. Soc.* **122**:8037, 2000.

152. Vergeles M, Keblinski P, Koplik J, Banavar JR. Stokes drag and lubrication flows: a molecular dynamics study. *Phys. Rev. E* **53**:4852, 1996.

153. Squires TM, Quake SR. Microfluidics: fluid physics at the nanoliter scale. *Rev. Mod. Phys.* **77**:977, 2005.

154. Choi CH, Westin KJA, Breuer KS. Apparent slip flows in hydrophilic and hydrophobic microchannels. *Phys. Fluids* **15**:2897, 2003.

155. Raschke TM, Levitt M. Nonpolar solutes enhance water structure within hydration shells while reducing interactions between them. *Proc. Natl. Acad. Sci. USA* **102**:6777, 2005.

156. Moulin F, Devel M, Picaud S. Molecular dynamics simulations of polarizable nanotubes interacting with water. *Phys. Rev. B* **71**:165401, 2005.

157. Gao J, Luedtke WD, Landman U. Nano-elastohydrodynamics: structure, dynamics and flow in nonuniform lubricated junctions. *Science* **270**:605, 1995.

158. Travis KP, Todd BD, Evans DJ. Departure from Navier–Stokes hydrodynamics in confined liquids. *Phys. Rev. E* **55**:4288, 1997.

159. Viswanath DS, Natarajan G. *Data Book on the Viscosity of Liquids*, Hemisphere Publishing, New York, 1989.

160. Kuzmenko I, Rapaport H, Kjaer K, Als-Nielsen J, Weissbuch I, Lahav M, Leiserowitz L. Design and characterization at crystalline thin film architectures at the air-liquid interface: simplicity to complexity. *Chem. Rev.* **101**:1659, 2001.

161. Purcell EM. Life at low Reynolds number. *Am. J. Phys.* **45**:3, 1977.

162. Woods N-B, Bottero V, Schmidt M, von Kalle C, Verma IM. Gene therapy: therapeutic gene causing lymphoma. *Nature* **440**:1123, 2006.

163. Lu DR, Oie S. *Cellular Drug Delivery: Principles and Practice*, Humana Press, Totowa NJ, 2004.

164. Akin A, et al. Bacteria-mediated delivery of nanoparticles and cargo into cells. *Nat. Nanotechnol.* **2**:441, 2007.

165. Dreyfus R, et al. Microscopic artificial swimmers. *Nature* **437**:2005 862.

166. Shirai Y, et al. Surface-rolling molecules. *J. Am. Chem. Soc.* **128**:4854, 2006.

167. Manghi M, Schlagberger X, Netz RR. Propulsion with a rotating elastic nanorod. *Phys. Rev. Lett.* **96**:068101, 2006.

168. Im H, Huang XJ, Gu B, Choi Y-K. A dielectric-modulated field-effect transistor for biosensing. *Nat. Nanotechnol.* **2**:430, 2007.

169. Král P, Sadeghpour HR. Laser spinning of nanotubes: a path to fast-rotating microdevices. *Phys. Rev. B* **65**:161401(R), 2002.

170. Plewa J, Tanner E, Mueth DM, Grier DG. Processing carbon nanotubes with holographic optical tweezers. *Opt. Expr.* **12**:1978, 2004.

171. Tan S, Lopez HA, Cai CW, Zhang Y. Optical trapping of single-walled carbon nanotubes. *Nano Lett.* **4**:1415, 2004.

172. Vacek J, Michl J. Molecular dynamics of a grid-mounted molecular dipolar rotor in a rotating electric field. *Proc. Natl. Acad. Sci. USA* **98**:5481, 2001.

173. Kottas GS, Clarke LI, Horinek D, Michl J. Artificial molecular rotors. *Chem. Rev.* **105**:1281, 2005.

174. Král P, Seideman T. Current-induced rotation of helical molecular wires. *J. Chem. Phys.* **123**:184702, 2005.

175. Kelly TR, De Silva H, Silva RA. Unidirectional rotary motion in a molecular system. *Nature* **401**:150, 1999.

176. Wang B, Vukovic L, Král P. Nanoscale rotary motors driven by electron tunneling. *Phys. Rev. Lett.* **101**:186808, 2008.

177. Forrest S. Genetic algorithms - principles of natural-selection applied to computation. *Science* **261**:872, 1993.

178. Pedersen JT, Moult J. Genetic algorithms for protein structure prediction. *Curr. Opin. Struct. Biol.* **6**:227, 1996.

179. Pokala N, Handel TM. Energy functions for protein design: adjustment with protein–protein complex affinities, models for the unfolded state, and negative design of solubility and specificity. *J. Mol. Biol.* **347**:203, 2005.

180. Hart GLW, Blum V, Walorski MJ, Zunger A. Evolutionary approach for determining first-principles Hamiltonians. *Nat. Mater.* **4**:391, 2005.

181. Tannor DJ, Kosloff R, Rice SA. Coherent pulse sequence induced control of selectivity of reactions: exact quantum mechanical calculations. *J. Chem. Phys.* **85**:5805, 1986.

182. Peirce AP, Dahleh MA, Rabitz H. Optimal control of quantum-mechanical systems: existence, numerical approximation, and applications. *Phys. Rev. A* **37**:4950, 1988.

183. Král P. Nanoscale processing by adaptive laser pulses. *Phys. Rev. B* **66**:241401, 2002.

184. Stephan O, Ajayan PM, Colliex C, Redlich P, Lambert JM, Bernier P, Lefin P. Doping graphitic and carbon nanotube structures with boron and nitrogen. *Science* **266**:1683, 1994.

185. Leopold PE, Montal M, Onuchic JN. Protein folding funnels: a kinetic approach to the sequence structure relationship. *P*roc. Natl. Acad. Sci. USA* **89**:8721, 1992.

186. Carlsson F, Hyltner E, Arnebrant T, Malmsten M, Linse P. Lysozyme adsorption to charged surfaces: a Monte Carlo study. *J. Phys. Chem. B* **108**:9871, 2004.

187. Synt K, Wang B, Král P. Selective ion passage through functionalized graphene nanopores. *J. Am. Chem. Soc.* **130**:16448, 2008.

6

MOLECULAR DYNAMICS MODELING OF NANODROPLETS AND NANOPARTICLES

D.T. Lussier, N.M.P. Kakalis, and Y. Ventikos
Department of Engineering Science, University of Oxford, Oxford, UK

6.1 INTRODUCTION

In the history of fluid mechanics, droplets and bubbles have proven to be a productive model system for research into diverse phenomena, such as surface tension, interface structure, two-phase flow, and phase change. Through this chapter we provide an overview of the application of direct simulation molecular dynamics to the study of nanodroplets. There are many different variants within this family of techniques. In general, the work surveyed here makes use of fully atomistic molecular dynamics. In addition to offering a rich platform for the study of fundamental fluid mechanics at the nanometer length scale, nanodroplets are also emerging as a fundamental and important building block for numerous practical applications. In some cases these are simply scaled-down versions of previous droplet applications, whereas in others, new behavior specific to the nanometric length scale plays a key role. Areas of potential application are diverse and include materials, manufacturing, drug delivery, and energy. Given the huge number of fluid-mediated processes, particularly in biology, it is reasonable to expect that the development of a greater understanding in nanoscale fluid behavior has the potential to underpin many other promising nanoscale technologies.

Multiscale Modeling of Particle Interactions: Applications in Biology and Nanotechnology,
Edited by Michael R. King and David J. Gee

Before going any further, it is worth pausing to define some of our terms; specifically, our use of the word or prefix *nano*. Building out from the initial sketch provided by Richard Feynman in his classic 1959 talk "There's Plenty of Room at the Bottom" [1], nanoscience and technology have developed into a large and varied domain ranging all the way from semiconducting devices, such as quantum dots, to the study of transport across biological membranes. In general, nanotechnology is understood to deal with systems with their important characteristic components under 100 nm, right down to single atoms at less than a single nanometer. Although it may appear pedantic, we believe that it is important to restrict use of the word *nano* to this range of length scales, where the behavior of the system can no longer obviously be looked at as a smooth continuum, at least a priori. At the nanoscale, the world is clearly made up of atoms.

Returning to nanodroplets in more specific terms, a study of their behavior and associated phenomena belongs to the subfield within fluid mechanics which over the past few years has acquired the name *nanofluidics*. The study of nanometric fluid systems is not new, but it is only recently that this type of work from a variety of fields has been bundled together under the nanofluidics banner. The name itself is a clear pointer to the existing trend of continued miniaturization of microfluidic applications, such as commercial lab-on-chip devices [2,3]. It is worth noting the potential for confusion with nanofluids, which is the terminology used to refer to colloidal suspensions of nanoparticles within a bulk liquid that can exhibit enhanced transport properties compared to the pure bulk fluid [4–8].

As the size of the system is reduced into the nanoscale, deviations from the well-understood models for phenomena in large macroscopic systems can emerge. These unexpected phenomena can be caused by a number of factors which gain increased importance as the size of the system is reduced into the nanoscale, such as surface forces, thermal fluctuations, the molecular nature of matter, and reduced dimensionality. The increased role of surface forces, for example, is most easily seen in the fact that surface forces scale as $O(L)$ with system dimensions while inertia scales as $O(L^3)$. As the system characteristic length is reduced many orders of magnitude from the scale of everyday life into the nanoscale, it is clear that surface forces will assume a much more significant role. Quantum effects can also have an important role in nanofluidics under certain conditions, such as very low temperature. We restrict our consideration here to systems that do not show significant long-range quantum effects (see Section 6.2.1). Despite all these factors, even when deviations from expected behavior are not observed, this remains an interesting finding, as it demonstrates the breadth of continuum mechanical models and provides evidence for the use of molecular dynamics in diverse applications.

6.2 MOLECULAR DYNAMICS SIMULATION

Molecular simulation can be broken down into two large families: *molecular dynamics*, which deterministically solves the mechanical equations of motion, and

Monte Carlo, which uses stochastic methods to explore the solution space. Both techniques are well used and mature, having numerous variants tailored to specific application areas. A complete description of either method alone would fill its own book. In this chapter we focus our discussion on providing an outline of the molecular dynamics method, which is used heavily in the study of dynamic time-dependent fluid phenomena.

Molecular dynamics (MD) simulation is one of the most common tools used for exploring the physics of materials at the atomic length scale. Since the pioneering work by Alder and Wainwright [9] on phase transitions of hard-sphere atomic systems in the 1950s, molecular dynamics has been used in many areas of science, including general statistical mechanics, fluid mechanics, material science, biochemistry, and molecular biology.

Molecular dynamics is well suited to problems driven by molecular or atomic phenomena because it can provide a detailed all-atom description of the system which allows for important phenomena such as thermal fluctuations to emerge naturally. Through the use of statistical thermodynamics, it is also possible to connect the atomistic behavior (i.e., positions and velocities) to the familiar thermodynamic parameters (ρ, T, P) and transport coefficients.

At the core of all MD simulations is the simultaneous numerical integration of the classical equations of motion for each atom in the system. A simplified block diagram for the overall molecular dynamics algorithm is shown in Fig. 6.1. In line with the high-level simplicity of the top-level MD algorithm, the implementation of MD simulation for a small number of atoms can be accomplished with relative ease. However, given that the size of systems of interest to fluid mechanics and materials science can range from 10^6 to 10^9 atoms [10–12], the efficient and accurate implementation of molecular dynamics simulation requires a sophisticated combination of physics, scientific computing, and computer science (hardware and software).

6.2.1 Quantum Atoms and Classical Assumptions

The molecular dynamics simulation technique is underpinned by two important assumptions:

1. Classical mechanics provides a good approximation to atomic trajectories.
2. Empirical potential-energy functions provide a good approximation of interatomic interaction.

Although atoms are fundamentally quantum mechanical in nature, under certain conditions atom–atom interactions do not require a full quantum mechanical description. For example, at high temperatures and low densities, where interatomic separation is much larger than the atomic diameter, atoms in a gas may be treated accurately as discrete, classical billiard ball–like objects. However, as the temperature is reduced, the classical treatment will no longer be appropriate and must be replaced with a quantum mechanical treatment where the atom is considered as a wave packet.

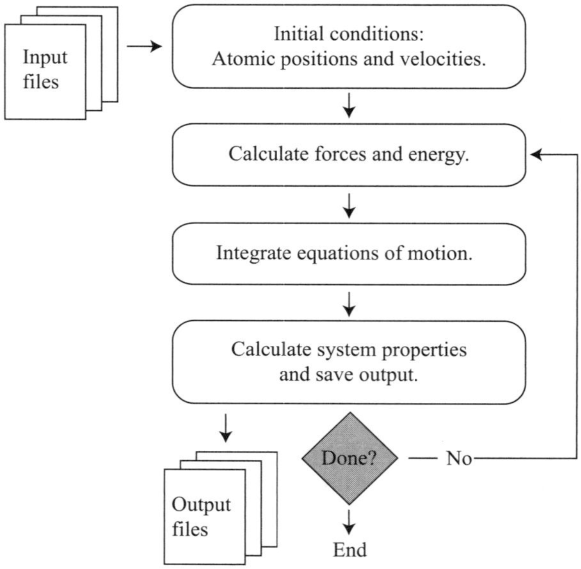

Figure 6.1 Molecular dynamics algorithm flowchart. Each block in the diagram represents a significant amount of algorithmic implementation and computation, which could be broken down into a number of smaller steps in a more detailed block diagram.

The spatial extent of the atomic wave packet can be approximated by the thermal wavelength, defined as

$$\lambda = \sqrt{\frac{2\pi\hbar^2}{mk_BT}}, \tag{6.1}$$

where $\hbar$ is the reduced Planck's constant, m is the mass, k_B is Boltzmann's constant, and T is the temperature. Using the thermal wavelength, it is possible to define a limit to the classical regime as the condition where $n\lambda = 1$, where n is the number density [13]. This limit defines a density–temperature curve

$$T = n^{2/3}\frac{2\pi\hbar^2}{mk_B} \tag{6.2}$$

at which the atomic spacing becomes approximately equal to the spatial extent of the atomic wave packet. Avoiding conditions near this limit is important to the accuracy of the classical treatment of atom trajectories but is relatively easily satisfied for most substances. For example, in the case of nitrogen at standard temperature and pressure, the limiting temperature is found to be less than 1 K.

In cases where the quantum mechanical nature of the system is important, such as low temperature or complex chemical interactions, molecular dynamics can be modified to include the necessary quantum mechanical calculations. These methods are generally referred to as *ab initio molecular dynamics methods*. One of the most important ab initio methods is the Car–Parrinello [14] method, which combines classical molecular dynamics ideas with density functional theory (DFT) to solve the electronic structure of the many-body system. As expected, the addition of quantum mechanical calculations can significantly increase the amount of computation necessary and restrict the size of tractable systems.

6.2.2 Equations of Motion

For the simplest case (i.e., a set of atoms treated as point masses), the equations of motion for the system are simply the atom-wise set of Newton's second law equations,

$$\mathbf{F}_i = m_i \frac{d^2\mathbf{r}_i}{dt^2}, \qquad i = 1, 2, \ldots, N. \tag{6.3}$$

This set of second-order linear ordinary differential equations defines a $3N$-dimensional initial value problem which can be solved numerically given the initial position and velocities of all atoms. Although this formulation of the equations of motion is intuitive, a deeper analysis can be obtained when the equations of motion are rewritten in the Hamiltonian formulation of classical mechanics as [15]

$$\dot{\mathbf{q}}_i = \frac{\partial \mathcal{H}}{\partial \mathbf{p}_i}, \qquad \dot{\mathbf{p}}_i = -\frac{\partial \mathcal{H}}{\partial \mathbf{q}_i}. \tag{6.4}$$

In this representation, $\mathcal{H}$ is the Hamiltonian of the system and $\mathbf{q}$ and $\mathbf{p}$ are the generalized position and conjugate momentum of the system within a now $6N$-dimensional phase space. The generalized position and momentum contain all the coordinates necessary to describe the system. In the simplest case of unbonded atoms, the generalized coordinates are simply the Cartesian position ($\mathbf{q} = [\mathbf{r}_1, \ldots, \mathbf{r}_N]$) and linear momentum ($\mathbf{p} = [\mathbf{p}_1, \ldots, \mathbf{p}_N]$). In the case of more complicated molecules, additional degrees of freedom such as bond stretching, bond rotation, and so on, will also be included in the generalized set of coordinates.

The Hamiltonian is, in general, a function of the generalized position, conjugate momentum, and time [i.e., $\mathcal{H} = \mathcal{H}(\mathbf{q}, \mathbf{p}, t)$]. In the case of a simple atomic system outlined here, the Hamiltonian does not have an explicit dependence on time and is equal to the total energy of the system expressed as

$$\mathcal{H}(\mathbf{p}, \mathbf{q}) = \mathcal{U}(\mathbf{q}) + \sum_{i=i}^{3N} \frac{p_i^2}{2m_i}, \tag{6.5}$$

where $\mathcal{U}$ is the total potential energy of the system. It can also be shown that under these conditions the Hamiltonian is a constant of motion for the system by taking the total time derivative of the Hamiltonian and utilizing the equations of motion,

$$\frac{d\mathcal{H}}{dt} = \sum_{i=1}^{3N} \left(\frac{\partial \mathcal{H}}{\partial \mathbf{r}_i} \dot{\mathbf{r}}_i + \frac{\partial \mathcal{H}}{\partial \mathbf{p}_i} \dot{\mathbf{p}}_i \right) = \sum_{i=1}^{3N} \left(\frac{\partial \mathcal{H}}{\partial \mathbf{r}_i} \frac{\partial \mathcal{H}}{\partial \mathbf{p}_i} + \frac{\partial \mathcal{H}}{\partial \mathbf{p}_i} \left(- \frac{\partial \mathcal{H}}{\partial \mathbf{r}_i} \right) \right) = 0.$$

Therefore, with a constant Hamiltonian the total system energy is also constant and the system evolution samples the constant NVE *microcanonical* statistical mechanical ensemble (i.e., constant $N \rightarrow$ number of atoms, $V \rightarrow$ volume, $E \rightarrow$ energy). Other statistical mechanical ensembles with different constants of motion are discussed in Section 6.2.5. The ability to identify constants of motion is one of the great advantages of using the Hamiltonian description of the system and assists in making linkages to system thermodynamics.

6.2.3 Interatomic Potentials

In any molecular dynamics simulation, the calculation of the force on each atom accounts for the vast majority of the total computational effort. Therefore, in addition to the accuracy of the potential in describing the atomic interaction, the speed and efficiency of the potential function are absolutely crucial to the success of molecular dynamics simulations. This fact is in keeping with the whimsical maxim that "as machines become more powerful, the efficiency of algorithms grows more important, not less" [16]. Despite the vast increases in computing power available, accurate, yet simple empirical potentials remain at the heart of molecular simulation.

Many families of empirical potential functions exist from which it is possible to calculate the interatomic force by $\mathbf{F}_i = -\nabla \mathcal{U}_i$. These empirical potentials are parameterized based on regressions performed from both experimental data and quantum mechanical calculations so that they describe accurately the interatomic interactions and system thermodynamics over a broad range of conditions.

The total potential energy can be described as the sum of a number of potential energies resulting from bonded (e.g., bond vibration, bending, rotation) and nonbonded interactions (e.g., long-range electrostatic and dispersion forces). In the strictest sense, the nonbonded potential energy of an atom is a function of all its neighbors, which can create a computationally expensive many-body function. It is, however, possible to achieve good results by reducing the complexity of atomic interaction to effective pairwise potential functions.

One of the oldest and still most important pairwise potentials used in molecular simulation is the 12-6 Lennard-Jones (LJ) potential,

$$u_{\mathrm{LJ}}(r_{ij}) = 4\epsilon \left[\left(\frac{r_{ij}}{\sigma} \right)^{-12} - \left(\frac{r_{ij}}{\sigma} \right)^{-6} \right], \tag{6.6}$$

where σ represents the distance where the potential is zero and ϵ is the depth of the potential-energy well. It was introduced to MD simulation by Rahman [17] and is

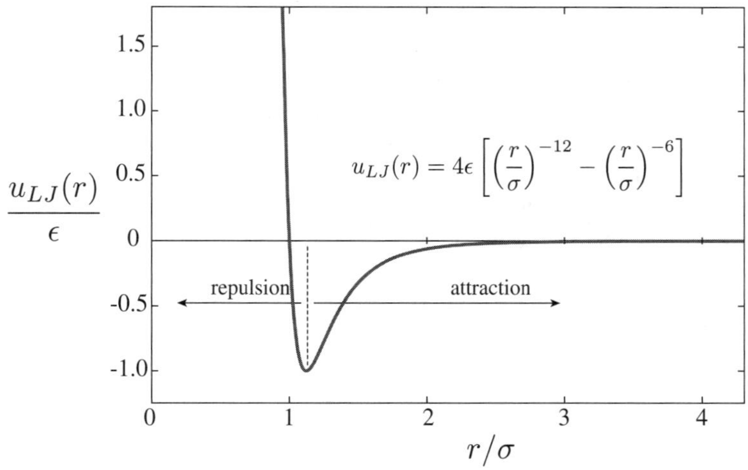

Figure 6.2 12-6 Lennard-Jones potential function. The shape of the Lennard-Jones potential is characteristic of most interatomic potentials for dispersion forces featuring an inner region of near hard-core repulsion and a long attractive tail.

shown graphically in Fig. 6.2. The Lennard-Jones potential remains popular due to its simplicity and accuracy in describing the dispersion force interaction between spherical, nonpolar molecules. It is, therefore, useful in providing a full description of simple spherically symmetrical atoms, such as the noble gases, and in acting as a component of a larger molecular model for more complex molecules such as water or biological macromolecules. The Lennard-Jones equation of state is well reviewed and parameterized by Johnson et al. [18,19]. Due to its frequent use, it is also common to see the results of MD simulation in nondimensional form, where the characteristic units are taken from the LJ potential to be σ, ϵ, and $\tau(=\sqrt{m\sigma^2/\epsilon})$ as the characteristic length, energy, and time. A set of LJ parameters for the noble gases is given in Table 6.1.

Given the huge proportion of the computational cost associated with the calculation of interatomic forces, potential functions are truncated at some cutoff radius r_c rather than continuing to calculate attractive interaction of the long tail as $r \to \infty$. Truncation serves to reduce the number of pairwise interactions acting on each atom and therefore can decrease the computational complexity of force evaluation from

Table 6.1 Lennard-Jones Parameters σ and ϵ for Noble Gases (Except Helium)

	σ (Å)	ϵ (J)	ϵ/k_b
Ne	2.74	0.50×10^{-21}	36.2
Ar	3.40	1.67×10^{-21}	121
Kr	3.65	2.25×10^{-21}	163
Xe	3.98	3.20×10^{-21}	232

Source: [20].

$O(N^2)$ to $O(N)$ when coupled with additional algorithms that track each atom's nearest neighbors [21]. Great care must be taken, however, when choosing the cutoff radius to avoid causing excessive changes in the system dynamics. In a study of the effect of potential truncation on the liquid–vapor interface, Trokhymchuk and Alejandre [22] showed important changes in simulated interfacial properties due to changes in the cutoff radius. They report that a cutoff radius of $r_c = 5.5\sigma$ is necessary to reproduce the results of the full LJ potential. This is larger than the cutoff radius of 2.5σ, which operated as a de facto standard throughout early molecular simulation works.

In addition to dispersion forces, electrostatic forces are another important type of nonbonded interaction. The electrostatic potential is given by the familiar coulombic equation,

$$u_{\mathrm{Elec}}(r_{ij}) = C\frac{q_i q_j}{r_{ij}}, \tag{6.7}$$

where q is charge and C is Coulomb's constant. Electrostatic potentials decay slowly at only $O(1/r)$, compared to short-range dispersion forces, which decay at $O(1/r^6)$, increasing the associated computational cost by greatly augmenting the number of pairwise interactions acting on each atom.

To simulate a complex molecule, it is necessary to use a number of different potential models to simulate each of the different interactions. A good example of this is provided by considering water, which is typically modeled as the combination of coulombic interactions and Lennard-Jones interactions. Given the uniqueness and importance of water, a huge amount of research has gone into the development of many competing models, starting with the early work by Bernal and Fowler [23].

In molecular dynamics simulation for fluid mechanics, the most common water models in use today date to the 1980s. In these models, the atoms are rigidly bonded (i.e., no bond vibration or rotation) but differ in the number of interaction sites and, correspondingly, in the parameterization. The simple point charge (SPC) model [24] features three charge sites located on the oxygen and two hydrogen atoms in addition to Lennard-Jones interactions between the oxygen atoms only. The parameterization for SPC is given in Table 6.2.

Another common family of water models includes the TIP3P and TIP4P models [25]. As their names suggest, these offer three- and four-site models where in the case of TIP4P, an additional charge is placed along the bisector of ∠HOH. A diagram of the TIP4P model is shown in Fig. 6.3, and the parameterization of the model is provided in Table 6.2. More recently, five-site TIP5P [26] and six-site TIP6P [27] models have also been developed. A recent review of the field has been presented by Guillot [28].

Another important type of material is metals, but due to the nature of metallic bonding they are not well modeled by pairwise potentials. Metallic bonding is produced by a sea of shared electrons which cannot be well modeled by simple pairwise potentials. One of the most successful families is of potentials in use for metals and alloys is the embedded-atom model (EAM), proposed by Daw and Baskes [29,30].

Table 6.2 SPC and TIP4P Water Model Parameters

	r_{OH} (Å)	∠HOH (deg)	σ_{OO} (Å)	ϵ_{OO}/k_B	r_{OM} (Å)	q_H (e)[a]	$q_{O\|M}$ (e)[a]
SPC	1.00	109.47	3.166	0.1553	—	0.410	−0.820
TIP4P	0.9572	104.52	3.1536	0.1550	0.15	0.520	−1.040

Source: [24,25].

[a]Charge of electron, e = 1.60219 × 10^{-19}C.

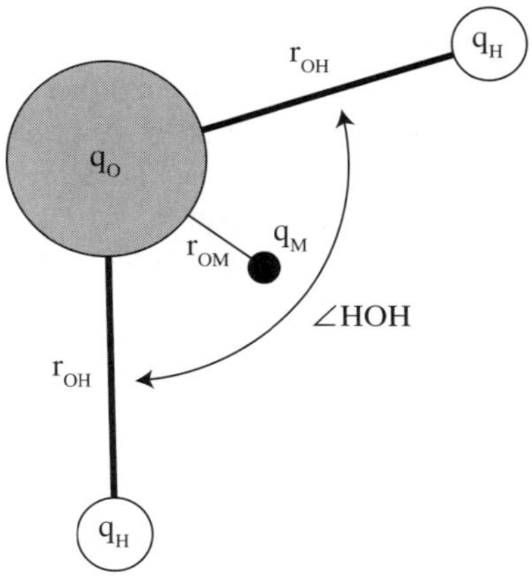

Figure 6.3 TIP4P water model showing the distribution of charges between the hydrogen atoms, oxygen atom, and dummy site located on the bisector of angle HOH.

The EAM potential retains a pairwise component which is supplemented by a further nonpairwise term such that

$$\mathcal{U}_{\mathrm{EAM},i} = \frac{1}{2}\sum_{j,i\neq j}^{N} u(r_{ij}) + F\left(\sum_{i\neq j} \rho(r_{ij})\right), \tag{6.8}$$

where $u(r_{ij})$ is the pairwise potential between atoms i and j and F is the embedding energy of atom i in the electron density ρ contributed by neighboring atoms.

6.2.4 Integrators

As we have emphasized repeatedly, the major computational bottleneck in molecular dynamics simulation is the calculation of interatomic forces and energy. Therefore, the actual speed of the integrator is not the most important criterion in choosing this component of the MD algorithm. Instead, the accuracy and ease of implementation tend to be more important deciding factors in the choice of integration methods.

The method of integration does, however, have an indirect impact on the speed of the simulation by setting the length of the time step, which, in turn, determines the number of time steps required to produce a given amount of simulation time. At each time step, a huge amount of computation takes place in the force calculation and, therefore, it is advantageous to choose as large a time step as possible while still resolving the fastest-moving degree of freedom in the system and maintaining stability of the simulation. Stability is manifested by good steady behavior in the constants of motion (i.e., total energy for simulations in the microcanonical ensemble). This effect is important because, in rough terms, the time step in MD simulations can be on the order of 1×10^{-14} to 1×10^{-15} s. Hundreds of thousands, if not millions, of time steps can be required to accumulate a significant amount of simulation time. This fact exposes the inherent difficulty in incorporating information derived from molecular dynamics simulation into multiscale simulation frameworks. Statistical sampling of MD results for material properties or boundary conditions, rather than a direct continuous coupling between scales, is commonly used in an attempt to negotiate the problematic separation of time scales.

The most common integrator used for molecular dynamics is the Verlet scheme [31]. There are many variants of the Verlet algorithm, but the most commonly used form in MD simulation is the velocity-Verlet method [32,33],

$$\mathbf{r}(t+\delta t) = \mathbf{r}(t) + \mathbf{v}(t)\,\delta t + \frac{1}{2}\mathbf{a}(t)\,\delta t^2 \tag{6.9}$$

$$\mathbf{v}(t+\delta t) = \mathbf{v}(t) + \frac{1}{2}\delta t[\mathbf{a}(t) + \mathbf{a}(t+\delta t)]. \tag{6.10}$$

Like all Verlet integration schemes, the velocity-Verlet integrator provides good accuracy with local error $O(\delta t^4)$ in position and $O(\delta t^2)$ in velocity. The velocity-Verlet, in particular, is preferred compared to other implementations in the Verlet family of integrators because it produces the updated atom positions and velocities at coincident times (i.e., $t+\delta t$) which is convenient for postprocessing.

Another commonly used integration scheme is the Gear predictor–corrector integrator, which can offer improved accuracy or longer time steps than to the velocity-Verlet scheme, at the expense of a more complex algorithm [34,35]. In the prediction step, the predicted position, velocity, and acceleration are generated using a Taylor series expansion,

$$\begin{aligned}
\mathbf{r}^P(1+\delta t) &= \mathbf{r}(t)+(\delta t)\mathbf{v}(t)+1/2(\delta t)^2\mathbf{a}(t)+1/6(\delta t)^3\mathbf{b}(t)+1/24(\delta t)^4\mathbf{c}(t)+O(\delta t^5)\\
\mathbf{v}^P(t+\delta t) &= \mathbf{v}(t)+(\delta t)\mathbf{a}(t)+1/2(\delta t)^2\mathbf{b}(t)+1/6(\delta t)^3\mathbf{c}(t)+O(\delta t^4)\\
\mathbf{a}^P(t+\delta t) &= \mathbf{a}(t)+(\delta t)\mathbf{b}(t)+1/2(\delta t)^2\mathbf{c}(t)O(\delta t^3)\\
\mathbf{b}^P(t+\delta t) &= \mathbf{b}(t)+(\delta t)\mathbf{c}(t)+O(\delta t^2)\\
\mathbf{c}^P(t+\delta t) &= \mathbf{c}(t)+O(\delta t).
\end{aligned} \tag{6.11}$$

The values predicted are then corrected using an updated value for the acceleration, $\mathbf{a}^c$, calculated from the interatomic forces at the position predicted $\mathbf{r}(t+\delta t)$,

$$\begin{aligned}
\mathbf{r}^C(t+\delta t) &= \mathbf{r}^P(t+\delta t)+c_0\,\Delta\mathbf{a}\\
\mathbf{v}^C(t+\delta t) &= \mathbf{v}^P(t+\delta t)+c_1\,\Delta\mathbf{a}\\
\mathbf{a}^C(t+\delta t) &= \mathbf{a}^P(t+\delta t)+c_2\,\Delta\mathbf{a}\\
\mathbf{b}^C(t+\delta t) &= \mathbf{b}^P(t+\delta t)+c_3\,\Delta\mathbf{a}\\
\mathbf{c}^C(t+\delta t) &= \mathbf{c}^P(t+\delta t)+c_4\,\Delta\mathbf{a},
\end{aligned} \tag{6.12}$$

where $\Delta\mathbf{a}=\mathbf{a}^C-\mathbf{a}^P$. The values for the coefficients depend on the order of the predictor–corrector method selected. The most commonly used predictor–corrector method in MD simulation is the five-term scheme with $c_0=19/20$, $c_1=3/4$, $c_2=1$, $c_3=1/2$, $c_4=1/12$. Although multiple correction steps are used in some applications, due to the computational expense of the force calculation required, a single correction step is typical in molecular dynamics.

In systems with widely varying time scales (i.e., high frequency bonded interaction), it can be advantageous to use different timesteps for the various length scales that make up the system. One such integration scheme is the reversible reference system propagator algorithm (rRESPA) popularized by Tuckerman et al. [36], which is available in many MD software packages.

6.2.5 Thermostats

Although the constant-energy microcanonical ensemble is the most natural statistical mechanical ensemble, it is much more common in experiments to have a system at constant temperature or constant pressure than at constant energy.

The simplest form of temperature control is temperature rescaling. At each time step, or at a specified frequency, the velocity of each atom is scaled according to a ratio of the instantaneous system temperature to the desired temperature according to

$$\mathbf{v}_i^{\text{scaled}} = \mathbf{v}_i\sqrt{\frac{T_0}{T}}. \tag{6.13}$$

Temperature rescaling is a successful way of controlling temperature, but it does not result in a phase-space trajectory that samples a well-formed statistical mechanical ensemble. Further, due to the artificial nature of the temperature rescaling intervention, unexpected consequences can result, such as improper redistribution of thermal kinetic energy into translational kinetic energy of the center of mass [37].

To produce a system that samples ensembles with constant temperature or pressure, the system's equations of motion must be modified. The Nosé–Hoover [38,39] thermostat achieves constant temperature by extending the system and adding an additional degree of freedom corresponding to a heat bath. Under these modifications, the equations of motion are now

$$\begin{aligned} \dot{\mathbf{r}}_i &= \frac{\mathbf{p}_i}{m_i} \\ \dot{\mathbf{p}}_i &= \mathbf{F}_i - \xi \mathbf{p}_i \\ \dot{\xi} &= \frac{1}{Q}\left(\sum_i^N \frac{p_i^2}{m_i} - 3Nk_B T_0\right), \end{aligned} \tag{6.14}$$

where Q is the mass of the thermal bath and T_0 is the desired temperature. These modified equations of motion can be shown to sample the *NVT* or *canonical ensemble* (i.e., constant $N \rightarrow$ number of atoms, $V \rightarrow$ volume, $E \rightarrow$ energy). In addition to the Nosé–Hoover thermostat, other well-known thermostats include the Berendsen [40] and Andersen thermostats [41]. Nosé–Hoover thermostatting remains popular because the Berendsen thermostat will not sample the canonical ensemble, while the Andersen thermostat can introduce nonphysical artifacts into transport phenomena.

6.2.6 Initialization

Initializing molecular dynamics simulations is a matter of providing the initial atom positions and velocities. For a liquid, it is tempting initially to consider placing atoms at random. Random placement of atoms, on its own, is unlikely to be successful because it can easily create atom pairs which are very close together, yielding unphysically high potential energies and huge repulsive forces. Therefore, it is more typical to initialize atoms on a crystalline lattice such as the face-centered-cubic (FCC) crystal structure. The initial atom velocities are assigned randomly by sampling from a Gaussian distribution for each component of the velocity vector such that, overall, the system achieves the desired initial temperature.

Once the simulation is started, the atoms will move off their initial lattice positions as the system quickly *melts* and moves toward thermodynamic equilibrium. During this equilibration phase, it is possible to observe significant fluctuations in measures of instantaneous thermodynamic parameters like temperature, which decay with time until they become statistically steady. This process can be viewed as part of the system preparation and it is imperative that sufficient time be allowed for complete equilibration before proceeding with the actual study.

6.2.7 Parallel Computing Implementation

In the practical implementation of molecular dynamics, parallel computing is almost always used due to the volume of computation required by MD simulation. As discussed in Section 6.2.3, the computational intensity of MD is caused primarily by the huge number of interatomic force calculations required at each time step. Molecular dynamics is well suited to parallel computation, as at each time step, computations for each molecule are independent from all others and thus can be performed simultaneously.

The goal of parallelization is to split up the computational work between a number of processors. In the case of molecular dynamics, this is accomplished principally by splitting up the force computation step of the algorithm. Three different approaches to parallelizing MD simulations with short-range forces are outlined by Plimpton [42]:

1. *Atom decomposition:* Subsets of atoms are assigned to each processor, and forces on these atoms are computed by that processor.
2. *Force decomposition:* Subsets of individual force calculations, rather than atoms, are assigned to each processor regardless of atom position.
3. *Spatial Decomposition:* Atoms within a subdomain of the simulation domain are assigned to a processor and the forces on these atoms are computed by the processor. An illustration of the spatial decomposition of a simulation domain into four rectangular cuboid subdomains is shown in Fig. 6.4.

The success of any parallelization scheme is dependent on the length scale of the interatomic interactions and the distribution of atoms in the system as a function of time. If the spatial distribution of atoms changes dramatically during the simulation, it may be advantageous to rebalance the load between processors at some interval. In any parallelization scheme there will always be some computational cost added due to the time spent in interprocessor communication and in allocating atoms to each processor.

There are many established and well-used software packages available for molecular dynamics simulation. Most packages are academic research codes and,

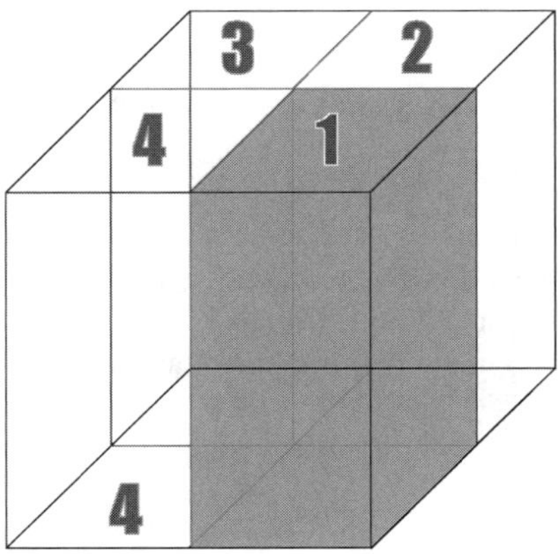

Figure 6.4 Sample spatial decomposition of a cubic simulation domain into four rectangular cuboid subdomains. The arrangement of subdomains is arbitrary in this case. In general, subdomains should be constructed such that they provide the best distribution of computational load across available resources while minimizing communication time and while, of course, spanning the computational domain. The nearest subdomain is shaded for visual purposes.

as such, are often free and open source. A short list of a few of the most used general-purpose molecular dynamics research codes is:

DL_POLY	http://www.cse.scitech.ac.uk/ccg/software/DL_POLY/
GROMACS	http://www.gromacs.org
LAMMPS	http://lammps.sandia.gov
NAMD	http://www.ks.uiuc.edu/Research/namd

Although each software package differs in implementation details, all use message-passing techniques (e.g., MPI) across a distributed memory computing cluster. The computing cluster hardware may be as simple as a handful of desktop workstations linked through standard networking to many hundreds or thousands of specifically designed high-performance computing (HPC) nodes linked through a specialized low-latency interconnect. A new trend in computing hardware which has the potential to have a great impact on molecular simulation is the use of highly powerful graphics processing units (GPUs) to perform calculations traditionally assigned to the central processing unit (CPU).

6.2.8 Reference Texts

Four excellent and highly cited reference texts on the molecular dynamics simulation technique are:

Allen MP, Tildesley DJ. *Computer Simulation of Liquids*, Clarendon Press, 1987.

Haile JM. *Molecular Dynamics Simulations: Elementary Methods*, Wiley-Blackwell, 1997.

Frenkel D, Smit B. *Understanding Molecular Simulation: From Algorithms to Applications*, Academic Press, 2001.

Rapaport DC. *The Art of Molecular Dynamics Simulation*, Cambridge University Press, 2004.

Each of these texts contains the core statistical mechanics and computing algorithms that underpin molecular dynamics simulation while also providing its own choice of emphasis. For example, Rapaport's monograph is unique in its greater focus on algorithm development and implementation and therefore features significant amounts of pseudocode.

6.3 NANODROPLETS

Although nanodroplets may not be directly observable in everyday experience, they are in fact the product or by-product of many natural or industrial processes and are, as such, almost omnipresent. In the natural world, nanodroplet aerosols with a diameter as small as 100 nm are known to be produced in ocean spray and play a role in the regulation of weather and climate [43]. In industrial processes, nanodroplets are

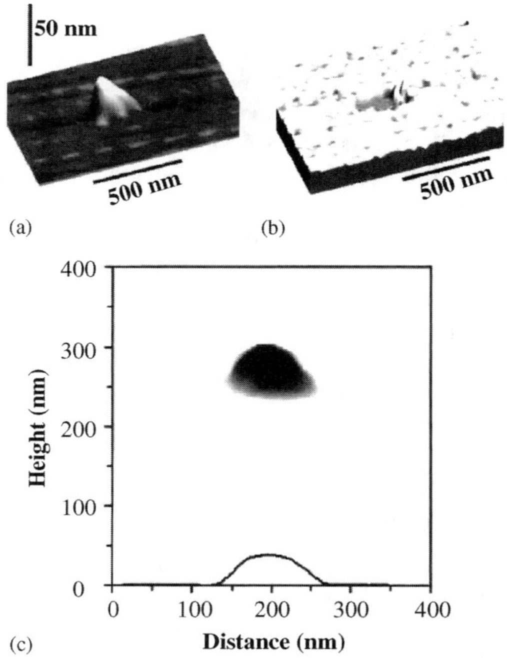

Figure 6.5 Tapping-mode atomic force microscopy of a nanodroplet formed on the surface of a hard disk platter: (a) distorted surface image; (b) phase image; (c) rotated surface image and droplet profile perpendicular to scan direction. (From [45], with permission. Copyright © 2004, Taylor & Francis Group.)

readily produced by rapidly expanding flows such as supersonic nozzles [44] and can also be found in unexpected places such as on the surface of computer hard disks, where they are unwanted contaminants that can cause decreased disk performance [45]. A tapping mode atomic force microscopy (AFM) image of a nanodroplet found on the surface of a hard disk is shown in Fig. 6.5.

In addition to providing a greater understanding of existing natural and industrial processes, the investigation of nanodroplet phenomena is also of great interest, due to their potential for use in a wide array of new and novel technological applications. As is the case for nanofluidics in general, interest in nanodroplets stems out of the larger trend toward device miniaturization across a huge variety of technologies. Device miniaturization has great advantages in many fields, such as high-throughput chemical or biological testing, where material is either scarce or expensive. Nanodroplet processes are also seen as having potential in the fabrication of new and novel materials. Some of these are thought to have application in medicine and biotechnology as nanometric drug-delivery vectors [46–48]. In semiconductor manufacturing, nanodroplet technology is already in place for cleaning away stubborn debris from in and around submicronal feature sizes between processing stages.

In the following section we discuss a number of important nanodroplet phenomena, such as evaporation, condensation, collision, and impact onto solid or liquid surfaces. Due to limitations in the total number of atoms that can be simulated by MD, it has been and remains common to treat a single or a limited number of droplets. This is not usually a significant limitation on the utility of MD simulation, as the strength of the technique is in resolving processes at the local molecular scale rather than at larger global scales. Information gained from MD simulation of a single droplet (i.e., fluid properties, boundary conditions, etc.) can then be integrated into larger-scaled models to treat larger, more complex phenomena such as aerosols or sprays.

For the nanodroplet processes treated here, we highlight key findings while illustrating the progress made recently in treating nanodroplet systems using molecular dynamics simulation. Simulation remains one of the primary options for the study of nanodroplets, due to the difficulty in performing experiments on fluid systems of this size, particularly with free surfaces. For example, small-angle neutron scattering (SANS) techniques and atomic force microscopy (AFM) techniques have been able to observe and characterize nanodroplets, but remain unable to study detailed dynamic nanodroplet phenomena [49,50].

6.3.1 Droplet Definition and Basic Properties

When discussing nanodroplets, it is important to make a distinction between a droplet and a simple cluster of atoms. Clusters are a small and typically temporary collection of loosely interacting molecules. Very small clusters, although not droplets, do, however, play important roles in many nanodroplet processes, such as condensation nucleation (see Section 6.3.2).

Once a cluster is large enough, there will be a region of *true liquid* at the center. Early results by Thompson et al. [51] show that depending on the temperature of the system, it is possible to achieve full liquid density at the core of the nanodroplet for even as few as several hundred atoms. A reproduction of the plot of reduced density (ρ^*) from their study is shown in Fig. 6.6, where it is possible to see that at 85 K ($T^* = 0.71$), near the triplepoint, a flat reduced density profile is achieved at the center of the droplet for droplets with as few as 250 or even 138 atoms. At higher temperatures, this threshold shifts upward toward larger droplets as the greater thermal motion at higher temperature serves to broaden the interface.

Unlike macroscale droplets, where the liquid–vapor interface is of negligible thickness compared to the droplet diameter, at the nanoscale the interface has a significant spatial extent which cannot be ignored. As seen from the density profile plots produced by Thompson et al. [51], even identifying the size of a nanodroplet is no longer trivial. At equilibrium, interface density profiles can be fitted with either a hyperbolic tangent formula or an error function formula,

$$\rho(r) = \frac{1}{2}(\rho_l + \rho_v) - \frac{1}{2}(\rho_l - \rho_v)\tanh\left(2\frac{r - R_O}{D}\right), \tag{6.15}$$

$$\rho(r) = \frac{1}{2}(\rho_l + \rho_v) - \frac{1}{2}(\rho_l - \rho_v)\mathrm{erf}\left(\sqrt{\pi}\frac{r - R_O}{D}\right), \tag{6.16}$$

where D is the interface thickness and R_O is the droplet radius. Although the hyperbolic tangent formula has been used traditionally, Sides et al. [52] recommend

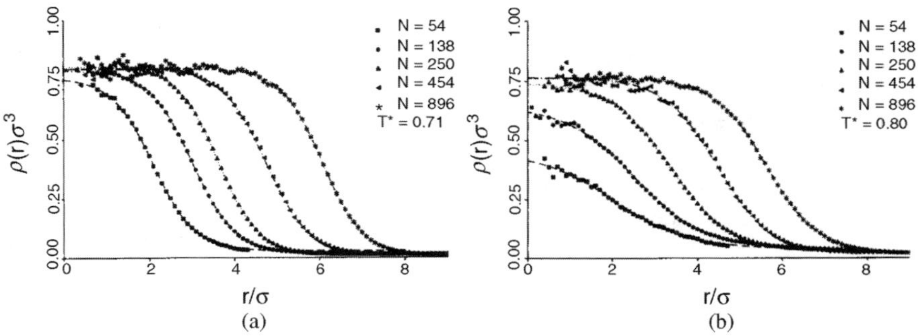

Figure 6.6 Density profiles for nanodroplets $N = 54$, 138, 250, 454, and 896 argon atoms: (a) $T^* = 0.71$; (b) $T^* = 0.80$. (From [51], with permission. Copyright © 1984, American Institute of Physics.)

using the error function to improve the accuracy of the surface thickness and tension predicted.

The case of an isolated droplet provides a good opportunity to observe the increasing complexity and sophistication in MD simulation that has developed to treat nanodroplet problems. Perhaps the earliest simulation of a droplet was performed by Rusanov ans Brodskaya in 1977 [53]. In this study, the authors were only able to treat droplets with a maximum of 256 atoms; only seven years later, Thompson et al. [51] were able to simulate systems of 2048 atoms. Although this increase is impressive, modern studies of nanodroplets often treat systems with hundreds of thousands or millions of atoms but, more importantly, systems exhibiting more complex and sophisticated dynamics. For example, Arcidiacono et al. [54] investigated the oscillation of a 330,000-atom, or approximately 30-nm-diameter nanodroplet, to find good agreement with both experiment and continuum models regarding surface tension. Droplet oscillation is an extremely common phenomenon, due to the constant interaction of droplets with their surroundings. For example, pendent drops dripping out of a faucet will undoubtably oscillate, due to stretching at the moment of detachment, and if this droplet then splashes down into a liquid pool, it has been found that the state of this oscillation at the moment of impact has an influence on the dynamics of impact [55]. Along a similar line, Moseler and Landman [56] have shown that instability in a nanojet will produce a stream of oscillating nanodroplets.

6.3.2 Evaporation and Condensation

Evaporation and condensation processes have been the subject of numerous MD studies. Beyond explicit studies of evaporation or condensation, it is worth noting that in the setup of most molecular dynamics simulations, these processes are more of a background fact (and potentially, afterthought) than the topic of the study. For example, when a liquid slab is initialized within a large, empty (i.e., low initial density) simulation cell, an evaporation process will occur as the system seeks thermodynamic equilibrium.

The earliest example of an explicit study of nanodroplet evaporation is that of Long et al. [57]. In that study, a relatively small system of 2048 Lennard-Jones atoms was used to study the complete evaporation of a nanodroplet with heating supplied from the surrounding vapor, and good agreement was found with the Knudsen aerosol theory for the evaporation rate. The simulations were limited, however, to a single system with $T = 120$ K and $P = 0.4$ MPa, which clearly limits the scope over which we may utilize their conclusions.

Walther and Koumoutsakos [58] simulate a subcritical nanodroplet at 300 K and 3 MPa. Although this study also treats a single case of system temperature and pressure, the increased computing resources available allowed for examination of the convergence of the method under the effect of changing potential cutoff, time-step size, simulation cell size, and heating frequency. Moreover, they find that with increasing droplet size from 8 nm (5768 atoms) to 17 nm (51,105 atoms), the discrepancy of the simulated evaporation coefficient with theoretical predictions can be reduced from 35% to 5%. This result makes a strong case for the utility of MD simulation for use in systems featuring evaporation, but also for the scalability of the continuum theory down to very small systems.

Consolini et al. [59] simulate a 6.88-nm (2000-atoms) xenon droplet evaporating in a nitrogen gas under a range of subcritical and supercritical conditions. In keeping with previous results, they find good qualitative agreement between the nanodroplet evaporation observed and expectations from continuum models or macroscale experiments. They are also able to show significant differences between the behavior of the system at subcritical and supercritical conditions. Subcritical droplets first undergo a heating period and then finish evaporating at approximately constant temperature, while supercritical droplets experience a constant increase in temperature. The droplet shape is also profoundly effected by the criticality of the system, as subcritical droplets initialized as ellipses are found to return quickly to a spherical shape, whereas supercritical droplets never attain a spherical shape. The time evolution of an initially elliptical droplet is shown in Fig. 6.7.

When considering condensation processes, a key difference is that condensation, unlike evaporation, is an activated process requiring the formation of small clusters of a critical size that act as the site of nucleation for the condensing droplet. The formation of clusters in a vapor is driven by thermal fluctuations, making it a problem that is ideally treated by molecular dynamics simulation in preference to continuum modeling. Wedekind and Regura [60] present simulation results which investigate different existing models for identifying the size of clusters. In this work they find that although the results from each cluster definition differ, all are successful in identifying the largest cluster correctly and therefore the most likely site for nucleation. Interestingly, they also show that assuming nucleation with a high energy barrier, it is easy to make use of the largest cluster size information taken from either definition in further modeling of nucleation without causing significant changes in the resulting nucleation rate. Wedekind et al. [61,62] have also recently presented further molecular dynamics work investigating the effects of thermostats and finite simulation cell size in the implementation of MD simulation, resulting in a systematic investigation of argon nucleation [63].

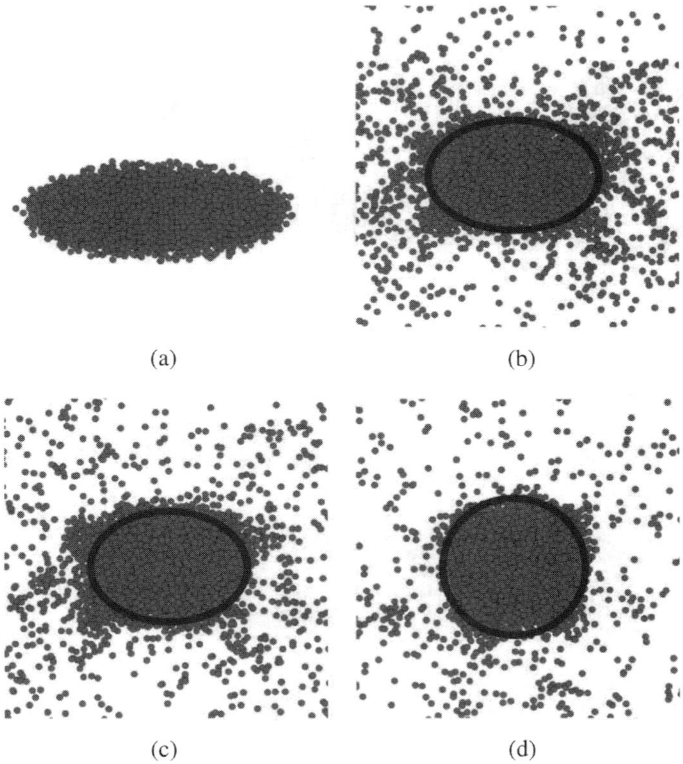

Figure 6.7 Evaporation of Xenon nanodroplet from nonspherical initial state under subcritical conditions: (a) $t = 0$ ps; (b) 10 ps; (c) 16 ps; (d) 36 ps. Surrounding noncondensing nitrogen atoms not shown. (From [59], with permission. Copyright © 2003, Elsevier.)

6.3.3 Collision, Impact, and Wetting

Collision of droplets with each other and droplet impact with solid or liquid surfaces exhibit many similarly complex and fast-moving dynamic phenomena. In both cases, much of the behavior can be understood through the balance between surface and inertial forces in the droplet, represented by the Weber number, defined as

$$We = \frac{\rho v^2 D}{\gamma}, \tag{6.17}$$

where ρ is the liquid density, v is the relative droplet speed, D is the droplet diameter, and γ is the liquid's surface tension.

An influential early work in both collision and impact on a surface is that of Koplik and Banavar [64], which presents simulation of the coalescence of droplets in a shear flow and the impact of a liquid cylinder with a planar liquid at a low Weber number. Snapshots from these simulations are shown in Fig. 6.8. In both cases, the critical

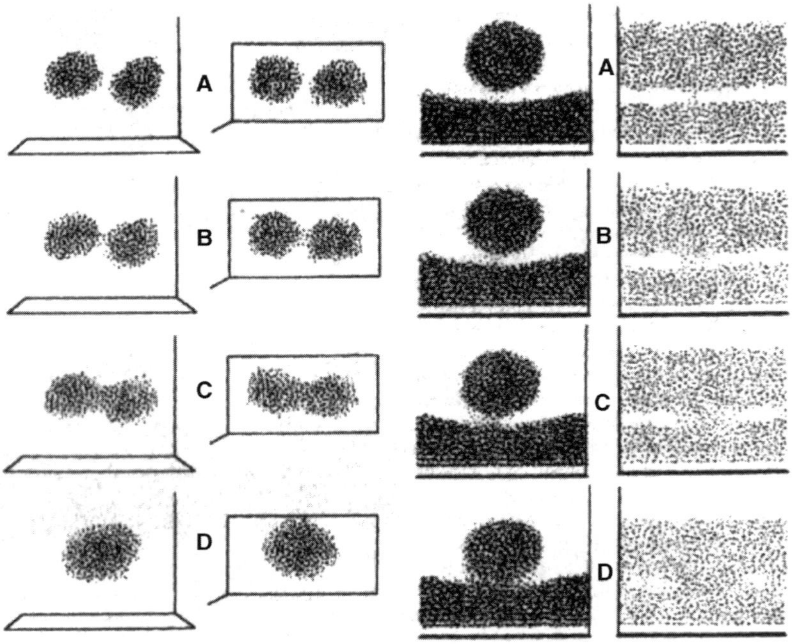

(a) (b)

Figure 6.8 (a) Coalescence of nanodrops under shear flow; (b) liquid cylinder into planar liquid surface flow shown from the front and side. Immiscisble background fluid is not shown. (From [64], with permission of AAAS.)

feature to note is the development of a liquid bridge between the liquid masses, which is central to achieving eventual coalescence. This bridging is produced as the two liquid masses are brought together and then, due to thermal fluctuations, some atoms from each move farther out into the remaining separation. At this point, the gap will be small enough that these atoms can have a strong attractive interaction which serves to draw more atoms into the gap between the liquid masses. This process is shown in Fig. 6.9. Once the bridge begins to form as such, the result is a cascade where the droplet atoms attract one another. Since it is now energetically favorable for continued and increased flow to occur, coalescence results. The decision by Koplik and Banavar to use a cylinder rather than a spherical droplet for studying the impact onto a liquid layer helps to further emphasize the thermally driven nature of the bridging process. In another realization of this simulation, the details of the thermal fluctuations will no doubt change and the location of bridging is likely to be elsewhere along the cylinder. In effect, the available bridging loci on the cylinder fall along a line, whereas for droplets, bridging will be localized around the closest point of approach.

As the mirror image of the droplet collision process, Koplik and Banavar [65] also simulate the process of droplet fission under the action of a strong shear. The evolution

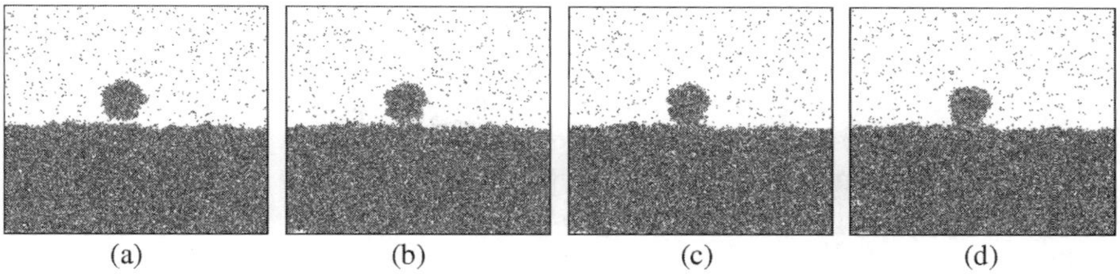

Figure 6.9 Impact of a 5000-atom Lennard-Jones argon nanodroplet onto a liquid surface with $v = 18\,\mathrm{m/s}$. The gap between the nanodroplet and the liquid surface is closed by the formation of a molecular bridge in frames (b) and (c).

of the shape of the liquid–vapor interface here bears qualitative resemblance to that seen in Fig. 6.8 for coalescence under shear, and to the case of the stretching-separation collision type (see Fig. 6.10).

Early work looking at the range of droplet collision behavior, given different impact conditions, was performed by Greenspan and Heath [66] using droplets of 2051 coarse-grained water molecules (see Section 6.4). Through variation in the magnitude and direction of the relative velocity of the colliding droplets, they were able to reproduce a number of previously observed types of collision with varying degrees of final cohesion (i.e., coalescence, stretching separation, and shattering), all readily seen in experiment. An example of a simulated stretching-separation collision is shown in Fig. 6.10.

Later, Svanberg et al. [67] expanded on these results and presented simulations for droplet collisions with varying droplet size, temperature, and velocity. They were able to make favorable comparisons with well-known experimental results by Ashgriz and Poo [68], finding good agreement with experiment for the evolution of collision morphology from coalescence to stretching-separation collisions with increasing Weber number. This type of agreement is impressive given that the nanodroplets in this study remain quite small, at 125 and 1000 SPC water molecules. It is worth noting

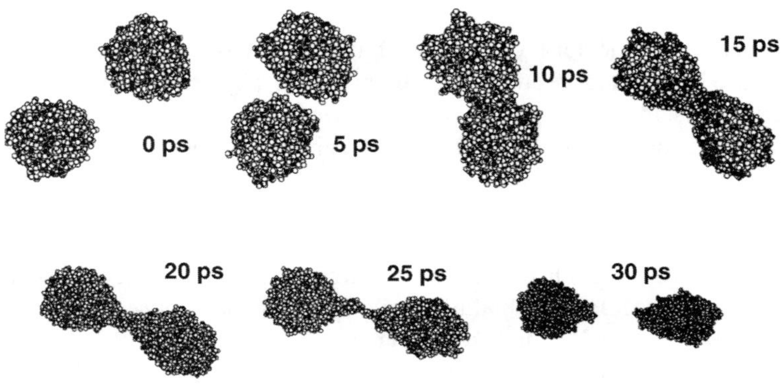

Figure 6.10 Stretching-separation collision for two 1000-molecule water nanodroplets at $T = 300\,\mathrm{K}$ and $v_{\mathrm{rel}} = 600\,\mathrm{m/s}$. (From [67], with permission. Copyright© 1998, American Institute of Physics.)

that despite the increased range of their simulations, Svanberg et al. [67] were unable to obtain a reflexive-separation type of coalescence. They propose that this type of coalescence may only be possible for larger droplets (diameter $\gg$ 5 nm), due to large amounts of dissipation in very small droplets, which effectively damps away the oscillations necessary to produce breakup of the temporarily joined nanodroplets. This theory is supported by a more recent study by Kalweit and Drikakis [69] in which they are also unable to produce reflexive separation for collisions of clusters with 9-nm diameters containing approximately 10,000 atoms. It should be noted, however, that in Kalweit and Drikakis's study, the clusters used are, in fact, solid prior to impact. Although the incident particles liquefy very quickly on impact, this fact may influence the result.

Droplet bouncing, another experimentally observed coalescence morphology, remained elusive in molecular dynamics simulation of droplet collision until an ambient gas was added by Murad and Law [70]. The presence of ambient gas had been shown in experimental work by Qian and Law [71] to play a key role in causing a bouncing collision. Poor drainage and the continued presence of this ambient gas in the gap between the droplets can act as a barrier to the development of the thermally driven molecular-scale bridging (see Fig. 6.9), reducing the potential for coalescence. The importance of the ambient gas was confirmed for nanoscale droplets, as bouncing was readily observed. Further, the importance of the molecular characteristics of the ambient gas is also highlighted here, as slight changes to the molecular potential, used for the ambient gas, produced coalescence from conditions that previously created a bouncing collision. Murad and Law [70] were also able to create a single instance of bouncing without ambient gas. They speculate, we believe correctly, that this result may be due to particularly significant vapor pressure at this specific condition which could have been missed in previous studies due to the limited number of trials attempted, given restrictions in the computing resources available at the time.

In a more recent work, Zhao and Choi [72] further demonstrate the importance of the molecular characteristics of the surrounding fluid in the development of bridging, and therefore the overall progress of coalescence. In their study, the coalescence of 100-molecule water nanodroplets surrounded by n-heptane is slowed significantly by the presence of the n-heptane, compared to the baseline case of coalescence simulated in vacuum. Here it is more intuitive, how given the physical extent of the flexible n-heptane molecule, its presence in the narrow gap between the approaching nanodroplets could inhibit formation of the molecular bridging. However, it is interesting to note that Zhao and Choi nevertheless find that once a bridge is formed successfully, coalescence takes place as quickly in the presence of the n-heptane as seen in the case of coalescence in vacuum. This result highlights the sensitivity of some fluidic processes, such as bridging, which are active at the molecular scale, to the details of their molecular surroundings and the corresponding insensitivity of others, such as the bulk flow of the linked droplets.

Moving away from droplet-on-droplet collision, we turn to the problem of droplet impact onto planar solid and liquid surfaces. Droplet impact is a hugely important process found in nature, such as splashing raindrops, as well as in industry, such as spray painting. Of particular interest in many applications and to fundamental

Figure 6.11 Low-inertia nanodroplet impact onto a solid surface for $v = 18\,\text{m/s}$ up to $t_f = 0.8$ ns. (From [75], with permission. Copyright © 2004, American Chemical Society.)

research are the boundaries and transitions between different regimes of impact behavior, which are described in detail in reviews by Rein [73] and Yarin [74].

Gentner et al. [75] simulate low-inertia impact of nanodroplets onto a flat solid surface. Given the low speed and small mass of the nanodroplet, inertia is low for impacts in this study, producing smooth nonsplashing behavior. The low-speed impact ($< 18\,\text{m/s}$) generates a slowly and smoothly changing liquid–vapor interface which allows the shape of the drop to be accurately modeled as an ellipsoid during the course of impact. Using this model shape, they find that while the spreading velocity of the droplet contact line is influenced by the inertia of impact immediately after impact, this effect quickly decays, and further spreading is controlled by the wetting characteristics of the solid–liquid interaction. Views of one impact from this study are shown in Fig. 6.11.

As discussed in Gentner et al. [75], the nature of the wetting interaction between the liquid and the solid surface plays a significant role in droplet impact on a solid surface. The importance of this effect for nanodroplet impact will be enhanced compared to larger length scales because of the strong scaling of surface-driven forces. It may therefore be possible to generate novel types of nanodroplet impact by modifying the nature of the liquid–solid wetting. One possible modification is fine-scale chemical patterning of the solid to produce regions of strong and weak liquid–solid interaction. Grest et al. [76] study the effect of a chemically striped surface on the outcome of nanodroplet spreading and show that it offers the ability to control the distribution of liquid along the surface by modifying the pitch of the striping, size of the droplet, or the fluid's composition.

The potential for distinctive droplet dynamics due to surface modification is highlighted by Cieplak et al. [77]. In their work, they simulate a 1800-atom nanodroplet sitting on an atomically smooth substrate with a checker board pattern of wetting and nonwetting squares. Unlike a droplet resting on a fully wetting surface, in this case when a force is applied parallel to the surface the droplet responds and moves easily along the surface, mimicking, to some degree, the lotus effect.

In addition to surface modifications, the nature of nanodroplet spreading can also be altered by changing the nature of the nanodroplet. Kim et al. [78] demonstrate this effect by showing an increase in the speed of spreading with the addition of a surfactant with strong affinity for the solid. Their results also show significant relocation of surfactant to the liquid–solid interface during spreading, which would give a driving force for Marangoni convection, thought to provide further enhancement of spreading velocity. A recent review of the wetting phenomena of simple fluids using MD simulations has been completed by De Coninck and Blake [79].

At the other end of the spectrum, very high speed [$O(10^3)$ m/s] impact onto solid surfaces has also been studied using molecular dynamics. One topic of interest in this area is surface modification applications, which were investigated by Kholmurodov et al. [80]. In this study, the small incident aluminum cluster (864 atoms) is initially a solid but behaves in a liquidlike fashion on impact. The amount of surface modification achieved was found to increase with impact speed, allowing for the identification of three main types of impact: soft landing, droplet spreading, and implantation. Tomsic et al. [81–84] have presented a sequence of works focusing on the fragmentation of water clusters striking a repulsive surface at high speed. Similar to the work of Kholmurodov et al. [80], the clusters used in these studies are initially solid nanoparticles; however, Tomsic et al. focus on tracking the behavior of the cluster and its fragments through impact rather than on surface modification.

In contrast to these impact studies, which use only initially solid nanoparticles, Germann [85] studies the high-speed impact of both solid nanoparticles and fully liquid nanodroplets. He simulates the impact of 51.2-nm copper nano-clusters containing approximately 5×10^6 atoms onto a rigid, but wetting solid surface using the embedded atom method (EAM) potential for both the cluster and solid-surface atoms. In considering the droplet impacts at impact velocities of 250 to 500 m/s, the nanodroplet is observed to spread out smoothly over the surface, which Germann suggests may provide a useful test case for existing continuum mechanics models of similar problems [86–88]. As the velocity of impact is increased to 750 m/s and beyond, hydrodynamic jetting parallel to the solid surface develops. The amount of material ejected increased with temperature and impact velocity, producing a distribution of fragment sizes that is consistent with existing models for cluster emission. The results of impacts at $v = 750$ m/s from this work are shown in Fig. 6.12. The ability to generate this diverse range of phenomena from a single model is a good example of the ability of MD simulation to allow for emergent or unexpected behaviors to develop without a need for model changes or additional assumptions.

Despite the relatively large amount of work done on nanodroplet (and nanoparticle) impact onto solid surfaces, there has not been a commensurate amount of work for impact onto liquid surfaces. The authors have simulated the impact of an argon nanodroplet approximately 9 nm in diameter (5300 to 5800 Lennard-Jones atoms) into a deep pool of liquid argon. We have tested impacts over a wide range of velocities ranging from 18 to 1100 m/s. Despite high impact velocities, coalescence-type impact is observed in all cases without the energetic splashing seen for macroscale droplets (i.e., secondary droplet ejection, jetting, etc.) [89]. As the impact velocity is increased, the size of the crater formed in the liquid layer increases dramatically. A snapshot taken at maximal crater depth for a high-speed impact is shown in Fig. 6.13. In contrast, for low-speed impact ($v < 100$ m/s) the liquid surface appears essentially undisturbed on nanodroplet impact, while the droplet assumes a shape similar to low-speed nanodroplet impact onto a solid seen in Gentner et al. [75] (see Fig. 6.11).

Coalescence is hugely favored for nanodroplets in this system, due to the fact that we have treated a completely homogeneous system with argon vapor filling the space

Figure 6.12 High-speed impact of copper nanoparticles ($T = 300$ and 1000 K for top and middle) and nanodroplet ($T = 1500$ K) onto a rigid copper surface at $v = 750$ m/s. (From [85], with permission. Copyright © 2006, Elsevier.)

around the argon nanodroplet and liquid layer. Under these conditions, where the liquid and surrounding vapor are of the same material, the formation of a molecular liquid bridge between the approaching droplet and the liquid layer is greatly enhanced. This is due to the consistently attractive forces between the liquid and vapor molecules, which extend the distance and time over which it is possible to initiate bridging. Bridging is clearly visible in the lower-speed impacts (see Fig. 6.9), while at higher impact speeds it is not as easily observed, due to the speed of the impact event relative to the frequency at which we visualize the molecular configurations.

6.4 MOLECULAR DYNAMICS AND MULTISCALE SIMULATION

It is becoming apparent that the greatest strength of molecular dynamics simulations, their capacity to probe at the most minute spatial and temporal scales, can also be their greatest weakness when the phenomena studied extend beyond the size that available

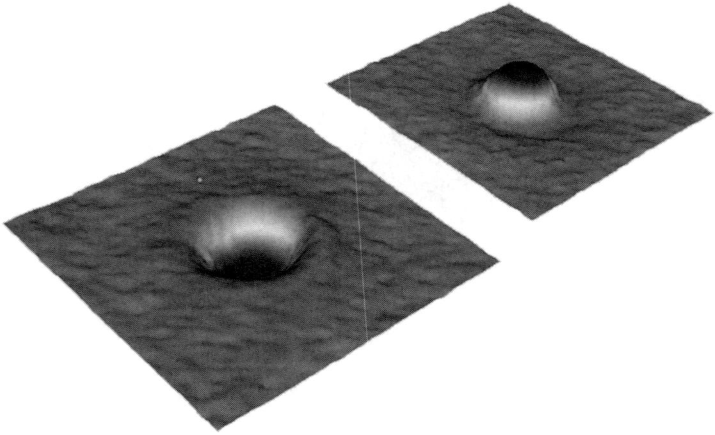

Figure 6.13 Crater formed from the impact of a 9-nm (5800 Lennard-Jones atoms) argon nanodroplet onto a liquid argon surface ($L_{x,z} = 108\,\text{nm}$, $L_y = 54\,\text{nm}$) with $v \approx 1100\,\text{m/s}$. Isosurface of density through the liquid–vapor interface plotted and shaded with respect to depth. The second image is an inverted version of the surface rotated to reveal the underside of the crater.

computational infrastructure can comfortably address. To state again the main problem of scale disparity, when crucial phenomena take place simultaneously at times or locations that are far apart in the scales spectrum, the requirement to resolve the smaller and larger scales simultaneously is often impossible to satisfy. This argument, which holds for all modeling approaches, is exacerbated further for MD techniques because they are intrinsically aimed at the nanoscale, whereas most of the technological problems that they address have, at least, some macroscopic component.

Sometimes, attacking this issue head-on by increasing drastically the number of molecules in a simulation is adequate. However, in many cases this is impossible either because a complex spatiotemporal discrepancy exists, or because it is simply not feasible computationally to include an adequate number of molecules in a simulation or to run a simulation for the desired physical time interval. Quite often, the physics of the problem hints that this might not be a sensible approach. For example, there are cases where the interesting or influential nanoscale effects cover only a small region but can inform the large-scale behavior of the system.

To address such issues, computational simulation techniques have evolved to account for scales disparity. A class of techniques that shows great interest and potential involves the coarse graining of elements that are used in the model. We shall not expand on this class of techniques, but we can briefly say that it involves the simulation of the interactions of more generalized entities and not individual molecules. As such, more comprehensive interaction potentials and geometric descriptions of such entities are necessary. Coarse-grained MD simulations have been used successfully in a wide range of applications. For example, the dispersion of carbon nanotubes in water is achieved by adsorbing lipids and surfactants on the nanotube surface. Wallace and Sansom [90] computationally studied the self-assembly properties of such nanotube populations in aqueous solutions using coarse-grained MD.

A class of techniques that are of particular interest are known as *hybrid methods*. They involve the simultaneous evaluation of processes at the larger and smaller scales, focusing appropriate simulation techniques on the spatial or temporal domain for which they are most suited. The nature of the coupling or exchange of information between the techniques used is one of the critical features and most pressing challenges in the implementation of hybrid techniques. For our purposes, molecular dynamics simulation can be used for phenomena at the smallest length and time scales, allowing for the incorporation of atomistic or molecular information into larger-scale simulations. Hadjiconstantinou [91] demonstrated the efficacy of this approach by combining a large-scale finite element/spectral method with molecular dynamics for problems involving fluid–fluid or fluid–solid contact. The moving contact problem is often used as a test case for hybrid techniques because it exhibits very clearly the difficulties in modeling problems with wide spatial disparities. Here the region near the contact line is governed by molecular processes which while only a small fraction of the total system, has a great impact on the bulk fluid motion away from the contact line. In Hadjiconstantinou's paper, in addition to the hybrid method solution, a more computationally expensive MD solution over the entire domain was also completed and was found to be in good agreement with the results obtained from the hybrid approach.

Within the hybrid framework, various transport phenomena can be studied, as, for example, in Liu et al. [92], where heat transfer characteristics are tested, or in Werder et al. [93], where transport and flow around a carbon nanotube are simulated. These and similar papers emphasize the importance of parameter continuity across the micro–macro interface (not always reaching the same conclusions regarding flux vs. velocity matching, however). Actually, in most examples of this nature, a coupling intermediate domain is utilized that effectively implies that the molecular dynamics and the continuum mechanics domains overlap. On the same note, Ren [94] has conducted a very detailed analytical and numerical study on the stability and convergence of such hybrid atomistic–continuum techniques. The importance of the overlapping region and of the various coupling approaches (e.g., velocity matching, flux matching) are emphasized, and a clear message regarding stability (and instability) is offered.

Recently, a new class of techniques of this nature is emerging that aims at addressing the fact that if the scale disparity is large enough, there is a range between the fully atomistic and fully continuum that is not addressed optimally by either methodology. An example of such a technique, coined *triple-decker*, has been presented by Fedosov and Karniadakis [95]. Here, MD is coupled with a mesoscopic and a continuum flow methodology. Molecular dynamics and classical continuous fluid dynamics are supplemented by an intermediate layer, that of the dissipative particle dynamics (DPD) method [96]. DPD is a coarse-grained transport technique where interactions are accounted for between fluid parcels. The methodology is tested on a series of reference flows (e.g., Couette, Poiseuille, lid-driven cavity) with very encouraging results.

6.5 CONCLUDING REMARKS

Droplet and particle phenomena at the nanometer range pose fundamental questions and are critical to a wide range of technological applications. The omnipresent trend for miniaturization makes the study of their properties and dynamics pivotal to future technological development. A first question that emerges naturally when one engages in the study on nanoscale phenomena is defining when the continuum assumptions (and consequently, the vast knowledge of such processes that we have accumulated) break down as sizes are reduced. A more practical question strongly linked with the first one is connected with the applicability and capacity of tools and techniques to provide relevant and accurate answers to design questions at these scales.

We believe that molecular dynamics simulation techniques can offer answers to both these issues. First, they provide a vehicle for minimum-assumptions hypothesis testing. For many problems, simulations of this nature have already delineated the boundaries where departure from macroscopic considerations start becoming important. It is of great interest that this does not always happen in an intuitive or predictable way. Similarly, the capacity to parametrize and probe in great spatial and temporal detail is putting this kind of modeling at the forefront of any nanotechnology design cycle. Finally, to supplement the inherent weakness of a nano-focused technique to span scales, MD has found its place in multiscale computational frameworks that can address a great range of problems.

Basic molecular dynamics is based on simple physical and algorithmic principles; however, proper implementation remains fundamental and the technique continues to be extremely computationally intensive. It is therefore critical to use MD simulation carefully in cases where the molecular-level insight it can provide is crucial and worthwhile. As we have highlighted in this chapter, phenomena that are particularly in need of molecular dynamics simulation are those of reduced dimensionality, those of dominant surface forces, or those in which thermal fluctuations are important. Although this seems like a restrictive list, such problems are found in many naturally occurring systems, in numerous biological processes, and in many synthetic engineered devices.

We expect that the rate of progress in nanodroplet research will depend not only on the undoubted increase in raw computational power available, but more directly on the ability of the community of interested researchers to leverage this power into interesting and tangible results.

REFERENCES

1. Feynman RP. There's plenty of room at the bottom. *J. Microelectromech. Syst.* **1**:60–66, 1992.
2. Gad-el Hak M. The fluid mechanics of microdevices: the Freeman Scholar Lecture *J. Fluid Eng. Trans. ASME* **121**:5–33, 1999.
3. Squires TM, Quake SR. Microfluidics: fluid physics at the nanoliter scale. *Rev. Mod. Phys.* **77**:977–1026, 2005.

4. Eapen J, Li J, Yip S. Beyond the Maxwell limit: thermal conduction in nanofluids with percolating fluid structures. *Phys. Rev. E* **76**:4, 2007.
5. Sankar N, Mathew N, Sobhan CB. Molecular dynamics modeling of thermal conductivity enhancement in metal nanoparticle suspensions. *Int. Comm. Heat Mass Transfer* **35**:867–872, 2008.
6. Sarkar S, Selvam RP. Molecular dynamics simulation of effective thermal conductivity and study of enhanced thermal transport mechanism in nanofluids. *J. Appl. Phys.* **102**:074302, 2007.
7. Xue L, Keblinski P, Phillpot SR, Choi SU-S, Eastman JA. Effect of liquid layering at the liquid–solid interface on thermal transport. *Int. J. Heat Mass Transfer* **47**:4277–4284, 2004.
8. Zhou XF, Gao L. Thermal conductivity of nanofluids: effects of graded nanolayers and mutual interaction. *J. Appl. Phys.* **103**:083503, 2008.
9. Alder BJ, Wainwright TE. Phase transition for a hard sphere system. *J. Chem. Phys.* **27**:1208–1209, 1957.
10. Abraham F, Walkup R, Gao H, Duchaineau M, De La Rubia TD, Seager M. Simulating materials failure by using up to one billion atoms and the world's fastest computer: brittle fracture. *Proc. Natl. Acad. Sci. USA* **99**:5777–5782, 2002.
11. Kadau K, Germann TC, Lomdahl PS. Molecular dynamics comes of age: 320 billion atom simulation on BlueGene/L. *Int. J. Mod. Phys. C* **17**:1755–1761, 2006.
12. Germann TC, Kadau K. Trillion-atom molecular dynamics becomes a reality. *Int. J. Mod. Phys. C* **19**:1315–1319, 2008.
13. Huang K. *Introduction to Statistical Physics*, CRC Press, Boca Raton, FL, 2001.
14. Car R, Parrinello M. Unified approach for molecular-dynamics and density-functional theory. *Phys. Rev. Lett.* **55**:2471–2474, 1985.
15. Kibble TWB, Berkshire FH. *Classical Mechanics*, Imperial College Press, London, 2004.
16. Trefethen LN. Maxims about numerical mathematics, computers, science and life. *SIAM News* **31**,1998.
17. Rahman A. Correlations in motion of atoms in liquid argon. *Phys. Rev. A* **136**:A405–A411, 1964.
18. Johnson JK, Zollweg JA, Gubbins KE. The Lennard-Jones equation of state revisited. *Mol. Phys.* **78**:591–618, 1993.
19. Nicolas JJ, Gubbins KE, Streett WB, Tildesley DJ. Equation of state for the Lennard-Jones fluid. *Mol. Phys.* **37**:1429–1454, 1979.
20. Maruyama S. Molecular dynamics method for microscale heat transfer. In: *Advances in Numerical Heat Transfer*, Vol. 2, Taylor & Francis, Philadelphia, 2000.
21. Rapaport DC. *The Art of Molecular Dynamics Simulation*, Cambridge University Press, New York, 2004.
22. Trokhymchuk A, Alejandre J. Computer simulations of liquid/vapor interface in Lennard-Jones fluids: some questions and answers. *J. Chem. Phys.* **111**:8510–8523, 1999.
23. Bernal JD, Fowler RH. A theory of water and ionic solution, with particular reference to hydrogen and hydroxyl ions. *J. Chem. Phys.* **1**:515–548, 1933.
24. Berendsen HJC, Postma JPM, van Gunsteren WF, Hermans J. Interaction models for water in relation to protein hydration. In: *Intermolecular Forces*, D. Reidel, Dordrecht, The Netherlands, 1981.

25. Jorgensen WL, Chandrasekhar J, Madura JD, Impey RW, Klein ML. Comparison of simple potential functions for simulating liquid water. *J. Chem. Phys.* **79**:926–935, 1983.

26. Mahoney MW, Jorgensen WL. A five-site model for liquid water and the reproduction of the density anomaly by rigid, nonpolarizable potential functions. *J. Chem. Phys.* **112**:8910–8922, 2000.

27. Nada H, van der Eerden JPM. An intermolecular potential model for the simulation of ice and water near the melting point: a six-site model of H2O. *J. Comput. Phys.* **118**:7401–7413, 2003.

28. Guillot B. A reappraisal of what we have learnt during three decades of computer simulations on water. *J. Mol. Liq.* **101**:219–260, 2002.

29. Daw MS, Baskes MI. Embedded-atom method: derivation and application to impurities, surfaces, and other defects in metals. *Phys. Rev. B* **29**:6443–6453, 1984.

30. Daw MS, Foiles SM, Baskes MI. The embedded-atom method: a review of theory and applications. *Mater. Sci. Rep.* **9**:251–310, 1993.

31. Verlet L. Computer experiments on classical fluids: I. Thermodynamical properties of Lennard-Jones molecules. *Phys. Rev.* **159**:98–103, 1967.

32. Swope WC, Andersen HC, Berens PH, Wilson KR. A computer-simulation method for the calculation of equilibrium-constants for the formation of physical clusters of molecules: application to small water clusters. *J. Chem. Phys.* **76**:637–649, 1982.

33. Haile JM. *Molecular Dynamics Simulations: Elementary Methods*, Wiley–Blackwell, New York, 1997.

34. Gear CW. *Numerical Initial Value Problems in Ordinary Differential Equations*, Prentice-Hall, Englewood Cliffs, NJ, 1971.

35. Allen MP, Tildesley DJ. *Computer Simulation of Liquids*, Clarendon Press, Oxford, UK, 1987.

36. Tuckerman M, Berne BJ, Martyna GJ. Reversible multiple time scale molecular-dynamics. *J. Chem. Phys.* **97**:1990–2001, 1992.

37. Harvey SC, Tan RK-Z, Cheatham TE. The flying ice cube: velocity rescaling in molecular dynamics leads to violation of energy equipartition. *J. Comput. Chem.* **19**:726–740, 1998.

38. Nosé S. A unified formulation of the constant temperature molecular-dynamics methods. *J. Chem. Phys.* **81**:511–519, 1984.

39. Hoover WG. Canonical dynamics: equilibrium phase-space distributions. *Phys. Rev. A* **31**:1695–1697, 1985.

40. Berendsen HJC, Postma JPM, van Gunsteren WF, DiNola A, Haak JR. Molecular-dynamics with coupling to an external bath. *J. Chem. Phys.* **81**:3684–3690, 1984.

41. Andersen HC. Molecular-dynamics simulations at constant pressure and/or temperature. *J. Chem. Phys.* **72**:2384–2393, 1980.

42. Plimpton SJ. Fast parallel algorithms for short-range molecular-dynamics. *J. Comput. Phys.* **117**:1–19, 1995.

43. Buzea C, Pacheco II, Robbie K. Nanomaterials and nanoparticles: sources and toxicity. *Biointerphases* **2**:MR17–MR71, 2007.

44. Streletzky KA, Zvinevich Y, Wyslouzil BE, Strey R. Controlling nucleation and growth of nanodroplets in supersonic nozzles. *J. Chem. Phys.* **116**:4058–4070, 2002.

45. Karis TE, Nayak UV. Liquid nanodroplets on thin film magnetic recording disks. *Tribol. Trans.* **47**:103–110, 2004.

46. Ackley DE. Nanodroplet technology targets drug delivery. Drug Deliv. Rep. Spring/ Summer, 2006.

47. Habenicht A, Olapinski M, Burmeister F, Leiderer P, Boneberg J. Jumping, nanodroplets. *Science* **309**:2043–2045, 2005.

48. Ferrari M. Nanovector therapeutics. *Curr. Opin. Chem. Biol.* **9**:343–346, 2005.

49. Heath CH, Streletzky KA, Wyslouzil BE, Wolk J, Strey R. Small angle neutron scattering from D_2O–H_2O nanodroplets and binary nucleation rates in a supersonic nozzle. *J. Chem. Phys.* **118**:5465–5473, 2003.

50. Wyslouzil BE, Wilemski G, Strey R, Heath CH, Dieregsweiler U. Experimental evidence for internal structure in aqueous–organic nanodroplets. *Phys. Chem. Chem. Phys.* **8**:54–57, 2006.

51. Thompson SM, Gubbins KE, Walton JPRB, Chantry RAR, Rowlinson JS. A molecular-dynamics study of liquid-drops. *J. Chem. Phys.* **81**:530–542, 1984.

52. Sides SW, Grest GS, Lacasse M-D. Capillary waves at liquid–vapor interfaces: a molecular dynamics simulation. *Phys. Rev. E* **60**:6708–6713, 1999.

53. Rusanov AI, Brodskaya EN. Molecular-dynamics simulation of a small drop. *J. Colloid. Interface Sci.* **62**:542–555, 1977.

54. Arcidiacono S, Poulikakos D, Ventikos Y. Oscillatory behavior of nanodroplets. *Phys. Rev. E* **70**:011505, 2004.

55. Rodriguez F, Mesler R. The penetration of drop-formed vortex rings into pools of liquid. *J. Colloid. Interface Sci.* **121**:121–129, 1988.

56. Moseler M, Landman U. Formation, stability, and breakup of nanojets. *Science* **289**:1165–1169, 2000.

57. Long LN, Micci MM, Wong BC. Molecular dynamics simulations of droplet evaporation. *Comp. Phys. Commu.* **96**:167–172, 1996.

58. Walther JH, Koumoutsakos P. Molecular dynamics simulation of nanodroplet evaporation. *J. Heat Transfer Trans. ASME* **123**:741–748, 2001.

59. Consolini L, Aggarwal SK, Murad S. A molecular dynamics simulation of droplet evaporation. *Int. J. Heat Mass Transfer* **46**:3179–3188, 2003.

60. Wedekind J, Reguera D. What is the best definition of a liquid cluster at the molecular scale? *J. Chem. Phys.* **127**:154516, 2007.

61. Wedekind J, Reguera D, Strey R. Finite-size effects in simulations of nucleation. *J. Chem. Phys.* **125**:214505, 2006.

62. Wedekind J, Reguera D, Strey R. Influence of thermostats and carrier gas on simulations of nucleation. *J. Chem. Phys.* **127**:064501, 2007.

63. Wedekind J, Woelk J, Reguera D, Strey R. Nucleation rate isotherms of argon from molecular dynamics simulations. *J. Chem. Phys.* **127**:154515, 2007.

64. Koplik J, Banavar JR. Molecular-structure of the coalescence of liquid interfaces. *Science* **257**:1664–1666, 1992.

65. Koplik J, Banavar JR. Molecular dynamics of interface rupture. *Phys. Fluids A* **5**:521–536, 1993.

66. Greenspan D, Heath LF. Supercomputer simulation of the modes of colliding microdrops of water. *J. Phys. D* **24**:2121–2123, 1991.

67. Svanberg M, Ming L, Markovic N, Pettersson JBC. Collision dynamics of large water clusters. *J. Chem. Phys.* **108**:5888–5897, 1998.

68. Ashgriz N, Poo JY. Coalescence and separation in binary collisions of liquid-drops. *J. Fluid Mech.* **221**:183–204, 1990.

69. Kalweit M, Drikakis D. Collision dynamics of nanoscale Lennard-Jones clusters. *Phys. Rev. B* **74**:235415, 2006.

70. Murad S, Law CK. Molecular simulation of droplet collision in the presence of ambient gas. *Mol. Phys.* **96**:81–85, 1999.

71. Qian J, Law CK. Regimes of coalescence and separation in droplet collision. *J. Fluid Mech.* **331**:59–80, 1997.

72. Zhao L, Choi P. Molecular dynamics simulation of the coalescence of nanometer-sized water droplets in *n*-heptane. *J. Chem. Phys.* **120**:1935–1942, 2004.

73. Rein M. Phenomena of liquid-drop impact on solid and liquid surfaces. *Fluid Dyn. Res.* **12**:61–93, 1993.

74. Yarin AL. Drop impact dynamics: splashing, spreading, receding, bouncing. *Annu. Rev. Fluid Mech.* **38**:159–192, 2006.

75. Gentner F, Rioboo R, Baland JP, De Coninck J. Low inertia impact dynamics for nanodrops. *Langmuir* **20**:4748–4755, 2004.

76. Grest GS, Heine DR, Webb EB, Liquid nanodroplets spreading on chemically patterned surfaces. *Langmuir* **22**:4745–4749, 2006.

77. Cieplak M, Koplik J, Banavar JR. Nanoscale fluid flows in the vicinity of patterned surfaces. *Phys. Rev. Lett.* **96**:114502, 2006.

78. Kim H-Y, Qin Y, Fichthorn KA. Molecular dynamics simulation of nanodroplet spreading enhanced by linear surfactants. *J. Chem. Phys.* **125**:174708, 2006.

79. De Coninck J, Blake TD. Wetting and molecular dynamics simulations of simple liquids. *Annu. Rev. Mater. Res.* **38**:1–22, 2008.

80. Kholmurodov K, Puzynin I, Smith W, Yasuoka K, Ebisuzaki T. MD simulation of cluster–surface impacts for metallic phases: soft landing, droplet spreading and implantation. *Comp. Phys. Commun.* **141**:1–16, 2001.

81. Tomsic A, Andersson PU, Markovic N, Piskorz W, Svanberg M, Pettersson JBC. Molecular-dynamics simulations of cluster–surface collisions: emission of large fragments. *J. Chem. Phys.* **115**:10509–10517, 2001.

82. Tomsic A, Gebhardt CR. A comparative study of cluster–surface collisions: molecular-dynamics simulations of (H_2O) and (SO_2)(1000). *J. Chem. Phys.* **123**:064704 2005

83. Tomsic A, Markovic N, Pettersson JBC. Molecular dynamics simulations of cluster-surface collisions: mechanisms for monomer emission. *Phys. Chem. Chem. Phys.* **3**:3667–3671, 2001.

84. Tomsic A, Schroder H, Kompa K-L, Gebhardt CR. Impact dynamics of molecular clusters on surfaces: fragmentation patterns and anisotropic effects. *J. Chem. Phys.* **119**:6314–6323, 2003.

85. Germann TC. Large-scale molecular dynamics simulations of hyperthermal cluster impact. *Int. J. Impact Eng.* **33**:285–293, 2006.

86. Haller KK, Poulikakos D, Ventikos Y, Monkewitz P. Shock wave formation in droplet impact on a rigid surface: lateral liquid motion and multiple wave structure in the contact line region. *J. Fluid Mech.* **490**:1–14, 2003.

87. Haller KK, Ventikos Y, Poulikakos D. Wave structure in the contact line region during high speed droplet impact on a surface: solution of the Riemann problem for the stiffened gas equation of state. *J. Appl. Phys.* **93**:3090–3097, 2003.

88. Haller KK, Ventikos Y, Poulikakos D, Monkewitz P. Computational study of high-speed liquid droplet impact. *J. Appl. Phys.* **92**:2821–2828, 2002.

89. Rein M. The transitional regime between coalescing and splashing drops. *J. Fluid Mech.* **306**:145–165, 1996.

90. Wallace EJ, Sansom MSP. Carbon nanotube self-assembly with lipids and detergent: a molecular dynamics study. *Nanotechnology* **20**:2009.

91. Hadjiconstantinou NG. Hybrid atomistic–continuum formulations and the moving contact-line problem. *J. Comput. Phys.* **154**:245–265, 1999.

92. Liu J, Chen S, Nie X, Robbins MO. A continuum–atomistic simulation of heat transfer in micro- and nano-flows. *J. Comput. Phys.* **227**:279–291, 2007.

93. Werder T, Walther JH, Koumoutsakos P. Hybrid atomistic–continuum method for the simulation of dense fluid flows. *J. Comput. Phys.* **205**:373–390, 2005.

94. Ren WQ. Analytical and numerical study of coupled atomistic–continuum methods for fluids. *J. Comput. Phys.* **227**:1353–1371, 2007.

95. Fedosov DA, Karniadakis GE. Triple-decker: interfacing atomistic– mesoscopic–continuum flow regimes. *J. Comput. Phys.* **228**:1157–1171, 2009.

96. Groot RD, Warren PB. Dissipative particle dynamics: bridging the gap between atomistic and mesoscopic simulation. *J. Chem. Phys.* **107**:4423–4435, 1997.

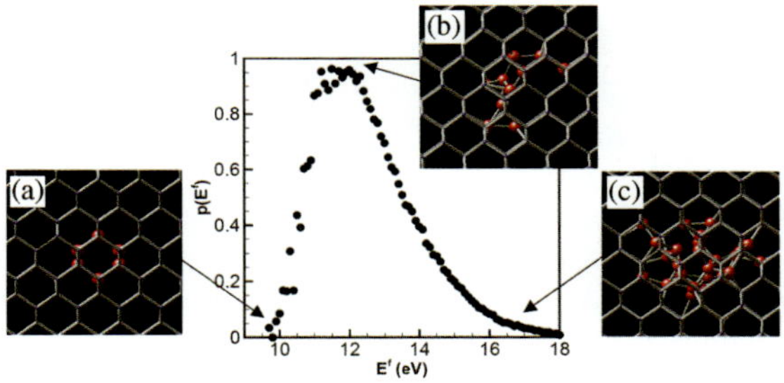

Figure 2.7 Probability distribution function for the formation energies of the six-vacancy cluster obtained by EDIP MD simulation at 1600 K. The insets are sample configurations; the large red spheres denote atoms that are displaced more than 10% of a bond length from their equilibrium positions.

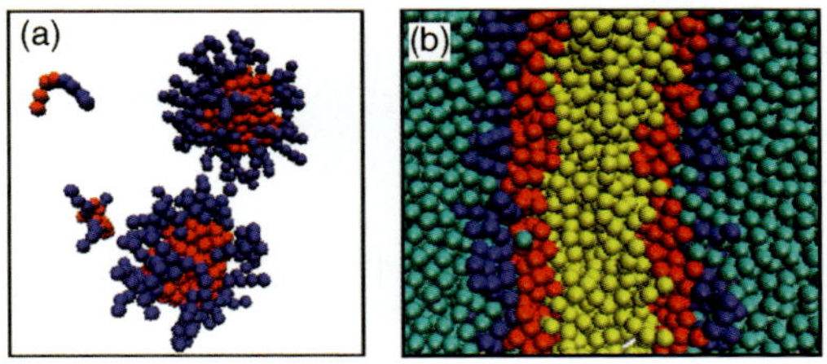

Figure 3.2 (a) Aqueous solution of H_4T_4 micelles and monomers; (b) H_3T_3 monolayers at hexadecane–water interfaces. Surfactant head and tail beads are shown by blue and red spheres, respectively. Solvent molecules are represented by cyan (water) and yellow (oil) beads. In plot (a), water molecules are omitted for clarity.

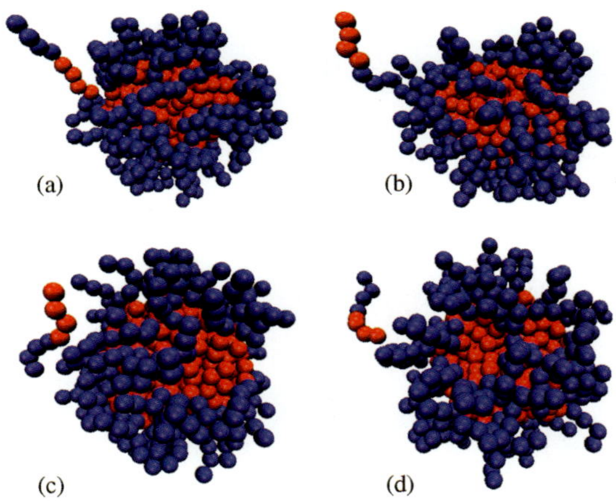

Figure 3.4 MD simulations of attempts of monomer entry into micelle A_{87}. Hydrophilic head beads and hydrophobic tail beads are shown by blue and red spheres, respectively. (a) Both the monomer and the micelle are in a favorable configuration leading to a successful entry; (b, c) the monomer, with an unfavorable orientation, approaches an exposed hydrophobic patch on the surface of the micelle; (d) the monomer, with a favorable orientation, approaches the hydrophilic micellar corona (i.e., the micelle is in an unfavorable configuration). The attempts at monomer entry fail in cases (b) to (d).

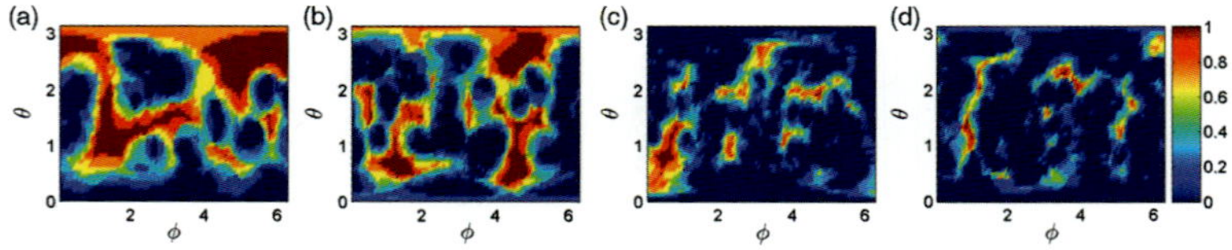

Figure 3.5 Fraction of time that cluster surfaces are occupied by hydrophobic patches during a period of 200 ps. Cluster aggregation numbers are (a) $N = 16$, (b) $N = 32$, (c) $N = 64$, and (d) $N = 88$. The surface of each cluster is parameterized by spherical angles θ and ϕ, with the origin of the system of coordinates located at the cluster center of mass.

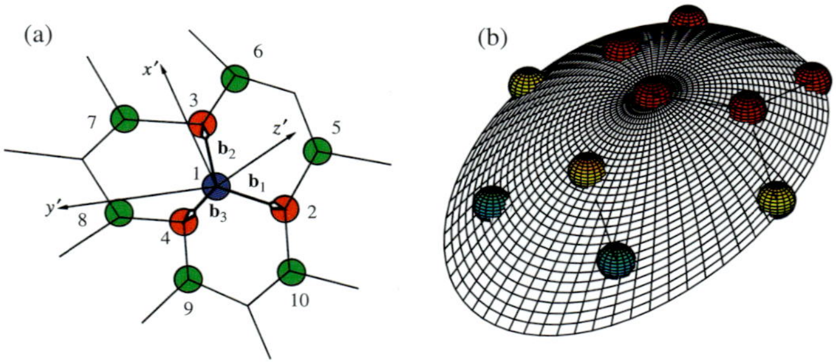

Figure 4.5 (a) Planar or curved hexagonal atomic lattice representing a graphene sheet; (b) a cubic surface can be defined passing through the 10 particles.

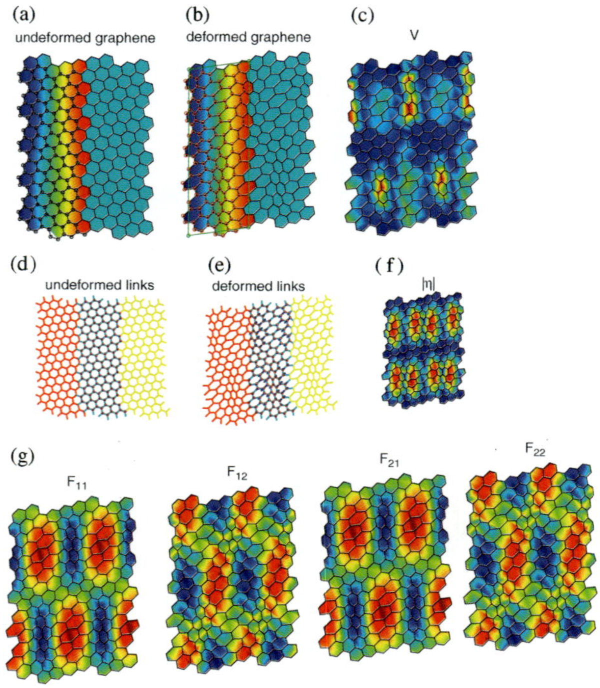

Figure 4.7 (a–c) Doubly periodic deformation of the planar graphene sheet in the $(10, 5)$ chiral mode, and induced Tersoff–Brenner potential; (d, e) bond network before and after deformation (the arrows indicate the exaggerated inner displacement, η); (f) plot of the magnitude of the inner displacement, $|\eta|$; (g) four components of the associated deformation gradient.

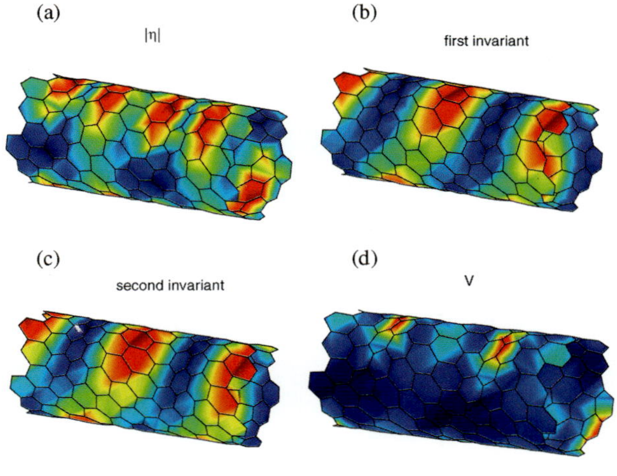

Figure 4.8 The graphene has been rolled up into a nanotube. (a) Plot of the magnitude of the inner displacement; (b,c) first and second invariants of the surface Cauchy–Green tensor; (d) atomic potential.

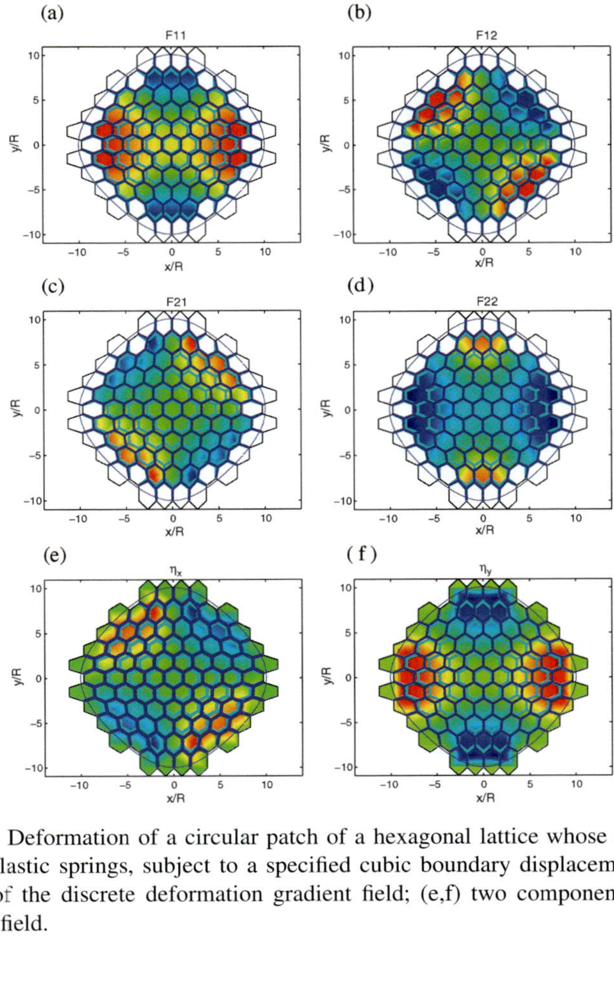

Figure 4.10 Deformation of a circular patch of a hexagonal lattice whose atoms are connected with elastic springs, subject to a specified cubic boundary displacement: (a–d) four components of the discrete deformation gradient field; (e,f) two components of the inner displacement field.

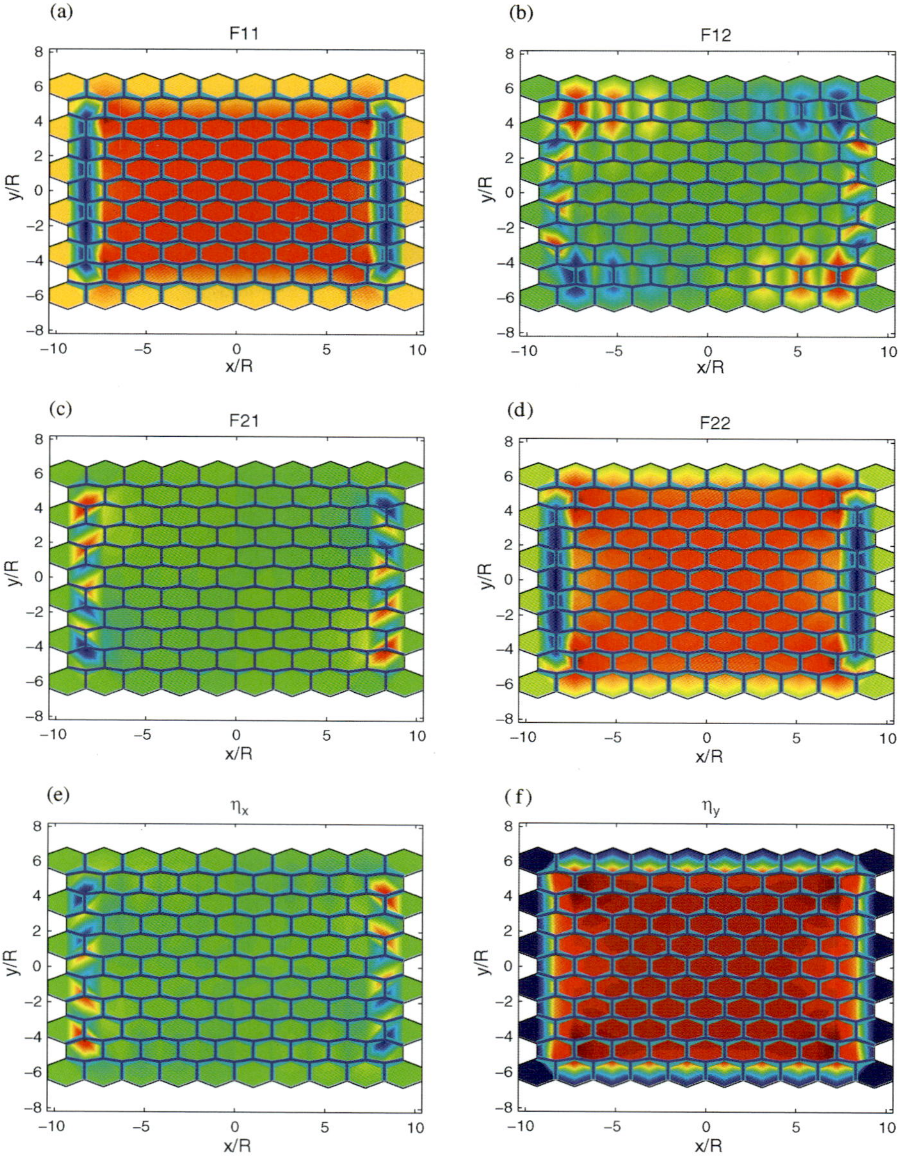

Figure 4.13 (a–d) Distribution of the four components of the discrete deformation gradient; (e,f) the inner displacement over a patch with side length $8R$ in extensional area-preserving boundary deformation with $\varepsilon = 0.2$.

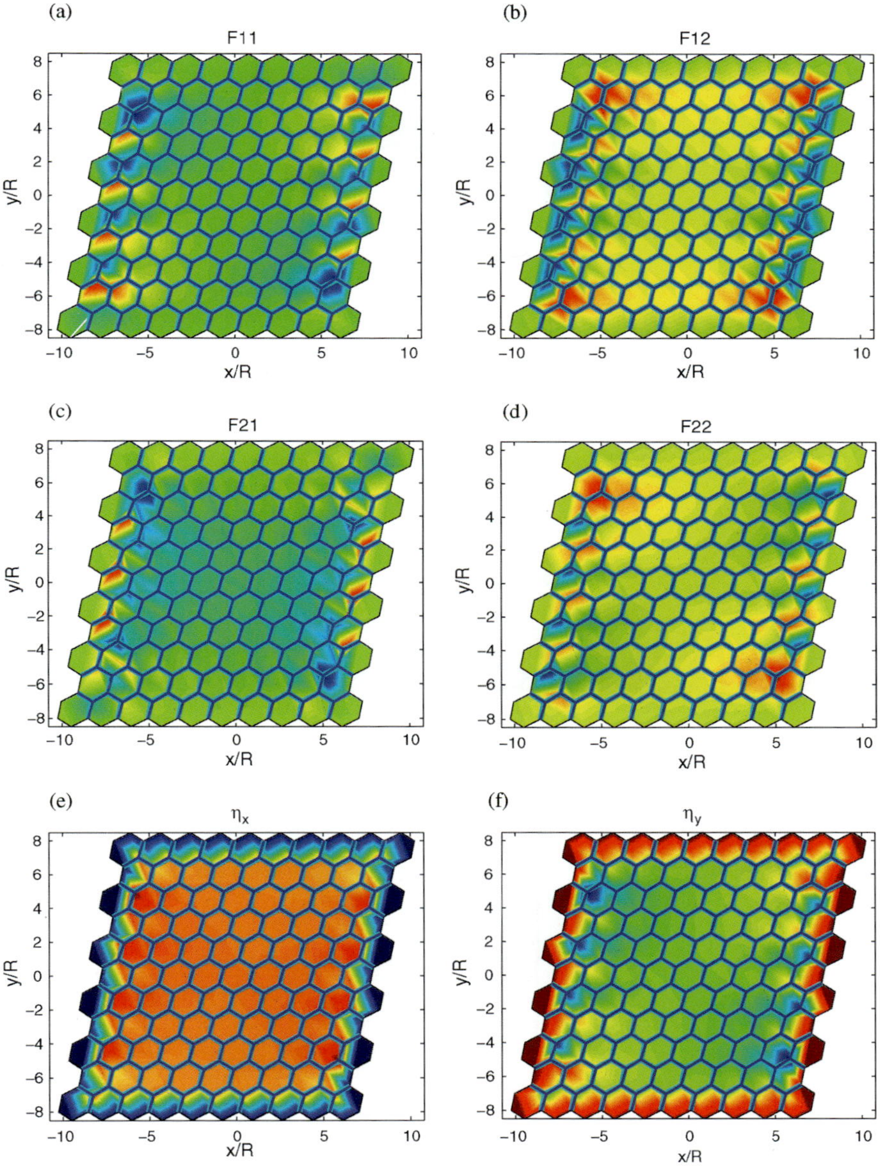

Figure 4.16 (a–d) Distribution of the four components of the discrete deformation gradient; (e,f) inner displacement over a patch with side length $8R$ in extensional area-preserving boundary deformation with $\varepsilon = 0.2$.

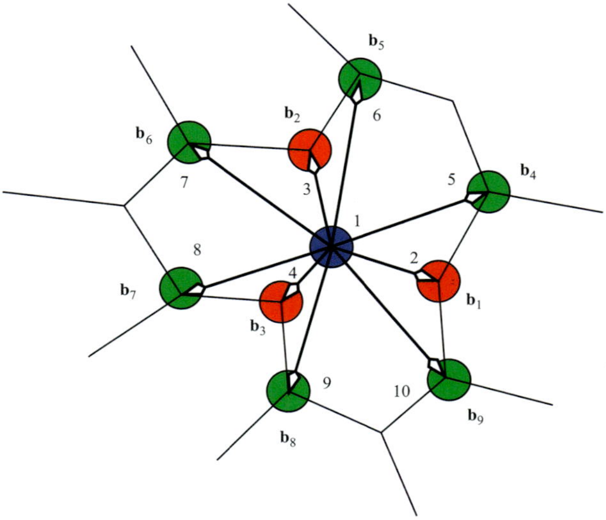

Figure 4.18 Nine bonds originate from the central atom up to the first and second neighbors.

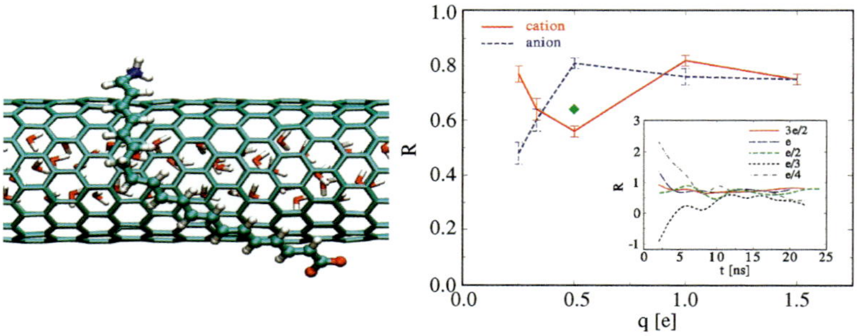

(a) (b)

Figure 5.2 See text on page 120 for full caption.

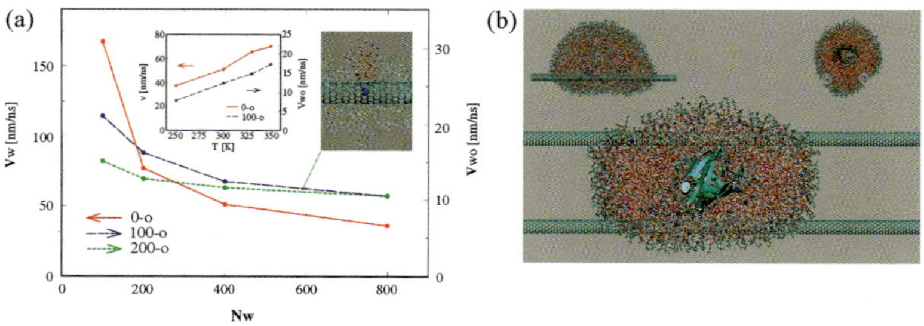

Figure 5.4 (a) Dependence of the nanodroplet velocity on the number of its water molecules N_W. Dragging of the droplet in the presence of $N_O = 100$ and 200 octane molecules is considered as well. Right inset: Visualization of the nanodroplet with $N_W = 50$ in $N_O = 200$ oil molecules. Left inset: Temperature dependence of the droplet velocity for $N_W = 400$. (b) Three views on a micelle formed by phospholipids and water molecules, which contains a GFP protein and five Na^+ ions neutralizing it [water is partly removed to see the protein (bottom and right up)]. It is driven by two Na^+ ions intercalated inside two CNTs.

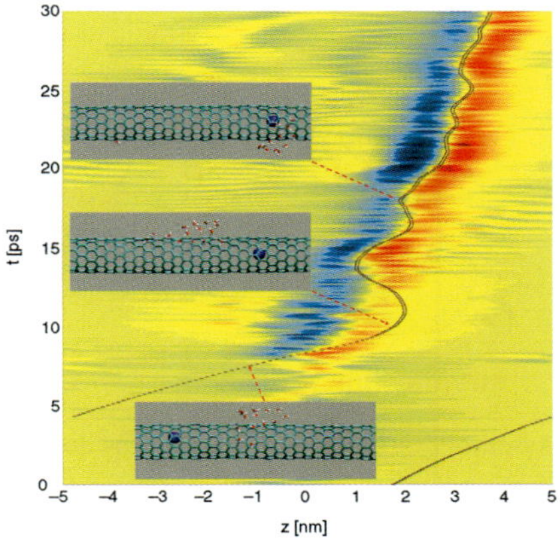

Figure 5.5 Trajectory of the Na^+ ion that is recaptured by a water nanodroplet with $N_W = 20$. The axial position and the time of ion motion are shown on the horizontal and vertical axes, respectively. The electric field generated by the water molecules along the CNT axis is plotted by contours.

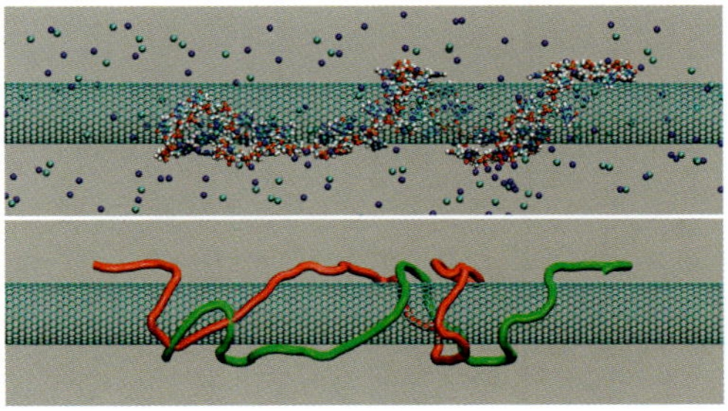

Figure 5.6 (top) A 50-base-long single-stranded DNA molecule, with the sequence CCTTCAGTGG CCGGTCATTG ATGAAGCCCT GAGGAACAAG GACACTCCGG, driven on the surface of a (25,0) CNT by the ionic solution flowing inside; (bottom) motion to the left of the DNA during a 7.4-ns simulation is clearly visible.

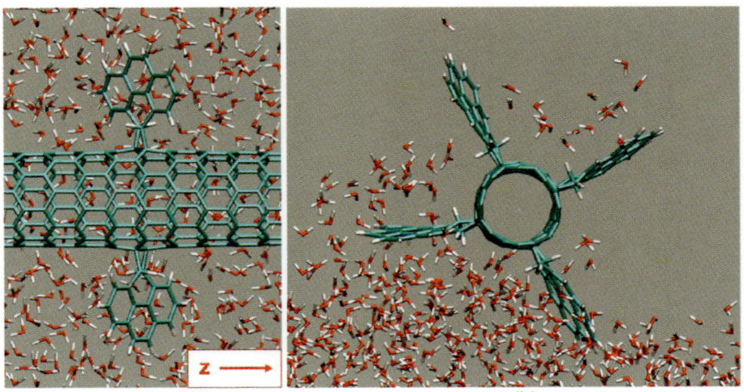

Figure 5.8 The model bulk (a) and surface (b) molecular propellers pump water along the tube (z) axis and orthogonal to it, respectively. Both systems are based on the (8,0) CNTs and have covalently attached aromatic (hydrophobic) blades. Water is partly removed from the front to uncover propeller detail.

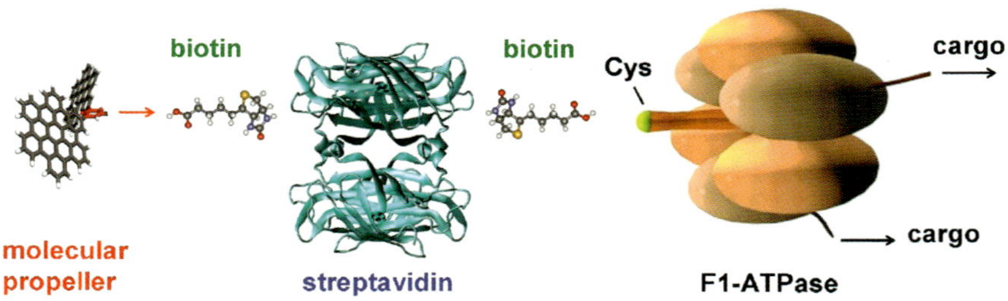

Figure 5.11 Assembly of the propelling unit, as described in the text.

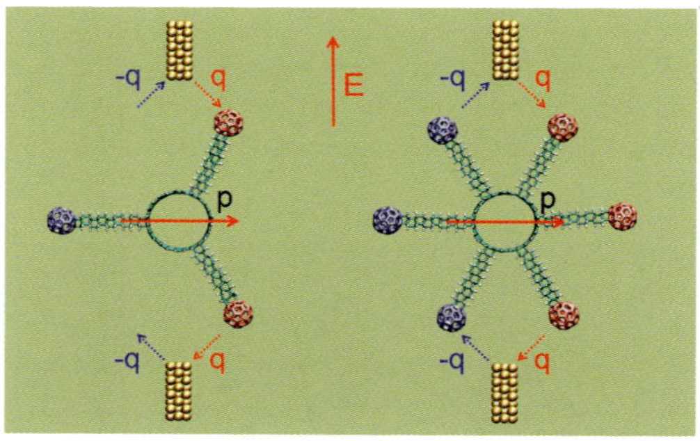

Figure 5.12 Tunneling-driven nanoscale motors with three (left) and six (right) fullerene blades. In an external homogeneous electric field E oriented along the vertical z direction, the electron tunneling from the neutral electrodes to the blades maintains an electric dipole p on the rotor, which is on average orthogonal to the field direction. This dipole is rotated unidirectionally by the electric field.

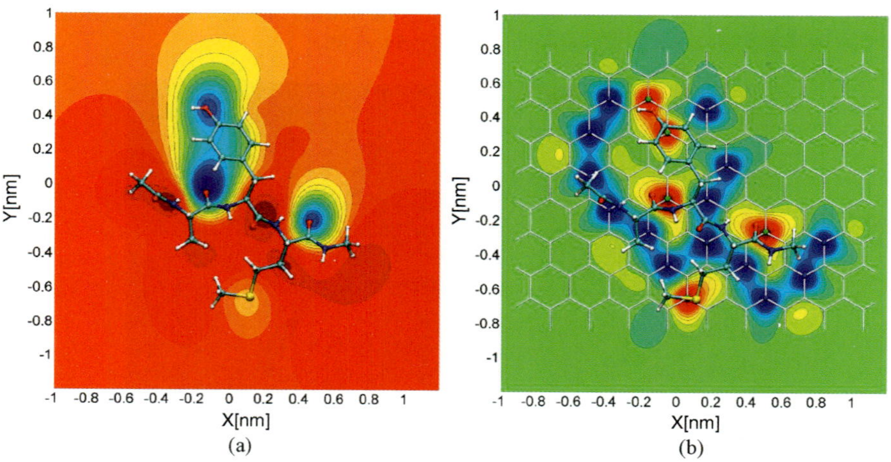

Figure 5.13 (a) Electrostatic potential distribution formed by the AYM peptide docked on the surface of the doped graphene sheet; (b) "complementary" distribution formed by the doped graphene sheet.

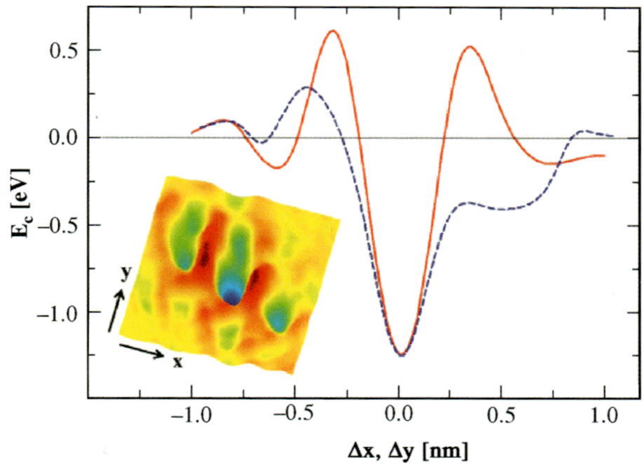

Figure 5.14(a) Dependence of the coulombic E_C potential energy of the system, when the peptide is shifted in the x (solid) and y (dashed) directions along the surface, away from its docking site. In the inset, the full two-dimensional potential surface of E_C is shown.

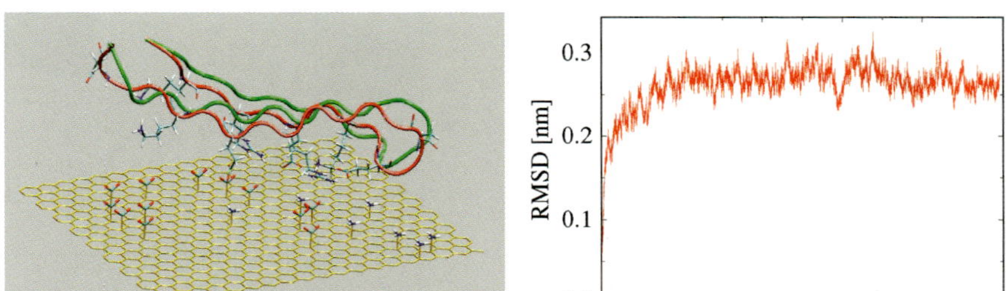

Figure 5.15 (a) Initial (red) and final (green) structures of the 105 to 130 residues from the GFP that are nested on the ligand-doped graphene layer. The peptide sequence is NYK-TRAEVKFEGDTLVNRIELKGIDF, where the 10 bold residues are charged and displayed in the initial structure. (b) RMSD of this peptide obtained during the simulation.

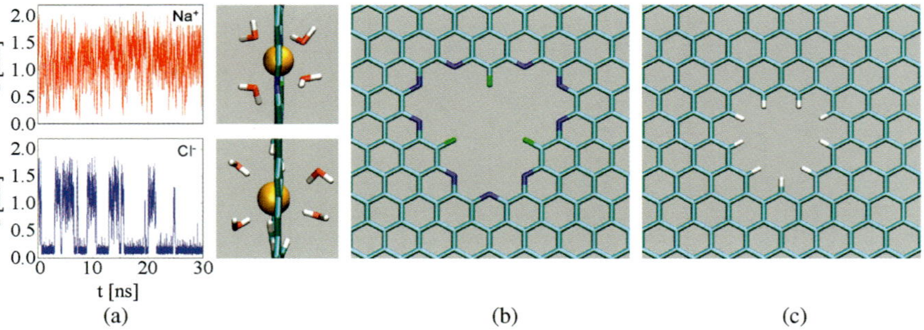

Figure 5.16 (a) Time-dependent distance d between the Na^+ and Cl^- ions and the centers of the F–N pore and H pore, respectively, at the field of $E = 6.25$ mV/nm. The dynamics of passage of these ions through the two pores is very different. (b) While both ions are surrounded by two water half-shells when passing through their pores, only the Cl^- ion has relatively stable binding to the H pore. (c) Detail of the F–N pore and the H pore, respectively, formed in the graphene monolayer.

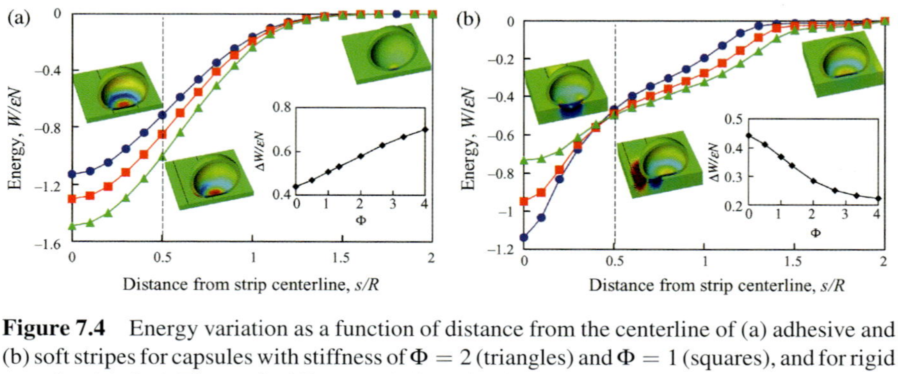

Figure 7.4 Energy variation as a function of distance from the centerline of (a) adhesive and (b) soft stripes for capsules with stiffness of $\Phi = 2$ (triangles) and $\Phi = 1$ (squares), and for rigid capsules (circles). The vertical lines mark the stripes' boundaries. From left to right, capsules with $\Phi = 2$ at $s = 0, 0.7$, and 2; colors indicate the strain as in Figs.7.5 and 7.6. The insets show the energy drop across the stripe versus the capsule stiffness.

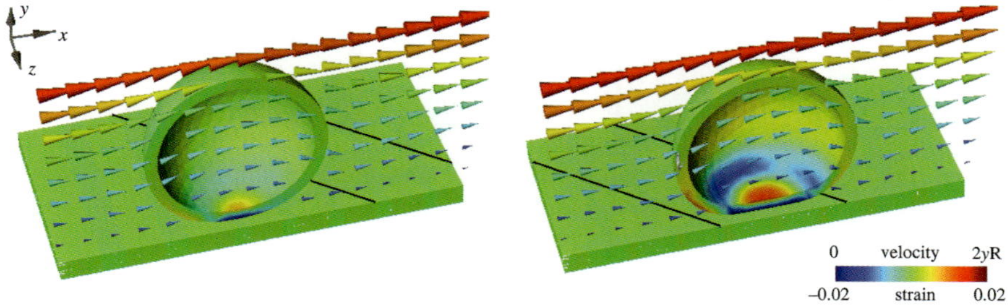

Figure 7.5 Elastic capsule rolling along a rigid substrate patterned with adhesive stripes in shear flow. The system is shown in cross section (made through the capsule's center of mass). The lines on the substrate indicate the boundaries of a sticky stripe. The colors in the capsule and substrate reveal the strain (see the color bar). The arrows indicate the flow direction and the arrow's color indicates the magnitude of the velocity. The capsule is located in front of a sticky stripe (left panel) and is crossing a sticky stripe (right panel). Note that the flattening of the capsule is due to a stronger adhesion with the sticky stripe. The capsule's deformation enhances the contact area with the substrate.

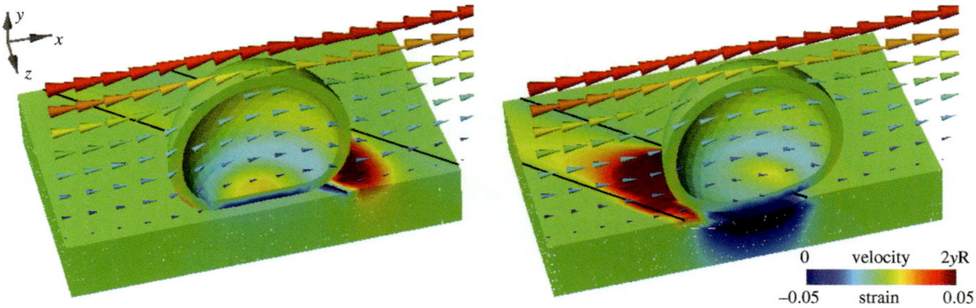

Figure 7.6 Elastic capsule rolling along an adhesive, mechanically patterned substrate in shear flow. The system is shown in cross section (made through the capsule's center of mass). The lines on the substrate mark the boundaries of a soft stripe. The characteristics of the strain and velocity are depicted in the same manner as in Fig. 7.5. Capsule approaches (left panel) and moves past (right panel) a soft stripe. Note the strong deformation of the soft stripe due to the adhesive interaction with the capsule, and consequently, an increase in contact area between the capsule and substrate.

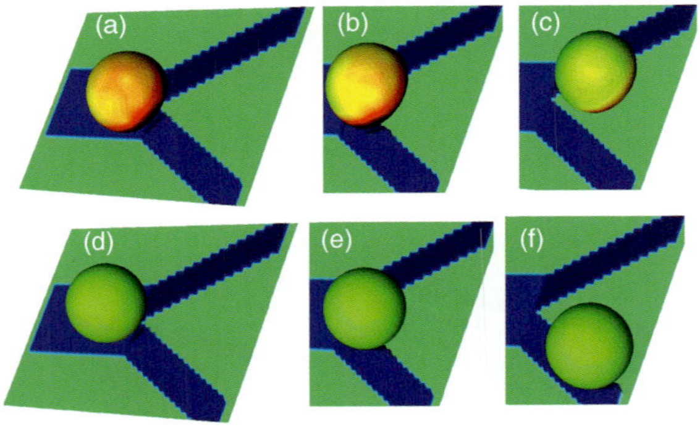

Figure 7.10 Capsule motion on the patterned substrate under an imposed shear flow with $\Lambda = 1.5 \times 10^3$. Images in parts (a) to (c) are for a soft capsule characterized by $\Phi = 6$ and images in parts (d) to (f) are for a stiff capsule characterized by $\Phi = 0.75$. The soft capsule deforms and moves toward the adhesive (sticky) branch, whereas the stiff capsule moves toward the soft branch. (The branches of the Y pattern appear to be jagged due to the discrete nature of the lattice-spring model.)

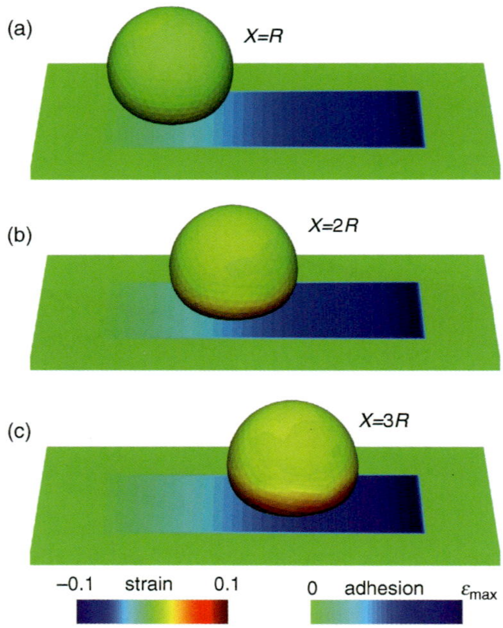

Figure 7.16 Compliant capsule on a substrate patterned with increasing adhesive strength. The capsule migrates along the pattern toward the region with the greatest adhesion. The migration occurs in the absence of any external flow and is due solely to the adhesive gradient. Part (a), (b), and (c) show three consequent moments when the capsule reaches the locations at $x = R$, $2R$, and $3R$, respectively. The width of the gradient pattern is $2R$ and its length is $5R$. Adhesion strength increases linearly with distance (i.e., $\varepsilon_a = \Delta\varepsilon_a x$). The color on the substrate indicates the local adhesion, while the color on the capsule's shell reveals the strain in the elastic shell due to the capsule's deformation. Note that the capsule deformation is enhanced when the compliant capsule moves along the gradient. Capsule stiffness is $\Delta\Phi = 20$, where $\Delta\Phi = \Delta\varepsilon_a R N_0 N_s / Eh\kappa^2$.

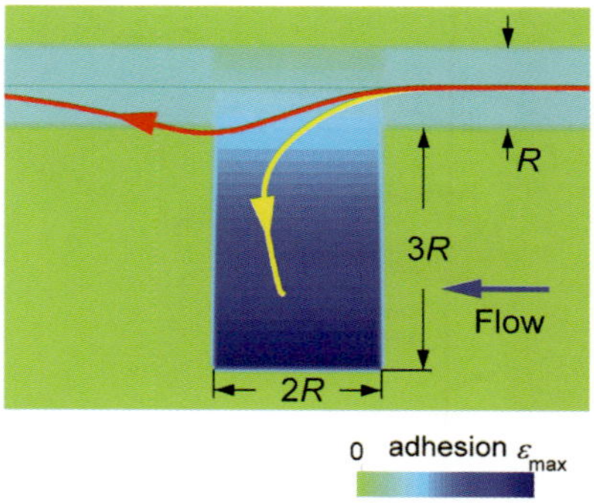

Figure 7.18 Trajectories of soft and rigid capsules on the patterned substrate. The pattern encompasses two parts: a delivery stripe with a uniform adhesion, and a side patch with adhesion increasing perpendicular to the flow direction. The color reveals the local adhesion on the substrate. The red and yellow lines show the trajectories of a rigid capsule and a soft capsule ($\Delta\Phi = 20$), respectively. The capsules are initially located at the same position and are driven to move by the imposed shear flow. The rigid capsule, although affected by the adhesion gradient on the side patch, escapes the trap and continues to move along the delivery stripe, whereas the soft capsule ends up at the side patch, entrapped due to the greater adhesive interaction.

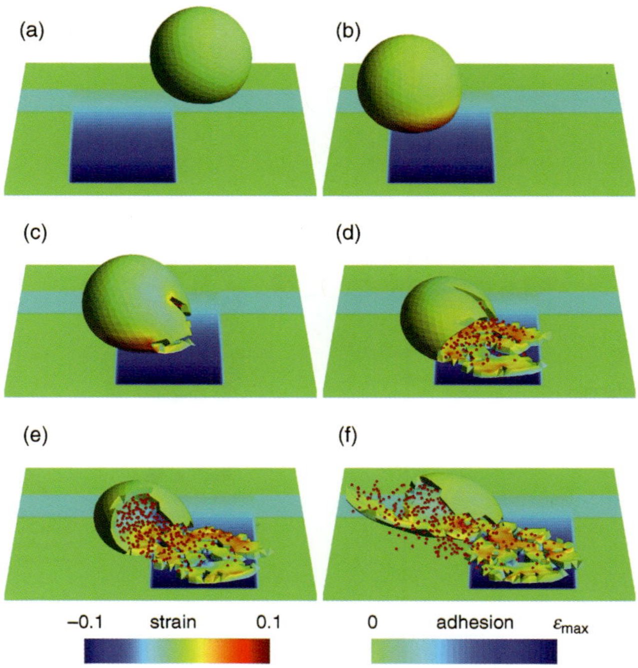

Figure 7.19 Soft capsule ($\Delta\Phi = 20$) as it becomes trapped and bursts on the patterned substrate. Time intervals between snapshots (a)–(b) and (e)–(f) are equal to $3\dot{\gamma}^{-1}$, and between snapshots (b)–(c), (c)–(d), and (d)–(e) are equal to $1.5\dot{\gamma}^{-1}$. The red dots indicate tracer particles that are initially enclosed inside the capsule and are released due to the capsule's bursting. The color at the substrate indicates the local adhesion; the color at the capsule's surface reveals the strain in the elastic shell.

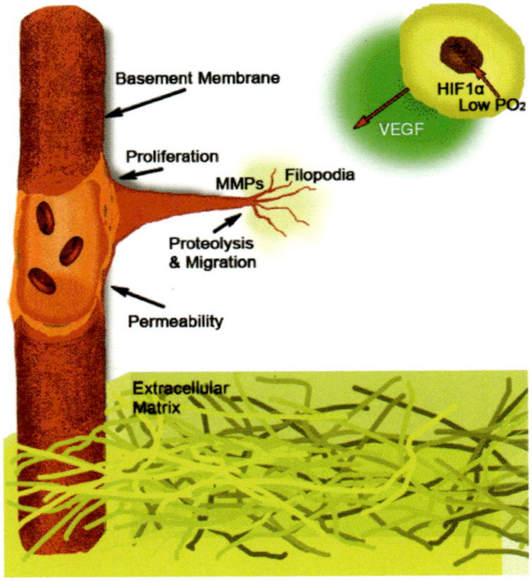

Figure 10.1 Initial steps in angiogenesis. Low oxygen in a neighboring cell (e.g., a cancer cell) triggers the expression of HIF1α, HIF1α up-regulates VEGF secretion, an adjacent vessel senses VEGF and a cell on the vessel becomes activated. This tip cell begins to migrate, using MMPs to proteolyze the surrounding matrix. Adjacent stalk cells proliferate behind the tip. HIF1α, hypoxia-inducible factor 1α. VEGF, vascular endothelial growth factor. MMPs, matrix metalloproteinases.

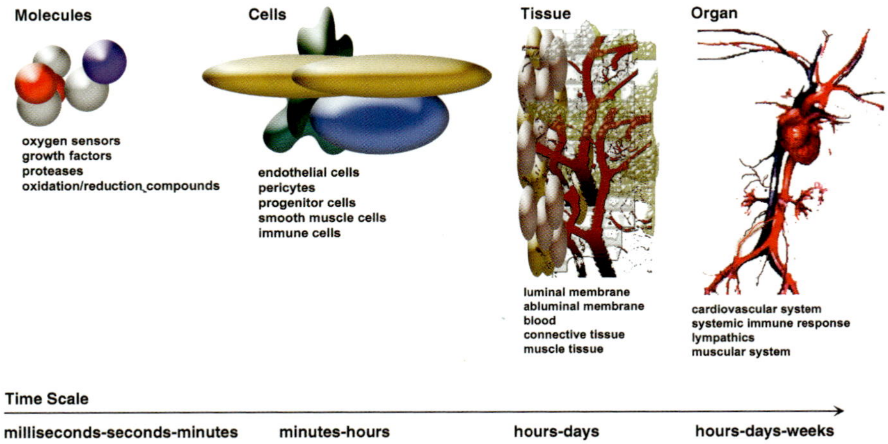

Figure 10.2 Time and spatial scales used in modeling processes in angiogenesis.

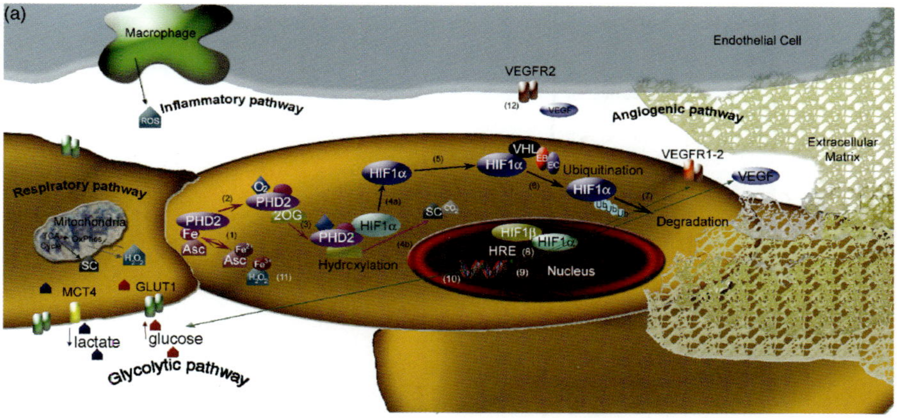

Figure 10.4(a) HIF1 system and the pathways it affects.

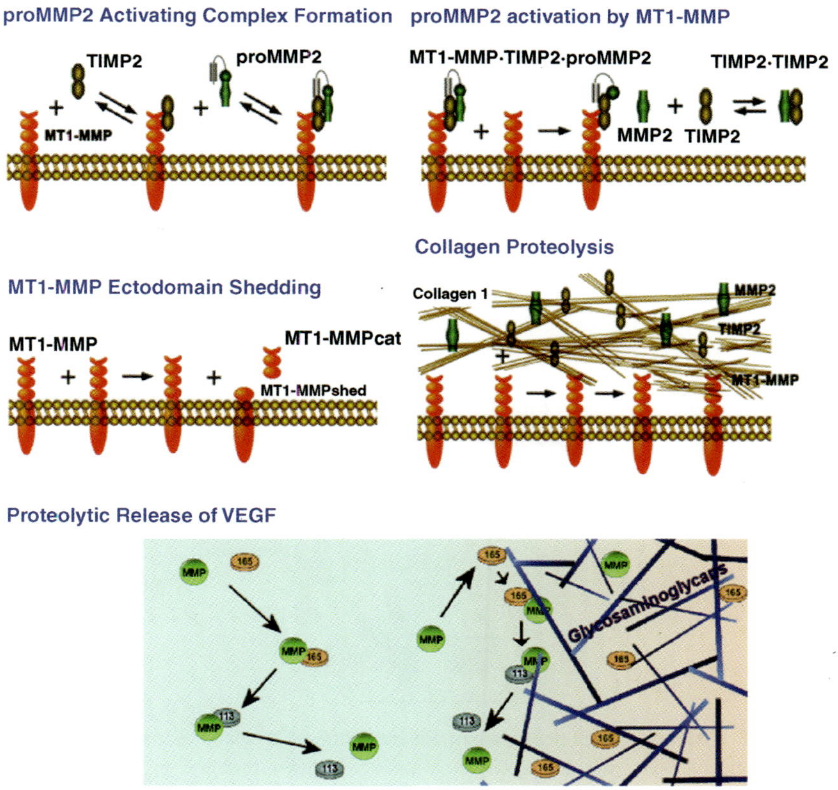

Figure 10.6 Schematic of proMMP2, MMP2, TIMP and MT-MMP interactions in the extracellular matrix (ECM), and subsequent proteolytic release of VEGF from the ECM.

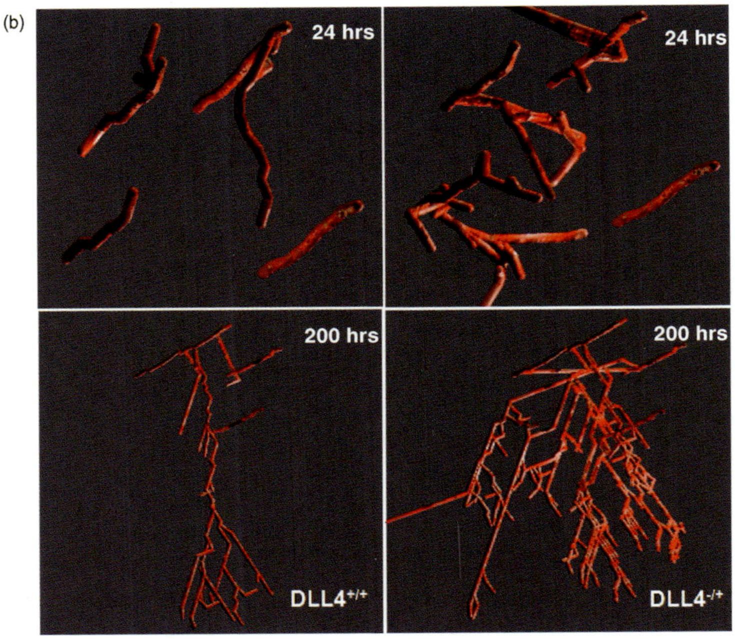

Figure 10.7(b) Model output showing vessel growth in time for control conditions (left panel) and haploinsufficiency of the Notch ligand Dll4 (right panel).

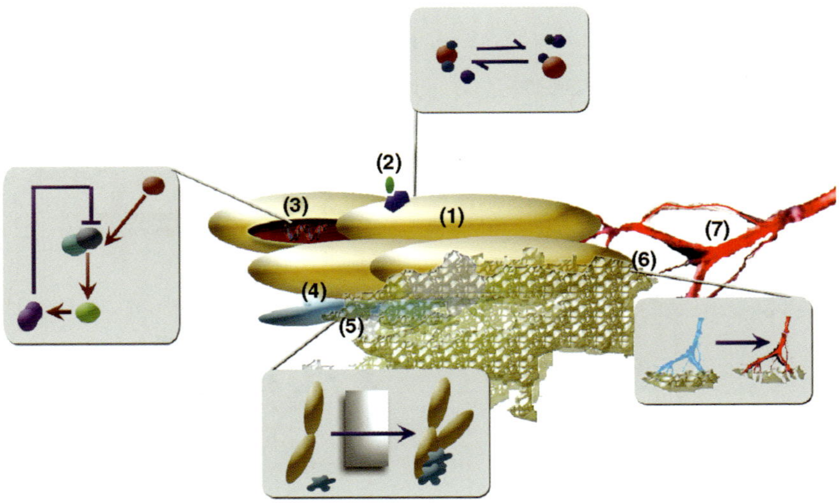

Figure 10.8 Examples of modeling areas and techniques that can be used to model angiogenesis at multiple levels, and advance angiogenic synthetic biology. (1) Intercellular and (2) membrane-based reactions modeled via chemical kinetic models; (3) transcriptional signaling modeled by biological networks; (4, 5) cell–cell and cell–matrix interactions modeled via agent-based programming techniques using rules; (6, 7) cell movement through the matrix and blood flow modeled via partial differential equations.

7

MODELING THE INTERACTIONS BETWEEN COMPLIANT MICROCAPSULES AND PATTERNED SURFACES

ANNA C. BALAZS
Department of Chemical and Petroleum Engineering, University of Pittsburgh, Pittsburgh, Pennsylvania

ALEXANDER ALEXEEV
The George W. Woodruff School of Mechanical Engineering, Georgia Institute of Technology, Atlanta, Georgia

7.1 INTRODUCTION

It has been said that the development of microreactors, which involve the manipulation of minute quantities of reagents in micrometer-scale "flasks," could "do for synthetic chemistry what the solid-state transistor has done for computing," vastly increasing the speed at which large amounts of new chemical information can be obtained [1]. To realize this potential, researchers have recently created droplet-based devices, where micrometer-sized aqueous droplets are propelled by electric fields to move on a planar substrate. The entire assembly is immersed in a chamber that is filled with air or a liquid that is immiscible with the droplet. By applying voltages across specific electrodes on the surface, droplets are driven to move, merge, mix their contents, and split into smaller drops (for a recent review, see ref. [2]).

Multiscale Modeling of Particle Interactions: Applications in Biology and Nanotechnology,
Edited by Michael R. King and David J. Gee

One limitation of the foregoing approach, however, is that the range of chemical components or reactions that can be confined within aqueous drops is fairly restricted. In contrast, microcapsules, which are composed of a thin polymeric shell and an encapsulated fluid, constitute robust and versatile carriers. These microcapsules can be fabricated by a layer-by-layer deposition method [3–6], involving the sequential deposition of oppositely charged polyelectrolytes; typically, poly(styrene sulfonate) is used as the polyanion, and poly(allylamine hydrochloride) is used as the polycation [6]. Such polymeric microcapsules offer a range of advantages over pure fluid drops. In particular, a vast variety of compounds can be encapsulated within these microcapsules [7–10] and the assembly has significant storage stability [11]. The chemical composition of both the exterior and interior of the polymeric shell is easily tailored, and reagents can even be grown within the interior through a "ship in a bottle" approach [12]. The permeability of the microcapsule can be controlled in a straightforward manner [13] by modifying the external solution (e.g., by varying pH [14–17] or adding organic solvents [18]). In addition, the Young's modulus of the capsule's shell can be altered dramatically [3, 5] and nanoparticles can be incorporated into these shells to further alter the mechanical behavior, as well as the optical, thermal, and chemical properties [19]. Finally, the microcapsules can be driven to undergo fusion, allowing the contents of different capsules to be intermixed [20]. For all these reasons, the capsules constitute promising micrometer-sized reaction flasks.

What is needed, however, is an effective way of directing the polymeric capsules through the microfluidic devices; the assembly of electrodes described above was designed to propel simple fluid drops and has not been adapted for microcapsules [2]. One means of addressing this need is to design "smart" surfaces that can effectively regulate the motion of the capsules in the microchambers. The combination of responsive microcapsules and smart surfaces could lead to the ideal system for realizing the goals promised in the first sentence.

The fabrication of such smart surfaces will also address another vital need. Namely, one of the current challenges in designing microfluidic devices for bioassays involving biological cells is to keep the system sufficiently simple that the entire device can readily be transported, allowing it to be used at point-of-care locations [21,22]. In other words, the goal is to fabricate a device that works precisely and reliably without a tremendous amount of supporting electrical or optical equipment.

Both of the foregoing issues are part of a more general, overarching goal in fabricating the next generation of microfluidic devices: creating systems that can "sense" local variations and respond to these changes in a useful manner without the application of considerable external control [23]. For example, it would be highly desirable to create systems that can discriminate between various soft particles (colloids, microcapsules, and cells) moving through the microchannels and to direct certain species "autonomously" to specific locations. These devices could potentially perform a variety of separation processes in a highly cost-effective and energy-efficient manner.

Recent computational modeling studies are revealing a new approach for manipulating capsules and cells in microchannels and thereby providing potential guidelines for meeting the challenges noted above [24–39]. The approach proposed exploits the fact that the chemical and mechanical properties of the capsule's shell can readily be tailored and that the inherent chemomechanical properties of cells serve as identifying features. In other words, the adhesiveness and compliance of the synthetic capsules and the biological cells constitute a *code*, which, in turn, can be deciphered through the use of chemically and mechanically patterned substrates. Thus, by appropriate design of the substrate patterns, one can effectively "program" microfluidic devices to carry out multistage processes; once the surface has been patterned, no external controls (other than an imposed fluid flow) are needed to direct the system and the steering of the microcarriers operates in an essentially autonomous manner. The approach can be harnessed to create microreactors in which capsules of a specific compliance are selectively diverted to a precise location and then caused to burst spontaneously, discharging the encapsulated compounds. The concepts can also be used to create devices for sorting cells by their mechanical properties and thus facilitate assays for various diseases [28,29,32,34,37,38].

Below, we describe these modeling studies, providing examples that illustrate the fundamental design principles in creating such responsive microchannels in microfluidic devices. To carry out these studies, we integrated two mesoscale approaches to model a fluid-filled elastic shell that is immersed in a host solution and driven to move on a compliant surface. In our hybrid approach, the fluid dynamics is captured via the lattice Boltzmann model (LBM) [40], an efficient solver for the Navier–Stokes equation. The behavior of the elastic shell is simulated via the lattice spring model (LSM) [41–43], which consists of a network of harmonic springs connecting regularly spaced masses. Our integrated LBM/LSM approach allows for a dynamic interaction between the elastic walls and the surrounding fluid. In other words, dynamically and interactively, the moving walls exert a force on the fluid and, in turn, the fluid reacts back on the walls. Through this model, we can probe the dynamic interactions among all the components in the system (i.e., the host solution, the encapsulated fluid, the bounding elastic shell, and the compliant surface) and determine how these interactions affect the movement of the capsules in the system.

As reviewed elsewhere in this book (e.g., see Chapter 12), other researchers have also modeled biological cells as fluid-filled elastic shells and carried out computational studies of the cell's motion along substrates [44–48]. We note that in our systems, however, we are not considering the behavior of cells in vivo; rather, we exploit the ease with which the substrate's mechanical and chemical properties can be varied via the LSM to determine the optimal surface structure for regulating the cell's movement in vitro. An important reason for focusing on mechanically compliant surfaces is that recent advances in soft lithography make it possible to tailor the surface of polymeric materials significantly, making them the ideal systems for creating smart substrates. As we show below, simple mechanical or topographical patterning of these soft surfaces permits significant control over the motion of capsules (i.e., synthetic microcapsules or biological cells).

We emphasize that the LSM captures the behavior of an elastic, deformable *solidlike material* (or with addition of a dissipative unit between the springs, a viscoelastic material [49]); thus, the shells in our systems are also considered to be solidlike, as opposed to being fluidlike. Consequently, the systems that are most appropriately modeled here are the layer-by-layer microcapsules and polymersomes [50,51], where the high-molecular-weight polymers used to create the vesicles are entangled and cross-linked to enhance the mechanical integrity of the entity. In the case of white blood cells (leukocytes), the cytoskeleton acts to reinforce the plasma membrane, and thus the cell's outer casing can also be modeled as an elastic layer [26]. Our approach, however, may not be appropriate for capturing the behavior of vesicles formed from short-chain lipids since the lipid layer is commonly considered to be a dynamic, fluidlike entity [26].

We describe the LBM/LSM model in more detail below. We also correlate our simulation parameters with experimental values for polymeric microcapsules and biological cells. We then present our findings on the effects of surface compliance and topography on the fluid-driven motion of the capsules on these surfaces.

7.2 METHODOLOGY

Our hybrid LBM/LSM approach involves a coupling of the lattice Boltzmann and lattice spring models. We first describe the individual components of the approach and then detail how these components are integrated to capture interactions among the elastic shell, the surrounding fluid, and the substrate. We also discuss how we validated the model.

7.2.1 Lattice Boltzmann Model

The lattice Boltzmann model is a lattice-based method for simulating hydrodynamic flows. The model consists of two processes, the first being the propagation of fluid "particles" to neighboring lattice sites, and the second being collisions between particles when they reach a site. The system is characterized by a single-particle velocity distribution function, $f_i(\mathbf{r}, t) \equiv f(\mathbf{r}, \mathbf{c}_i, t)$, describing the mass density of fluid particles with velocity $\mathbf{c}_i$ at a lattice node $\mathbf{r}$ at time t. Here, $\mathbf{c}_i$, $\mathbf{r}$, and t are discrete variables, while the distribution function itself is a continuous variable. The hydrodynamic quantities are moments of the distribution function: the mass density $\rho = \sum_i f_i$, the momentum density $\mathbf{j} = \rho\mathbf{u} = \sum_i \mathbf{c}_i f_i$, with $\mathbf{u}$ being the local fluid velocity, and the momentum flux $\Pi = \sum_i \mathbf{c}_i \mathbf{c}_i f_i$.

The time evolution of the distribution function is governed by the discretized Boltzmann equation

$$f_i(\mathbf{r} + \mathbf{c}_i \Delta t,\ t + \Delta t) = f_i^*(\mathbf{r}, t) = f_i(\mathbf{r}, t) + \Omega[(\mathbf{f}(\mathbf{r}, t)]. \tag{7.1}$$

To illustrate the two steps (collisions and propagation), we define $f_i^*(\mathbf{r}, t)$ as the postcollision distribution function. The collision operator, $\Omega[(\mathbf{f}(\mathbf{r}, t)]$, accounts for the change in f_i due to instantaneous collisions at the lattice nodes; its action depends on all the f_i's at a node, denoted collectively by $\mathbf{f}(\mathbf{r}, t)$. We adopt a multi-relaxation-time collision operator, which in contrast to the more widely employed single-relaxation-time collision operator enables us to assign independent values to the shear and bulk viscosities. This collision operator conserves mass and momentum and relaxes the momentum flux (or stress) toward local equilibrium.

The velocity $\mathbf{c}_i$ in the ith direction is chosen such that fluid particles propagate from one lattice site to the next in exactly one time step, Δt. The 19 velocities of the three-dimensional model correspond to movement to the nearest- and next-nearest-neighbor directions of a simple cubic lattice and to rest particles. In the ensuing discussion, all dimensional parameters are expressed in terms of the lattice Boltzmann units for the lattice spacing, Δx, and the time step,Δt, where both values are taken to be unity.

7.2.2 Lattice Spring Model

The solid elastic material is represented by a lattice spring model which consists of a network of harmonic springs that connect regularly spaced mass points or nodes. The elastic energy associated with a node at position $\mathbf{r}_i$ is given by

$$E(\mathbf{r}_i) = \frac{1}{2}\sum_j k_j \left(r_{ij} - r_{ij}^{\mathrm{eq}}\right)^2, \tag{7.2}$$

where the summation runs over all nearest- and next-nearest-neighbor nodes. Here, $r_{ij} = \left|\mathbf{r}_i - \mathbf{r}_j\right|$ is the length of the spring between two nodes with positions $\mathbf{r}_i$ and $\mathbf{r}_j$, r_{ij}^{eq} is its equilibrium length, and k_j is its spring constant. This results in a spring force $\mathbf{F}_S$ on node $\mathbf{r}_i$ of the form

$$\mathbf{F}_S(\mathbf{r}_i) = -\sum_j k_j \left(\frac{r_{ij} - r_{ij}^{\mathrm{eq}}}{r_{ij}}\right)\mathbf{r}_{ij}. \tag{7.3}$$

To capture the dynamics of the solid material, we assign a mass M_i to each node $\mathbf{r}_i$ and integrate Newton's equation of motion, $\mathbf{F}(\mathbf{r}_i) = M_i(\partial^2\mathbf{r}_i/\partial t^2)$, where $\mathbf{F}$ is the total force acting on the node. A friction term could be introduced within (7.3) to account for dissipation. The total force $\mathbf{F}$ consists of the spring force in Eq. (7.3) and the force exerted by the fluid on the solid at the solid–fluid boundary, discussed in more detail below. We utilize the velocity-Verlet algorithm to integrate Newton's equation of motion. This is a well-known molecular dynamics algorithm that updates the positions, velocities, and accelerations of each node in discrete time steps.

For small deformations, this system of equations can be shown to obey linear elasticity theory and results in a Young's modulus $E = 5k/2\Delta x_{\mathrm{LS}}$ and a speed of sound $c_s = \Delta x_{\mathrm{LS}}\sqrt{3k/M}$, where k is the spring constant, M is the mass of the node, and Δx_{LS} is the lattice spacing in the LSM. The solid density is then given by $\rho_s = M/\Delta x_{\mathrm{LS}}^3$. This simple model is restricted to a Poisson's ratio of $\nu = \frac{1}{4}$, although more complicated many-body interactions can be included in order to vary ν [43].

Our system has different solid domains: the compliant underlying substrate and the compliant shells of the individual microcapsules (see Fig. 7.1). The capsule's three-dimensional shell is constructed from two concentric layers of LSM nodes; using the Delaunay triangulation technique [52], we distribute nodes in a regular manner on each surface. These two concentric surfaces are separated by a distance that is equal to the average size of a triangular bond and are connected by springs between the nearest- and next-nearest-neighbor nodes. The spring constant for springs located on the capsule surfaces and normal to the surfaces is k, while the diagonal springs (omitted for the sake of clarity) have spring constants of $2/3k$.

The velocity-Verlet algorithm that we use to integrate Newton's equation of motion is explicit in time. For stability purposes, this requires that we choose a lattice spacing, Δx_{LS}, and time step, Δt_{LS}, in the lattice spring part of the code such that the Courant number $\mathrm{Cr} = c_s\, \Delta t_{\mathrm{LS}}/\Delta x_{\mathrm{LS}}$ is smaller than 1. Furthermore, we must maintain $\Delta x_{\mathrm{LS}} \geq \Delta x$, which prevents large fluctuations in the force of the fluid on the lattice spring nodes due to the discretization. In our applications, we set $\Delta t_{\mathrm{LS}} = \Delta t = 1$ (equal to the time step in the lattice Boltzmann part of the code), while $\Delta x_{\mathrm{LS}} \geq 1.25\Delta x$ and $\Delta x = 1$. This allows us to satisfy the Courant condition for the entire range of c_s that was used to obtain our results.

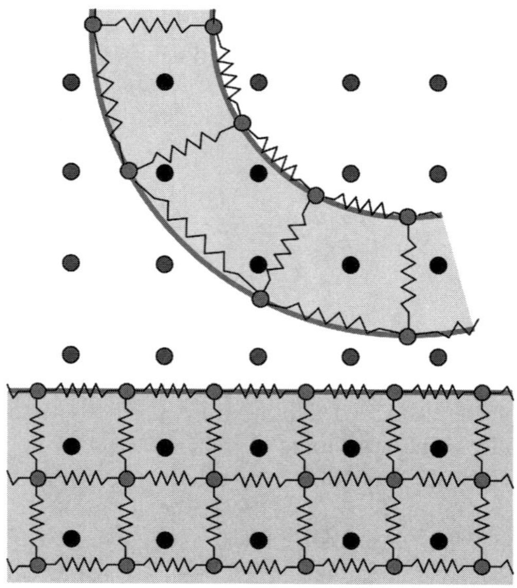

Figure 7.1 Schematic of the system. The nodes that are connected by springs form the lattice spring grid (for clarity, we omitted the diagonal springs that connect each node with all its next-nearest neighbors). The remaining nodes represent the lattice Boltzmann grid. The solid lines are the solid–fluid interfaces.

7.2.3 Solid–Fluid Boundary Condition

In order to capture the interactions between the solids and fluids, lattice spring nodes that are situated at the solid–fluid interface must impose velocities on the surrounding fluids through boundary conditions and, in turn, experience forces due to the fluid pressure and viscous stresses. Put concisely, the simulation proceeds through the sequential update of both the lattice spring and the lattice Boltzmann systems. The LSM system is updated by first calculating the forces that are acting on the LSM nodes due to the LSM springs, the enclosed fluid, and the surrounding solvent. New positions, velocities, and accelerations of the LSM nodes are then calculated using the velocity-Verlet algorithm. In updating the LBM system, we first establish which LBM links intersect the solid–fluid interface. We then obtain the velocities at these points of intersection from neighboring LSM nodes. Next, we propagate the distribution function by streaming fluid particles to their neighboring nodes whenever these nodes are in the fluid domain, and otherwise, we apply the appropriate boundary condition (described below). Finally, we modify the distribution functions at the LBM nodes to account for the collision step. We then repeat the entire cycle.

Before describing these processes in more detail below, we note that the values for the masses of nodes at the solid–fluid interface and the spring constants of the corresponding interconnecting springs are half the values of their bulk counterparts. The latter specifications facilitate the implementation of the solid–fluid boundary condition by placing the solid–fluid interface at the actual nodal positions.

Fluid particles that are moving on a link that intersects the solid–fluid interface are bounced back into the fluid phase at the intersection point (or boundary node) $\mathbf{r}_b$, as illustrated in Fig. 7.2. Here, $\delta_1 = |\mathbf{r}_1 - \mathbf{r}_b| / |\mathbf{r}_1 - \mathbf{r}_2|$, with $\mathbf{r}_1$ being the fluid node at which the fluid particles originate and $\mathbf{r}_2 = \mathbf{r}_1 + \mathbf{c}_i\,\Delta t$ being the neighboring solid node in the direction of the fluid particles' motion. For $\delta_1 = \frac{1}{2}$, these particles will arrive back at node $\mathbf{r}_1$ after precisely one time step, with a velocity that is opposite in direction to their original motion. This situation represents the well-known link bounce-back rule. However, for $\delta_1 \neq \frac{1}{2}$, fluid particles will end up at positions that do not coincide with a regular lattice node, and some sort of interpolation is needed. We follow the scheme that is developed by Bouzidi et al. [53] and described in detail in ref. [26].

First we consider the case of a stationary interface. For $\delta_1 < \frac{1}{2}$, we obtain $f_i^*(\mathbf{r}_0, t)$, the postcollision distribution at position $\mathbf{r}_0$, by linear interpolation between $f_i^*(\mathbf{r}_1 - \mathbf{c}_i \Delta t, t)$ and $f_i^*(\mathbf{r}_1, t)$ (Fig. 7.2b). After traveling for one time step and being reflected at the boundary node, the interpolated postcollision distribution at $\mathbf{r}_0$ will end up precisely at node $\mathbf{r}_1$ with a velocity in the opposite direction (Fig. 7.2b). Hence,

$$f_k(\mathbf{r}_1, t+\Delta t) = f_i^*(\mathbf{r}_0, t) = (1-2\delta_1) f_i^*(\mathbf{r}_1 - \mathbf{c}_i\,\Delta t, t) + 2\delta_1 f_i^*(\mathbf{r}_1, t), \quad \delta_1 < \tfrac{1}{2}. \tag{7.4}$$

Here the subscript k stands for fluid particles with a velocity in the direction opposite that of the incoming particles (i.e., $\mathbf{c}_k = -\mathbf{c}_i$). For $\delta_1 \geq \frac{1}{2}$ we propagate the fluid particles at node $\mathbf{r}_1$ such that they end up at the position $\mathbf{r}_0$ [i.e.,

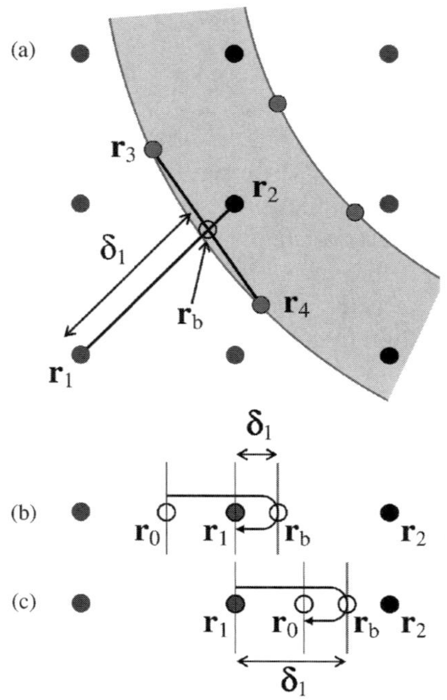

Figure 7.2 Implementation of solid–fluid boundary conditions (see the text for details).

$f_k(\mathbf{r}_0, t+\Delta t) = f_i^*(\mathbf{r}_1, t)]$ (Fig. 7.2c). We then obtain $f_k(\mathbf{r}_1, t+\Delta t)$ by linear interpolation between $f_k(\mathbf{r}_1-\mathbf{c}_i\,\Delta t, t+\Delta t)$ and $f_k(\mathbf{r}_0, t+\Delta t)$ as follows:

$$\begin{aligned} f_k(\mathbf{r}_1, t+\Delta t) &= \frac{2\delta_1-1}{2\delta_1} f_k(\mathbf{r}_1-\mathbf{c}_i\,\Delta t, t+\Delta t) + \frac{1}{2\delta_1} f_k(\mathbf{r}_0, t+\Delta t) \\ &= \frac{2\delta_1-1}{2\delta_1} f_k^*(\mathbf{r}_1, t) + \frac{1}{2\delta_1} f_i^*(\mathbf{r}_1, t), \qquad \delta_1 \geq \tfrac{1}{2} \end{aligned} \tag{7.5}$$

For a moving interface, we have to account for the velocity $\mathbf{u}_b$ of the solid material at the intersection point $\mathbf{r}_b$. This leads to additional terms in Eqs. (7.4) and (7.5) that are proportional to the component of the velocity $\mathbf{u}_b$ in the direction of the fluid particle's velocity:

$$\begin{aligned} f_k(\mathbf{r}_1, t+\Delta t) = {} & (1-2\delta_1) f_i^*(\mathbf{r}_1-\mathbf{c}_i\Delta t, t) + 2\delta_1 f_i^*(\mathbf{r}_1, t) \\ & - \frac{2\alpha_i}{c_f^2}\rho(\mathbf{r}_1, t)\mathbf{c}_i\cdot\mathbf{u}_b(\mathbf{r}_b, t), \qquad \delta_1 < \tfrac{1}{2} \end{aligned} \tag{7.6a}$$

$$f_k(\mathbf{r}_1, t+\Delta t) = \frac{2\delta_1-1}{2\delta_1} f_k^*(\mathbf{r}_1, t) + \frac{1}{2\delta_1} f_i^*(\mathbf{r}_1, t) - \frac{\alpha_i}{\delta_1 c_f^2}\rho(\mathbf{r}_1, t)\mathbf{c}_i\cdot\mathbf{u}_b(\mathbf{r}_b, t), \qquad \delta_1 \geq \tfrac{1}{2} \tag{7.6b}$$

Here, $\alpha_i = \frac{1}{18}$ for the orthogonal directions and $\frac{1}{36}$ for the diagonal links. The velocity $\mathbf{u}_b$ is obtained through a straightforward interpolation between the velocities of the lattice spring nodes around the intersection point (Fig. 7.2a). This implementation gives a no-slip boundary condition at the solid–fluid interface, with errors that are of second-order accuracy in the spatial discretization.

As a result of the bounce-back process, the fluid exerts a force on the solid–fluid interface. This force is taken to be equal to the rate of exchange in momentum that takes place as the fluid particles are reflected at the interface:

$$\mathbf{F}_b(\mathbf{r}_b, t + \tfrac{1}{2}\Delta t) = \left\{ \left[f_i^*(\mathbf{r}_1, t) - \alpha_i\,\rho(\mathbf{r}_1, t) \right] \mathbf{c}_i - [f_k(\mathbf{r}_1, t + \Delta t) - \alpha_i\,\rho(\mathbf{r}_1, t)] \mathbf{c}_k \right\} \frac{\Delta x^3}{\Delta t}. \tag{7.7}$$

Here the terms $\alpha_i \rho(\mathbf{r}_1, t)$ compensate for the ambient pressure, ensuring that the force on the interface is zero when the entire system is at rest. $\mathbf{F}_b$ is then distributed to the lattice spring nodes near the intersection point, while conserving the normal and tangential force on the interface.

7.2.4 Capsule–Substrate Adhesive Interaction

The interaction between the capsule and the substrate is modeled through a Morse potential, which is given by

$$\phi(r) = \varepsilon \left[1 - \exp\left(\frac{r_0 - r}{\kappa} \right) \right]^2. \tag{7.8}$$

Here ε and κ characterize the respective strength and range of the interaction potential, and r is the distance between a pair of LS nodes, where one node lies on the capsule's outer surface and the other lies at the substrate–fluid interface (Fig. 7.3). The parameter r_0 is the equilibrium distance where the force due to the potential is zero

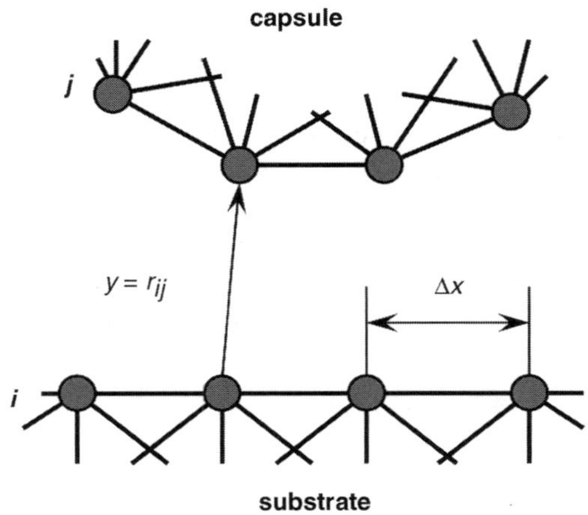

Figure 7.3 Implementation of the capsule–substrate adhesive interaction (see the text for details).

and is set to 1.8 in the ensuing simulations. To prevent overlap between the capsule and the substrate, we fix $\varepsilon = \varepsilon_r = 0.005$ for the repulsive part of the potential ($r < r_0$), while the values for the adhesive part ($r > r_0$) $\varepsilon = \varepsilon_a$ are specified below. In addition, we set $\kappa = 1$. Note that all the values above are in LBM units.

It is worth noting that interaction potentials similar to the Morse potential have been used to model the adhesion between vesicles and a substrate [54–56]. In general, in order to capture physically realistic behavior, it is important to utilize an interaction potential that includes a strongly repulsive short-range interaction and a weakly attractive long-range interaction [54]. Numerically, the Morse potential is particularly suitable for these calculations because it has a nearly harmonic repulsive part, which simplifies coupling with the LSM. Although the specific form of the potential may depend on particular experimental conditions, our findings will hold at least qualitatively as long as the general shape of the potential is preserved.

7.2.5 Rupture of Capsule

High-velocity gradients in the shear flow and strong capsule–substrate adhesion leads to significant strain in the capsule's shell and, ultimately, to rupturing of the capsule [36]. The strain in a lattice bond connecting sites i and j is equal to $s_{ij} = \left(r_{ij} - r_{ij}^{\mathrm{eq}}\right)/r_{ij}^{\mathrm{eq}}$, where r_{ij}^{eq} is the equilibrium length of the bond. To model capsule rupture, we selectively remove LS bonds having a strain s_{ij} exceeding the yield value, $s_{\mathrm{crit}} = 0.1$. Through this process, new surfaces are created on the capsule's shell, modeling the formation and propagation of cracks. These cracks expose new fluid–solid interfaces and we apply the boundary conditions described above at these new interfaces.

To identify the location of these new interfaces, we divide the entire capsule's shell into volume elements formed from tetrahedrons; the vertices of these tetrahedrons are LS nodes, and edges are LS bonds. We note that some of the bonds are associated with two elementary tetrahedrons. If a breaking LS bond is located on an edge, we eliminate all associated tetrahedrons and define the exposed surfaces of the neighboring tetrahedrons to be new solid–fluid interfaces. (If the breaking LS bond is not an edge, we simply remove it.) The volume of the tetrahedrons eliminated is filled with a fluid whose local properties are extrapolated from the surrounding LBM and LSM nodes. If, due to this process of eliminating elementary tetrahedrons, a bond happens not to be associated with any volume element, we remove this bond; for example, this situation occurs when a bond connects two otherwise separate fragments of the shell. When the capsule bursts, the resulting shell fragments interact through the repulsive part of the Morse potential (with $\varepsilon = \varepsilon_r$ and $r_e = 0.9$) and thus are prevented from overlapping.

7.2.6 Validation

We validated our LBM/LSM model by calculating the drag force on a periodic array of spheres [26]. We fixed the density and the shear viscosity of the fluid to

$\rho_f = 1$ and $\mu = \frac{1}{6}$, respectively. We constructed a stationary spherical shell and varied the outer diameter of the sphere to study the effect of volume fraction $\varphi = \frac{4}{3}\pi R^3/L^3$ on the accuracy of our numerical method; here R is the radius of the sphere and L the length of the cubic simulation box. We varied the volume fractions from 0.05 to 0.5 and then compared our findings to the theoretical predictions [26]. We found that the error is less than 1% for $\varphi < 0.2$ and does not exceed 6% for larger volume fractions.

To further verify our model, we calculated the breathing mode oscillations of empty and fluid-filled shells in vacuum and compared our results with theory [24,26]. We again fixed the density and viscosity of the encapsulated fluid. Two concentric layers of LSM nodes represented the shell. We fixed the shell density at $\rho_s = 1$ and varied the spring constants k to determine the effect of the shell stiffness on the frequency of the breathing mode oscillations. For the empty shell and for the first-mode oscillations of the fluid-filled shell, the difference between the numerical and analytical models is less than 0.5%. For the second-mode oscillations (which reflect the oscillations occurring within the solid shell of the capsule), the error is less than 5%. Note that in our model, the shell is represented by only two layers of LSM nodes; nonetheless, our model can accurately capture the second-mode oscillations of the fluid-filled shell in vacuum.

7.2.7 Simulation Parameters

As noted above, we set the density and viscosity of both the host and the encapsulated fluids to $\rho_f = 1$ and $\mu = \frac{1}{6}$, respectively. Unless specified otherwise, we chose the following parameters for our simulations. The outside radius of the undeformed microcapsule is $R = 10$ and the capsule density is $\rho_c = 1$. The number of LSM nodes on this outer radius is $N_c = 642$; the average spacing between nodes is $\Delta x_c \approx 1.4$ and the distance between the outer and inner layers (thickness of the capsule's shell) is $h = 1.4$. Our simulation box has length L_x, height L_y, and width L_z. The top wall (at $y = L_y$) moves with a constant velocity $(U_0, 0, 0)$ and serves to drive the flow with a uniform shear rate $\dot{\gamma} = U_0/L_y$. Periodic boundary conditions are applied in the x and z directions. The stationary bottom wall (at $y = 0$) represents the floor of our substrate. The spacing between the LSM nodes in the substrate is $\Delta x_s = 1.25$; the substrate thickness is $h_s = 10$ and density $\rho_s = 1.5$.

7.2.8 Comparison of Simulation Parameters with Experimental Values

For the majority of the scenarios considered here, an imposed fluid flow drives the capsules to move along the bottom wall of a microchannel. We can characterize the behavior of the system by the dimensionless capillary number,Ca $= \dot{\gamma}\mu R/E_c h$, where $\dot{\gamma}$ is the shear rate, μ is the shear viscosity of the host fluid, R is the radius of the undeformed capsule, E_c is the Young's modulus of the capsule's shell, and h is the shell thickness. The capillary number represents the relative importance of the viscous stress in the surrounding fluid and elastic stress in the capsule's shell. Another dimensionless number that characterizes the system is the Reynolds number, $\mathrm{Re} = \rho\dot{\gamma}R^2/\mu$, where ρ is the density of the host fluid. The Reynolds number is a measure of the relative importance of inertial and viscous effects. In our simulations, we set the shear rate such that $\mathrm{Re} \sim 0.1$, which is still sufficiently small that inertial effects in the fluid can be neglected.

We also define a dimensionless interaction strength, $\Phi = \varepsilon N/E_c h\kappa^2$, where ε and κ characterize the interaction potential (i.e., Morse potential) between the capsule and the substrate and N is the number of interacting nodes; thus, εN characterizes the total adhesion interaction of the capsule. The parameter Φ represents the ratio of the interaction strength to the membrane stiffness $E_c h$. For $\Phi \gg 1$, the interaction leads to significant deformation of the capsule, while for $\Phi \ll 1$, the effect on its shape is small. Finally, to characterize the importance of the adhesive force relative to the viscous force acting on a capsule, we introduce the dimensionless parameter $\Lambda = \varepsilon_a N/\dot{\gamma}\mu R\kappa^2$. Note that $\Lambda = \Phi/\mathrm{Ca}$.

By expressing the data in terms of such dimensionless numbers, we can provide more flexible guidelines for potential experimental studies. When using the capillary number, the experimentalist is free to choose, for instance, a combination of capsule sizes, Young's modulus, and shell thickness that will result in the ratio specified. In the studies presented here, we examined phenomena that span the Ca range 10^{-3} to 10^{-4} since this range yielded the most intriguing behavior and, just as important, is relevant to a number of experimental systems that are described below.

We first consider the case of polyelectrolyte microcapsules which are propelled by an applied shear in an aqueous solution whose viscosity $\mu \approx 10^{-3}$ kg/s · m and density $\rho \approx 10^3$ kg/m^3 [26]. The elasticity of the microcapsule's shell is on the order of 0.1 to 1 GPa [5,57,58], and these elastic properties are found to be independent of the speed of deformation [57]. (We note that Young's modulus of the membrane can be tailored by varying the salt concentration in the solution [59] or the solvent temperature [60].) In typical flow experiments within a microchannel, the shear rate can be up to $\dot{\gamma} \approx 10^3$ to 10^4 s^{-1} [61]. If we take experimentally realistic values of the membrane thickness h to be about 50 nm for a 50-μm capsule [58], we obtain $\mathrm{Ca} \sim 10^{-4}$, which is in the range considered in the simulations.

A polyelectrolyte coating on the substrate can be utilized to produce an adhesive interaction between the microcapsule and the substrate. The total adhesion energy can be estimated to be on the order of $w \sim 10^{-14}$ J/capsule [62], while the interaction range is roughly $\kappa \sim 10^{-8}$ to 10^{-7} m [63]. If we set w equal to εN in the expression for Φ and take $E_c h$ from the example above (0.1 to 1 GPa × 50 nm), we find that Φ is on the order of unity. The latter value is in the range of the parameters considered in our simulations.

The model is also appropriate for modeling the interactions between leukocytes and a substrate. Leukocytes have a cytoskeleton that serves to maintain the spherical shape of the cell, and these cells have in fact been modeled as fluid-filled elastic shells [44–48,64,65]. In such modeling studies, researchers typically use an effective membrane stiffness that lies in the range 0.01 to 0.3 N/m. The shear rate in the blood vessels is about $\dot{\gamma} \approx 100$ to 500 s^{-1}. Thus, for leukocytes with radii on the order of 5 μm in an aqueous solution, one obtains typical values of the capillary number

$\text{Ca} \sim 10^{-4}$, which again is in the range of our numerical simulations. We therefore expect that qualitatively similar behavior can be observed in experimental studies on the interactions between leukocytes and the endothelium, or in studies where ex vivo leukocytes are used as the fluid-driven capsules (and the imposed shear rate is comparable to $\dot{\gamma} \approx 100$ to $500\ \text{s}^{-1}$, so that Ca remains in the range cited above). For micronmeter-scale biological particles interacting with substrates, the adhesion energy is in the range $w \sim 10^{-19}$ to 10^{-15} J [34,55,56,66,67] and $\kappa \sim 10$ nm [34,55,56,66]. Thus, values of Φ on the order of unity lie within experimentally realistic values.

7.3 RESULTS AND DISCUSSION

In the following sections we discuss how to harness patterned substrates to modify the free-energy landscape in the system and thereby control the motion of fluid-driven capsules in microfluidic channels. We show how the introduction of simple surface patterns permits the separation of microcapsules according to their compliance, the routing of capsules to specific locations in the system, and the selective trapping and bursting of specified microcarriers. We note that for both biological cells and synthetic microcapsules, mechanical compliance is a key parameter, since it can reveal the presence of disease in the former case and the quality of the fabricated product in the latter case. To date, however, assessing the mechanical properties of such micrometer-scale particles in an efficient, cost-effective means remains a critical challenge. Recent advances in soft lithography, however, make it possible to significantly tailor the chemical, mechanical, and topological features of polymeric substrates. As we describe below, we exploit these advances to design low-cost devices that can ultimately perform continuous on-the-fly separation processes.

7.3.1 Stripes That Segregate

In microchannels with adhesive or "sticky" substrates, fluid-driven capsules experience not only the actions of a hydrodynamic drag force, but also forces due to interactions with the sticky surface. On a *uniformly* adhesive substrate, capsules simply travel in the direction of the imposed flow, being drawn by the viscous drag; the adhesive force just points toward the surface and keeps the capsules localized on the substrate by balancing the hydrodynamic lift. When the surface adhesion is *nonuniform*, however, the adhesive force is not directed solely toward the substrate, but also includes an in-plane component, which directs the capsule toward the location with the strongest adhesion and thereby reduces the free energy of the system. In other words, there is a free-energy minimizing thermodynamic force that affects the motion of capsules along substrates that exhibit spatially nonuniform adhesion. Moreover, elastic forces due to the deformation of compliant capsules and substrates also contribute to the total free energy of the system and alter the thermodynamic force.

To illustrate how patterned substrates can modify the free-energy landscape, we first consider the properties of the system in the absence of imposed flow and focus on two types of patterning: chemical and mechanical [32]. In the former, the substrate is rigid and includes an adhesive stripe. In the latter, the surface is uniformly adhesive but includes a stripe that is softer than the rest of the substrate.

Figure 7.4 reveals the changes in the energy W associated with the adhesive interaction between the capsule and the substrate, and the deformations due to this interaction for the two patterns described above. The adhesive interaction energy is measured directly from the Morse potential acting between the substrate and the capsules, while the energy associated with the elastic deformation is calculated from

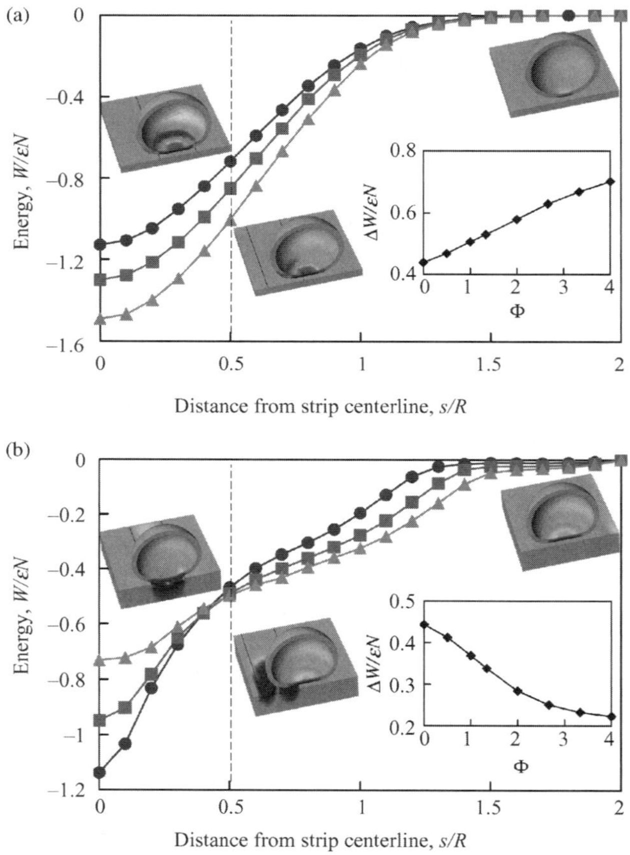

Figure 7.4 Energy variation as a function of distance from the centerline of (a) adhesive and (b) soft stripes for capsules with stiffness of $\Phi = 2$ (triangles) and $\Phi = 1$ (squares), and for rigid capsules (circles). The vertical lines mark the stripes' boundaries. From left to right, capsules with $\Phi = 2$ at $s = 0, 0.7$, and 2; colors indicate the strain as in Figs.7.5 and7.6. The insets show the energy drop across the stripe versus the capsule stiffness. (From[32], with permission. Copyright © 2007, American Chemical Society.) (*See insert for color representation of figure.*)

the strain within the spring network of our LSM. To calculate W, we fix s, the distance between the capsule's center of mass and the stripe's centerline, while allowing the capsule to equilibrate in the vertical direction. We increase s and repeat this calculation to obtain the dependence of W on s (Fig. 7.4), where W is measured relative to the energy of a capsule on the homogeneous bulk of the substrate. In both cases, W has a minimum at $s = 0$ and increases with increasing s, meaning that the interaction energy depends on the change in the local properties of the patterned substrate.

The findings in Fig. 7.4 demonstrate that the stripes introduce local free-energy minima into the system and thereby alter the energy landscape. The inherent energy minimization gives rise to a force on the capsules, $F_w \sim \partial W/\partial s$, which acts along the surface and is directed toward the centerline. The superposition of F_w on the drag force due to the flow, F_d, controls the capsule's motion along the patterned substrate and alters the capsule trajectory.

Although both types of surface patterning result in similar behavior for W, the reasons for the energy variation with s are different. For the chemically patterned substrate with a sticky stripe, the decrease in energy at $s = 0$ is due primarily to the following: (1) a stronger interaction within the more adhesive strip and (2) the increase in contact area A between the capsule and substrate due to a greater deformation of the capsule within the stripe (see Fig. 7.4a). The value of A depends on the capsule's elasticity; the softer shell experiences a larger change in the contact area and thus a larger variation in W when the capsule moves onto the adhesive stripe (inset, Fig. 7.4a, where $\Delta W$is the amplitude of the energy variation over the stripe). Because ΔW depends on the capsule's mechanical properties, the force F_w attains different magnitudes for soft and stiff capsules.

For the mechanically patterned, chemically uniform substrate, the largest contribution to ΔW comes from changes in the contact area A; however, the energy of elastic deformations within the substrate and capsule's shell also contribute to ΔW. The change in A is due primarily to deformations that occur in the soft stripe as it "envelops" the capsule (see Fig. 7.4b). The most dramatic change in A, and consequently W, occurs for a rigid capsule (inset, Fig. 7.4b). Thus, on the mechanically patterned substrate, F_w is lower for softer capsules.

As demonstrated above, patterned surfaces can yield different magnitudes of the interaction forces for capsules with different stiffness. This observation allows us to design a patterned substrate that can discriminate between microcapsules of differing mechanical compliance [32]. To this end we introduce a series of stripes on an otherwise uniform substrate. A key element here is that the stripes are oriented at an angle of $\alpha = 45°$ relative to the flow direction. Thus, the force due to interaction with the stripe, F_w, is not aligned with the flow direction and therefore will tilt the capsule's trajectory as it crosses the stripes. Moreover, since F_w depends on the capsule's elasticity, different capsules will follow different trajectories and effectively separate themselves by their mechanical properties.

Figure 7.5 shows snapshots of the capsule motion across a rigid, chemically patterned substrate that contains two stripes of width R, which are oriented $\alpha = 45°$ relative to the flow direction. The stripes display an adhesive strength of

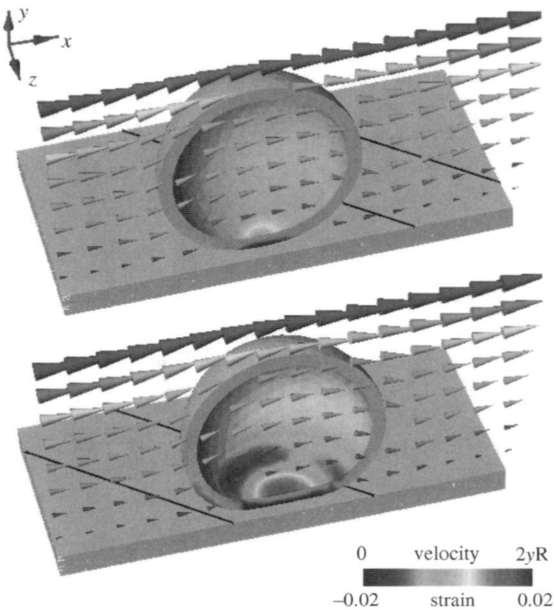

Figure 7.5 Elastic capsule rolling along a rigid substrate patterned with adhesive stripes in shear flow. The system is shown in cross section (made through the capsule's center of mass). The lines on the substrate indicate the boundaries of a sticky stripe. The colors in the capsule and substrate reveal the strain (see the color bar). The arrows indicate the flow direction and the arrow's color indicates the magnitude of the velocity. The capsule is located in front of a sticky stripe (upper panel) and is crossing a sticky stripe (lower panel). Note that the flattening of the capsule is due to a stronger adhesion with the sticky stripe. The capsule's deformation enhances the contact area with the substrate. (From [32], with permission. Copyright © 2007, American Chemical Society.) (*See insert for color representation of figure.*)

$\varepsilon_a = 5.43 \times 10^{-4}$, whereas the bulk of the surface exhibits a fivefold decrease in the adhesive interaction, with $\varepsilon_a = 1.09 \times 10^{-4}$. The latter interaction strength is sufficiently strong to prevent the capsule's detachment by a lift force but exerts only a minor influence on the capsule's shape. In contrast, adhesion within the patterned stripes is sufficiently strong to cause a substantial deformation of the capsule (Fig. 7.5).

Figure 7.6 displays the capsule's motion along a flat, mechanically patterned substrate that contains two stripes of material with a mechanical stiffness that is 100 times lower than the rest of the substrate. In this case, the capsule–substrate adhesion is characterized by $\varepsilon_a = 5.43 \times 10^{-4}$ along the entire surface. Note that the rolling capsules do not distort the stiffer regions, but they can significantly deform the compliant stripes (Fig. 7.6).

Figure 7.7 shows the trajectories of the capsule's center of mass motion on the chemically (Fig. 7.7a) and mechanically (Fig. 7.7b) patterned substrates. The distance between the centerlines of the two stripes is $3R$. We define Δz_2 as the total lateral

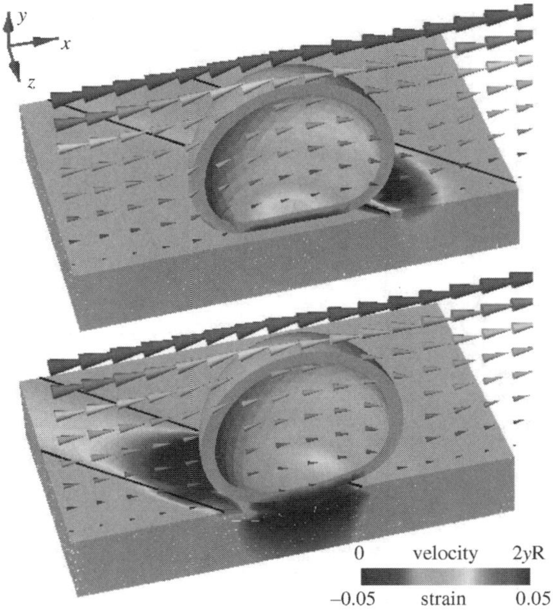

Figure 7.6 Elastic capsule rolling along an adhesive, mechanically patterned substrate in shear flow. The system is shown in cross section (made through the capsule's center of mass). The lines on the substrate mark the boundaries of a soft stripe. The characteristics of the strain and velocity are depicted in the same manner as in Fig. 7.5. Capsule approaches (upper panel) and moves past (lower panel) a soft stripe. Note the strong deformation of the soft stripe due to the adhesive interaction with the capsule, and consequently, an increase in contact area between the capsule and substrate. (From [32], with permission. Copyright © 2007, American Chemical Society.) (*See insert for color representation of figure.*)

displacement of the capsule in the direction transverse to the fluid flow, after it has traversed both stripes (Fig. 7.7a). Although located at the same initial position, capsules with different E_c values attain remarkably different Δz_2 values, indicating that the capsule's stiffness dictates its response to the patterned substrate. With chemical patterning, the softest capsule exhibits the largest lateral displacement (Fig. 7.7a), whereas with mechanical patterning, the rigid capsule attains the largest displacement Δz_2 (Fig. 7.7b).

The trajectories for capsules with different E_c nonetheless show some common trends, which we describe using the example of a rigid capsule on the chemically patterned surface (Fig. 7.7a). Driven by the fluid flow, the capsules initially roll on a homogeneous part of the substrate, and their trajectories are parallel to the flow direction, corresponding to section a–b in Fig. 7.7a. As the capsules interact with the strip, the adhesive force F_w pushes the capsule toward the centerline of the stripe, as indicated by the displacements in section b–c. Upon reaching this centerline, the capsules' lateral displacements change direction (section c–d); this is because F_w changes its direction as the capsules cross the centerline. Now, F_w tends to keep the

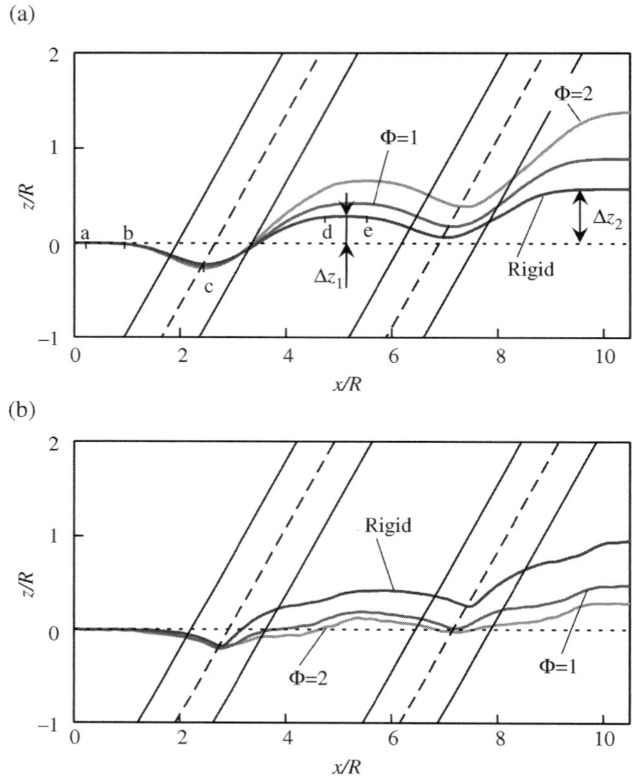

Figure 7.7 Trajectories of the center-of-mass motion for capsules with different compliance on substrates with (a) adhesive and (b) soft diagonal stripes. The substrates encompass two consequent stripes, which are oriented 45° relative to the flow direction. Starting from the same initial location, the capsules then move from left to right. As a result of the interaction with the stripes, capsules of different stiffness gain different lateral displacements. (From [32], with permission. Copyright © 2007, American Chemical Society.)

capsules within the stripe, as they are propelled by the imposed flow. As a result of traversing the entire stripe, the capsules gain a net lateral displacement Δz_1. The capsules continue to move in a straightforward manner (section d–e) until they reach the following stripe. Repeating the dynamic behavior described above, the capsules attain an additional displacement of Δz_1 such that the total displacement over two stripes yields $\Delta z_2 \approx 2\Delta z_1$. These results reveal that mechanical or chemical patterning can indeed be harnessed to segregate capsules of different stiffness. Note that multiple stripes will enhance this effect.

An important feature of an effective sorting device is the ability to tune the sensitivity of the apparatus. In the design proposed, the sensitivity of the device can be controlled simply by varying the flow rate within the chamber. This point is illustrated in Fig. 7.8, where we plot the dependence of Δz_1 on the capsule's stiffness for different

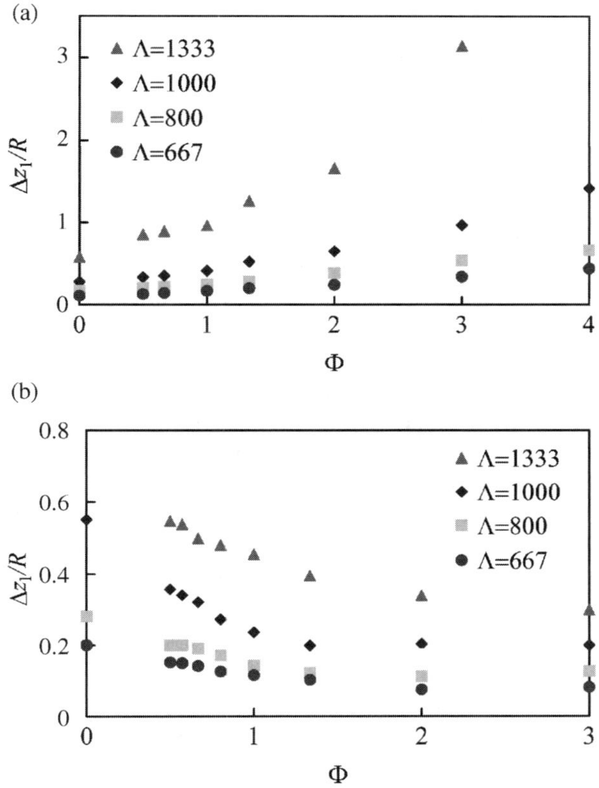

Figure 7.8 Capsule's lateral displacement after moving over (a) adhesive or (b) soft stripe as a function of capsule stiffness for different shear rates, $\dot{\gamma}$. Here we specify the shear rate in terms of the dimensionless parameter $\Lambda = \varepsilon_a N/\dot{\gamma}\mu R\kappa^2$.

values of the shear rate $\dot{\gamma}$. Note that the data in Fig. 7.8 can be used to extract the capsules stiffness $E_c h$ by measuring Δz_1, which is directly related to Φ. As shown in Fig. 7.8, a decrease in shear rate $\dot{\gamma}$ leads to an increase in Δz_1, indicating that reducing the flow rate enhances the sensitivity of the device. Thus, at relatively low $\dot{\gamma}$, even minute variations in capsule stiffness can be registered, due to the relatively large differences in the respective values of Δz_1. (Recall that such differences in the lateral excursions can be further amplified by introducing additional stripes.)

By comparing Fig. 7.8a and b, it is clear that the chemically patterned substrate produces a larger variation in Δz_1 for softer capsules (Fig. 7.8a), while for the mechanical patterning, Δz_1 grows rapidly for $\Phi < 1$ (Fig. 7.8b). This indicates that the chemical patterning is optimal for separating softer capsules, whereas the mechanical patterning is more effective for segregating stiffer capsules. In either case, patterned substrates can readily be incorporated into microfluidic devices, and capsules of differing compliance can be streamed continuously over the surfaces, leading to a continuous sorting of these particles by their mechanical properties.

We also note that since the energy variation ΔW depends on the local interaction strength between the capsule and the substrate, this approach can be used to separate capsules with identical mechanical properties but different adhesiveness or binding energies. This is an important attribute that can be utilized to characterize or identify the chemical composition of a cell's surface.

7.3.2 Taking the Fork in the Road

We can build on the observation above to design a smart patterned surface that effectively routes mechanically distinct capsules to specific regions and thus regulates traffic in a microfluidic device. To this end we introduce a simple Y- shaped pattern on the floor of the microchannel (Fig. 7.9). With respect to the stem, one branch of this Y is relatively soft, whereas the other branch is relatively sticky. In Section 7.3.1, we found that mechanically different capsules are deflected differently when they cross soft or sticky diagonal stripes. Indeed, the soft, mechanically deformable stripes are more effective at diverting the stiffer microcapsules, while the sticky stripes are more effective at diverting the softer ones. As discussed above, for soft and sticky stripes this behavior is due to the different mechanisms involved in minimizing the system's free energy. The soft stripe is more deformed by the stiffer capsules. This leads to a greater change in the contact area between the hard capsule and attractive surface and a larger force for diverting the stiffer capsules to move along the stripe. The higher

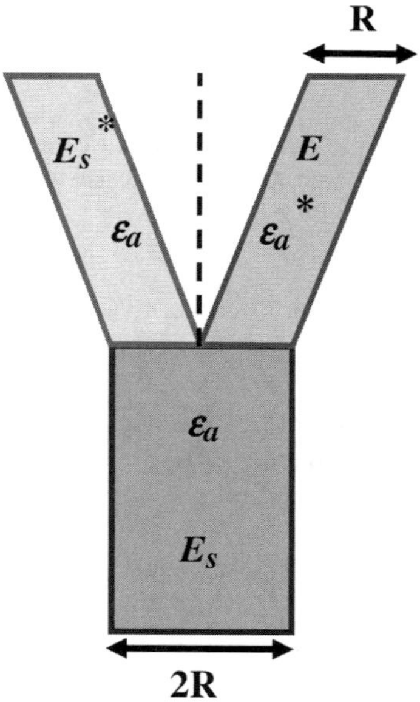

Figure 7.9 Schematic of the Y-shaped pattern where the branch on the right is more adhesive than the stem and the branch on the left is mechanically softer than the stem. The rest of the substrate is nonadhesive and has the same stiffness as the stem. Note that the patches are not drawn to scale.

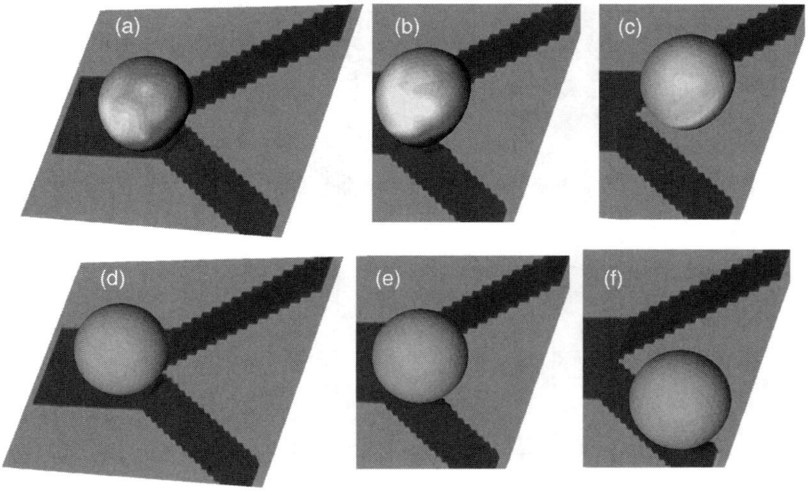

Figure 7.10 Capsule motion on the patterned substrate under an imposed shear flow with $\Lambda = 1.5 \times 10^3$. Images in parts (a) to (c) are for a soft capsule characterized by $\Phi = 6$ and images in parts (d) to (f) are for a stiff capsule characterized by $\Phi = 0.75$. The soft capsule deforms and moves toward the adhesive (sticky) branch, whereas the stiff capsule moves toward the soft branch. (The branches of the Y pattern appear to be jagged due to the discrete nature of the lattice-spring model.) (From [37], with permission. Copyright © 2007, American Chemical Society.) (*See insert for color representation of figure.*)

deformability of softer capsules on the sticky stripe leads to the greater particle–surface contact and a greater force acting on the softer capsules that displaces them along the sticky stripe. We can expect, therefore, that the Y pattern with both soft and sticky branches could separate capsules of different stiffness.

Our simulation results indeed demonstrate that the particles segregate in a manner consistent with the arguments above: a soft capsule moves toward the adhesive branch and a stiff capsule moves toward the deformable branch [37,38]. Figure 7.10 shows views from the simulations, illustrating the separation achieved. In Fig. 7.11a we present trajectories for a number of examples to illustrate the behavior for a range of capsule compliances.

As indicated above, there is a thermodynamic driving force that contributes to this particle separation. To gain insight into these thermodynamic forces, we construct the energy landscape for each capsule, employing the approach described in Section 7.3.1. Figure 7.11b and c reveal the free energy of the system for soft and stiff particles, respectively. The dashed lines in Fig. 7.11a indicate the specific crosssections where the energy is measured. As can be seen for different positions along the flow direction, the total free energy of the system reaches a local minimum when a soft capsule is localized over the most adhesive branch (Fig. 7.11b), whereas in the case involving a rigid particle, the free energy of the system reaches a local minimum over the mechanically patterned surface (Fig 7.11c). Therefore, the fluid-driven capsules are

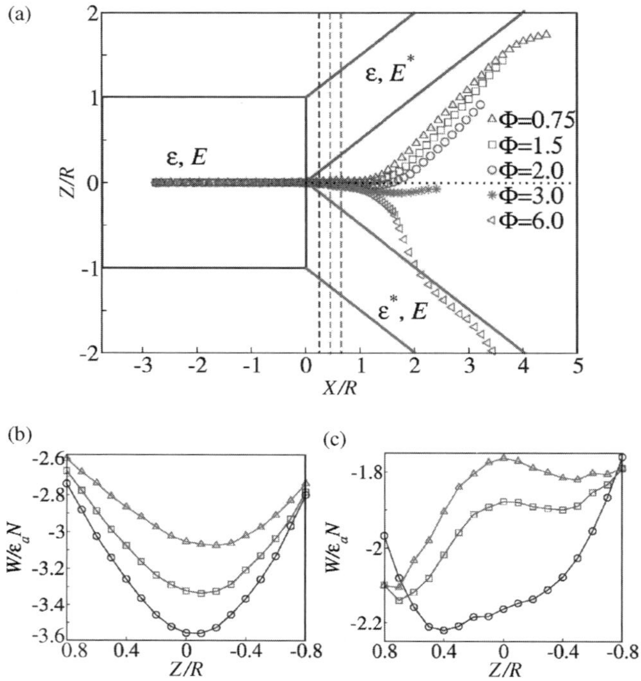

Figure 7.11 (a) Trajectories for capsules of different mechanical properties overlaid on the basic geometry used in this study. Here $\Delta\varepsilon/\varepsilon_a = 0.166$, $E_s^*/E_s = 0.01$, and $\Lambda = 1.5 \times 10^3$. Total energy of the system for (b) a soft ($\Phi = 6$) and (c) a stiff capsule ($\Phi = 0.75$) as a function of *Z/R*, for different *X/R*. The three slices for the energy shown are at $X/R = 0.25$ (circles), $X/R = 0.45$ (squares), and $X/R = 0.65$ (triangles). The dashed lines in (a) indicate the location of these slices. (From [37], with permission. Copyright © 2007, American Chemical Society.)

shifted toward the respective branches according to the steepest descent in the energy landscape.

It is noteworthy that for the shear rates examined here, the outcome of the particle segregation is independent of the value of $\dot{\gamma}$ (Fig. 7.12). The shear rate does, however, affect the particular details of the capsule trajectory and introduces small variations in the capsule motion. For a relatively soft capsule, the differences between trajectories for various shear rates are shown in the inset of Fig 7.12. At very low shear rates, the soft capsule initially moves toward the soft substrate, but it eventually deforms and moves toward the most adhesive branch. These subtle variations are due to differences in the relative time scales for the motion of the capsule, the deformation of the capsule, and the deformation of the substrate [37,38]. On the other hand, when the flow is too fast, the capsules escape the effects of the diverging branches, due to the strong drag force.

It is important to delineate the range of surface parameters for which we observe capsule separation (for the range of capsule compliances considered here). The two crucial parameters are the modulus of the soft branch with respect to the rest of the

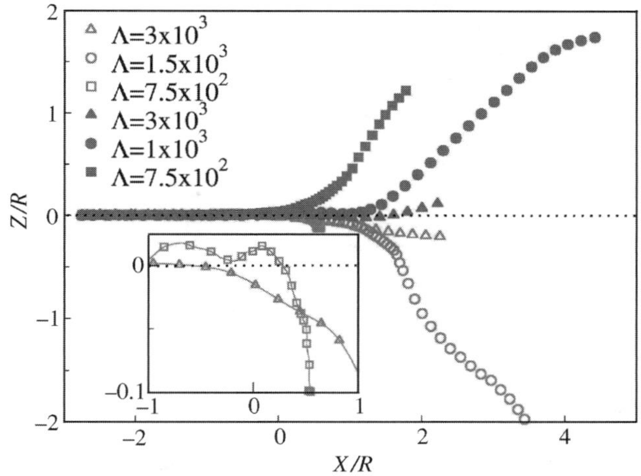

Figure 7.12 Trajectories for a soft ($\Phi = 6$, hollow symbols) and a stiff capsule ($\Phi = 0.75$, solid symbols), for different imposed shear rates. Here $\Delta\varepsilon/\varepsilon_a = 0.166$ and $E_s^*/E_s = 0.01$. The inset details the trajectories of the soft particles near the junction point. The capsule separation is achieved at all shear rates studied here, although as illustrated in the inset, the dynamic details are different due to the change in relative times cales of the deformation and the motion of the capsules. (From [37], with permission. Copyright © 2007, American Chemical Society.)

surface, E_s^*/E_s, and the adhesion difference between the two branches, $\Delta\varepsilon = \varepsilon_a^* - \varepsilon_a$ (where ε_a^* is the parameter that characterizes the stickier domain). Figure 7.13 reveals that when the two branches have identical adhesion coefficients ($\Delta\varepsilon = 0$), both capsules move toward the soft branch (which is on the left in this setup and indicated by the L in the phase map) since the deformation of this soft region around the particles increases the particle–surface contact area and thereby lowers the free energy of the system. Thus, a critical value of $\Delta\varepsilon$ is necessary for the particle separation to occur; the particles are segregated by the surface for the range of parameters marked by a V in the phase map. If, however, $\Delta\varepsilon$ is significantly above this critical value, the deformation of the soft branch will not be a sufficient driving force even for the rigid capsules, and both capsules will now move toward the most adhesive branch (which is indicated by an R in the phase map). It is important to note that this critical value of $\Delta\varepsilon$ also depends on the modulus of the soft branch, as shown in Fig. 7.13.

The foregoing findings reveal that microcapsules of different stiffness can be directed along different, targeted paths by using both mechanical and chemical patterning of the surface. The role of the surface patterns is to introduce an asymmetry into the energy landscape, which then modifies the capsules' movement due to the imposed flow. The greater this asymmetry, the more robust and stable the separation between the different capsules will be.

From the observation above, we can deduce that by implementing a more asymmetric pattern than the Y-shaped design, we could enhance the energy asymmetry and therefore improve the segregation efficiency. To illustrate this point, we

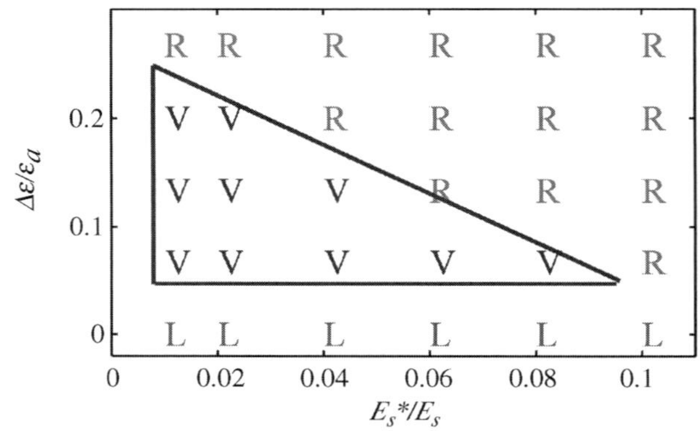

Figure 7.13 Separation diagram for stiff ($\Phi = 0.75$) and soft ($\Phi = 6$) capsules in terms of branch stiffness, E_s^*/E_s, and relative adhesion, $\Delta\varepsilon/\varepsilon_a$. The symbols in this graph signify the following: for R, both capsules move toward the most adhesive branch; for L, both capsules move toward the soft, mechanically patterned branch; and for V, capsules are separated. (From [37], with permission. Copyright © 2007, American Chemical Society.)

consider the geometry in Fig. 7.14, where two parallel stripes are placed asymmetrically with respect to the delivery patch. Here, we also replace the mechanically patterned branch with a patch that is identical to the delivery stripe. Thus, this motif utilizes only differences in the relative adhesiveness of the patches and their geometry

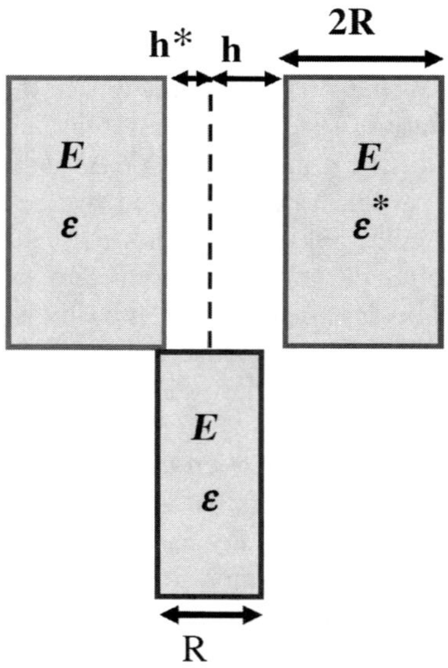

Figure 7.14 Schematic of a pattern with two parallel branches. The substrate has a uniform stiffness, while the branches have different adhesive strengths and display a spatial asymmetry around the centerline.

to accomplish the separation and, consequently, might be the simplest design to fabricate.

The plots in Fig. 7.15 reveal the behavior of the system above for soft ($\Phi = 6$) and stiff ($\Phi = 0.75$) capsules. From Fig. 7.15a we find that both stiff and soft capsules are driven by the imposed flow to move to the patch that is identical to the delivery patch. The soft capsule, however, ultimately moves onto the stickier (right) stripe, while the stiff capsule continues to move along the left strip. In this scenario, the soft capsule is initially compressed by the adhesive force on the delivery stripe and, due to a larger footprint, can "feel" the edges of the sticker stripe. The capsule deforms to maximize its contact with the sticky stripe and subsequently is "dragged" onto this region by the thermodynamic forces. Again, we see that the free energy of the system is decreased when the softer capsule is on the stickier patch (Fig. 7.15c). The stiff capsule is constrained to remain on the left (i.e., positive Z) patch since it could not deform to a sufficient degree to experience the attraction of the sticky patch on the right (i.e., negative Z). Note that the free energy of the stiff capsule increases as it is moved from the left stripe since it must traverse a region where $\varepsilon = 0$. There is, therefore, an energy barrier that prevents the stiff capsule from migrating toward the stickier patch and, in this manner, induces separation between soft and stiff capsules.

7.3.3 Selectively Trapping and Bursting Capsules

In addition to the selective routing of microcapsules to specific locations, another desirable function for microfluidic or pharmaceutical applications is controlled release of the capsules' payload. In the context of microfluidic devices, there is a need to design smart devices that can be programmed effectively to carry out multistage processes. The program could be embedded into the microfluidic devices via appropriately designed patterns, while the values of "variables" could be encoded and stored by the mechanical and chemical properties of capsules' shells. More generally, the capsules could serve to encode certain information, while the surface patterning could function as a "processor" that decodes this information and handles the particles dynamically according to the unraveled instructions.

In this section we exploit the chemical and physical interactions between the microcapsules' shells and patterned substrates to design a microfluidic system that is programmed to perform a multistage process in an essentially autonomous manner [36]. More specifically, we design a device that can select capsules of a specific compliance, direct the movement of these capsules to a precise location in the microchannels, and then cause the capsules to burst spontaneously, thereby discharging the encapsulated compounds.

A key component of this system is shown in Fig. 7.16; here, the substrate is decorated with a specific pattern that exhibits a gradient in the adhesive interaction with the capsule's surface. Such adhesion gradients can readily be inscribed onto substrates through microcontact printing [68,69]. Before describing the operation of the entire system, we discuss the behavior of compliant capsules on such gradient surfaces in the absence of an imposed flow because these results provide insight into the factors that control the system's performance.

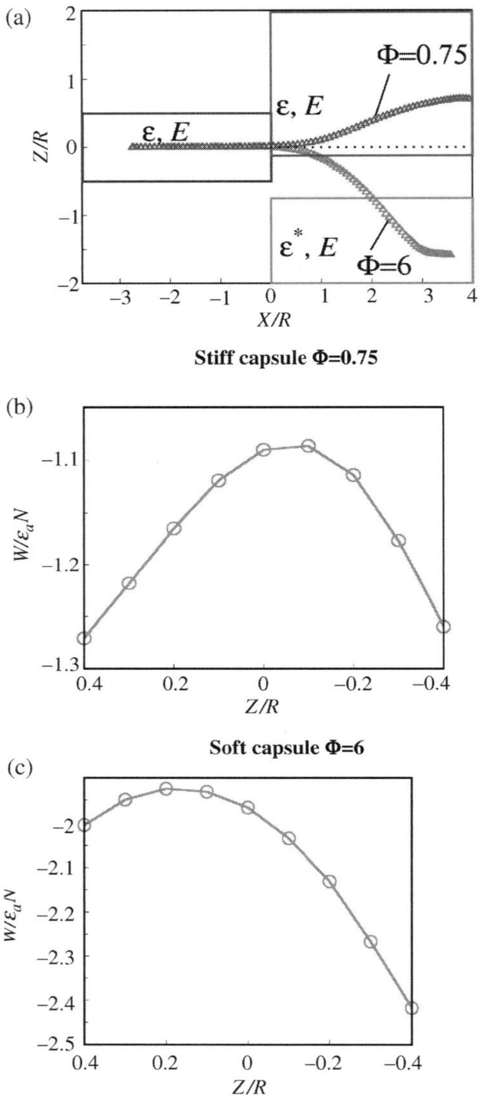

Figure 7.15 (a) Trajectories for a stiff capsule ($\Phi = 0.75$) and a soft capsule ($\Phi = 6$) at $\Lambda = 1.5 \times 10^3$ on the structure in Fig. 7.5 with $h = 0.625R$. The total energy of the system is shown as a function of Z/R for (b) a stiff capsule and (c) a soft capsule. The energy is measured at $X/R = 2.5$, where the capsules do not feel the effect of the delivery patch. The abscissas in parts (b) and (c) are inverted for easy transformation and superposition on the geometry layout in part (a).

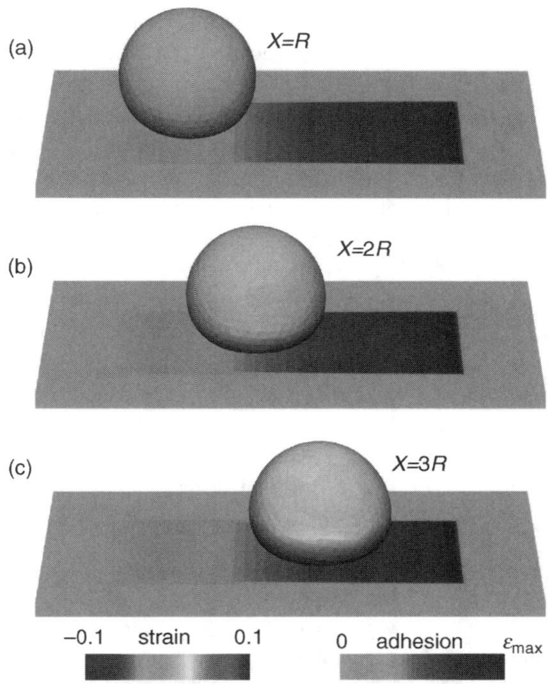

Figure 7.16 Compliant capsule on a substrate patterned with increasing adhesive strength. The capsule migrates along the pattern toward the region with the greatest adhesion. The migration occurs in the absence of any external flow and is due solely to the adhesive gradient. Part (a), (b), and (c) show three consequent moments when the capsule reaches the locations at $x = R$, $2R$, and $3R$, respectively. The width of the gradient pattern is $2R$ and its length is $5R$. Adhesion strength increases linearly with distance (i.e., $\varepsilon_a = \Delta\varepsilon_a x$). The color on the substrate indicates the local adhesion, while the color on the capsule's shell reveals the strain in the elastic shell due to the capsule's deformation. Note that the capsule deformation is enhanced when the compliant capsule moves along the gradient. Capsule stiffness is $\Delta\Phi = 20$, where $\Delta\Phi = \Delta\varepsilon_a R N_0 N_s / Eh\kappa^2$. (From [36], with permission of the Royal Society of Chemistry.) (*See insert for color representation of figure.*)

When a capsule is placed on a substrate with an adhesion gradient, a net force arises due to differences in the adhesive strength around the area of contact between the capsule and the substrate. The force induced by the adhesion gradient, F_g, sets the capsule into motion even in the absence of an imposed flow (Fig. 7.16) and drives the capsule to the location with the highest adhesive strength, since this minimizes the free energy of the system. We can estimate the magnitude of this force to be $F_g \approx \Delta w\, A_c A_s$, where Δw is the change in the adhesion per unit area per unit length, A_c is the contact area on the capsule's shell that interacts with the substrate, and A_s is the contact area on the substrate that interacts with a unit area of the shell. While Δw and A_s are characteristics of the specific pattern and interaction potential, the magnitude of A_c is defined by the capsule elasticity. The softer the capsule, the larger the area of contact, A_c, it has and thus a greater force F_g that the capsule experiences when it is placed on the gradient pattern. In other words, the specific magnitude of the force F_g is prescribed by the elastic properties of the capsule shell. We note that the capsule is surrounded by a viscous fluid and therefore the force F_g is balanced by the viscous drag force, which, in turn, is proportional to the velocity of the capsule along the gradient, U. Thus, the gradient pattern can induce capsule migration along the pattern with a velocity that is defined by the capsule compliance. This is how the gradient pattern can be used to discriminate between capsules of different compliance.

In Fig. 7.17a we plot the magnitude of the capsule's velocity along the adhesive gradient for capsules with different elasticity. In agreement with the arguments above, Fig. 7.17a indeed shows that soft capsules move sufficiently faster than rigid ones. Moreover, as softer capsules move toward regions of greater adhesion, they experience larger deformations and higher velocities, even though the magnitude of the adhesion gradient is kept constant. Again, this is related directly to the capsule's mechanical stiffness, which eventually dictates the speed of capsule displacement.

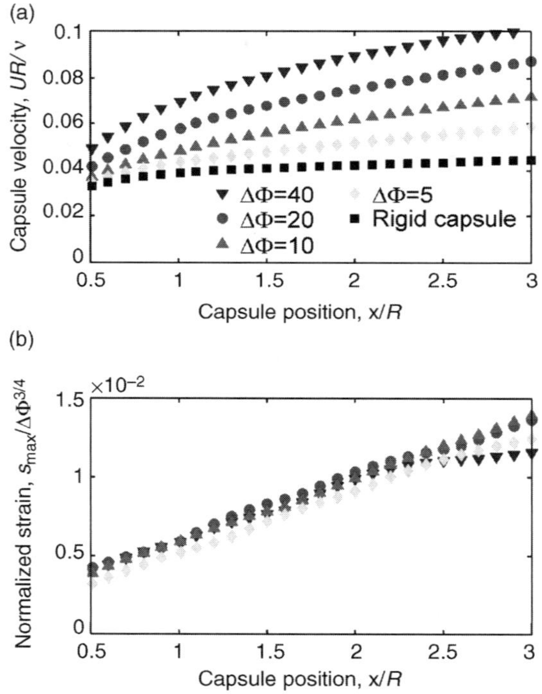

Figure 7.17 Capsule velocity and strain on a surface patterned with an adhesion gradient. (a) Velocity of capsules with different stiffness. The characteristics of the pattern are the same as in Fig. 7.16. Capsule stiffness is expressed via the dimensionless parameter $\Delta\Phi = \Delta\varepsilon_a R N_0 N_s / Eh\kappa^2$. Note that softer capsules migrate substantially faster than stiffer ones due to an enhanced contact with the gradient patterns. (b) Maximum strain $s_{\max}$ within the capsule's shell.

The adhesive interaction not only deforms the capsules, but also induces a strain in the capsules' shell. Figure 7.17b shows the magnitude of the maximum strain within the shell, $s_{\max}$, as a function of the capsule's position along the gradient stripe. The maximum strain increases linearly with distance. Furthermore, our results for capsules of different stiffness indicate that $s_{\max}$ scales with $\Delta\Phi^{3/4}$, where $\Delta\Phi = \Delta\varepsilon_a R N_0 N_s / Eh\kappa^2$, or equivalently, $s_{\max} \sim (Eh)^{-3/4}\Delta\varepsilon_a^{7/4}x$. This finding suggests that one can predict the adhesive strength that is needed to induce a critical level of strain within the shell and thus control the location where the adhesion triggers the capsule to burst.

We can use as a design principle the fact that the softer capsules exhibit larger deformations and move faster than their more rigid counterparts. In particular, we now construct patterned substrates that can selectively direct and burst capsules with specific mechanical properties. Such a patterned substrate is illustrated in Fig. 7.18. The nonadhesive rigid substrate is decorated with an adhesive pattern that incorporates two essential parts: a delivery track with uniform adhesion, and a side patch with increasing adhesion. The delivery track is oriented along the direction of the imposed fluid flow that propels the capsules through the system. The adhesion on the delivery track is sufficiently strong to keep the capsules localized on the surface by

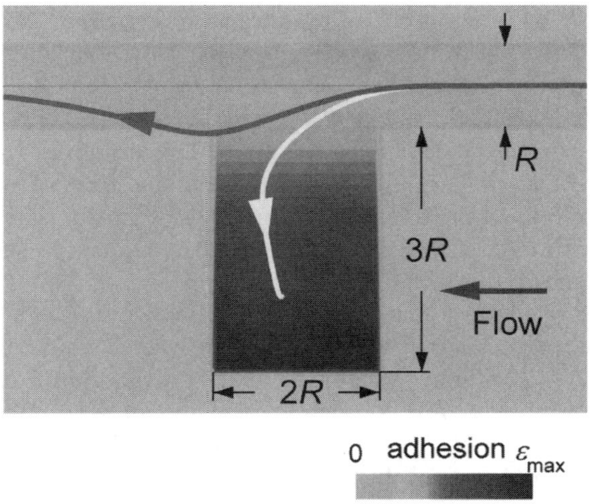

Figure 7.18 Trajectories of soft and rigid capsules on the patterned substrate. The pattern encompasses two parts: a delivery stripe with a uniform adhesion, and a side patch with adhesion increasing perpendicular to the flow direction. The color reveals the local adhesion on the substrate. The red and yellow lines show the trajectories of a rigid capsule and a soft capsule ($\Delta\Phi = 20$), respectively. The capsules are initially located at the same position and are driven to move by the imposed shear flow. The rigid capsule, although affected by the adhesion gradient on the side patch, escapes the trap and continues to move along the delivery stripe, whereas the soft capsule ends up at the side patch, entrapped due to the greater adhesive interaction. (From [36], with permission of the Royal Society of Chemistry.) (*See insert for color representation of figure.*)

counteracting the lift force. The delivery track guides the capsules toward the junction with the side patch, which encompasses the adhesion gradient (Fig. 7.18). The adhesive strength along the gradient increases linearly in the direction away from the delivery track (i.e., the adhesion gradient is directed perpendicular to the fluid flow).

We illustrate the operation of the device above by considering the behavior of soft and rigid capsules in the system with the trajectories of the capsules on the patterned substrate being shown in Fig. 7.18. Initially, capsules are located on the centerline of the delivery track and are driven to move along the track with a velocity $U_d \sim \dot{\gamma}R$ by an imposed flow with shear rate $\dot{\gamma}$. When the capsules reach the junction between the delivery and the side patches, a force due to the adhesion gradient induces a velocity U_g in the lateral direction (i.e., perpendicular to the flow direction). The magnitude of U_g depends on the mechanical properties of the capsule. As a result, capsules with different properties follow different trajectories as they cross the junction.

Specifically, softer capsules gain a larger lateral displacement. The magnitude of the displacement at the moment when the capsule is about to leave the junction defines the further behavior of the capsules. For stiff capsules, the lateral displacement is relatively small and the capsule continues to move down the track, in the direction of the shear flow. If the capsule is relatively soft, however the lateral velocity U_g can be sufficiently high to move the capsule onto the side patch. In this situation, the capsule moves perpendicular to the flow due to the adhesion gradient; the viscous drag is balanced by the force that arises from the adhesion drop between the downstream side of the gradient patch and the nonadhesive substrate.

The sticky side patch not only serves to divert capsules selectively with specific properties, but can also be harnessed to destroy those capsules by inducing their bursting. Indeed, if the strain exceeds a critical yield value, cracks form and propagate throughout the capsule's shell. Eventually, under the mutual action of adhesion and shear flow, the capsule's shell disintegrates and the capsule releases its contents.

We illustrate the bursting process in Fig. 7.19, where a soft capsule is shown being trapped and destroyed due to interaction with the increasingly adhesive side patch. When the capsule reaches a location on the substrate where the adhesion is sufficiently strong, an initial crack is formed near the bottom of the capsule, on the upstream side. The first crack serves as a stress concentrator and promotes the development of new cracks that originate from this initial defect on the shell. The newborn cracks rapidly propagate along the capsule's shell, resulting eventually in complete disintegration of the capsule.

The parts of the capsule's shell that are closer to the substrate stick to the adhesive surface, while fragments from the upper part of the capsule are carried away by the fluid flow. The contents of the capsule are thus released and intermix with the external solvent. To illustrate the dispersion of the capsule's contents, we introduce inertial-less tracer particles which follow the streamlines within the flow. These particles, initially located inside the capsule shell, demonstrate how the contents spread after the capsule's breakdown.

As discussed above, the discriminating behavior of the device is governed by the surface pattern and the interplay between the capsule's velocity due to the shear flow

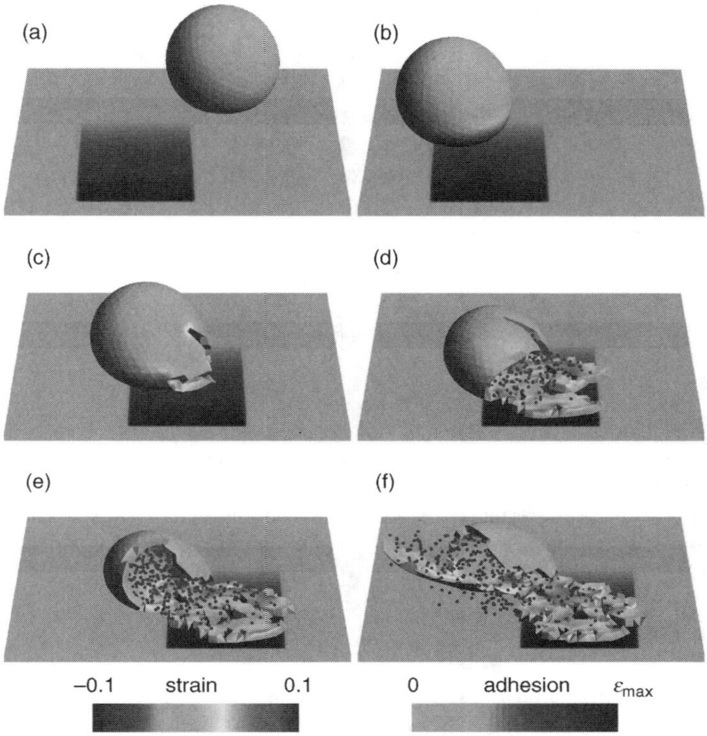

Figure 7.19 Soft capsule ($\Delta\Phi = 20$) as it becomes trapped and bursts on the patterned substrate. Time intervals between snapshots (a)–(b) and (e)–(f) are equal to $3\dot{\gamma}^{-1}$, and between snapshots (b)–(c), (c)–(d), and (d)–(e) are equal to $1.5\dot{\gamma}^{-1}$. The red dots indicate tracer particles that are initially enclosed inside the capsule and are released due to the capsule's bursting. The color at the substrate indicates the local adhesion; the color at the capsule's surface reveals the strain in the elastic shell. (From [36], with permission of the Royal Society of Chemistry.) (*See insert for color representation of figure.*)

and the velocity induced by the adhesion gradient. Although the gradient velocity is defined solely by the mechanical properties of the capsule and its interaction with the adhesive substrate, the fluid velocity can readily be modified by changing the flow rate through the microfluidic device. In this manner, the sensitivity of a device with a given patterned substrate can easily be adjusted to trap capsules with specified mechanical properties. To demonstrate this aspect, we constructed a phase diagram in terms of the imposed shear rate and capsule stiffness, as shown in Fig. 7.20. The diagram indicates the specific regimes in which a capsule with certain mechanical properties is either trapped by the adhesive side patch (marked by the crosses) or escapes by continuing in a straight motion along the delivery track (marked by the dots). As seen, by choosing the appropriate shear rate, one can regulate the range of capsules that are trapped.

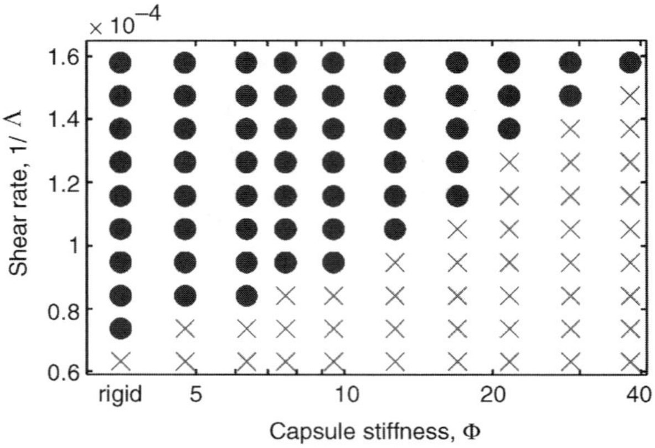

Figure 7.20 Phase diagram showing the flow regimes where capsules with specific stiffness are either trapped or escape. The crosses indicate the parameters for which the capsule is entrapped, while the circles are for the situations where the capsule escapes the trap. The capsule stiffness and the shear rate are expressed in terms of Φ and Λ, respectively, where ε_a is that for the delivery stripe. (From [36], with permission of the Royal Society of Chemistry.)

7.4 CONCLUSIONS

We have shown how patterned substrates can be harnessed to regulate the motion of capsules (synthetic and biological) in microchannels. Continuing advances in microfluidics will require the devices to perform in a more "intelligent" manner, taking inputs from the environment and effectively "making decisions" about their operation. In this manner, the devices can carry out a broad array of functions or regulate multistage reactions in a relatively autonomous manner. Additionally, there is a significant push to create microfluidic platforms that require less electronics for their operation and thus would be more portable. These needs impose significant design challenges. Our results point to one way of meeting these challenges: By patterning substrates with the appropriate design, one can create a smart, robust device for autonomously routing, sorting, and bursting soft particles.

In our discussion, we focused primarily in separating capsules by mechanical properties. This is a particularly important goal because there are a number of diseases (e.g., malaria and various cancers) that alter the elasticity of biological cells [70] and in some instances the stiffness of the cell indicates the stage of infection [71]. Although researchers are currently developing sophisticated approaches (involving optical tweezers) to measure the mechanical characteristics of diseased cells [72], there remains a critical need for facile methods to sort cells by their stiffness and thereby enable rapid, low-cost assays.

As we showed above, a capsule's path on these heterogeneous substrates is highly dependent on its stiffness: that is, mechanically different particles that start at the same position eventually roll to separate locations on the substrate and thus are effectively sorted by the surface. Since this approach exploits the particle's inherent response to the substrate, it does not involve explicit measurement and assessment and thus could prove to be a highly efficient method for carrying out continuous, on-the-fly separation processes.

Along these lines, we note that scientists are now generating new types of polymeric microcapsules [5] and polymersomes [50,51] with a range of tailored compositions and flexibilities. For example, relatively rigid capsules are produced by incorporating nanoparticles into the bounding shells [19,73], while more elastic species are formed using block copolymers [50]. To utilize these polymeric capsules as robust microcarriers or microreactors, it becomes necessary to isolate species with the desired mechanical properties. Thus, here too the proposed device could provide an effective, low-cost means of carrying out quality control studies, isolating the more robust carriers from the less optimal units.

The approach could be adapted equally well to sort capsules by their adhesiveness, that is, the chemical composition on their surface. This too provides useful capabilities for applications in the chemical and pharmaceutical industries. Here, the capsules could sort themselves on the fly into separate bins based on their intrinsic chemical interactions with the surface.

As we described in the example involving the simple striped patterns, the sensitivity of the device could be tuned by varying the imposed shear rate, with lower $\dot{\gamma}$ increasing the sensitivity and hence the selectivity. Further selectivity could be achieved by taking capsules that have been sorted into a particular bin and then running this sample again so that the variance in each bin becomes smaller and smaller.

The examples above also reveal that the intrinsic physical properties of synthetic and biological capsules constitute a code, and this code can be deciphered through the use of chemically and mechanically patterned substrates. The coupling between the encoded particles and the decoding surfaces opens up new methods for regulating the motion of microcarriers in microchannels. This design principle could be harnessed to construct logic gates and more complex circuits, where cells or microcapsules carry out the logic operations, in analogy with the functions carried out by droplets and bubbles in microfluidic networks [74,75].

Finally, we mention that topological patterning of the surfaces, by introducing posts or corrugations, also permits the separation of the capsules by mechanical properties [28,29,34]. The introduction of such topological features can readily be achieved by soft lithography, and thus such architectural patterning could also be exploited in microfluidic devices. With these recent advances in the fabrication of patterned polymeric surfaces with highly regular, microscopic features and chemically tailored polymeric microcapsules, our findings on the behavior of microcapsules on a variety of substrates can be tested experimentally and can provide the needed design rules for creating the optimal low-cost, portable devices for crucial bioassays and fundamental chemical studies.

Acknowledgments

The authors gratefully acknowledge financial support from the U.S. Department of Energy and the Office of Naval Research.

REFERENCES

1. Haswell SJ, Middleton RJ, O'Sullivan B, Skelton V, Watts P, Styrine P. The application of micro-reactors to synthetic chemistry. *Chem. Commun.* **5**:391–398, 2001.
2. Teh S-Y, Lin R, Hung L-H, Lee AP. Droplet microfluidics. *Lab Chip* **8**:198–220, 2008.
3. Gao CY, Leporatti S, Moya S, Donath E, Möhwald H. Stability and mechanical properties of polyelectrolyte capsules obtained by stepwise assembly of poly(styrenesulfonate sodium salt) and poly(diallyldimethyl ammonium) chloride onto melamine resin particles. *Langmuir* **17**:3491–3495, 2001.
4. Lulevich VV, Radtchenko IL, Sukhorukov GB, Vinogradova OI. Mechanical properties of polyelectrolyte microcapsules filled with a neutral polymer. *Macromolecules* **36**(8):2832–2837, 2003.
5. Fery A, Dubreuil F, Möhwald H. Mechanics of artificial microcapsules. *New J. Phys.* **6**:18, 2004.
6. Yow HN, Routh AF. Formation of liquid core–polymer shell microcapsules. *Soft Matter* **2**:940–949, 2006, and references therein.
7. Caruso F, Trau D, Möhwald H, Renneberg R. Enzyme encapsulation in layer-by-layer engineered polymer multilayer capsules. *Langmuir* **16**:1485–1488, 2000.
8. Moya S, Sukhorukov GB, Auch M, Donath E, Möhwald H. Microencapsulation of organic solvents in polyelectrolyte multilayer micrometer-sized shells. *J. Colloid Interface Sci.* **216**:297–302, 1999.
9. Shchukin DG, Radtchenko IL, Sukhorukov GB. Micron-scale hollow polyelectrolyte capsules with nanosized magnetic Fe_2O_3 inside. *Mater. Lett.* **57**(11):1743–1747, 2003.
10. Voigt A, Buske N, Sukhorukov GB, Antipov AA, Leporatti S, Lichtenfeld H, Bäumler H, Donath E, Möhwald H. Novel polyelectrolyte multilayer micro- and nanocapsules as magnetic carriers. *J. Magn. Magn. Mater.* **225**:59–66, 2001.
11. Ibarz G, Dähne L, Donath E, Möhwald H. Smart micro- and nanocontainers for storage, transport, and release. *Adv. Mater.* **13**:1324–1327, 2001.
12. Dähne L, Leporatti S, Donath E, Möhwald H. Fabrication of micro reaction cages with tailored properties. *J. Am. Chem. Soc.* **123**:5431–5436, 2001.
13. Antipov AA, Sukhorukov GB, Leporatti S, Radtchenko IL, Donath E, Möhwald H. Polyelectrolyte multilayer capsule permeability control. *Colloids Surf. A* **198**:535–541, 2002.
14. Volodkin DV, Balabushevitch NG, Sukhorukov GB, Larionova NI. Model system for controlled protein release: pH-sensitive polyelectrolyte microparticles. *STP Pharma Sci.* **13**(3):163–170, 2003.
15. Antipov AA, Sukhorukov GB, Möhwald H. Influence of the ionic strength on the polyelectrolyte multilayers' permeability. *Langmuir* **19**(6):2444–2448, 2003.
16. Petrov AI, Gavryushkin AV, Sukhorukov GB. Effect of temperature, pH and shell thickness on the rate of Mg^{2+} and Ox^{2-} release from multilayered polyelectrolyte shells deposited onto microcrystals of magnesium oxalate. *J. Phys. Chem. B* **107**(3): 868–875, 2003.
17. Kim B-S, Vinogradova OI. pH-Controlled swelling of polyelectrolyte multilayer microcapsules. *J. Phys. Chem. B* **108**:8161–8165, 2004.

18. Kim B-S, Lebedeva OV, Koynov K, Gong H, Glasser G, Lieberwith I, Vinogradova OI. Effect of organic solvent on the permeability and stiffness of polyelectrolyte multilayer microcapsules. *Macro* **38**(12):5214–5222, 2005.
19. Gil PR, del Mercato LL, del Pino P, Muñoz-Javier A, Parak WJ. Nanoparticle-modified polyelectrolyte capsules. *Nanotoday* **3**:12–21, 2008.
20. Park M-K, Deng S, Advincula RC. Sustained release control via photo-cross-linking of polyelectrolyte layer-by-layer hollow capsules. *Langmuir* **21**:5272–5277, 2005.
21. Yager P, Edwards T, Fu E, Helton K, Nelson K, Tam MR, Weigl BH. Microfluidic diagnostic technologies for global public health. *Nature* **442**:412–418, 2006.
22. Martinez AW, Phillips ST, Butte MJ, Whitesides GM. Patterned paper as a platform for inexpensive, low-volume, portable bioassays. *Angew. Chem. Int. Ed.* **46**(8):1318–1320, 2007.
23. Eddington DT, Beebe DJ. Flow control with hydrogels. *Adv. Drug Deliv. Rev.* **56** (2):199–210, 2004.
24. Buxton GA, Verberg R, Jasnow D, Balazs AC. A Newtonian fluid meets an elastic solid: coupling lattice Boltzmann and lattice spring models. *Phys. Rev. E* **71**:056707, 2005.
25. Alexeev A, Verberg R, Balazs AC. Designing compliant substrates to regulate the motion of vesicles. *Phys. Rev. Lett.* **96**:148103, 2006.
26. Alexeev A, Verberg R, Balazs AC. Modeling the motion of capsules on compliant polymeric surfaces. *Macromolecules* **28**:10244–10260, 2005.
27. Alexeev A, Verberg R, Balazs AC. Modeling the interactions between deformable capsules rolling on a compliant surface. *Soft Matter* **2**:499–509, 2006.
28. Smith KA, Alexeev A, Verberg R, Balazs AC. Designing a simple ratcheting system to sort microcapsules by mechanical properties. *Langmuir* **22**:6739–6742, 2006.
29. Alexeev A, Verberg R, Balazs AC. Motion of compliant capsules on corrugated surfaces: a means of sorting by mechanical properties. *J. Polym. Sci. B* **44**:2667–2678, 2006.
30. Verberg R, Alexeev A, Balazs AC. Modeling the release of nanoparticles from mobile microcapsules. *J. Chem. Phys.* **126**:224712, 2006.
31. Verberg R, Dale AT, Kumar P, Alexeev A, Balazs AC. Healing substrates with mobile, particle-filled microcapsules: designing a "repair and go" system. *J. R. Soc. Interface* **4**:349–357, 2007.
32. Alexeev A, Verberg R, Balazs AC. Patterned surfaces segregate compliant microcapsules. *Langmuir* **23**:983–987, 2007.
33. Zhu G, Alexeev A, Kumacheva E, Balazs AC. Modeling the interactions between compliant microcapsules and pillars in microchannels. *J. Chem. Phys.* **127**:034703, 2007.
34. Zhu G, Alexeev A, Balazs AC. Designing constricted microchannels to selectively entrap soft particles. *Macromolecules* **40**:5176–5181, 2007.
35. Balazs AC. New approaches for designing 'programmable' microfluidic devices. *Polym. Int.* **57**(5):669–671, 2008.
36. Alexeev A, Balazs AC. Designing smart systems to selectively entrap and burst microcapsules. *Soft Matter* **3**:1500–1505, 2007.
37. Usta OB, Alexeev A, Balazs AC. Fork in the road: patterned surfaces direct microcapsules to make a decision. *Langmuir* **23**(22):10887–10890, 2007.
38. Usta OB, Nayhouse M, Alexeev A, Balazs AC. Designing patterned substrates to regulate the movement of capsules in microchannels. *J. Chem. Phys.* **128**:235102, 2008.

39. Usta OB, Alexeev A, Zhu G, Balazs AC. Modeling microcapsules that communicate through nanoparticles to undergo self-propelled motion. *ACS Nano* **2**(3):471–476, 2008.

40. Succi S, *The Lattice Boltzmann Equation for Fluid Dynamics and Beyond*, Clarendon Press, Oxford, UK, and Oxford University Press, New York, 2001.

41. Ladd AJC, Kinney JH, Breunig TM. Deformation and failure in cellular materials. *Phys. Rev. E* **55**:3271–3275, 1997.

42. Ladd AJC, Kinney JH. Elastic constants of cellular structures. *Phys. A* **240**:349–360, 1997.

43. Buxton GA, Care CM, Cleaver DJ. A lattice spring model of heterogeneous materials with plasticity. *Model. Simul. Mater. Sci. Eng.* **9**:485–497, 2001.

44. Dong C, Cao J, Struble EJ, Lipowsky HH. Mechanics of leukocyte deformation and adhesion to endothelium in shear flow. *Ann. Biomed. Eng.* **27**:298–312, 1999.

45. Jadhav S, Eggleton CD, Konstantopoulos K. A 3-D computational model predicts that cell deformation affects selectin-mediated leukocyte rolling. *Biophys. J.* **88**:96–104, 2005, and references therein.

46. Marella SV, Udaykumar HS. Computational analysis of the deformability of leukocytes modeled with viscous and elastic structural components. *Phys. Fluids* **16**(2):244–264, 2004.

47. Krasik EF, Hammer DA. A semianalytic model of leukocyte rolling. *Biophys. J.* **87**:2919–2930, 2004.

48. Khismatullin DB, Truskey GA. A 3D numerical study of the effect of channel height on leukocyte deformation and adhesion in parallel-plate flow chambers. *Microvasc. Res.* **68**(3):188–202, 2004, and references therein.

49. Buxton G, Balazs AC. Lattice spring model of filled polymers and nanocomposites. *J. Chem. Phys.* **117**(16):7649–7658, 2002.

50. Discher BM, Won Y-Y, Ege DS, Lee JC-M, Bates FS, Discher DE, Hammer DA. Polymersomes: Tough vesicles made from diblock copolymers. *Science* **284** (5417):1143–1146, 1999.

51. Discher DE, Eisenberg A. Polymer vesicles. *Science* **297**(5583):967–973, 2002, and references therein.

52. Delaunay BN. Sur la sphere vide. *Bull. Acad. Sci. USSR: Class. Sci. Math.* **7**:793–800, 1934.

53. Bouzidi M, Firdaouss M, Lallemand P. Momentum transfer of a Boltzmann–lattice fluid with boundaries. *Phys. Fluids* **13**:3452–3459, 2001.

54. Sukumaran S, Seifert U. Influence of shear flow on vesicles near a wall: a numerical study. *Phys. Rev. E* **64**(1):011916, 2001.

55. Cantat I, Misbah C. Lift force and dynamical unbinding of adhering vesicles under shear flow. *Phys. Rev. Lett.* **83**(4):880–883, 1999.

56. Cantat I, Misbah C. Dynamics and similarity laws for adhering vesicles in haptotaxis. *Phys. Rev. Lett.* **83**(1):235–238, 1999.

57. Elsner N, Dubreuil F, Weinkamer R, Wasicek M, Fischer FD, Fery A. Mechanical properties of freestanding polyelectrolyte capsules: a quantitative approach based on shell theory. *Prog. Colloid Polym. Sci.* **132**:117–123, 2006.

58. Dubreuil F, Elsner N, Fery A. Elastic properties of polyelectrolyte capsules studied by atomic-force microscopy and RICM. *Eur. Phys. J. E* **12**(2):215–221, 2003.

59. Heuvingh J, Zappa M, Fery A. Salt softening of polyelectrolyte multilayer capsules. *Langmuir* **21**(7):3165–3171, 2005.

60. Mueller R, Kohler K, Weinkamer R, Sukhorukov G, Fery A. Melting of PDADMAC/PSS capsules investigated with AFM force spectroscopy. *Macromolecules* **38**(23):9766–9771, 2005.

61. Stone HA, Stroock AD, Ajdari A. Engineering flows in small devices: microfluids toward a lab-on-a-chip. *Annu. Rev. Fluid Mech.* **36**:381–411, 2004.

62. Nolte M, Fery A. Coupling of individual polyelectrolyte capsules onto patterned substrates. *Langmuir* **20**(8):2995–2998, 2004.

63. Bosio V, Dubreuil F, Bogdanovic G, Fery A. Interactions between silica surfaces coated by polyelectrolyte multilayers in aqueous environment: comparison between precursor and multilayer regime. *Colloids Surf. A* **243**(1–3):147–155, 2004.

64. King MR, Hammer DA. Multiparticle adhesive dynamics: hydrodynamic recruitment of rolling leukocytes. *Proc. Natl. acad. Sci. USA.* **98**(26):14919–14924, 2001.

65. Hodges RS, Jensen OE. Spreading and peeling dynamics in a model of cell adhesion. *J. Fluid Mech.* **460**:381–409, 2002.

66. Leckband D, Israelachvili J. Intermolecular forces in biology. *Q. Rev. Biophys.* **34** (2):105–267, 2001.

67. Zhang XH, Chen A, De Leon D, Li H, Noiri E, Moy VT, Goligorsky MS. Atomic force microscopy measurement of leukocyte–endothelial interaction. *Am. J. Physiol. Heart Circ. Physiol.* **286**(1):H359–H367, 2004.

68. Chaudhury MK, Whitesides GM. How to make water run uphill. *Science* **256** (5063):1539–1541, 1992.

69. Xia YN, Whitesides GM. Inorganic/block copolymer-dye composites and dye doped mesoporous materials for optical and sensing applications. Soft lithography. *Angew. Chem. Int. Ed.* **37**:551–575, 1998.

70. Suresh S. Mechanical response of human red blood cells in health and disease: some structure–property–function relationships. *J. Mater. Res.* **21**(8):1871–1877, 2006.

71. Suresh S, Spatz J, Mills JP, Micoulet A, Dao M, Lim CT, Beil M, Sefferlein T. Connections between single-cell biomechanics and human disease states: gastrointestinal cancer and malaria. *Acta Biomater.* **1**(1):15–30, 2005.

72. Van Vliet KJ, Bao G, Suresh S. The biomechanics toolbox: experimental approaches for living cells and biomolecules. *Acta Mater.* **51**(19):5881–5905, 2003.

73. Shchukin DG, Sukhorukov GB, Mohwald H. Smart inorganic/organic nanocomposite hollow microcapsules. *Angew. Chem. Int. Ed.* **42**:4472–4475, 2003.

74. Fuerstman MJ, Garstecki P, Whitesides GM. Coding/decoding and reversibility of droplet trains in microfluidic networks. *Science* **315**(5813):828–832, 2007.

75. Prakash M, Gershenfeld N. Microfluidic bubble logic. *Science* **315**(5813):832–835, 2007.

PART II

APPLICATIONS IN BIOLOGY

8

COARSE-GRAINED AND MULTISCALE SIMULATIONS OF LIPID BILAYERS

Harry A. Stern
Department of Chemistry, University of Rochester, Rochester, New York

8.1 INTRODUCTION

Cell membranes are an essential component of cells. The outer plasma membrane maintains the integrity of the cell and the distinct environments of the cytosol and extracellular fluid. Organelles within the cell, such as the endoplasmic reticulum, Golgi apparatus, and outer-segment disks in visual rod cells, are also bound by membranes. Membranes are complex and heterogeneous structures consisting of a bilayer of phospholipid molecules and other components, mostly proteins such as receptors or channels. The fluid-mosaic picture of membranes suggests that the lipid bilayer is effectively a medium or solvent in which other membrane components, such as receptor proteins or ion channels, are embedded. The membrane may be considered to be a two-dimensional fluid in which proteins and other constituents can diffuse freely in the plane of the bilayer but whose motion perpendicular to the bilayer is restricted. The hydrophobic interior of the bilayer serves as a barrier for transport of water-soluble molecules across the membrane; such transport is accomplished via specialized proteins and may either be active (requiring the expenditure of free energy in the form of ATP hydrolysis) or passive.

Multiscale Modeling of Particle Interactions: Applications in Biology and Nanotechnology,
Edited by Michael R. King and David J. Gee

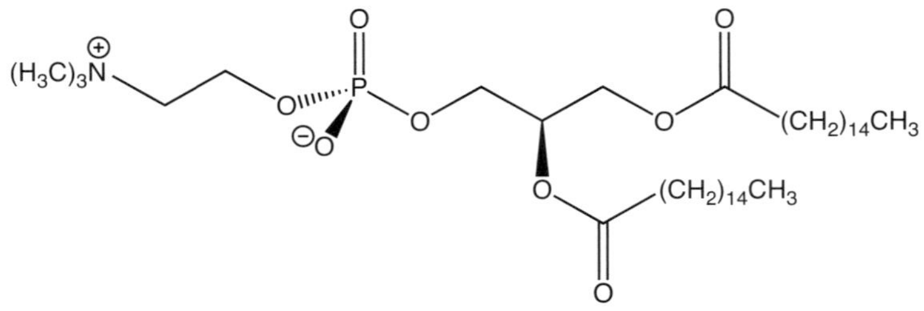

Figure 8.1 Phospholipid dipalmitoylphosphatidylcholine, which consists of partially charged choline and phosphate groups, glycerol, and ester linkages to two saturated hydrocarbon tails.

Biological membranes contain three primary types of lipid molecules: phospholipids, cholesterol, and glycolipids. Phospholipids are the most abundant type, and membranes consisting of a single kind of phospholipid are often used as models for biological membranes. They consist of a hydrophilic (charged, or highly polar—essentially zwitterionic) head group made up of choline and phosphate moieties connected via glycerol linkages to hydrocarbon tails. The tails are of varying length and may be completely saturated or contain one or more double bonds. Figure 8.1 shows one of the most prevalent phospholipids, dipalmitoylphosphatidylcholine (DPPC), whose hydrocarbon side chains are completely saturated. Glycolipids contain one or more sugar moieties in the head-group region. The distribution of glycolipids is often extremely asymmetric between the bilayer leaflets of a cellular membrane; they are much more prevalent in the outer (noncytosolic) leaflet, reflecting their functional importance in cell recognition, protection against harsh extracellular conditions (e.g., in intestinal epithelial cells), and neurotransmission [1]. Cholesterol is a smaller, more rigid molecule that makes bilayers less deformable and fluid, but also inhibits aggregation of hydrocarbon chains of the other membrane constituents, thereby preventing crystallization. For in vitro or computational studies, model systems consisting of only one or a few types of lipid molecules, such as DPPC, in water are used. Such model systems exhibit many of the same physical and chemical properties as those of more complicated biological membranes.

In addition to their biological functions, lipid bilayers are of interest as examples of a complex mechanical system formed by self-assembly. In general, the driving force for bilayer formation is the tendency of the amphiphilic membrane constituents to expose their polar head groups to water and to bury their hydrophobic tails. Depending on the temperature, pressure, and composition, they may form micelles (inverted or ordinary), planar bilayers, bilayer vesicles, or more complex lamellar or columnar structures [2] (Fig. 8.2). Many biological processes, such as membrane fusion, vesicle budding, and protein insertion, depend on specific mechanical properties of the bilayer phase, such as surface tension, flexibility and elasticity, lateral pressure profile, and diffusion.

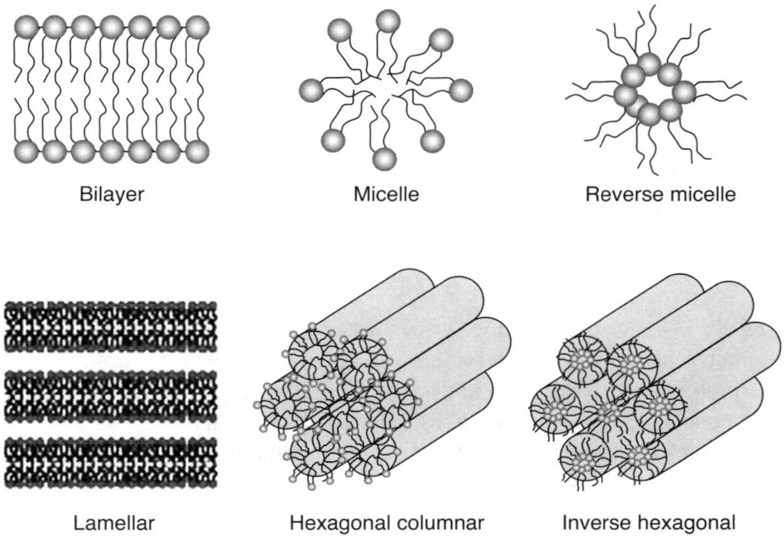

Figure 8.2 Different aggregates and phases formed by amphiphilic lipids.

In the past several years there have been many computational studies of lipid bilayers using models at several different levels of detail, from atomistic molecular dynamics simulations to continuum studies [3,4]. Simulations using an all-atom potential energy function (force field) in addition to explicit representation of water molecules provide the finest level of detail and, in principle, are the most accurate. However, because the time step in such simulations must not exceed the period of the highest-frequency motion in the system (covalent bond vibrations to hydrogen, on the order of femtoseconds), available computational resources often limit the simulation time to several nanoseconds, far too short to observe processes of biological interest or spontaneous self-assembly.

As such, there has been a great deal of recent interest on coarse-grained or multiscale simulations. Like all-atom simulations, these use discrete sites or particles to represent phospholipid and water molecules and water, with similar functional forms to evaluate interaction energies. However, each particle no longer represents a single atom, but several (e.g., three or four) heavy atoms. In this way, computations can be made orders of magnitude faster, both from the reduced number of particle interactions that need to be calculated and by the effect that when individual atoms and molecules are replaced by coarse-grained sites, there is less friction, so that particles diffuse more quickly.

In this chapter we will focus on recent models proposed for coarse-grained and multiscale models of lipid bilayers in aqueous solution. Variations and details in the energy functions for describing particle interactions, parameterization methods, and applications toward understanding biological membranes and self-assembly will be discussed.

8.2 METHODS

The essential idea behind a coarse-grained model is to replace several heavy atoms with a single interaction site. In the coarse-grained model proposed by Shelley and co-workers [5,6], a three-to-one mapping is used. For aliphatic chains, triplets of carbons and attached hydrogens are represented by a single site. These sites interact via a Lennard-Jones (LJ)-type 9-6 potential:

$$U_{\mathrm{LJ}}(r) = \frac{15}{4}\varepsilon_{ij}\left[\left(\frac{\sigma_{ij}}{r}\right)^9 - \left(\frac{\sigma_{ij}}{r}\right)^6\right], \tag{8.1}$$

where r is the distance between sites, σ_{ij} is an interaction radius for types i and j, and ε_{ij} is a parameter describing the interaction strength. For the phospholipid dimyristoyl-phosphatidylcholine (DMPC), additional spherical sites are used to represent the choline, phosphate, glycerol, and ester groups (Fig. 8.3). Finally, a "W" site is used to represent, roughly, three water molecules. The interaction potential between these coarse-grained sites is not given by a simple functional form but, instead, is tabulated based on fitting data (described below). Intramolecular bonded interactions are given by harmonic potentials of the form

$$U_{\mathrm{bond}}(r) = \tfrac{1}{2}K_{\mathrm{bond}}(r - r_{\mathrm{bond}})^2, \tag{8.2}$$

where r is the bond length, r_{bond} is the equilibrium value, and K_{bond} is a force constant. Bond-angle potentials are also used to maintain the linearity of hydrophobic chains, of the form

$$U_{\mathrm{angle}}(\theta) = K_{\mathrm{angle}}[1 - \cos(\pi - \theta)]^2. \tag{8.3}$$

Similar mapping is used in the coarse-grained models of Ayton, Izvekov, and co-workers [7–13]. To describe DMPC, single interaction sites are associated with

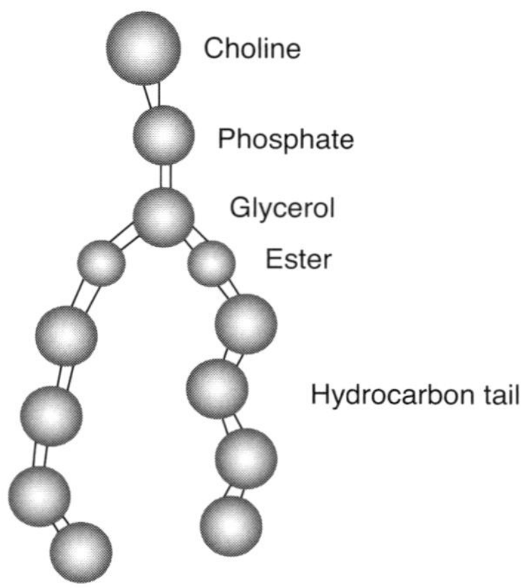

Figure 8.3 Coarse-grained model for dimyristoylphosphatidylcholine.

the centers of mass of the choline, phosphate, glycerol, and ester groups and triplets of carbon atoms.

In the MARTINI model of Marrink and co-workers [14,15], a four-to-one mapping is used. Four different basic interaction site types are considered: polar (P), nonpolar (N), apolar (C), and charged (Q). Within these basic site types, subtypes are assigned (five, four, five, and four different subtypes for each of the types above, respectively). Sites interact via a Lennard-Jones (LJ) 6-12 potential:

$$U_{\mathrm{LJ}}(r) = 4\varepsilon_{ij}\left[\left(\frac{\sigma_{ij}}{r}\right)^{12} - \left(\frac{\sigma_{ij}}{r}\right)^{6}\right], \tag{8.4}$$

the same functional form that is used to represent dispersion and repulsion interactions in all-atom force fields. Here σ_{ij} is taken to be 0.47 nm for all interactions, with a few exceptions. The interaction strength ε_{ij} varies from 2.0 to 5.6 kcal/mol, depending on the type and subtype of both of the sites. Interactions between charged sites (Q) also include a coulombic force, depending on the product of the charges and the reciprocal of the distance between them:

$$U_{\mathrm{coulomb}}(r) = \frac{q_i q_j}{4\pi\varepsilon_0\varepsilon_r r}. \tag{8.5}$$

Here q_i and q_j are charges on sites i and j, ε_0 is the permittivity of the vacuum, and ε_r is the effective relative permittivity, taken to be 15.

Different molecules are built out of combinations of the coarse-grained sites. Four water molecules are represented by a single P-type site. Lipids are represented by several coarse-grained sites connected by bonded interactions. For example, DPPC is represented by two Q-type sites (the choline and phosphate moieties), two N-type sites (the ester linkages), and eight C-type sites (four representing each aliphatic chain). Bonds between sites are described by a harmonic potential,

$$U_{\mathrm{bond}}(r) = \tfrac{1}{2}K_{\mathrm{bond}}(r - r_{\mathrm{bond}})^2 \tag{8.6}$$

with an equilibrium distance r_{bond} taken to be 0.47 nm (the same value as σ) and a force constant K_{bond} of 1250 kJ/mol·nm^2. Bond angles are also maintained near an equilibrium distance with a similar harmonic potential:

$$U_{\mathrm{angle}}(\theta) = \tfrac{1}{2}K_{\mathrm{angle}}(\cos\theta - \cos\theta_0)^2. \tag{8.7}$$

For the coarse-grained sites in aliphatic chains, the force constant is chosen to be $K_{\mathrm{angle}} = 25$ kJ/mol, while the equilibrium angle θ_0 is taken to be 180° to keep the sites close to linear. Linkages between sites corresponding to unsaturated double bonds are assigned an angle-bending potential with a stiffer force constant (45 kJ/mol) and an equilibrium angle of 120° for cis double bonds.

The exception to the four-to-one heavy atom mapping in the MARTINI model is ring systems. Phenyl rings are represented by three coarse-grained sites and the rings, in cholesterol, by a single site. Because of the different scale of mapping, the interaction parameters for rings were adjusted from those for other sites; in particular, σ_{ij} in the Lennard-Jones potential for interactions between ring sites is reduced to 0.43 nm.

Many different molecules may be represented by combinations of the basic site types; alkanes are modeled by a linear chain of C-type sites, cyclohexane as three ring particles, ions as single Q-type sites (in which the site is assumed to represent both the ion and the immediate hydration layer), and polar solvents such as octanol as a combination of C- and P-type sites.

A similar coarse-grained model for DMPC has been proposed by Lyubartsev [16]. The head group is represented by a choline (N) and phosphate (P) group and there are two ester groups (CO), each connected to a chain of three CH sites, representing the hydrocarbon tails. The four types of sites interact with 10 different pairs of intermolecular potentials and four different intramolecular bond potentials. Water molecules are not included; the effect of water is modeled implicitly via the interaction potentials between the lipid sites.

Another, more minimalist, coarse-grained model for lipids has been proposed by Stevens [17] and makes use of only two site types, hydrophilic and hydrophobic. The head group is composed of three hydrophilic sites and each chain by four hydrophobic sites. Solvent molecules are represented by a single hydrophilic site equivalent to those comprising the lipid head group. Sites interact with a Lennard-Jones potential, and the sites in a lipid molecule are connected with bond and angle interactions.

A different approach to coarse-grained models has been proposed by Essex and coworkers [18,19] and makes use of the Gay–Berne potential [20] (which describes interactions between ellipsoidal bodies) and has been used for several decades to model liquid crystals:

$$U_{\text{Gay-Berne}}(\boldsymbol{r}_{ij}, \hat{\boldsymbol{u}}_i, \hat{\boldsymbol{u}}_j) = 4\varepsilon(\boldsymbol{r}_{ij}, \hat{\boldsymbol{u}}_i, \hat{\boldsymbol{u}}_j)\left[\left(\frac{\sigma_0}{r_{ij}-\sigma(\boldsymbol{r}_{ij}, \hat{\boldsymbol{u}}_i, \hat{\boldsymbol{u}}_j)+\sigma_0}\right)^{12} - \left(\frac{\sigma_0}{r_{ij}-\sigma(\boldsymbol{r}_{ij}, \hat{\boldsymbol{u}}_i, \hat{\boldsymbol{u}}_j)+\sigma_0}\right)^{6}\right], \tag{8.8}$$

where the well depth and radius parameters depend on the orientations $\hat{\mathbf{u}}_i$ and $\hat{\mathbf{u}}_j$ of the ellipsoids. In this model, a water molecule is represented by a Lennard-Jones site, a dipole moment, and a tetrahedral hydrogen-bonding interaction (the soft–sticky dipole model) [21–23]. Octane is represented by two Gay–Berne ellipsoids connected by a harmonic spring. The coarse-grained models for octane and water have been combined with all-atom models for solutes in multiscale simulations used to predict partition coefficients. A coarse-grained model for DMPC represented as an assembly of spheres and ellipsoids has been proposed; specifically, two charged Lennard-Jones spheres representing the choline and phosphate, ellipsoids with embedded dipole moments representing the glycerol–ester linkages, and three nonpolar ellipsoids representing each of the two aliphatic chains. The electrostatic interactions between charges and dipoles are given by Coulomb's law with the relative permittivity set to 1.

The model proposed by Michel and Cleaver [24] also makes use of the Gay–Berne potential. This model is a very simple generalized model of an assembly of amphiphilic molecules where solvent particles are modeled with a Lennard-Jones sphere and amphiphiles by a Gay–Berne ellipsoid. To introduce amphiphilic behavior, anisotropy is included in the sphere–ellipsoid interaction to make interactions with one end of the amphiphilic particle (the *solvophilic* end) more favorable, and interactions with the other solvophobic end less favorable.

All of the coarse-grained models proposed above are fairly similar. However, different strategies have been employed for parameterization. For the model of Shelley and co-workers [5,6], the water model (a single site interacting with a LJ 6-4 potential, intended to represent a loose grouping of roughly three water molecules) was parameterized so as to reproduce the density of water at 303.15 K and the melting temperature. The alkane-bonded parameters were fit to reproduce appropriate bond distances and angles obtained from atomic-detail simulations. The Lennard-Jones parameters for alkanes were chosen such that the density and vapor pressure from simulations of dodecane and nonane were in good agreement with experimental values. The tabulated potentials used for interactions between the coarse-grained sites used to represent the polar head groups of lipid molecules were fit so as to reproduce radial distribution functions of these groups, again obtained from atomistic simulations.

The coarse-grained models of Voth and co-workers [7–13] also are parameterized using atomic-detail simulations. However, a force-matching procedure is used rather than adjusting parameters to obtain agreement with geometries or radial distribution functions. Each configuration of coarse-grained sites corresponds to many possible configurations of atomic sites. For any such coarse-grained configuration, the average total force (i.e., the sum of the forces for the atoms comprising the coarse-grained site) can be calculated from a long equilibrium atomic-detail molecular dynamics simulation. The interactions between coarse-grained sites are then adjusted so as to minimize the difference between the coarse-grained forces and the averaged atomic-detail forces.

The MARTINI model of Marrink and co-workers [14,15] is parameterized by direct fitting to experimental quantities rather than reproducing data from atomic-detail simulations. Model small molecules and ions (e.g., ethanolamine, phosphate, ethanol, hydrocarbons) are constructed making use of each coarse-grained site type. For each model small molecule, the free energy of hydration, the free energy of vaporization, and partition coefficients between water and organic phases such as hexadecane, chloroform, ether, and octanol are computed. Parameters describing the interaction strengths are then adjusted so as to obtain the best agreement with experimental values.

As in the approach of Shelley, the coarse-grained model of Lyubartsev [16] is fit to reproduce data from atomic-detail simulations: in particular, radial distribution functions. An inverse Monte Carlo (MC) algorithm is used to extract interaction potentials from RDFs (i.e., an iterative procedure with a series of MC runs in which successive corrections to the potential are taken from the logarithm of the ratio of the calculated coarse-grained and target atomic-detail RDF).

The model for DPPC [18,19], which makes use of Gay–Berne ellipsoids [20], is parameterized so as to reproduce the experimental volume and area per lipid and the amount of order in the tail as measured by nuclear magnetic resonance [25]. The generalized amphiphile model of Michel and Cleaver [24] makes use of only two adjustable parameters: the anisotropy of amphiphile–solvent interactions (i.e., the ratio of the well depths for the interactions between the solvent and the amphiphile head, and the amphiphile side), and the relative size of the solvophobic region of the amphiphile. The parameters are not fit to experimental properties for a particular lipid system; rather, the behavior of the model is examined as a function of the parameters.

A multiscale model in which atomic-detail and mesoscopic membrane systems are bridged has been presented by Ayton, Chang, and co-workers [9,26–28]. The mesoscopic model, called EM2, is a discretization of the Hamiltonian for an elastic membrane of a given thickness, bending modulus, bulk modulus, local curvature, and local in-plane strain. The membrane is represented by a set of quasiparticles with a position $\mathbf{r}_i$ and orientation Ω_i, and effective Hamiltonian

$$H_{\text{eff}}(\mathbf{r}_{ij}, \check{\mathbf{z}}_i, \check{\mathbf{z}}_j) = H_{k_c,\text{eff}}(\mathbf{r}_{ij}, \check{\mathbf{z}}_i, \check{\mathbf{z}}_j) + H_{\lambda,\text{eff}}(r_{ij}), \tag{8.9}$$

where

$$H_{k_c,\text{eff}}(\mathbf{r}_{ij}, \check{\mathbf{z}}_i, \check{\mathbf{z}}_j) = \frac{1}{2}\sum_{i,j} \frac{8k_c}{\rho_A N} \frac{\left[(\check{\mathbf{z}}_i \cdot \mathbf{r}_{ij})^2 + (\check{\mathbf{z}}_j \cdot \mathbf{r}_{ij})^2\right]}{r_{ij}^2} \tag{8.10}$$

and

$$H_{\lambda,\text{eff}}(r_{ij}) = \frac{1}{2}\sum_{i,j} \frac{2\pi (r_{ij}^0)^2 h\lambda}{N^2} \left[2\left(\frac{r_{ij} - r_{ij}^0}{r_{ij}^0}\right)\right]^2. \tag{8.11}$$

Here k_c is the bending modulus, ρ_A is a reduced density, N is the number of particles, h is the membrane thickness, λ is the bulk modulus, and r_{ij}^0 is the initial distance between two quasiparticles when the membrane is in a completely flat, undeformed state. These parameters are taken from atomic-detail simulations. The membrane model also interacts with explicit mesoscopic solvent models.

8.3 RESULTS

The primary advantage of coarse-grained models is the ability to simulate processes far beyond the time scale accessible to atomic-detail simulation, such as self-assembly. As a preliminary check of consistency, Shelley and co-workers performed simulations of a bilayer, calculated the density profiles for the various constituents of the lipid, and obtained reasonable agreement with atomistic simulation [5]. They subsequently performed simulation of a system containing 50 DMPC molecules and 428 W sites, started from an entirely random configuration. After about 10^6 Monte Carlo moves per site, the DMPC self-assembled into a rough bilayer structure containing defects. Additional molecular dynamics simulation for 6 ns resulted in a defect-free bilayer structure. Further simulations examined the self-assembly of a reverse hexagonal phase. The lipids diheptadecanoylphosphatidylcholine (DHPC) and nonane were simulated with water: specifically, 369 DHPC molecules, 984 nonane molecules, and 7530 W sites. After about 10 ns of molecular simulation starting from a random initial structure, a columnar reverse-hexagonal structure was obtained. One conclusion from these simulations is that molecular dynamics is a far more effective sampling method than Monte Carlo for enabling the system to perform the collective moves necessary for it to find an equilibrium ordered state.

A more recent simulation from the Klein group using the same lipid model examined the effects of anesthetics on membrane properties [29]. Simulations were performed for halothane modeled with a single coarse-grained site in a bilayer of 512 DMPC molecules and 4384 W sites. Six simulations were performed, with the number of halothane molecules varied so as to correspond to concentrations ranging from infinitely dilute to 1 : 1 mixtures of halothane and lipid molecules. Systems were equilibrated for 2 ns and data collected for an additional 5 ns. Several bilayer properties were monitored to study the effects of increasing halothane concentration. The average area per lipid increased monotonically from 70 Å to 78 Å over the range of halothane concentrations. The interlamellar spacing underwent a small decrease, but this was ascribed to redistribution of water molecules with increasing area per lipid, not to a property of the bilayer itself. The distributions of water and lipid group profiles throughout the bilayer, the diffusion constant of lipids and halothane, and the orientational probability distribution of the angle between the vector connecting the head-group choline and phosphate and the bilayer normal are not strongly affected by the halothane concentration. However, there is a strong monotonic increase in the segmental order parameters of the lipid chains as halothane concentration increases, in agreement with results from NMR. The ordering effect is strongest for bonds closest to the membrane surface.

The mesoscopic model of Ayton and Voth has been employed in a multiscale simulation to examine changes in protein structure due to membrane deformations [27]. An ensemble of atomic-detail simulations of a transmembrane protein, influenza A virus M2 proton channel, in a DMPC bilayer was coupled to a mesoscopic model of a DMPC bilayer in explicit mesoscopic solvent. The transmembrane domain exists as a homotetramer. Analysis focused on the four His37 residues which are postulated to form a gate controlling whether the channel is open or closed. In general, coupling the channel to large-wavelength fluctuations of the mesoscale simulation resulted in a greater effective volume for the His37 residues. The effect was observed to be larger for the open state of the channel than for the closed state.

The coarse-grained model of Marrink and co-workers [14,15] (most recently called MARTINI) has been used for several applications. In the original parameterization, several properties were calculated and compared with experiment as a check of the model validity. The water model was used to compute the density, isothermal compressibility, self-diffusion constant, and freezing point, all of which are in reasonable agreement with experiment. Bulk properties of liquid alkanes (butane through eicosane) were calculated using chains of hydrophobic C particles. Densities and compressibilities were well-reproduced (within 10%); however, self-diffusion of alkanes was calculated to be two to three times slower than experiments would indicate. A simulation of 1600 coarse-grained DPPC molecules and 60,000 water sites started from a random configuration spontaneously formed a defect-free bilayer after about 200 ns of simulation time at 323 K. Simulations of smaller numbers of lipids formed bilayers more quickly. The equilibrium area per lipid was 0.64 nm^2, in excellent agreement with experiment. To examine the mechanical properties of the bilayer, a large simulation of a patch containing 6400 lipid molecules was run for 250 ns. The calculated bending modulus was about 4×10^{-20} J, in good agreement with the experimental value of 5.6×10^{-20} J. The area compressibility modulus was found to be between 0.26 and 0.40 N/m, in reasonable agreement with the reported value of 0.231 N/m. The lateral self-diffusion rates of lipids and the permeation rate of water molecules through the bilayer were calculated and were of the same order of magnitude as experimental values.

Other phases were examined. When a simulation was run in which the bilayer model was cooled to 283 K, a freezing transition to a crystalline state was observed over the course of a few nanoseconds. In this phase, the lipids packed in a hexagonal lattice with an increased area per head group of 0.47 nm^2, in very good agreement with that observed for the crystalline L_c phase that forms experimentally for DPPC at temperatures below 273 K. A model for the lipid dioleoylphosphatidylethanolamine (DOPE) was constructed and used to examine phase preferences at temperatures of 273 and 318 K. Simulations of 1024 lipids and 4224 water sites starting from random initial conditions were performed. At the lower temperature, the lipids aggregated spontaneously into a lamellar phase, while at the higher temperature a reverse hexagonal phase formed, in agreement with experiment. Micellar formation was also studied using a model for dodecylphosphatidylcholine (DPC). A simulation of 400 lipids and 125,000 coarse-grained water sites was performed, giving an effective lipid concentration well above the experimental critical micelle concentration. Within a microsecond, lipids aggregated into micelles, although the simulation time was not long enough for the micellar size distribution to have converged.

A subsequent reparameterization of the coarse-grained model [15] did not change properties addressed previously and, in addition, resulted in smaller spontaneous curvature of bilayers and a preference for micelles over a bilayer for shorter-chained lipids, in agreement with experiment. Parameters for cholesterol were also introduced and the effect of cholesterol concentration on the area per lipid was calculated and shown to be in good agreement with atomic-detail simulations.

The MARTINI force field has been used by several other groups in order to examine lipid bilayer behavior. Dickey and Faller [30] investigated the effects of *n*-butanol on DPPC bilayer properties using both atomic-detail and coarse-grained simulations. Increasing alcohol concentration caused a monotonic increase in the area per lipid and decrease in chain ordering, in agreement with experiment. Another study from Faller examined mixtures of DPPC and dipalmitoylphosphatidylethanolamine (DPPE) at

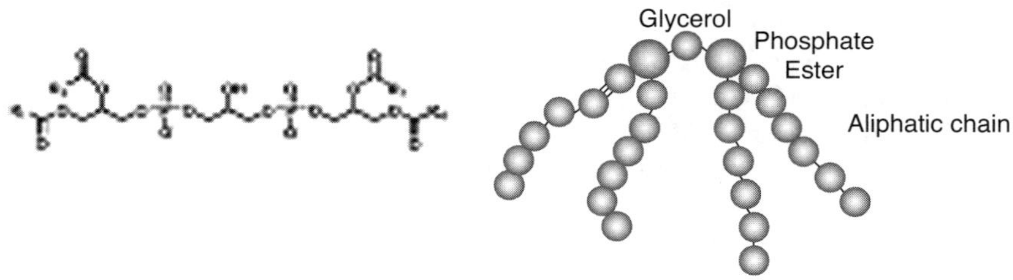

Figure 8.4 Molecular formula and coarse-grained model for cardiolipin.

various concentrations and temperatures. Simulations of 128 lipids were run with 0, 25, 50, 75, and 100% DPPE, and at temperatures of 285 to 400 K for 1 μs. At higher temperatures there is ideal mixing between lipids, and the area per lipid increases linearly with increasing DPPC concentration, in agreement with experiment. However, at the lowest temperature there is a minimum of the area per lipid when there are equal concentrations of each lipid and, in general, a weak, nonmonotonic dependence of area per lipid on concentration, indicating nonideal mixing.

Based on the MARTINI site parameters, Dahlberg [31] constructed a coarse-grained model for cardiolipin derivatives. Cardiolipin, a lipid with four acyl chains and negatively charged phosphate groups (Fig. 8.4), is found in bacterial and mitochondrial membranes and accounts for the negative charge in the inner membrane of human mitochondria. Cardiolipin derivatives containing various numbers of aliphatic chains have been shown to aggregate into either lamellar or inverted hexagonal phases, depending on the pH and the number of chains per head group. Models for cardiolipin derivatives containing two to five aliphatic chains (CL2 to CL5) were constructed and simulated along with coarse-grained sites representing water and sodium ions to maintain overall system neutrality. The primary goal of the simulations was to examine the phase preferences of the derivatives as a function of pH (the effect of acidic pH was modeled by neutralizing the charged phosphate head groups). For CL2, a micellar phase formed spontaneously at neutral pH (i.e., with negatively charged phosphates and sodium counterions) for both a system of 500 lipids and about 20,000 water sites run for 200 ns, and a larger system of 8000 lipids and 160,000 water sites run for 60 ns, both started from isotropic random initial conditions. A low-pH system of 1000 lipids and 18,000 water sites formed a lamellar phase, in agreement with experiments. For CL3, the lamellar phase was most stable, independent of the pH. At neutral pH, CL4 prefers the lamellar phase; however, at acidic pH the bilayers spontaneously make a transition to an inverse hexagonal phase. CL5 lipids form an inverse hexagonal phase with defects, independent of the pH. In general, the simulated phase behavior of the CL-type lipids, with respect to acidity and the number of aliphatic chains, was in good agreement with that observed experimentally.

Recently, Catte and co-workers [32] published a study comparing atomic-detail and coarse-grained simulations of spheroidal high-density lipoprotein (HDL) particles. Spheroidal HDL particles are formed by esterification of cholesterol, creating a core of cholesteryl ester molecules embedded in a phospholipid bilayer. The model particle simulated by Catte et al. was constructed from a "double belt" of apolipoprotein A-I monomers surrounded by a bilayer consisting of 80 palmitoyloleoylphosphatidylcholine (POPC) molecules, and with an inserted core of 16 cholesteryl oleate (CO) molecules. Atomic-detail simulations were run for 25 ns and compared with coarse-grained simulations, using parameters based on the MARTINI model [14,15], for 1 μs. The structures resulting from both scales of simulation were consistent with each other; the HDL particle assumed a prolate ellipsoidal shape, with a major axis of about 40 to 50 Å, in good agreement with measurements of in vitro reconstituted HDL particles from nondenaturing gradient gel electrophoresis and electron microscopy. The fraction of solvent-accessible surface area (SASA) of the core of CO molecules is a very small fraction of the total SASA; the protein and lipid molecules shield the core from solvent. There is significant intercalation of CO molecules with lipids and apolipoproteins, which the authors suggest might play an important role in the stability of the HDL particle.

Wong-Ekkabut and co-workers [33] recently published a study of simulations of fullerenes in dioleoylphosphatidylcholine (DOPC) and DPPC bilayers. Fullerene was modeled as an assembly of 16 hydrophobic particles, each representing roughly four atoms. In the same spirit as the MARTINI parameterization, the interaction parameters for the fullerene sites were fit to reproduce experimental transfer free energies between different solvents (ethanol, acetone, benzene, and cyclohexane). In molecular dynamics simulations started from configurations in which fullerene was placed in surrounding water, it entered the bilayer spontaneously on a nanosecond time scale. Free-energy calculations for the transfer of a fullerene to the interior gave a value of –50 kJ/mol, further indicating that the process is spontaneous. In general, results were not sensitive to the identity of the side chains on the lipids composing the bilayer; similar properties were observed for DOPC and DPPC. Aggregation of fullerenes was studied by performing three sets of simulations of 16 fullerenes and a bilayer, started from different initial configurations. For one set, initial conditions were chosen such that the fullerenes started in water. On a time scale of about 100 ns, they formed aggregates that penetrated the bilayer within 1 μs. For another set, fullerenes were started near the lipid head groups, rapidly formed small clusters which penetrated the bilayer interior within a few hundred nanoseconds, and then dispersed. For the third set, fullerenes were started in center of the bilayer, where they remained without forming aggregates. Bilayer perturbation was examined by molecular dynamics simulations in which a single fullerene was restrained at different positions within the bilayer; the fullerene was shown to perturb structural parameters of the bilayer only slightly; perturbation was greatest when the fullerene was near the center of the bilayer. Larger-scale simulations of about 1000 lipid molecules with various numbers of fullerenes giving molar concentration ratios up to one fullerene for every nine lipids showed increases in the area per lipid and bilayer thickness, and decreases in lipid lateral diffusion, area compressibility modulus, and bending modulus. Despite this "softening" of the membrane, no disruption or introduction of defects was observed. The primary biological conclusion was that mechanical damage of lipid bilayers is an unlikely mechanism for fullerene toxicity.

Baoukina and co-workers studied the mechanism of lipid monolayer collapse due to lateral compression using coarse-grained simulations with the MARTINI model [34].

Binary lipid mixtures of DPPC and palmitoylphosphatidylglycerol (POPG) at two concentration ratios, 4:1 and equal parts, were examined. The model system consisted of a water slab, bounded by a vacuum, with two symmetric monolayers at the water–vacuum interfaces, containing a total of 8192 lipids and about 300,000 water sites. Lateral compression was modeled in two different ways: (1) by applying a lateral pressure of 5 to 50 bar in a constant lateral-pressure simulation with fluctuating box dimensions, and (2) by decreasing the simulation box dimensions parallel to the monolayers while keeping constant the length in the direction normal to the monolayers. Upon compression, undulations form in the monolayers, becoming large buckling deformations. The monolayer then folds to form a bilayer with a semi-elliptical shape, in agreement with theoretical predictions. Bilayer folds can also transform into vesicles that detach from the monolayer. The qualitative features of monolayer collapse were independent of the particular method used to model compression.

Another recent study using the MARTINI coarse-grained model examined the interactions of lipid bilayers with supports [35]. Supported bilayers may be formed by fusion with solid substrates such as glass, aerogels, xerogels (i.e., dried gels forming a solid without shrinkage), or mica, and have attracted a great deal of interest from experimentalists. To model a supported bilayer, Xing and Faller [35] performed simulations of a DPPC bilayer on top of a fixed lattice of type N (intermediate hydrophilic) coarse-grained sites. Supports having either a smooth or a rough surface topology were examined. It has been hypothesized that bilayers are separated from supports by a thin layer of water molecules. In the current simulations, two layers of water sites were placed between the bilayer and the support, in initial configurations. Simulations of the supported bilayer were compared with simulations of a free DPPC bilayer in water. Overall lateral diffusion constants were smaller by about a factor of 2 for lipids in supported bilayers, compared with lipids in unsupported bilayers; furthermore, there was significant asymmetry in the bilayer leaflets. Lipids in the inner leaflet closest to the support diffused more slowly, by around a factor of 10. The support also had a strong effect on the local ordering of lipids in the inner leaflet. Diffusive slowing due to the support was less significant for a rough surface than for a smooth surface.

Lyubartsev [16] used his coarse-grained potential for DMPC to examine bilayers as well as spontaneous formation of bicelles and vesicles. Preliminary simulations of 392 lipids were initiated from a bilayer configuration. The bilayer was stable and the distribution of coarse-grained sites was in good agreement with atomic-detail simulations. The final structure was then simulated without periodic boundary conditions and formed a bicelle. After a further 0.5 ns of molecular dynamics simulation, the bicelle underwent a spontaneous transition to a spherical vesicle shape. Longer simulations were run to examine self-assembly, starting from a configuration in which about 1000 lipids were distributed randomly within a periodic simulation box. Bicelles and vesicles formed spontaneously after a simulation time on the order of several nanoseconds.

Using a minimalist coarse-grained model, Stevens performed simulations at different temperatures and with lipids containing different numbers of aliphatic chains, in order to study the liquid–gel transition [17]. At the melting temperature, a discontinuity in the area per lipid was observed. The melting temperature increased with the number of carbon atoms in the lipid tail, in agreement with experimental data for DPPC. The bending modulus was also calculated as a function of temperature and was observed to increase with lower temperature (i.e., a bilayer becomes stiffer as the gel transition temperature is approached), again in agreement with experiment. The bending modulus was also observed to increase with the length of the hydrophobic chain.

As discussed above, Essex and co-workers use a different approach toward constructing a coarse-grained lipid model, based on Gay–Berne ellipsoids rather than spheres. They used this model to calculate octanol–water partition coefficients of small molecules [18] using multiscale simulations in which the small molecules were modeled with an atomic-detail force field. Such partition coefficients are important as an estimate of drug bioavailability. Partition coefficients were calculated using free–energy perturbation for a set of 15 solutes representing analogs of neutral amino acid side chains. The results were in good agreement with experimental partition coefficients for water–hexane (which are expected to be very similar to those for octane). The same coarse-grained approach was used to study a system of 128 DMPC lipids and 3400 water molecules [24]. Starting from an initial random configuration and periodic boundary conditions, the system self-assembled into a defect-free bilayer on a time scale of about 100 ns. The electron density profile, area per lipid, and order parameters of the hydrocarbon tails inferred from the simulation were in good agreement with experiments. The compressibility and bending moduli calculated were also observed to lie within the range of values reported by experimentalists. The lateral diffusion constant of lipid molecules, calculated over a time scale spanning about 100 ns, was calculated to be about $12\,\text{nm}^2/\mu\text{s}$, consistent with the value of $9\,\text{nm}^2/\mu\text{s}$ reported from NMR experiments. The permeability of water molecules through the bilayer was also estimated from the simulation; it was about 90 μm/s, in reasonable agreement with the experimental estimate of 70 μm/s.

The model of Michel and Cleaver [24] is also based on the Gay–Berne ellipsoid model, but is more abstract and is not intended to represent a particular lipid molecule. Solvent molecules are represented by Lennard-Jones spheres, and lipids, by Gay–Berne ellipsoids. Amphilicity is introduced by making the ellipsoid–sphere interaction anisotropic so that the interaction of spheres with one end of the ellipsoid (the solvophilic end) is stronger than with the other (the solvophobic end). The interaction is modulated by two parameters which are analogous to ε and σ for the Lennard-Jones interaction: the depth of the energy well for the interaction with the solvophilic end, and the extent of the head group. Several molecular dynamics runs were performed starting from an isotropic, randomized configuration of spheres and ellipsoids at various concentration ratios. With increasing concentrations of ellipsoids, the spontaneous formation of spherical and cylindrical micelles, lamellar phases, and inverse micelles was observed. Simulations were also performed in which the parameters describing the amphiphilic interaction were varied. When the extent of the solvophilic head group was reduced, fewer phases were observed to form (e.g., a true lamellar phase was not observed). Gradually switching off the deeper energy well for the solvophilic end of the ellipsoid led to a shift from a spherical micelle, to a cylindrical micelle, to a rod-shaped aggregate, and eventually, to complete disruption of structure and return to an isotropic random configuration.

8.4 CONCLUSIONS AND FUTURE DIRECTIONS

From the large number of simulation studies that have been carried out in the last few years, it is clear that coarse-grained and multiscale particle-based models of lipid bilayers are a good compromise between accuracy and computational efficiency. Such models demonstrate good agreement with both more detailed and expensive atomic-detail models, and with physical properties of bilayers that have been measured experimentally, such as compression and bending moduli and order parameters. Coarse-grained models can account for the rich phase behavior of lipid systems as the concentration, temperature, and composition are varied, and spontaneous formation of micelles, vesicles, and bilayers has been observed in simulations. There does not appear to be a compelling reason to prefer one particular functional form (e.g., interactions between Lennard-Jones spheres or Gay–Berne ellipsoids). As long as parameterization is done carefully, good results may be obtained for a variety of model potential functions.

It is expected that future work and further increases in computational power will stimulate additional studies focusing on more complex membrane behavior, encompassing both larger length scales and longer time scales, such as vesicle budding and membrane rupture and fusion. Future studies incorporating coarse-grained models of proteins, and examining insertion into membranes, aggregation, and effects of membrane composition and mechanical strain on protein structure, can also be expected. Finally, additional work focusing on simulations combining coarse-grained models for lipid bilayers and atomic-detail models for proteins or other membrane-bound biomolecules are expected. In general, In the coming years, coarse-grained and multiscale models of lipid bilayers should provide a great deal of insight into our understanding of cellular membranes.

REFERENCES

1. Alberts B, Johnson A, Lewis J, Raff J, Roberts K, Walter P. *Molecular Biology of the Cell*, 4th ed., Garland Science, New York, 2002.
2. Chaikin PM, Lubensky TC. *Principles of Condensed Matter Physic*. Cambridge University Press, Cambridge, UK, 1995.
3. Saiz L, Klein ML. Computer simulation studies of model biological membranes. *Acc. Chem. Res.* **35**(6):482–489, 2002.
4. Klein ML, Shinoda W. Large-scale molecular dynamics simulations of self-assembling systems. *Science* **321**(5890):798–800, 2008.
5. Shelley JC, Shelley MY, Reeder RC, Bandyopadhyay S, Klein ML. A coarse grain model for phospholipid simulations. *J. Phys. Chem. B* **105**(19):4464–4470, 2001.

6. Shelley JC, Shelley MY, Reeder RC, Bandyopadhyay S, Moore PB, Klein ML. Simulations of phospholipids using a coarse grain model. *J. Phys. Chem. B* **105**(40):9785–9792, 2001.
7. Ayton GS, Izvekov S, Noid WG, Voth GA. Multiscale simulation of membranes and membrane proteins: connecting molecular interactions to mesoscopic behavior. *Comp. Mod. Membr. Bilayers* **60**:181–225, 2008.
8. Ayton GS, Noid WG, Voth GA. Systematic coarse graining of biomolecular and soft-matter systems. *MRS Bull.* **32**(11):929–934, 2007.
9. Ayton GS, Noid WG, Voth GA. Multiscale modeling of biomolecular systems: in serial and in parallel. *Curr. Opin. Struct. Biol.* **17**(2):192–198, 2007.
10. Izvekov S, Voth GA. A multiscale coarse-graining method for biomolecular systems. *J. Phys. Chem. B* **109**(7):2469–2473, 2005.
11. Noid WG, Chu JW, Ayton GS, Krishna V, Izvekov S, Voth GA, Das A, Andersen HC. The multiscale coarse-graining method: I. A rigorous bridge between atomistic and coarse-grained models. *J. Chem. Phys.* **128**:244114, 2008.
12. Noid WG, Chu JW, Ayton GS, Voth GA. Multiscale coarse-graining and structural correlations: connections to liquid-state theory. *J. Phys. Chem B* **111**(16):4116–4127, 2007.
13. Noid WG, Liu P, Wang Y, Chu JW, Ayton GS, Izvekov S, Andersen HC, Voth GA. The multiscale coarse-graining method: II. Numerical implementation for coarse-grained molecular models. *J. Chem. Phys.* **128**:244115, 2008.
14. Marrink SJ, de Vries AH, Mark AE. Coarse grained model for semiquantitative lipid simulations. *J. Phys. Chem. B* **108**(2):750–760, 2004.
15. Marrink SJ, Risselada HJ, Yefimov S, Tieleman DP, de Vries AH. The MARTINI force field: coarse grained model for biomolecular simulations. *J. Phys. Chem. B* **111**(27):7812–7824, 2007.
16. Lyubartsev AP. Multiscale modeling of lipids and lipid bilayers. *Eur. Biophys. J. Biophys. Lett.* **35**(1):53–61, 2005.
17. Stevens MJ. Coarse-grained simulations of lipid bilayers. *J. Chem. Phys.* **121** (23):11942–11948, 2004.
18. Michel J, Orsi M, Essex JW. Prediction of partition coefficients by multiscale hybrid atomic-level/coarse-grain simulations. *J. Phys. Chem. B* **112**(3):657–660, 2008.
19. Orsi M, Haubertin DY, Sanderson WE, Essex JW. A quantitative coarse-grain model for lipid bilayers. *J. Phys. Chem. B* **112**(3):802–815, 2008.
20. Gay JG, Berne BJ. Modification of the overlap potential to mimic a linear site–site potential. *J. Chem. Phys.* **74**(6):3316–3319, 1981.
21. Fennell CJ, Gezelter JD. On the structural and transport properties of the soft sticky dipole and related single-point water models. *J. Chem. Phys.* **120**(19):9175–9184, 2004.
22. Liu Y, Ichiye T. The static dielectric constant of the soft sticky dipole model of liquid water: Monte Carlo simulation. *Chem. Phys. Lett.* **256**(3):334–340, 1996.
23. Liu Y, Ichiye T. Soft sticky dipole potential for liquid water: a new model. *J. Phys. Chem.* **100**(7):2723–2730, 1996.
24. Michel DJ, Cleaver DJ. Coarse-grained simulation of amphiphilic self-assembly. *J. Chem. Phys.* **126**:034506, 2007.
25. Douliez JP, Leonard A, Dufourc EJ. Restatement of order parameters in biomembranes: calculation of C–C bond order parameters from C–D quadrupolar splittings. *Biophys. J.* **68** (5):1727–1739, 1995.

26. Chang R, Ayton GS, Voth GA. Multiscale coupling of mesoscopic- and atomistic-level lipid bilayer simulations. *J. Chem. Phys.* **122**:244716, 2005.
27. Ayton GS, Voth GA. Multiscale simulation of transmembrane proteins. *J. Struct. Biol.* **157** (3):570–578, 2007.
28. Ayton GS, McWhirter JL, Voth GA. A second generation mesoscopic lipid bilayer model: Connections to field-theory descriptions of membranes and nonlocal hydrodynamics. *J. Chem. Phys.* **124**:064906, 2006.
29. Pickholz M, Saiz L, Klein ML. Concentration effects of volatile anesthetics on the properties of model membranes: a coarse-grain approach. *Biophys. J.* **88**(3): 1524–1534, 2005.
30. Dickey AN, Faller R. Investigating interactions of biomembranes and alcohols: A multiscale approach. *J. Polym. Sci. B* **43**(8):1025–1032, 2005.
31. Dahlberg M. Polymorphic phase behavior of cardiolipin derivatives studied by coarse-grained molecular dynamics. *J. Phys. Chem. B* **111**(25):7194–7200, 2007.
32. Catte A, Patterson JC, Bashtovyy D, Jones MK, Gu F, Li L, Rampioni A, Sengupta D, Vuorela T, Niemela P, Karttunen M, Marrink SJ, Vattulainen I, Segrest JP. Structure of spheroidal HDL particles revealed by combined atomistic and coarse-grained simulations. *Biophys. J.* **94**(6):2306–2319, 2008.
33. Wong-Ekkabut J, Baoukina S, Triampo W, Tang IM, Tieleman DP, Monticelli L. Computer simulation study of fullerene translocation through lipid membranes. *Nat. Nanotechnol.* **3** (6):363–368, 2008.
34. Baoukina S, Monticelli L, Risselada HJ, Marrink SJ, Tieleman DP. The molecular mechanism of lipid monolayer collapse. *Proc. Natl. Acad. Sci. USA* **105**(31): 10803–10808, 2008.
35. Xing CY, Faller R. Interactions of lipid bilayers with supports: a coarse-grained molecular simulation study. *J. Phys. Chem. B* **112**(23):7086–7094, 2008.

9

STOCHASTIC APPROACH TO BIOCHEMICAL KINETICS

Ian J. Laurenzi

Department of Chemical Engineering, Lehigh University, Bethlehem, Pennsylvania

9.1 INTRODUCTION

The ability of organisms to respond to their environments often hinges on speed. That is, "how often" and "how quickly" are often as important as "how many" or "how much." Insofar as molecular processes define life in all its forms, one may often quantify temporal adaptation via the kinetics of biochemical reactions or networks underlying macroscopic phenomena.

For example, consider the process of fermentation whereby yeasts (e.g., *Saccharomyces cerevisiae*) converts sugars into ethanol and carbon dioxide. The process is facilitated by the glycolysis pathway of the cell to generate energy for homeostasis, growth, and so on. The overall reaction of pertinence to brewers and winemakers is

$$\text{sugar} \rightarrow 2\,\text{ethanol} + 2\,\text{CO}_2, \tag{9.1}$$

where "sugar" represents either glucose, fructose, or other six-carbon sugars. The qualities of beer or wine, defined by the amount of ethanol and other products of metabolism ("how much"), explicitly depend on the duration of the fermentation process. From the cellular perspective, the rates of cellular growth (via metabolism), response to stress, and transmission of information (e.g., via electrical signals or the interaction of cellular receptors) are essential for survival. The molecule-driven

Multiscale Modeling of Particle Interactions: Applications in Biology and Nanotechnology,
Edited by Michael R. King and David J. Gee

dynamic qualities of both unicellular and multicellular organisms define their fitness in a Darwinian sense and thus are of profound interest to scientists and engineers.

When considering the time evolutions of the populations of biochemical species, biochemical researchers customarily employ chemical rate equations: for example,

$$\begin{aligned}\frac{dc_1}{dt} &= R_1(c_1, c_2, \ldots, c_M)\\ \frac{dc_2}{dt} &= R_2(c_1, c_2, \ldots, c_M)\\ &\vdots\\ \frac{dc_M}{dt} &= R_M(c_1, c_2, \ldots, c_M),\end{aligned} \tag{9.2}$$

where c_i, $i \in [1, M]$ are the concentrations of reactive species—be they reactants, products, or both—and R_i are mathematical expressions for the rates of generation and consumption of each. Given a set of initial conditions, Eq. (9.2) specifies the deterministic time evolution of the biochemical process; that is, the concentrations c_i are known with absolute precision at any time t.

However, when the populations of reactive biomolecular species are small, the probabilistic nature of chemical reaction is manifested. Consider, for example, the association of a biological receptor with its ligand in a system containing one molecule of each. The two molecules will physically encounter each other by way of Brownian motion—itself random. Furthermore, reaction is not guaranteed upon collision since chemical bond formation is governed by the laws of quantum mechanics. There also remains the matter of interpreting Eq. (9.2) as well: If there is either one molecule of species i, or none at all, then either $c_i = 1/V$ or $c_i = 0/V$, where V is the volume of the reaction vessel, cytosol, nucleus, and so on. How can one define the derivative of a concentration that is discrete and, therefore, discontinuous?

One is thus forced to consider that quantitative description of the dynamics of small systems—which naturally includes any "nanosystem"—requires something other than the classical approach to chemical kinetics [e.g., Eq. (9.2)]. Moreover, this alternative must account for the random fluctuations of chemical populations in accordance with the predictions of statistical mechanics. Most important, it must predict experimental observations of small systems. The stochastic approach to chemical kinetics demonstrably satisfies all of these criteria.

In his famous book, Joseph L. Doob defined a *stochastic process* as "a mathematical abstraction of an empirical process whose development is governed by probabilistic laws" [1]. Quantitative descriptions of the chemical dynamics of biomolecular species certainly meet this criterion for the reasoning we have expounded in our discussion of receptor–ligand interactions: The physical laws governing both molecular encounters and the electronic transitions that define chemical reactions are probabilistic in nature. Furthermore, since chemical reaction rates depend only on the current state of the system and not on its prior states, they may also be formulated as Markov chains. Thus, the mathematical foundations of Markov chains may be utilized to describe the probabilistic time evolution of any chemical reaction or pathway. As we shall see, the transition probabilities defining a Markov chain for a chemical process are closely related to the aforementioned reaction rates, R_i.

Given definitions (or hypotheses) for these transition probabilities, the stochastic approach to chemical kinetics is customarily applied in one of two ways: (1) stochastic simulation of biochemical pathways and (2) solution of the chemical master equation. The first of these approaches was developed in the late 1970s by Daniel Gillespie and has become a mainstay of computational biology and systems biology [2,3]. In this approach, one uses the transition probabilities to construct a reaction probability density function,

$$\begin{aligned}\mathscr{P}(\mu,\tau)d\tau = {} & \Pr(\text{the next reaction event will be a reaction of type } \mu \in [1,N] \\ & \text{and will take place on the time interval } (\tau, \tau + d\tau)).\end{aligned} \tag{9.3}$$

One then repetitively selects reaction events and quiescence times via Monte Carlo, allowing the chemical system to follow a particular outcome of the process. Thus, each such simulation is akin to a particular experiment conducted on the virtual system. If one wishes to obtain a robust estimate of the average population of some species i, $E(X_i)$, or its variance $V(X_i)$, one would need to perform many simulations. One may estimate the joint distribution for the molecular populations $\{X_i\}$ similarly, given results from a sufficiently large number of simulations.

However, when applying the stochastic approach to estimate biophysical parameters from experimental data, one often requires an exact mathematical expression for the *grand probability distribution*,

$$P_{\mathbf{x}}(t) = \Pr(X_1 = x_1, X_2 = x_2, \ldots, X_N = x_N). \tag{9.4}$$

In this expression, the variables X_1, X_2, and so on, are integer-valued random variables for the populations of each species and the vector $\mathbf{x} = (x_1, x_2, \ldots, x_N)$ describes the state of the system in terms of the populations of each biomolecular species. Thus, $P_{\mathbf{x}}(t)$ is the probability that the chemical system will be in state $\mathbf{x}$ at some time t.

This time-dependent distribution is specified by the *chemical master equation* (CME), which is a birth–death equation resulting from a stochastic process for a chemical reaction. Birth–death equations are encountered in a variety of fields [4] and have been studied in detail [5,6]. In its most general form, a univariate birth–death equation may be written as

$$\frac{dP_x(t)}{dt} = \lambda_{x-1}P_{x-1}(t) - (\lambda_x + \mu_x)P_x(t) + \mu_{x+1}P_{x+1}(t), \tag{9.5}$$

where λ_x is the *birth rate* (i.e., the rate of transition from state x to state $x+1$) and μ_x is the corresponding *death rate*. The complexity of birth–death equations and the level of difficulty associated with their solution follow from the way that the state is defined as well as the mathematical forms of the birth and death rates.

Master equations for unimolecular reactions, such as $A \rightarrow B$ and $A \rightarrow B + C$, were the first to be considered by physical chemists. The first application is typically attributed to Delbrück [7], although Smoluchowski [8] described the elements of the underlying stochastic processes as early as 1917. Delbrück used the master equation to describe the kinetic fluctuations in an autocatalytic reaction $A \xrightarrow{B} 2A$, although his approach was mechanistically incorrect.[1] The subsequent developments quite appropriately focused on biochemical reactions. A series of analytical solutions of the master equation for unimolecular reactions (as well as the Michaelis–Menten reaction mechanism with one enzyme molecule) were published by Anthony Bartholomay between 1958 and 1962 [9–11]. Subsequently, Donald McQuarrie published solutions of the master equation for the reversible reaction $A \rightleftharpoons B$ (initiated completely with either A or B molecules) and the coupled reactions $A \rightarrow B$ and $A \rightarrow C$. In 1966, Arnold Fredrickson considered more generalized unimolecular reaction networks [12,13].

The first probabilistic treatment of bimolecular reactions (e.g., the associations of receptors with ligands) is properly attributed to Smoluchowski, who formulated the transition probabilities [8]. However, the first formulation of a master equation for such reactions was conducted by Alfréd Rényi in 1953 [14]. Using the master equation for the reaction $A + B \rightleftharpoons C$, Rényi identified the general form of the law of mass action, revealing it to be accurate only in the large-population limit and providing the correction factor for small populations. However, Rényi did not solve the master equation to obtain the grand probability distribution.

Solutions of the master equation for bimolecular reactions such as the associations of receptors with ligands first appeared in the literature in 1964. That year, Donald McQuarrie et al. [15] and Kenji Ishida [16] published the grand probability distributions for the reactions $A + B \rightarrow C$ and $2A \rightarrow B$. In 1966, Ivan Darvey and co-workers [17] published solutions of the master equations for most of the physically relevant reversible bimolecular reactions[2] in the limit $t \rightarrow \infty$, but the first complete solution of the master equation for the reaction $A + B \rightleftharpoons C$ was not published until 2000 [18]. In 2008, Erdem Arslan and Ian Laurenzi [19] published the grand probability distribution for the autocatalytic reaction $A + B \rightleftharpoons 2A$, which describes the propagation of prions etiological to Creutzfeldt–Jakob disease and bovine spongiform encephalopathy.

[1] Delbrück assumed that the number of noncatalytic molecules in such a process are not consumed, transforming the truly autocatalytic reaction $A + B \rightarrow 2A$ into $A \rightarrow 2A$. The two reaction mechanisms exhibit dramatically different time evolutions.

[2] Darvey and co-workers produced the grand probability distributions for the reactions $A + B \rightleftharpoons C$, $A + B \rightleftharpoons C + D$, $2A \rightleftharpoons C$, and $2A \rightleftharpoons C + D$ at equilibrium.

9.2 IRREVERSIBLE UNIMOLECULAR TRANSFORMATION

Thus far, we have discussed transition probabilities and the master equation without application. Let us illustrate the general procedure of formulating and solving a chemical master equation for the simplest of reactions, the elementary isomerization reaction

$$\mathrm{A} \xrightarrow{k_1} \mathrm{B}. \tag{9.6}$$

Equation (9.6) represents the spontaneous and direct conversion of an A molecule into a B molecule without catalysis or additional reaction steps. For examples, "A" might represent a protein in one state and "B" might represent the protein in the alternative state. We consider the case where the reaction occurs in a constant volume V held at a constant temperature T, which are typical experimental conditions. Moreover, let us consider the process initiated with N A molecules [i.e., $X_{\mathrm{A}}(0) = N$].

9.2.1 The Stochastic Process

Initially, let's consider the situation where we have just one A molecule. How long will it take for a "reaction event" to occur? Since chemical reactions occur randomly, we write

$$k_1\delta t + o(\delta t) = \Pr(\text{a specific A molecule will react within the next time interval } \delta \mathrm{t}), \tag{9.7}$$

where $o(\delta t)$ represents a sum of higher-order terms,

$$o(\delta t) = a_2(\delta t)^2 + a_3(\delta t)^3 + a_4(\delta t)^4 + \cdots. \tag{9.8}$$

The left side of Eq. (9.7) is just a Taylor series excluding a constant term [i.e., the term with $(\delta t)^0$]. But why exclude this term? Put simply, it would imply that the reaction could occur without the passage of any time, which does not make sense at a physical level.

Now let's consider a scenario whereby the population of A molecules, X_{A}, is greater than 1. The probability of a reaction event is modified only slightly: We multiply the left side of Eq. (9.7) by the total number of ways that the reaction could occur. As there are X_{A} A molecules, an $\mathrm{A} \rightarrow \mathrm{B}$ reaction can occur X_{A} ways. Thus,

$$k_1 X_{\mathrm{A}}\delta t + o(\delta t) = \Pr(\text{any A molecule will react within the next time interval } \delta t). \tag{9.9}$$

Our reasoning is similar to that employed in the course of the derivation of the binomial distribution[3] (Fig. 9.1),

[3] http://mathworld.wolfram.com/BinomialDistribution.html.

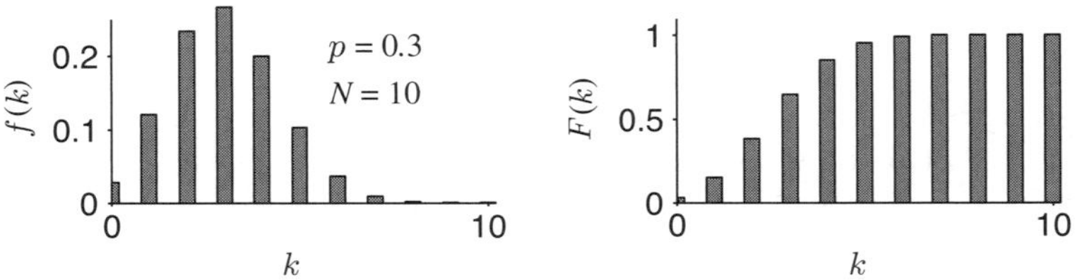

Figure 9.1 Binomial distribution. The distribution density, Eq. (9.10), and its cumulative distribution function $F(k|p, N)$ are provided for the case with $p = 0.3$ and $N = 10$.

$$f(k|p,N) = \binom{N}{k} p^k (1-p)^{N-k}. \tag{9.10}$$

Recall that $f(k|p, N)$ is the probability that there will be k successes and $N-k$ failures over the course of N games of chance, where p is the probability of a success. Since each game is independent of the others, the quantity p^k is the probability of k successes, and the quantity $(1-p)^{N-k}$ is the probability of $N-k$ failures. However, there are

$$\binom{N}{k} = \frac{N!}{(N-k)!\,k!} \tag{9.11}$$

ways that one could have obtained those k successes and $N-k$ failures. Thus, one must multiply these terms by the binomial coefficient, Eq. (9.11), to obtain the correct probability.

The failure to account properly for combinatorics in this way is exceptionally common in many areas of science. In fact, it is so common that it is given a name: the *base rate fallacy*. Fortunately, one may often check for it readily: The sum of any probability distribution density over the complete range of the random variable must be 1; that is, it is normalized. A probability distribution density derived in error by way of the base rate fallacy will often fail to be normalized.

In summary, Eq. (9.9) is the basis of a stochastic process describing this chemical reaction, with the quantity $k_1 X_A$ serving as a death rate since each reaction results in the reduction of the population of A molecules, X_A.

9.2.2 Derivation of the Master Equation

If we define the state of the reacting system by the populations of A and B molecules (X_A and X_B, respectively), the temperature T, and the volume V, then Eq. (9.9) represents the probabilities of transition between states. Defining the grand probability distribution in terms of the populations of A and B molecules,

$$P_{x_A,x_B}(t) = \Pr(X_A = x_A \cap X_B = x_B | t), \tag{9.12}$$

we then utilize the total probability theorem[4] to obtain

[4] http://mathworld.wolfram.com/TotalProbabilityTheorem.html.

$$
\begin{aligned}
P_{x_A,x_B}(t+\delta t) = & \ [k_1(x_A+1)\delta t + o(\delta t)]P_{x_A+1,x_B-1}(t) \\
& + [1-k_1 x_A \delta t + o(\delta t)]P_{x_A,x_B}(t),
\end{aligned}
\tag{9.13}
$$

where, from Eq. (9.9), we recognize that $k_1(x_A+1)\delta t + o(\delta t)$ is the probability that the state will transition from (x_A+1, x_B-1) to (x_A, x_B), and $1-k_1 x_A \delta t + o(\delta t)$ is the probability that the state will remain in the state (x_A, x_B) throughout the imminent time interval δt.

Next, we extract $o(\delta t)$ from the terms on the right of Eq. (9.13), transpose $P_{x_A,x_B}(t)$ from the right-hand side, and divide throughout by δt to obtain

$$
\begin{aligned}
\frac{P_{x_A,x_B}(t+\delta t)-P_{x_A,x_B}(t)}{\delta t} = & \ k_1(x_A+1)P_{x_A+1,x_B-1}(t) \\
& - k_1 x_A P_{x_A,x_B}(t) + \frac{o(\delta t)}{\delta t}.
\end{aligned}
\tag{9.14}
$$

In the limit $\delta t \to 0$, the left-hand side of Eq. (9.14) becomes the time derivative of $P_{x_A,x_B}(t)$, and the term $o(\delta t)/\delta t$ vanishes [see Eq. (9.8)]. One obtains

$$
\frac{dP_{x_A,x_B}}{dt} = k_1(x_A+1)P_{x_A+1,x_B-1} - k_1 x_A P_{x_A,x_B}. \tag{9.15}
$$

Equation (9.15) is the CME for the elementary reaction $A \xrightarrow{k_1} B$. It is a differential-difference equation, so called because it features a time derivative of P_{x_A,x_B} as well as terms with different population indices (e.g., P_{x_A,x_B} and P_{x_A+1,x_B-1}). Strictly speaking, it is really a set of equations, one for every stoichiometrically possible combination of x_A and x_B.

The master equation is fully specified by the initial condition. If $X_A = N$ and $X_B = 0$ at $t = 0$, then $P_{N,0}(0) = 1$ and $P_{x_A,x_B}(0) = 0$ for all other states (x_A, x_B). Using the Kronecker delta function

$$
\delta_{ij} = \begin{cases} 0, & i \neq j \\ 1, & i = j, \end{cases} \tag{9.16}
$$

we may give the initial condition of Eq. (9.15) as

$$
P_{x_A,x_B}(0) = \delta_{x_A,N}\delta_{x_B,0}. \tag{9.17}
$$

9.2.3 Solution of the Master Equation

The solution of the CME is facilitated by the stoichiometry of the process. Recall that the reaction is initiated with N A molecules and that the stoichiometry of Eq. (9.6) demands that

$$
X_B + X_A = N. \tag{9.18}
$$

This effectively introduces a Kronecker delta function into the double sums of the preceding equations, $\delta_{x_A,N-x_B}$, converting P_{x_A,x_B} into a univariate distribution. If we define

$$P_x = \Pr(X_A = x \cap X_B = N - x | t), \tag{9.19}$$

we may rewrite Eq. (9.15) as

$$\frac{dP_x}{dt} = k_1(x+1)P_{x+1} - k_1 x P_x, \tag{9.20}$$

which has the initial condition

$$P_x(0) = \delta_{N,x}. \tag{9.21}$$

In principle, this differential-difference equation is no more difficult to solve than a partial differential equation, a partial difference equation, or a linear system of equations. In fact, these three analogies suggest a few different strategies.

9.2.3.1 Derivation of the Generating Function A common approach to solving birth–death equations is the method of generating functions,[5] which is analogous to the *Z*-transform approach[6] to solving difference equations. In this approach one transforms the differential-difference equation [i.e., Eq. (9.5) or (9.20)] into a partial difference equation that may be solved using standard techniques. The generating function is defined as

$$G(s,t) = \sum_{x=0}^{\infty} s^x P_x(t), \tag{9.22}$$

where $s \in (-1, 1)$ is a continuous variable that does not depend on t. By consequence, one may write the s-derivatives of Eq. (9.22) directly:

$$\frac{\partial G}{\partial s} = \sum_{x=0}^{\infty} x s^{x-1} P_x(t), \tag{9.23}$$

$$\frac{\partial^2 G}{\partial s^2} = \sum_{x=0}^{\infty} x(x-1) s^{x-2} P_x(t), \tag{9.24}$$

which will become useful in subsequent steps. Note that Eqs. (9.23) and (9.24) are entirely general; that is, they do not depend on the details of the underlying stochastic process.

[5] http://mathworld.wolfram.com/GeneratingFunction.html.

[6] http://mathworld.wolfram.com/Z-Transform.html.

One transforms Eq. (9.20) by multiplying it by s^x and then summing over the complete state space, in accordance with Eq. (9.22). One obtains

$$\sum_{x=0}^{N} s^x \frac{dP_x}{dt} = \sum_{x=0}^{N} s^x (k_1(x+1)P_{x+1} - k_1 x P_x),$$
$$\frac{dG}{dt} = k_1 \left[\sum_{x=0}^{N} s^x (x+1) P_{x+1} - \sum_{x=0}^{N} s^x x P_x \right]. \tag{9.25}$$

Using Eqs. (9.23) and (9.24), we may evaluate the sums on the right-hand side of Eq. (9.25). By judicious reindexing of the sums and elimination of nonphysical terms (e.g., P_{-1} and $P_{x>N}$, both of which violate conservation of mass) arising from reindexing, one obtains

$$\frac{\partial G}{\partial \tau} = (1-s)\frac{\partial G}{\partial s}, \tag{9.26}$$

where we have introduced the dimensionless variable $\tau = k_1 t$. The initial condition may be obtained from Eq. (9.21) directly,

$$G(0, s) = \sum_{x=0}^{N} \delta_{x,N} s^x = s^N. \tag{9.27}$$

This partial differential equation may be solved via the method of separation of variables. By substituting $G(s, \tau) = T(\tau)S(s)$ into Eq. (9.26) and separating the variables, it is easy to show that

$$T(\tau) = A' \exp(C\tau) \tag{9.28}$$

and

$$S = B'(1-s)^{-C}, \tag{9.29}$$

where A' and B' are constants to be determined from the initial condition [Eq. (9.27)]. Recognizing that these results may hold for many different constants $\{C_n\}$, we may express the general solution as

$$G = \sum_n A_n((1-s)e^{-\tau})^{-C_n}. \tag{9.30}$$

The constants $\{A_n\}$ and $\{C_n\}$ may be obtained by taking the s-derivatives of the initial condition and setting them equal to the s-derivatives of Eq. (9.30). We then evaluate the resulting expressions at $s = 1$, at which point these quantities are related to the moments of the distribution $P_x(0)$. That is,

$$\left.\frac{(N)!}{(N-k)!}s^{N-k}\right|_{s=1} = \frac{d^kG}{ds^k}\sum_n A_n[(1-s)e^{-\tau}]^{-C_n}\Bigg|_{s=1}. \tag{9.31}$$

For example, for the first derivative

$$\begin{aligned}\left.\frac{dG}{ds}\right|_{s=1} &= \sum_n(-C_n)A_n(1-s)^{-C_n-1}\frac{d}{ds}(1-s)\\ &= -\sum_n(-C_n)A_n(1-s)^{-C_n-1} \\ N &= -\lim_{s\to 1}\sum_n(-C_n)A_n(1-s)^{-C_n-1}.\end{aligned} \tag{9.32}$$

The only term in the sum that survives in this limit is the one for which $C_n = -1$. If we set $n = 1$, we may write $C_1 = -1$ and, by extension, $A_1 = -N$. Similar considerations of the second derivatives [$k = 2$ in Eq. (9.31)] yields $C_2 = -2$ and

$$A_2 = \frac{N(N-1)}{(2)\cdot(2-1)} = \binom{N}{2}.$$

The third derivatives yield $C_3 = -3$ and

$$A_3 = \frac{N(N-1)(N-2)}{-(3)\cdot(2)\cdot(1)} = -\binom{N}{3}.$$

By induction, one may show that

$$C_n = -n \tag{9.33}$$

and

$$A_n = (-1)^n\binom{N}{n}. \tag{9.34}$$

Collecting terms and simplifying, the solution to Eq. (9.26) is

$$G(s,t) = \sum_{n=0}^{N}\binom{N}{n}[(s-1)e^{-k_1t}]^n, \tag{9.35}$$

where we have reexpressed the generating function in terms of t and established the limits of the series from the values of n permissible with the binomial coefficients. If we employ the binomial theorem,[7]

[7] http://mathworld.wolfram.com/BinomialTheorem.html.

$$(a+b)^n = \sum_{k=0}^{n} \binom{n}{k} a^{n-k} b^k, \tag{9.36}$$

we may express Eq. (9.35) in closed form,

$$G(s,t) = [1 + (s-1)e^{-k_1 t}]^N. \tag{9.37}$$

Note that $G(1,t) = \sum_{x=0}^{N} P_x(t) = 1$, demonstrating that the distribution $P_x(t)$ is normalized, as we would expect.

9.2.3.2 Inversion of the Generating Function The distribution $P_x(t)$ may be inverted from this expression using Taylor's theorem,

$$P_x(t) = \frac{1}{x!} \frac{\partial^x G(s,t)}{\partial s^x}\bigg|_{s=0} \tag{9.38}$$

since $P_x(t)$ are the Taylor coefficients in Eq. (9.22). In this case, the derivatives are easily calculated:

$$\begin{aligned}
\frac{\partial G(s,t)}{\partial s} &= N[1 + (s-1)e^{-k_1 t}]^{N-1} e^{-k_1 t} \\
\frac{\partial^2 G(s,t)}{\partial s^2} &= N(N-1)[1 + (s-1)e^{-k_1 t}]^{N-2} e^{-2k_1 t} \\
\frac{\partial^3 G(s,t)}{\partial s^3} &= N(N-1)(N-2)[1 + (s-1)e^{-k_1 t}]^{N-3} e^{-3k_1 t} \\
&\vdots \\
\frac{\partial^x G(s,t)}{\partial s^x} &= N(N-1) \cdots (N-x)[1 + (s-1)e^{-k_1 t}]^{N-x} e^{-xk_1 t}.
\end{aligned}$$

Collecting terms, one obtains

$$P_x(t) = \binom{N}{x} (1-e^{-k_1 t})^{N-x} e^{-xk_1 t}. \tag{9.39}$$

This is illustrated in Fig. 9.2 with $\tau = k_1 t$. Equation (9.39) is a binomial distribution with a probability of failure of $\exp(-k_1 t)$ and a probability of success of $1-\exp(-k_1 t)$. These probabilities, too, should be familiar: The quantity $1-\exp(-k_1 t)$ is the cumulative exponential distribution [i.e., the probability that an event will occur within the next time interval t (Fig. 9.3)]. More generally, $1-\exp(-k_1 t)$ is the probability that an A molecule will survive for a time interval of length t. So $(1-e^{-k_1 t})^{N-x}$ is the probability that $N-x$ A molecules have survived until t, and $e^{-xk_1 t}$ is the probability that the remaining x A molecules would not. Since the

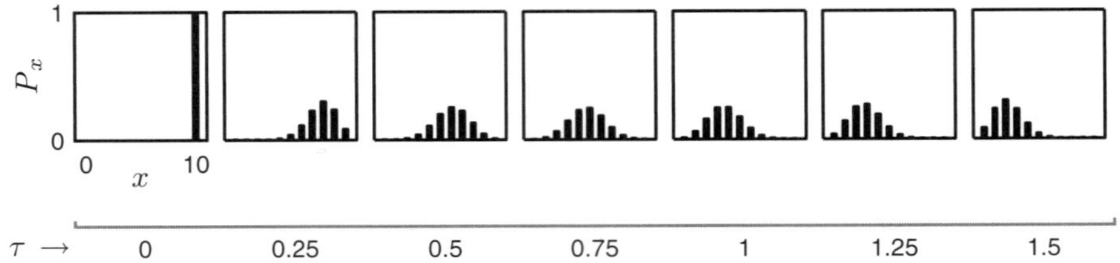

Figure 9.2 Grand probability distribution for the reaction A $\rightarrow$ B. The first frame illustrates the initial condition, Eq. (9.21). As $t \rightarrow \infty$, all A molecules will become B molecules, at which point the leftmost frame will be reflected (i.e., $\lim_{t\rightarrow\infty} P_x = \delta_{x,0}$).

total number of A molecules is initially N, there are $\binom{N}{x}$ ways that this could occur. In some sense, one might have been able to arrive at this distribution from heuristic argumentation alone! However, this will rarely be the case.

9.2.4 Fluctuations

The average population and its variance may be calculated directly by differentiating the generating function. It is straightforward to show that the expectation value and variance of the population of A molecules may be calculated from

$$E(x) = \sum_{x=0}^{\infty} x P_x(t) = \left.\frac{\partial G(s,t)}{\partial s}\right|_{s=1} \tag{9.40}$$

and

$$\begin{aligned} V(x) &= \sum_{x=0}^{N} [x - E(x)]^2 P_x(t) \\ &= \left.\frac{\partial^2 G(s,t)}{\partial s^2}\right|_{s=1} + E(x) - [E(x)]^2, \end{aligned} \tag{9.41}$$

respectively. These two expressions are entirely general.

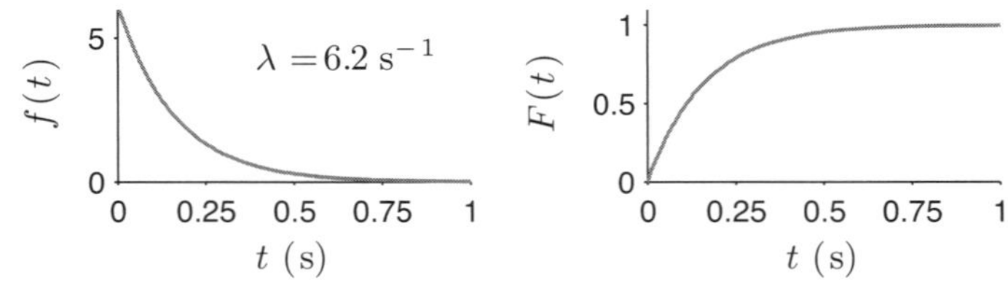

Figure 9.3 Exponential distribution. Both the distribution density, $f(t) = \lambda e^{-\lambda t}$, and cumulative distribution, $F(t) = 1 - e^{-\lambda t}$, are illustrated.

However, the fact that $x \sim \text{Binomial}(N, e^{-k_1 t})$ for the reaction $A \longrightarrow B$ allows us to exploit the well-known properties of the binomial distribution. We immediately obtain

$$\bar{X}_A = E(X_A) = Ne^{-k_1 t}, \tag{9.42}$$

with variance

$$V(X_A) = Ne^{-k_1 t}(1 - e^{-k_1 t}). \tag{9.43}$$

Both are illustrated in Fig. 9.4.

9.2.5 Relationship to Classical Chemical Kinetics

Equation (9.15) is considerably different from the governing equation from classical chemical kinetics,

$$\frac{dc_A}{dt} = -k_1 c_A \tag{9.44}$$

where c_A is the concentration of A molecules. However, the two expressions may be related by considering the average population of A molecules,

$$\bar{X}_A = \sum_{x_B=0}^{N} \sum_{x_A=0}^{N} x_A P_{x_A, x_B}(t). \tag{9.45}$$

We begin by multiplying both sides of Eq. (9.15) by x_A and summing over the possible ranges of x_A and x_B, as in Eq. (9.45). After some algebra, one obtains

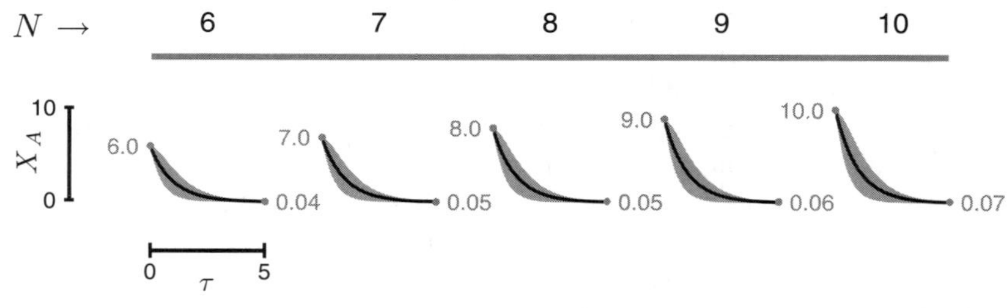

Figure 9.4 Fluctuations of the first-order unimolecular transformation reaction $A \rightarrow B$ initiated with $N = 6, \ldots, 10$ A molecules. The black line illustrates the time evolution of the population mean, Eq. (9.42). The shaded region represents $E(X) \pm \sigma$, where $\sigma^2 = V(X)$. Note that the variance is zero at $t = 0$, reflecting the initial condition, Eq. (9.21). As $t \rightarrow \infty$, the variance returns to zero as the A molecules are consumed completely.

$$\begin{aligned}\frac{d}{dt}\bar{X}_{\mathrm{A}} = {} & k_1 \sum_{x_{\mathrm{B}}=0}^{N} \sum_{x_{\mathrm{A}}=0}^{N} x_{\mathrm{A}}(x_{\mathrm{A}}+1) P_{x_{\mathrm{A}}+1, x_{\mathrm{B}}-1} \\ & - k_1 \sum_{x_{\mathrm{B}}=0}^{N} \sum_{x_{\mathrm{A}}=0}^{N} x_{\mathrm{A}}^2 P_{x_{\mathrm{A}}, x_{\mathrm{B}}}.\end{aligned} \tag{9.46}$$

The first of these sums may be simplified using the reaction stoichiometry, Eq. (9.18), and judicious reindexing of the sum:

$$\begin{aligned}\sum_{x_{\mathrm{B}}=0}^{N} \sum_{x_{\mathrm{A}}=0}^{N} x_{\mathrm{A}}(x_{\mathrm{A}}+1) P_{x_{\mathrm{A}}+1, x_{\mathrm{B}}-1} = {} & \sum_{y_{\mathrm{B}}=0}^{N} \sum_{y_{\mathrm{A}}=0}^{N} y_{\mathrm{A}} y_{\mathrm{A}} P_{y_{\mathrm{A}}, y_{\mathrm{B}}} \\ & - \sum_{y_{\mathrm{B}}=0}^{N} \sum_{y_{\mathrm{A}}=0}^{N} y_{\mathrm{A}} P_{y_{\mathrm{A}}, y_{\mathrm{B}}}.\end{aligned}$$

Reinserting this expression into Eq. (9.46), one obtains

$$\begin{aligned}\frac{d}{dt}\bar{X}_{\mathrm{A}} &= -k_1 \sum_{y_{\mathrm{B}}=0}^{N} \sum_{y_{\mathrm{A}}=0}^{N} y_{\mathrm{A}} P_{y_{\mathrm{A}}, y_{\mathrm{B}}} \\ &= -k_1 \bar{X}_{\mathrm{A}}.\end{aligned} \tag{9.47}$$

Recognizing that the average concentration is defined by $\bar{c}_{\mathrm{A}} = \bar{X}_{\mathrm{A}}/V$, we may rewrite Eq. (9.47) as

$$\frac{d\bar{c}_{\mathrm{A}}}{dt} = -k_1 \bar{c}_{\mathrm{A}}. \tag{9.48}$$

This equation is identical to Eq. (9.44). That is, the time evolution of the average population of A molecules evolves in accordance with the classical deterministic approach, despite the fact that the time evolution of the reaction $\mathrm{A} \to \mathrm{B}$ is random. This is generally true for unimolecular reaction networks; however, it is only approximately true when chemical species undergo bimolecular reactions [14].

This result informs our understanding of chemical kinetics in three ways. First, it shows that classical chemical kinetics is consistent with the stochastic approach to chemical kinetics, and thus any rate constants estimated from bulk experiments can be used in the stochastic formalism to predict the dynamics of biomolecular processes in small systems. Second, the identity of Eqs. (9.44) and (9.48) shows us that the stochastic formalism underlies the classical formalism. That is, the stochastic approach to chemical kinetics is the more fundamental of the two. Finally, we recognize that Eq. (9.2) is valid to the extent that the c_i are interpreted as the average concentrations of species $i \in [1, M]$, which will generally be close but not equal to the actual concentrations. When the populations of chemical species are large (e.g., $X \sim 10^{23}$), the fluctuations tend to insignificance, although they cannot be dismissed for certain chemical systems (e.g., polymerization and aggregation [20,21]).

9.3 REVERSIBLE UNIMOLECULAR TRANSFORMATION

Solution of the chemical master equation becomes considerably more difficult when there are multiple reactions due, in part, to the increased number of "choices" that a system has. For example, let us consider the reverible reaction

$$\mathrm{A} \underset{k_1}{\overset{k_2}{\rightleftharpoons}} \mathrm{B}. \tag{9.49}$$

Here a state defined by X_{A} (with X_{B} defined by stoichiometry) can arise from either a forward or reverse reaction, which substantially complicates the master equation.

In 2004 and 2005, a team of researchers from Petar M. Djuric's group published two papers that ostensibly give a general solution of the master equation for any unimolecular chemical reaction network [22,23]. We apply their approach to obtain the grand probability distribution for Eq. (9.49), which represents the spontaneous and reversible transformation of A molecules into B molecule (and vice versa) without catalysis or additional reaction steps. As in Section 9.2, we consider the case where the reaction occurs in a constant volume V held at a constant temperature T. However, in this case we start with N_{A} A molecules and N_{B} B molecules at $t = 0$. Naturally, we could simplify this process by choosing either $N_{\mathrm{A}} = 0$ or $N_{\mathrm{B}} = 0$ such that the process is initiated with just one of the two species, but let us consider the most general case first.

9.3.1 The Stochastic Process

Once again, let us define the states of the chemical system by the volume V, temperature T, and the populations of A and B molecules, X_{A} and X_{B}, respectively. We have already demonstrated that

$$k_1 X_{\mathrm{A}} \delta t + o(\delta t) = \Pr(\text{any A molecule will react within the next time interval } \delta t).$$

in our formulation of the stochastic process for the reaction $\mathrm{A} \rightarrow \mathrm{B}$, where X_{A} is the population of the A molecules. By the same arguments we may write

$$k_2 X_{\mathrm{B}} \delta t + o(\delta t) = \Pr(\text{any B molecule will react within the next time interval } \delta \mathrm{t}). \tag{9.50}$$

Since changes in the system are brought about one reaction at a time, Eqs. (9.9) and (9.50) represent the probabilities of transition among all states of the system.

9.3.2 The Master Equation

Once again, we may fully quantify the process using the total probability theorem. In this case, it yields a somewhat more complex equation than Eq. (9.13),

$$P_{x_A,x_B}(t+\delta t) = [k_1(x_A+1)\delta t + o(\delta t)]P_{x_A+1,x_B-1}(t) + [k_2(x_B+1)\delta t + o(\delta t)]P_{x_A-1,x_B+1}(t) + [1-k_1 x_A \delta t - k_2 x_B \delta t + o(\delta t)]P_{x_A,x_B}(t). \quad (9.51)$$

Note that we have two types of transitions to the state defined by (x_A, x_B): one from the state (x_A-1, x_B+1) (via the reverse reaction), and one from the state (x_A+1, x_B-1) (via the forward reaction).

Once again, we transpose the term $P_{x_A,x_B}(t)$ to the left-hand side, divide throughout by δt, and allow $\delta t \to 0$ to obtain the CME:

$$\frac{dP_{x_A,x_B}}{dt} = k_1(x_A+1)P_{x_A+1,x_B-1} + k_2(x_B+1)P_{x_A-1,x_B+1} - (k_1 x_A + k_2 x_B)P_{x_A,x_B}. \quad (9.52)$$

As before, we may employ stoichiometry to our advantage. Since the reaction is initiated with N_A A molecules and N_B B molecules,

$$X_A + X_B = N_A + N_B = N, \quad (9.53)$$

we may employ the definition of Eq. (9.19),

$$P_x = \Pr(X_A = x \cap X_B = N-x|t)$$

and rewrite Eq. (9.52) as

$$\frac{dP_x}{dt} = k_1(x+1)P_{x+1} + k_2(N-x+1)P_{x-1} - (k_1 x + k_2(N-x))P_x. \quad (9.54)$$

Reflecting our specification that the initial populations of A molecules are N_A and N_B, the initial condition for this process is

$$P_x(0) = \delta_{N_A,x}. \quad (9.55)$$

That is, the number of A molecules x is certainly equal to N_A at $t = 0$.

9.3.3 Solutions of the Master Equation for One Molecule

Although we could try the method of generating functions, an alternative mathematical technique will be much more effective in this case. Let us begin by considering an experiment initiated with just one molecule. As a result, there are only two states of the system, which correspond to the cases with no A molecules ($X_A = x = 0, X_B = 1$) and one A molecule ($X_A = x = 1, X_B = 0$). Since $N_A + N_B = N = 1$, Eqs. (9.54) may be simplified to

$$\begin{aligned}\frac{dP_0}{dt} &= -k_2 P_0 + k_1 P_1,\\ \frac{dP_1}{dt} &= k_2 P_0 - k_1 P_1.\end{aligned} \tag{9.56}$$

The relations may be reexpressed in matrix form as

$$\frac{d}{d\tau}\mathbf{P}(t) = \mathbf{AP}(t), \tag{9.57}$$

where, again, we have defined $\tau = k_1 t$,

$$\mathbf{P}(t) = \begin{bmatrix} P_0(t) \\ P_1(t) \end{bmatrix} \tag{9.58}$$

and

$$\mathbf{A} = \begin{bmatrix} -K & 1 \\ K & -1 \end{bmatrix}. \tag{9.59}$$

Furthermore, we have defined $K = k_2/k_1$, which, conveniently, is the equilibrium constant for the reaction.

The system of equations may be solved by applying a Laplace transform,

$$P_x(s) = \int_0^{\infty} e^{-s\tau} P_x(\tau), \tag{9.60}$$

from which we obtain

$$s\mathbf{P}(s) - \mathbf{P}_0 = \mathbf{AP}(s), \tag{9.61}$$

where

$$\mathbf{P}_0 = \lim_{t\to 0} \begin{bmatrix} P_0(t) \\ P_1(t) \end{bmatrix} \tag{9.62}$$

and

$$\mathbf{P}(s) = \begin{bmatrix} P_0(s) \\ P_1(s) \end{bmatrix}. \tag{9.63}$$

Having placed the Laplace-transformed CME into the form of Eq. (9.61), we may solve it using the customary approaches of linear algebra,

$$\mathbf{P} = [s\mathbf{I} - \mathbf{A}]^{-1}\mathbf{P}_0. \tag{9.64}$$

The inverse of the 2×2 matrix $[s\mathbf{I}-\mathbf{A}]^{-1}$ is easily obtained as

$$[s\mathbf{I}-\mathbf{A}]^{-1} = \frac{1}{s(s+(1+K))}\begin{bmatrix} s+1 & 1 \\ K & s+K \end{bmatrix}.$$

9.3.3.1 Case I: Process Initiated with One A Molecule If $N_A = 1$, then $\mathbf{P}_0 = [0, 1]^T$ and we obtain

$$\mathbf{P}(s) = \frac{1}{s(s+(1+K))}\begin{bmatrix} s+1 & 1 \\ K & s+K \end{bmatrix}\begin{bmatrix} 0 \\ 1 \end{bmatrix}$$
$$= \frac{1}{s(s+(1+K))}\begin{bmatrix} 1 \\ s+K \end{bmatrix}$$

with solution

$$P_0(s) = \frac{1}{s(s+(1+K))} \tag{9.65}$$

$$P_1(s) = \frac{1}{s+(1+K)} + \frac{K}{s(s+(1+K))}. \tag{9.66}$$

After inverse Laplace transformation, we obtain the grand probability distribution for the reversible reaction $A \rightleftharpoons B$ initiated with one A molecule,

$$P_0(t) = \frac{1}{1+K}(1-e^{-(1+K)\tau}), \tag{9.67}$$

$$P_1(t) = \frac{1}{1+K}(K+e^{-(1+K)\tau}). \tag{9.68}$$

9.3.3.2 Case II: Process Initiated with One B Molecule On the other hand, if $N_B = 1$, then $\mathbf{P}_0 = [1, 0]^T$ and we obtain

$$\mathbf{P} = \frac{1}{s(s+(1+K))}\begin{bmatrix} s+1 & 1 \\ K & s+K \end{bmatrix}\begin{bmatrix} 1 \\ 0 \end{bmatrix}$$
$$= \frac{1}{s(s+(1+K))}\begin{bmatrix} s+1 \\ K \end{bmatrix}$$

with solution

$$P_0(s) = \frac{1}{s+(1+K)} + \frac{1}{s(s+(1+K))}, \tag{9.69}$$

$$P_0(s) = \frac{K}{s(s+(1+K))}, \tag{9.70}$$

Table 9.1 Solutions of the Master Equation for the Reaction A$\rightleftharpoons$B Initiated with Either One A Molecule ($N_A = 1$) or One B Molecule ($N_B = 1$)[a]

	$N_A = 1$	$N_B = 1$
P_0	$\frac{1}{1+K}\left(1-e^{-(1+K)\tau}\right)$	$\frac{1}{1+K}\left(1+Ke^{-(1+K)\tau}\right)$
P_1	$\frac{1}{1+K}\left(K+e^{-(1+K)\tau}\right)$	$\frac{K}{1+K}\left(1-e^{-(1+K)\tau}\right)$

[a] $P_x = \Pr(X_A = x \cap X_B = N-x|t = \tau/k_1)$, where $N = N_A + N_B$ and $K = k_2/k_1$. Note that the sum of the probabilities for each case is equal to 1.

from which we obtain the grand probability distribution for the reversible reaction A$\rightleftharpoons$B initiated with one B molecule,

$$P_0(t) = \frac{1}{1+K}\left(1+Ke^{-(1+K)\tau}\right), \tag{9.71}$$

$$P_1(t) = \frac{K}{1+K}\left(1-e^{-(1+K)\tau}\right). \tag{9.72}$$

These solutions of the CME for the reaction A$\rightleftharpoons$B initiated with a single molecule are summarized in Table 9.1. Quite clearly, the time evolutions of both processes differ as a result of the initial conditions. However, as $t \to \infty$, the distributions are exactly the same. One should expect this to be the case; this is the equilibrium state.

From a probabilistic point of view, if the initial molecule is an A molecule ($N_A = 1$), the number of A molecules at time t may be said to be Bernoulli distributed[8] with parameter

$$p = \frac{1}{1+K}\left(K+e^{-(1+K)\tau}\right). \tag{9.73}$$

Similarly, if the initial molecule is a B molecule ($N_B = 1$), then the number of A molecules at time t is Bernoulli distributed with parameter

$$q = \frac{K}{1+K}\left(1-e^{-(1+K)\tau}\right). \tag{9.74}$$

9.3.4 Solution of the Master Equation for Many Molecules

One may obtain a closed-form expression for the generating function corresponding to the grand probability distribution for the reaction A$\rightleftharpoons$B [Eq. (9.54)]. However, the resulting expression,

[8] http://mathworld.wolfram.com/BernoulliDistribution.html.

$$G(s,\tau) = \left[\frac{-(1-s)e^{-(1+K)\tau}+1+Ks}{K(1-s)e^{-(1+K)\tau}+1+Ks}\right]^{N_A} \times \left[\frac{K(1-s)e^{-(1+K)\tau}+1+Ks}{1+K}\right]^{N} \tag{9.75}$$

does not yield the "facile differentiability" of Eq. (9.37) that affords a simple inversion to $P_x(\tau)$ [23].

Therefore, let us consider an alternative approach to solving the CME for this process to obtain the grand probability distribution $P_x(\tau)$. Since the reactions of Eq. (9.49) do not involve molecular interaction, an A molecule will turn into a B molecule (and vice versa) independent of all of the other molecules in the system. This is an intrinsic quality of any unimolecular reaction.

Now let us consider a reaction process in which there are N_A A molecules at $\tau = 0$. Because the states of each of the N_A molecules are independent of each other, the probability that there will be $l \leq N_A$ A molecules at a dimensionless time $\tau = k_1 t$ is just

$$P_l(\tau|N_A) = \underbrace{p\cdots p}_{\Pr(l \text{ are A})}\ \underbrace{(1-p)\cdots(1-p)}_{\Pr(N_A-l \text{ are not A})}\binom{N_A}{l}$$

or more concisely,

$$P_l(\tau|N_A) = N_A\binom{N_A}{l}p^l(1-p)^{N_A-l}, \tag{9.76}$$

where p is given by Eq. (9.73). Note that Eq. (9.76) is a binomial distribution which is also the convolution of the Bernoulli distribution. A similar argument yields the probability that there are $m \leq N_B$ A molecules if the process is initiated with N_B B molecules,

$$P_m(\tau|N_B) = \binom{N_B}{m}q^m(1-q)^{N_B-m}, \tag{9.77}$$

where q is given by Eq. (9.74).

It remains to determine the probability that there are $x = \ell + m$ A molecules given that the reaction is initiated with N_A molecules and N_B molecules. This state may arise if one of the x A molecules comes from one of the N_A initial A molecules and the balance come from the N_B initial B molecules. Alternatively, it may arise if two of the x A molecules come from the N_A initial A molecules and the balance come from the N_B initial B molecules. More concisely, we may write

$$P_x(\tau) = \sum_{x=l+m} P_k(\tau, N_A)P_{x-k}(\tau, N_B). \tag{9.78}$$

If $x > N_A$ or $x > N_B$, many of these terms in Eq. (9.78) may be zero or will be undefined. For example, if $\ell > N_A$, then $P_l(\tau|N_A) = 0$. Thus, the complete solution for such cases will depend on the relative values of N_A, N_B, and x. In the case where $N_A \geq x > N_B$, we lose all terms $P_m(\tau|N_B)$ in Eq. (9.78) with $m > N_B$ such that

$$P_x(\tau) = \sum_{k=x-N_B}^{x} P_k(\tau, N_A)P_{x-k}(\tau, N_B) \qquad (N_A \geq x > N_B). \tag{9.79}$$

In the situation where $x > N_A \geq N_B$, Eq. (9.78) becomes

$$P_x(\tau) = \sum_{k=x-N_B}^{N_A} P_k(\tau, N_A)P_{x-k}(\tau, N_B) \qquad (x > N_A \geq N_B). \tag{9.80}$$

Finally, if $N_A \geq N_B \geq x$, we may express the solution to the CME as

$$P_x(\tau) = \sum_{k=0}^{x} P_k(\tau, N_A)P_{x-k}(\tau, N_B) \qquad (N_A \geq N_B \geq x). \tag{9.81}$$

Equations (9.79)–(9.81) constitute the general solution to the CME for the reaction A$\rightleftharpoons$B [Eq. (9.54)]. The equations are summarized in Table 9.2, and the time evolution of the process with $K = 10$, $N_A = 10$, and $N_B = 5$ is illustrated in Fig. 9.5.

In the special case where $N_A = N$ (the process is initiated with only one of the two molecules), the solutions summarized in Table 9.2 reduce to

$$P_x(\tau) = \binom{N}{x}\left(\frac{1}{1+K}\right)^N \left(e^{-(1+K)\tau} + K\right)^x \left(1 - e^{-(1+K)\tau}\right)^{N-x}, \tag{9.82}$$

Table 9.2 Solutions of the Master Equation for the Reaction A$\rightleftharpoons$B Initiated with Exactly N_A A Molecules and N_B B Molecules[a]

Range	$P_x(\tau)$
$N_A \geq N_B \geq x$	$\sum_{k=0}^{x} \binom{N_A}{k} p^k (1-p)^{N_A-k} \binom{N_B}{x-k} q^{x-k} (1-q)^{N_B-x+k}$
$N_A \geq x > N_B$	$\sum_{k=x-N_B}^{x} \binom{N_A}{k} p^k (1-p)^{N_A-k} \binom{N_B}{x-k} q^{x-k} (1-q)^{N_B-x+k}$
$x > N_A \geq N_B$	$\sum_{k=x-N_B}^{N_A} \binom{N_A}{k} p^k (1-p)^{N_A-k} \binom{N_B}{x-k} q^{x-k} (1-q)^{N_B-x+k}$

[a] $P_x(t)$ is the probability that there are x A molecules at time $t = \tau/k_1$ (the balance of the molecules being of type B). The quantities p and q are specified by Eqs. (9.73) and (9.74), respectively.

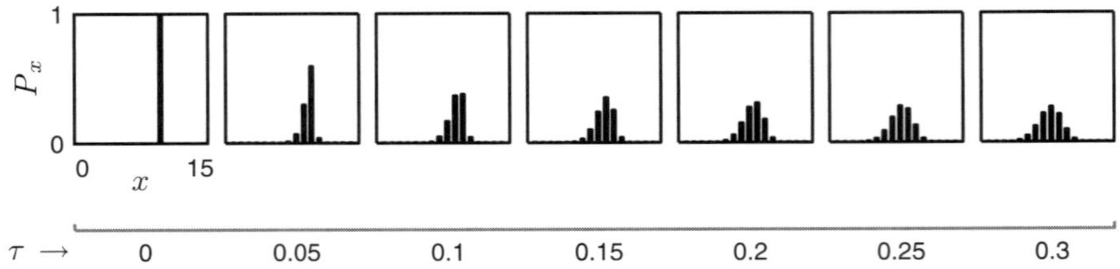

Figure 9.5 Time evolution of the grand probability distribution for reversible isomerization $A \rightleftharpoons B$ initiated with $N_A = 10$ A molecules and $N_B = 5$ B molecules, where $\tau = k_1 t$. The equilibrium constant $K = 0.25$ reflects the fact that B molecules will be more prevalent than A molecules as $t \rightarrow \infty$.

which may be inverted from the generating function [23], Eq. (9.75)). If we consider Eq. (9.82) more closely, after a little algebra we obtain

$$P_x(t) = \binom{N}{x} p^x (1-p)^x, \tag{9.83}$$

which is a binomial distribution with p given by Eq. (9.73).

Comparing this result with the grand probability distribution for the unidirectional reaction $A \rightarrow B$ [Eq. (9.39)], we recognize that if a biochemical reaction of the type $A \rightleftharpoons B$ is initiated with N A molecules and no B molecules—the experimentally expedient initial condition—the distribution of A-populations will always be binomial. The form of the distribution will depend on the value of the equilibrium constant $K = k_2/k_1$,

$$x \sim \text{Binomial}(p(t), N) \qquad p(t) = \begin{cases} e^{-k_1 t}, & K = 0, \\ \dfrac{1}{1+K}(K + e^{-(1+K)k_1 t}), & K > 0. \end{cases} \tag{9.84}$$

We may apply this distribution to characterize a variety of single-molecule experiments, including patch-clamp and adhesion flow assays for the characterization of ion-channel dynamics and receptor–ligand interactions, respectively.

9.4 RECEPTOR-MEDIATED ADHESION

Stochastic approaches for the description of receptor-mediated adhesion began to appear in the scientific literature in the early 1990s. The first of these may be attributed to Cindi Cozens-Roberts and co-workers [24,25], but were rapidly followed by others [26–28]. By the mid-1990s, these developments had led to experimental techniques for the characterization of single tether bonds between adhesion receptors on opposing surfaces, and investigations into the force dependence of their rates of dissociation [29–35]. By the early 2000s, these single-molecule techniques had advanced to the point where they were being used to characterize blood adhesion pathologies at the kinetic level [36–38].

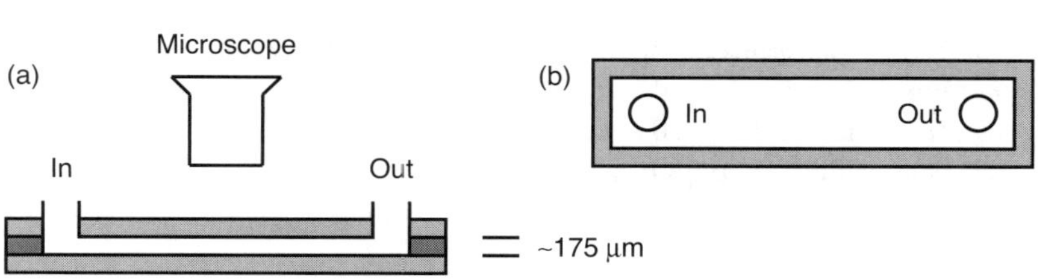

Figure 9.6 One of many experimental configurations for an adhesion flow assay: (a) side view; (b) top view.

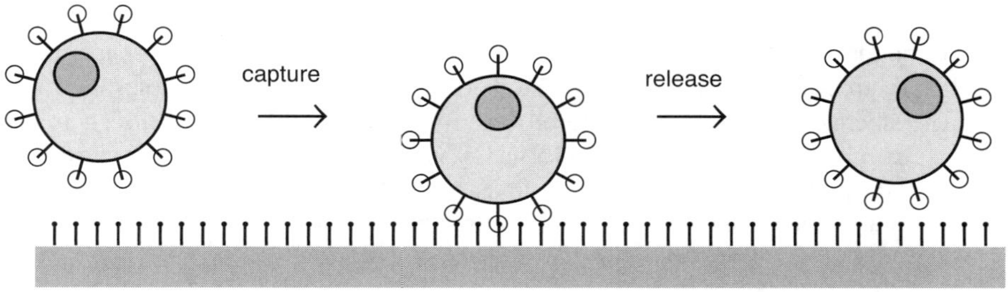

Figure 9.7 Receptor-mediated adhesion. Cells adhere to surfaces or other cells via interactions between surface-resident receptors. Leukocytes are not the only cells used in these experiments; platelets and ligand-coated beads are often used. Moreover, receptors need not be attached directly to the (silanized) glass slide. Often, a monolayer of cells provides the surface-bound receptors.

The common feature of these investigations is the *adhesion flow assay*, illustrated in Figs. 9.6 and 9.7 The experimental setup consists of a glass flow chamber attached to a microscope with a high-speed videocamera. The flow chamber is typically coated with receptor-expressing cells or the receptors themselves. Then, ligand-expressing cells (or ligand-coated beads) are drawn through the chamber under well-defined flow conditions. As cells pass through the flow chamber, they will transiently adhere to the surface via specific receptor–ligand interactions. The "pause times" of the cells are then calculated a posteriori from the video recordings of the experiment.

9.4.1 The Stochastic Process

Common to all receptor-mediated adhesion processes is the formation of a *tether bond*, a complex of a cell-bound receptor with a cell or surface-bound ligand. Representing the receptor by R, the ligand by L, and the tether bond by C, the process of tether bond formation and dissociation may be expressed as

$$\mathrm{R} + \mathrm{L} \underset{k_2}{\overset{k_1}{\rightleftharpoons}} \mathrm{C}, \tag{9.85}$$

where k_1 and k_2 are the chemical rates of association and dissociation, respectively. Hence, all receptor-mediated capture of cells in flow must consist of the four following steps:

1. The formation of one complex between a receptor, located on the surface of a moving cell, and its ligand, attached to the nonmoving surface or the surface of an immobilized cell. As soon as the tether bond is formed, the formerly moving cell is considered captured.
2. The (possible) formation of additional tether bonds between the receptors and ligands in the cell–surface contact area.
3. The dissociation of all tether bonds.
4. The motion of the unbound cell away from the binding site due to fluid flow.

Although all of these events may occur, the second is not necessary; single tether bonds are sufficient for transient cellular capture, although receptor-mediated rolling requires the formation of several tether bonds between a cell and a surface [37].

The stochastic description of the process is defined by the rates of the chemical reactions underlying these events. We begin by defining the state of the system in terms of the populations of receptors, ligands, and tether bonds (X_R, X_L, X_C) within the cross-sectional area shared by the cell and surface. Then

$$\frac{k_1}{V}\delta t+o(\delta t)=\Pr((1,1,0)\rightarrow(0,0,1)\text{ within the imminent time interval }\delta t) \quad (9.86)$$

$$k_2\delta t+o(\delta t)=\Pr((0,0,1)\rightarrow(1,1,0)\text{ within the imminent time interval }\delta t). \quad (9.87)$$

As discussed previously, these probabilities must be multiplied by the total number of ways in which the events can occur. When the populations of the reactive species are greater than 1, we have

$$\frac{k_1}{V}X_RX_L\delta t+o(\delta t)=\Pr((X_R,X_L,X_C)\rightarrow(X_R-1,X_L-1,X_C+1)\text{ within the imminent time interval }\delta t), \quad (9.88)$$

$$k_2X_C\delta t+o(\delta t)=\Pr((X_R,X_L,X_C)\rightarrow(X_R+1,X_L+1,X_C-1)\text{ within the imminent time interval }\delta t). \quad (9.89)$$

Note that Eqs. (9.86) and (9.88) feature the volume of the system in the denominator of the expression for the reaction rate. This is an essential quality of bimolecular reaction rates and we refer the interested reader to Gillespie [2,3,39] for a comprehensive discussion.

Equations (9.86)–(9.89) constitute the stochastic description of steps in the transient adhesion process with the exception of one detail: If the population of tether bonds becomes zero, the process must terminate. Upon the loss of the last tether bond, the cell will move with the fluid (step 4), thereby terminating the pause. Finally, the initial state is defined by the adhesion event itself: Since the adhesion event must begin with the formation of one bond, $X_C(0) = 1$.

9.4.2 Pause-Time Distribution

The pause-time distribution is particularly useful for experimental characterization and may be derived using the master equation. If there are N_R receptors and N_L ligands in the contact area, we may define the grand probability distribution as

$$P_x(t) = \Pr(X_R = N_R - x, X_L = N_L - x, X_C = x, t). \tag{9.90}$$

Once again, we may derive the master equation using the total probability theorem. In this case, we obtain

$$\begin{aligned}\frac{d}{dt}P_x = {} & k_2(x+1)P_{x+1} + k_1(N_R - x + 1)(N_L - x + 1))P_{x-1} \\ & -(k_2 x + k_1(N_R - x)(N_L - x))P_x.\end{aligned} \tag{9.91}$$

Next, let us consider an experimental design that reduces N_L or N_R to 1. This might, for instance, be accomplished by dilution of the surface concentration or by using beads in lieu of solution-phase cells and diluting their surface concentrations. This reduces the range of x to (0,1). Further, eliminating the probability of bond formation from the state with $X_C = 0$, we obtain

$$\frac{d}{dt}P_x = k_2(x+1)P_{x+1} - k_2 x P_x, \tag{9.92}$$

which has the initial condition $P_x(0) = \delta_{x,1}$.

Upon closer examination, we recognize that Eq. (9.92) has the same form as Eq. (9.54), including the same initial condition (with $N = 1$). Therefore, Eq. (9.92) has the same solution as Eq. (9.54) [i.e., Eq. (9.39), replacing k_1 with k_2].

We are not interested in the grand probability distribution per se; however, we may use it to obtain the pause-time distribution. Since the adhesion event terminates when the number of tether bonds is zero, the cumulative pause-time distribution must necessarily be $P_0(t)$. Therefore,

$$t \sim \text{Exponential}(k_2). \tag{9.93}$$

This distribution is essential for correct estimation of the dissociation rate constant from experimentally measured pause times. We discuss the procedure for doing so in Section 9.7.

9.5 SIMPLE ION CHANNELS

The patch-clamp technique developed by Erwin Neher and Bert Sakmann in the early 1970s permitted the measurement of current through a single ion channel [40,41]. The

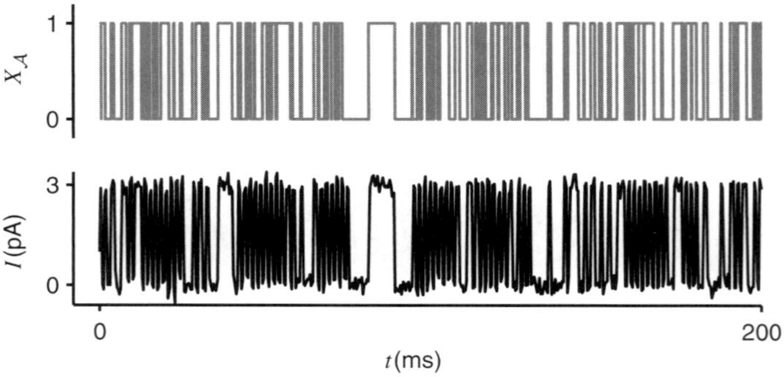

Figure 9.8 Current fluctuations resulting from the opening and closing of a single ion channel. The process illustrated here is the simple ion-channel blocking mechanism. If the population of $\mathscr{A}$ (open) channels is 1 ($X_{\mathscr{A}} = 1$), current flows. However, if the population of $\mathscr{A}$ ions is zero (by consequence of a transformation to another state), current is impeded.

power and novelty of the procedure has so transformed the field of physiology that Neher and Sakmann were awarded the Nobel Prize in Physiology or Medicine in 1991.

The patch-clamp technique entails the formation of a "tight seal" between an micropipette electrode and a cell membrane. Generally, one (or many) ion channel(s) will be captured within the cross-sectional area. As ions pass through the channel, the corresponding current may be recorded, thereby revealing whether the channel is in a closed or an open state (see Fig. 9.8). The rates of opening and closing of ion channels had been estimated by Alan Lloyd Hodgkin and Andrew Huxley [42–44] in 1952, using a "bulk" physical-chemical technique.[9] However, patch-clamp measurements revealed the randomness inherent in the transformations themselves, thereby establishing the need for a stochastic approach to the quantification of the kinetics of these transformations.

One of the first applications of the stochastic master equation to biochemical experimentation was conducted by Colquhoun and Hawkes, whose landmark papers revolutionized the biochemical characterization of ion channels [45–47].

Interestingly, the advances of Colquhoun and Hawkes appear to have been made independently of those in the chemical research community. Their 1977 paper does not cite the works of Delbrück [7], Rényi [14], Bartholomay [9–11], McQuarrie [15,48,49], Ishida [16,50], Darvey [17], or their co-workers. However, Colquhoun and Hawkes do cite Cox and Miller's The *Theory of Stochastic Processes* [51], which is also cited by McQuarrie in his well-cited review article of 1967 [49]. It would appear that in the 1960s and 1970s, many areas of Western science were undergoing a stochastic processes moment.

Let us first consider an ion channel that exhibits two states: one that permits the passage of ions (open) and another that does not (closed). The direction of

[9] Hodgkin and Huxley used their technique to characterize the action potential of the giant axon of the squid *Loligo*, for which they earned the Nobel Prize in Physiology or Medicine in 1963.

ionic transport is not important here, nor does the effect of voltage, although both most certainly would be important considerations in the design of the patch-clamp experiments. An open channel may close by consequence of voltage or some other perturbation, but we note that such openings will generally be random on account of the fact that the transitions themselves are molecular and, thus, random. That is,

$$\underset{(\text{open})}{\mathscr{A}} \underset{\beta'}{\overset{\alpha}{\rightleftharpoons}} \underset{(\text{closed})}{\mathscr{C}} . \tag{9.94}$$

We recognize immediately that this is the mechanism discussed in Section 9.3.

Fluctuations in current akin to those observed in real patch-clamp experiments are illustrated in Fig. 9.8. We immediately recognize that current flows only when the ion channel is in the open ($\mathscr{A}$) state, and thus analysis of the open and closed intervals can provide information on the kinetics of the processes of opening and closing.

Let us consider the intervals of time during which ions flow between the cytosol and media. During this interval, the ion channel is necessarily in the open state. Furthermore, as long as the ion channel is open, the rate of opening (β') is not a part of the stochastic process. Thus, the stochastic process describing the dynamics of open intervals is akin to the analysis of the irreversible reaction $\mathscr{A} \xrightarrow{\alpha} \mathscr{C}$. Similarly, the stochastic process describing the dynamics of closed intervals is akin to the analysis of the irreversible reaction $\mathscr{C} \xrightarrow{\beta'} \mathscr{A}$.

In these special cases, there is only one state that permits the flow of current ($\mathscr{A}$) and one that does not ($\mathscr{C}$). Reflecting on the similarities of these "two-state" processes with the irreversible reaction (Section 9.2) and the dissociation of single tether bonds during transient cellular adhesion (Section 9.4), we realize that

$$t \sim \exp(\beta') \qquad (\text{interval of ionic quiescence}), \tag{9.95}$$

and

$$t \sim \exp(\alpha) \qquad (\text{Interval of ionic flux}). \tag{9.96}$$

These distributions may be used to estimate the opening and closing rate constants β' and α from experimental data. Indeed, one is hard-pressed to identify another way of obtaining these chemical rate constants from measurements of the type illustrated in Fig. 9.8.

9.6 SIMPLE OPEN ION-CHANNEL BLOCKING MECHANISM

Several such mechanisms considered by Colquhoun and Hawkes address the actions of channel-unlocking and channel-blocking drugs [45]. All such mechanisms feature more than one "closed" state, which potentially confounds the analysis of the intervals of ionic flux and quiescence.

Therefore, let us consider one such mechanism of pharmacological relevance: the simple open ion-channel block mechanism. In this mechanism, a ligand blocks an ion channel's open state:

$$\underset{\text{(closed)}}{\mathscr{C}} \underset{\alpha}{\overset{\beta'}{\rightleftharpoons}} \underset{\text{(open)}}{\mathscr{A}}, \tag{9.97}$$

$$L + \mathscr{A} \underset{k_{-1}}{\overset{k_{+1}}{\rightleftharpoons}} \underset{\text{(blocked)}}{\mathscr{B}}. \tag{9.98}$$

In these expressions, $\mathscr{C}$ represents an ion channel in the closed state, $\mathscr{A}$ represents an ion channel in the open state, $\mathscr{B}$ represents the channel in the blocked state, and L is the ligand. Note that this mechanism covers both of the following cases: (1) the ligand physically blocks the passage of ions through the channel; and (2) the ligand allosterically modifies the ion channel, thereby precluding the passage of ions. This model features four reaction steps and, thus, four rate constants: k_{+1}, k_{-1}, α, and β. Each of these may significantly influence the propensity of the channel to be in the open or closed states. However, one must not neglect a fifth important quantity: the population of the ligand, X_L. Given these parameters at a designated temperature, T, and constraining the "control volume" to the patch, one may fully define a stochastic process for the time evolution of a single ion channel.

Let us consider a typical patch-clamp experiment, in which we have only one channel within the patch. Experimentally, this may be verified by monitoring the current through the patch: If there are clearly only two electronic states corresponding to the open and closed states of a single ion channel (in lieu of several "steps" in current), one may reasonably conclude that the observations result from ions passing through one channel, alone.

9.6.1 The Stochastic Process

The probability that this single ion channel will go from a closed state to an open state is defined by Eq. (9.97),

$$\beta' X_{\mathscr{C}} \delta t + o(\delta t) = \text{Pr(the closed ion channel will open } (\mathscr{C} \to \mathscr{A}) \\ \text{with in the imminent time interval } \delta t), \tag{9.99}$$

,where $X_{\mathscr{C}}$ is the population of closed ion channels within the control volume. By restricting our experimental consideration to single channels, this is either zero or one.

Considering both the reverse reaction of Eq. (9.97) and the reactions of Eq. (9.98), we have

$$\alpha X_{\mathscr{A}} \delta t + o(\delta t) = \text{Pr(any open ion channel will close } (\mathscr{A} \to \mathscr{C}) \\ \text{within the imminent time interval } \delta t), \tag{9.100}$$

,

$$\frac{k_{+1}}{V} X_L X_{\mathscr{A}} \delta t + o(\delta t) = \text{Pr(any open ion channel will be bound by a ligand}$$

$$\text{and be blocked } (\mathscr{A} \rightarrow \mathscr{B}) \text{ with in the imminent time interval } \delta\text{t}), \qquad (9.101)$$

$$k_{-1} X_{\mathscr{B}} \delta t + o(\delta t) = \text{Pr(any blocked ion channel will dissociate}$$

$$\text{from the ligand } (\mathscr{B} \rightarrow \mathscr{A}) \text{ within the imminent time interval } \delta\text{t}), \qquad (9.102)$$

where $X_{\mathscr{A}}$ and $X_{\mathscr{B}}$ are the populations of open and blocked ion channels, respectively. Again, these populations are restricted to values of zero and one, whereas X_L, the population of the ligand, is unconstrained.

When we considered the reactions of the type A $\rightarrow$ B, we defined the state in terms of the temperature, the volume of the system, and the population of A molecules alone. The choice of temperature is convenient for chemical investigations since isothermal experiments are almost universally more reproducible than nonisothermal (e.g., adiabatic) ones. Moreover, by setting the temperature, we ensure that the rate constants do not change throughout the experiment.

In this single-molecule experiment, we have multiple states of the ion channel ($\mathscr{C}$, $\mathscr{A}$, $\mathscr{B}$), effectively serving as multiple chemical species. Moreover, we have an additional chemical species, the drug, on which the rate of blocking depends. Were there multiple ion channels in the patch, it would be necessary to employ more than one discrete variable (e.g., X_A in the preceding discussions) to describe the state. As it is, the state of this process may be characterized by the three states of the single ion channel.

9.6.2 The Master Equation

Again, we employ the total probability theorem to obtain

$$\begin{aligned}
P_{\mathscr{C}}(t+\delta t) &= (\alpha \delta t + o(\delta t)) P_{\mathscr{A}}(t) + (1 - \beta' \delta t + o(\delta t)) P_{\mathscr{C}}(t), \\
P_{\mathscr{A}}(t+\delta t) &= (\beta' \delta t + o(\delta t)) P_{\mathscr{C}}(t) + (k_{-1} \delta t + o(\delta t)) P_{\mathscr{B}}(t) \\
&\quad + \left(1 - \alpha \delta t - \frac{k_{+1}}{V} X_{\mathrm{B}} \delta t + o(\delta t) \right) P_{\mathscr{A}}(t), \\
P_{\mathscr{B}}(t+\delta t) &= \left(\frac{k_{+1}}{V} X_{\mathrm{B}} \delta t + o(\delta t) \right) P_{\mathscr{A}}(t) \\
&\quad + (1 - k_{-1} \delta t + o(\delta t)) P_{\mathscr{B}}(t).
\end{aligned} \qquad (9.103)$$

Again, $o(\delta t)$ may be extracted from each line of Eq. (9.103). Applying the steps executed in previously discussed cases and defining $c_L = X_L/V$ as the concentration of the ligand, we obtain the following master equations:

$$\frac{dP_{\mathscr{A}}}{dt} = -(\alpha + k_{+1}c_B)P_{\mathscr{A}} + k_{-1}P_{\mathscr{B}} + \beta' P_{\mathscr{C}},$$

$$\frac{dP_{\mathscr{B}}}{dt} = k_{+1}c_L P_{\mathscr{A}} - k_{-1}P_{\mathscr{B}}, \tag{9.104}$$

$$\frac{dP_{\mathscr{C}}}{dt} = \alpha P_{\mathscr{A}} - \beta' P_{\mathscr{C}}.$$

We may rewrite Eq. (9.104) in matrix form as

$$\frac{d}{dt}\mathbf{P} = \mathbf{AP}, \tag{9.105}$$

where

$$\mathbf{P}(t) = \begin{bmatrix} P_{\mathscr{A}}(t) \\ P_{\mathscr{B}}(t) \\ P_{\mathscr{C}}(t) \end{bmatrix} \tag{9.106}$$

and

$$\mathbf{A} = \begin{bmatrix} -(\alpha + k_{+1}c_L) & k_{-1} & \beta' \\ k_{+1}c_L & -k_{-1} & 0 \\ \alpha & 0 & -\beta' \end{bmatrix}. \tag{9.107}$$

Note that this system of equations specifies a *conservation of probability*, the amount of which is specified by the initial condition.

As we have discussed, one typically measures the intervals of time during which the ion channel is open or closed. Often, one does not know the precise state of the ion channel during intervals of electronic quiescence (absence of current). However, a hypothesized mechanism will typically delineate the initial and final states corresponding to such intervals. By formulating stochastic processes and solving the corresponding master equations for these specific cases, one may obtain probability distributions from which one may make statistical point estimates of the kinetic parameters (e.g., α, β', k_{+1}, and k_{-1}).

9.6.3 Intervals of Ionic Flux

Let us first consider the uninterrupted interval of time during which ions flow. Once again, the process is initiated with the open state $\mathscr{A}$ so that $P_{\mathscr{A}}(0) = 1$ and $\mathbf{P}_0 = [1, 0, 0]^T$. Moreover, during an uninterrupted interval of ion flux, the closed and blocked states, $\mathscr{C}$ and $\mathscr{B}$, are never encountered. Thus, there is no opportunity for the dissociation of the ligand L nor the conventional opening of the channel. Hence,

$$\beta' = k_{-1} = 0, \tag{9.108}$$

and we obtain

$$[s\mathbf{I}-\mathbf{A}] = \begin{bmatrix} s+(\alpha+k_{+1}c_L) & 0 & 0 \\ -k_{+1}c_L & s & 0 \\ -\alpha & 0 & s \end{bmatrix}. \tag{9.109}$$

This matrix and initial condition completely specify the stochastic process for the cessation of current, and Cramer's rule may be used to solve the master equations in Laplace space. The solution of the probability that the channel will be in the open state is

$$\begin{aligned} P_{\mathscr{A}}(s) &= \frac{1}{\det[s\mathbf{I}-\mathbf{A}]} \det \begin{bmatrix} 1 & 0 & 0 \\ 0 & s & 0 \\ 0 & 0 & s \end{bmatrix} \\ &= \frac{1}{s+(\alpha+k_{+1}c_L)}, \end{aligned} \tag{9.110}$$

where

$$\det[s\mathbf{I}-\mathbf{A}] = s^2(s+(\alpha+k_{+1}c_L)). \tag{9.111}$$

This is easily inverted to yield

$$P_{\mathscr{A}}(t) = \exp[-(\alpha+k_{+1}c_L)t]. \tag{9.112}$$

We recognize this as the cumulative probability of closure, implying that the time until closure is exponentially distributed; that is,

$$t \sim \exp(\alpha+k_{+1}c_{\mathrm{B}}) \qquad \text{(interval of ionic flux, blocking)}. \tag{9.113}$$

Note that the expectation value of the open time is

$$\bar{t} = \frac{1}{\alpha+k_{+1}c_L}. \tag{9.114}$$

In this case, the open state may transition to two distinct states, whereas the simple ion channel permits only one escape from the open state. Note that the rates of escape are summed in the denominator of Eq. (9.114) and that the open time is a function of the ligand L in this case.

One might be tempted to estimate k_{+1} and α by performing a linear regression of Eq. (9.114); that is, $\bar{t}^{-1}$ versus c_L. However, this is not a valid approach, due to the fact that linear regression features key assumptions about the error associated with measurements, including homoskedasticity. The distribution of times is exponential, and thus these assumptions are not valid. We discuss the statistical point estimation of k_{+1} and α in Section 9.7.

9.6.4 Intervals of Ionic Quiescence

The distribution of the interval of electronic quiescence may be derived in similar fashion, but with some noteworthy differences. In this case we consider uninterrupted quiescence such that the open ($\mathscr{A}$) state is never encountered until the end of the interval. As a direct consequence, the rates of opening by ligand action or conventional transition are irrelevant and we may write

$$\alpha = k_{+1}x_{\mathrm{B}} = 0, \tag{9.115}$$

and we obtain

$$[s\mathbf{I}-\mathbf{A}] = \begin{bmatrix} s & -k_{-1} & -\beta' \\ 0 & s+k_{-1} & 0 \\ 0 & 0 & s+\beta' \end{bmatrix}. \tag{9.116}$$

To fully define the stochastic process, one must define the initial condition. At first glance, it might appear that this is unknown since the process might begin in either the blocked ($\mathscr{B}$) or closed ($\mathscr{C}$) states. This isn't quite the case, but definition of the initial state requires an alternative way of viewing the stochastic process.

Thus, let's consider for a moment the state immediately preceding the indeterminate closed state. The ion channel is in the open state and, as we have shown, will transition to one of the two closed states in a random exponential fashion. This result may be derived in a similar way from a slightly different approach. Consider a finite interval τ immediately preceding the reaction event to come, be it a "closing" or "locking". During this interval, the rates of closure via either reaction do not change by consequence of the fact that we have defined the interval by its imminence; that is, it precedes the event to come. As a direct consequence of Eq. (9.100), the probability that the event to come following the interval of τ will be a blocking reaction is

$$\alpha\, d\tau \exp[-(\alpha + k_{+1}c_{\mathrm{B}})\tau], \tag{9.117}$$

where we have utilized Eq. (9.112). Furthermore, Eq. (9.101) mandates that the probability that the event to come following the interval of τ will be a "closing" reaction is

$$k_{+1}c_{\mathrm{B}}\, d\tau \exp[-(\alpha + k_{+1}c_{\mathrm{B}})\tau]. \tag{9.118}$$

We may condition these two distributions into two distinct distributions: one for the time until the imminent event, and one for the reaction to come. We obtain

$$\begin{aligned} P(\mu, \tau) &= \frac{a_\mu}{a_0} a_0 \exp(-a_0 \tau)\, d\tau \\ &= P_2(\mu|\tau) P_1(\tau)\, d\tau, \end{aligned} \tag{9.119}$$

where $a_1 = \alpha, a_2 = k_{+1}c_L$, and $a_0 = a_1 + a_2$. Equation (9.119) is commonly denoted the "reaction probability density function" to distinguish it from the grand probability density function one obtains by solving the master equation. The reaction probability density function is the basis of Gillespie's Monte Carlo procedure for the simulation of the stochastic time evolution of chemical reactions.

Equation (9.119) answers the question of initial specification of the stochastic process for the interval of electronic quiescence. Because $\mu \in \mathbb{N}$ and $\tau \in \mathbb{R}$ are both random numbers, $P_2(\mu|\tau)$ gives the probability that the "state to come" will be a blocked ($\mathscr{B}$) or closed ($\mathscr{C}$) state. We obtain

$$\frac{\alpha}{\alpha + k_{+1}c_L} = \Pr(\text{the state following the open state } (\mathscr{A}) \tag{9.120}$$

$$\text{is the closed } (\mathscr{C}) \text{ state}), \tag{9.121}$$

$$\frac{k_{+1}c_B}{\alpha + k_{+1}c_L} = \Pr(\text{the state following the open state } (\mathscr{A}) \tag{9.122}$$

$$\text{is the blocked } (\mathscr{B}) \text{ state}). \tag{9.123}$$

As a direct consequence, the initial condition for our stochastic process for the uninterrupted interval of electronic quiescence is exactly

$$\mathbf{P}_0 = \left[0, \frac{k_{+1}c_L}{\alpha + k_{+1}c_L}, \frac{\alpha}{\alpha + k_{+1}c_L}\right]^T. \tag{9.124}$$

Solution of the master equation follows as before:

$$P_{\mathscr{A}}(s) = \frac{1}{\alpha + k_{+1}c_L}\left[\frac{\alpha\beta'}{s(s+\beta')} + \frac{k_{+1}c_L k_{-1}}{s(s+k_{-1})}\right]. \tag{9.125}$$

This is easily inverted to yield

$$P_{\mathscr{A}}(t) = \frac{1}{\alpha + k_{+1}c_L}\int_0^t \alpha\beta' \exp(-\beta' t)dt + k_{+1}c_B k_{-1}\exp(-k_{-1}t)dt. \tag{9.126}$$

Note that this is the weighted sum of two cumulative exponential distributions; that is,

$$P_{\mathscr{A}}(t) = \int_0^t \begin{bmatrix} \dfrac{\alpha}{\alpha + k_{+1}c_L} \\ \dfrac{k_{+1}c_L}{\alpha + k_{+1}c_L} \end{bmatrix} \left[\beta' \exp(-\beta' t)dt \quad k_{-1}\exp(-k_{-1}t)dt\right], \tag{9.127}$$

where each exponential distribution has a parameter associated with the opening of the channel and the weighting term is associated with the corresponding closing step.

9.7 STATISTICAL ANALYSIS OF SINGLE-MOLECULE EXPERIMENTS

In two key papers in the early twentieth century, Ronald Fisher introduced the method of maximum likelihood and proved that it yields the most robust estimates of model parameters from data [52,53]. As the name suggests, the method is based on the maximization of a likelihood function, defined as

$$L(\theta) = \prod_{i=1}^{N} f(t_i; \theta), \tag{9.128}$$

where $f(t_i; \theta)$ is a hypothesized distribution evaluated at data points $(t_1, t_2, \ldots, t_N)$ and θ is a vector of the parameters of the distribution. Since the logarithm is a monotonic function, maximization of the log-likelihood function[10]

$$\ell(\theta) = \ln L(\theta) \tag{9.129}$$

gives the same result.

The method of maximum likelihood is *the* bridge between the stochastic approach to chemical kinetics and biophysical characterization. Given a distribution for an experimental observable (e.g., pause time, intervals of ionic flux through an ion channel) and a sufficient amount of experimental data, one may obtain the most robust estimates of rate constants and, in turn, their distributions, allowing one to estimate the confidence intervals for the parameters and employ hypothesis testing to distinguish these parameters scientifically.

9.7.1 Applications for Receptor-Mediated Adhesion

To illustrate, let us utilize the method of maximum likelihood to estimate the rate constant k_2 associated with the dissociation of tether bonds and concomitant transient adhesion events. We begin by performing an experiment wherein the transient

[10] This is often referred to as the *score function*.

adhesion of cells (or beads) to a ligand-coated surface is monitored via high-speed video microscopy. From the image time series, we observe N transient adhesions, from which we record N pause-times $(t_1, t_2, \ldots, t_N)$. Recalling that $t \sim \exp(k_2)$ [i.e., $f(t) = k_2 \exp(-k_2 t)$], we may write the likelihood function concisely,

$$L(k_2) = k_2^N \exp\left(-k_2 \sum_{i=1}^{N} t_i\right). \tag{9.130}$$

As a result, the corresponding log-likelihood function is

$$\ell(k_2) = N \ln k_2 + \left(-k_2 \sum_{i=1}^{N} t_i\right). \tag{9.131}$$

9.7.1.1 Maximum Likelihood Statistical Point Estimate Using Eq. (9.131), we obtain a statistical point estimate of k_2 as follows:

$$\frac{d}{dk_2}\ell(k_2)\bigg|_{k_2=\hat{k}_2} = 0. \tag{9.132}$$

Taking the derivative, evaluating the resulting expression at $k_2 = \hat{k}_2$, and solving for $\hat{k}_2$, we obtain

$$\hat{k}_2 = \frac{N}{\sum_{i=1}^{N} t_i}. \tag{9.133}$$

Equation (9.133) is the maximum likelihood statistical point estimate of k_2. It is not the actual value of k_2, but it is indubitably the best estimate of the true rate constant insofar as it was obtained via the method of maximum likelihood. Note also that $\hat{k}_2$ is constructed from the random variables $\{t_i\}$ and thus it, too, is a random variable. One may show that the expectation value of the point estimate is equal to the true value [i.e., $E(\hat{k}_2) = k_2$].

9.7.1.2 Confidence Intervals Because the estimate of the dissociation rate constant $\hat{k}_2$ is itself a random variable, one is often interested in obtaining the range of values within which one can be reasonably certain that the true value k_2 resides. This confidence interval is not merely an afterthought—it is the measure by which differences in the off-rate may be quantified (e.g., due to some experimental perturbation, such as the introduction of a drug or application of force).

In this case, we benefit from the form of Eq. (9.133). Recognizing that the sum of the (exponentially distributed) pause-times is an Erlang random variable,

$$\Pr\left(X < \sum_{i=1}^{N} t_i\right) = P\left(N, \frac{k_2}{\hat{k}_2} N\right), \qquad 0 \le x < \infty, \tag{9.134}$$

where $P(a, x)$ is a regularized incomplete gamma function [54]. Further noting the relationship between the Erlang and chi-square distributions, we may write

$$2N\frac{k_2}{\hat{k}_2} \sim \chi^2_{2N}, \tag{9.135}$$

which means that the quantity $2Nk_2/\hat{k}_2$ is a chi-square random variable.

This cumulative distribution, in turn, establishes the upper and lower limits of the confidence interval for k_2,

$$\left(\frac{\hat{k}_2}{2N}\chi^2_{\alpha'/2,2N} < k_2 < \frac{\hat{k}_2}{2N}\chi^2_{1-\alpha'/2,2N}\right), \tag{9.136}$$

where α' is the type I error rate and is typically taken as 0.05. It should not be confused with the rate constant for an opening ion channel, discussed previously.

9.7.1.3 *Hypothesis Testing*

Equation (9.135) is the basis of statistical quantification of differences in rate constants. The simplest case involves the hypothesis that k_2 is a particular value, $k_{2,0}$. If $k_{2,0}$ is not within the confidence interval established by Eq. (9.136), the hypothesis fails.

A more common hypothesis is that the dissociation rate constant for a receptor–ligand complex is affected by force [29–31] or another factor (e.g., venom [38]). Let us define the rate constant under control conditions as $k_{2,1}$ and the rate constant under experimental conditions as $k_{2,2}$. We then measure N_1 pause times for control interactions to obtain a point estimate of the off-rate, $\hat{k}_{2,1}$, and N_2 pause times for experimental interactions to obtain $\hat{k}_{2,2}$. Employing Eq. (9.135) and recognizing that the ratio of two chi-squared variates divided by their respective degrees of freedom is F-distributed, we obtain

$$\frac{k_{2,1}}{k_{2,2}}\frac{\hat{k}_{2,2}}{\hat{k}_{2,1}} \sim F(2N_1, 2N_2). \tag{9.137}$$

Thus, we may test the hypothesis H_0: $k_{2,1} = k_{2,2}$ by setting $k_{2,1} = k_{2,2}$ and evaluating the cumulative F-distribution appropriately.

9.7.2 Ion-Channel Blocking Mechanism

In this case, $t \sim \exp(\alpha + k_{+1}c_{\mathrm{B}})$ [Eq. (9.113)] and to estimate the parameters α and k_{+1}, we require more information. Immediately, we recognize that this additional information can be introduced via variation of c_{B} in our experimental design.

Thus, let us consider the measurement of intervals of ionic flux under different drug concentrations. We define t_{ij} as the jth interval length ($j = 1, \ldots, N_i$) for the experimental blocker concentrations $c_{B,i}$, $i \in [1, M]$. The likelihood function is then

$$L(k_{+1},\alpha) = \prod_{i=1}^{M}\prod_{j=1}^{N_i}(\alpha + k_{+1}c_{\mathrm{B},i})\exp(-(\alpha + k_{+1}c_{\mathrm{B},i})t_{ij}) \tag{9.138}$$

and the corresponding log-likelihood function is

$$\ell(k_{+1},\alpha) = \sum_{i=1}^{M}\left(N_i \ln(\alpha + k_{+1}c_{\mathrm{B},i}) - \sum_{j=1}^{N_i}(\alpha + k_{+1}c_{\mathrm{B},i})t_{ij}\right). \tag{9.139}$$

9.7.2.1 Estimation of the Rates of Blocking and Opening The maximum likelihood statistical point estimates of $\hat{k}_{+1}$ and $\hat{\alpha}$ are obtained in the customary way, where

$$\left.\frac{\partial}{\partial k_{+1}}\ell(k_{+1},\alpha)\right|_{k_{+1}=\hat{k}_{+1}} = 0 \tag{9.140}$$

and

$$\left.\frac{\partial}{\partial \alpha}\ell(k_{+1},\alpha)\right|_{\alpha=\hat{\alpha}} = 0. \tag{9.141}$$

The first derivatives are

$$\frac{\partial \ell}{\partial k_{+1}} = \sum_{i=1}^{M} c_{\mathrm{B},i}\left(\frac{N_i}{\alpha + k_{+1}c_{\mathrm{B},i}} - \sum_{j=1}^{N_i} t_{ij}\right) \tag{9.142}$$

and

$$\frac{\partial \ell}{\partial \alpha} = \sum_{i=1}^{M}\left(\frac{N_i}{\alpha + k_{+1}c_{\mathrm{B},i}} - \sum_{j=1}^{N_i} t_{ij}\right). \tag{9.143}$$

From these we obtain two equations in the two unknowns, $\hat{\alpha}$ and $\hat{k}_{+1}$:

$$0 = \sum_{i=1}^{M} c_{\mathrm{B},i}\left(\frac{N_i}{\hat{\alpha} + \hat{k}_{+1}c_{\mathrm{B},i}} - \sum_{j=1}^{N_i} t_{ij}\right) \tag{9.144}$$

$$0 = \sum_{i=1}^{M}\left(\frac{N_i}{\hat{\alpha} + \hat{k}_{+1}c_{\mathrm{B},i}} - \sum_{j=1}^{N_i} t_{ij}\right). \tag{9.145}$$

These equations must be solved numerically to obtain $\hat{\alpha}$ and $\hat{k}_{+1}$. In principle, this is not a matter of difficulty since there are many software packages that may solve systems of nonlinear equations.

9.7.2.2 Fisher Information Lacking closed-form expressions for the distributions for $\hat{\alpha}$ and $\hat{k}_{+1}$ [e.g., Eq. (9.133)], we require an alternative approach for the estimation of their confidence intervals. Fortunately, the *Fisher information* can be used to estimate the standard errors associated with these rate constants, thereby permitting the calculation of confidence intervals and formulation of hypotheses tests.

The Fisher information for a model with many parameters θ (e.g., α and k_{+1}) is defined as

$$\mathcal{I}_{ij}(\theta) = -E\left[\frac{\partial^2}{\partial\theta_i\,\partial\theta_j}\ln L(q)\right] \tag{9.146}$$

and can be related to the covariances of the parameters [55]. The standard error of the estimate of each parameter is defined by the equation

$$se(\hat{\theta}_i) = \sqrt{(\mathcal{I}^{-1})_{ii}}, \tag{9.147}$$

where $(\mathcal{I}^{-1})_{ii}$ is the ith diagonal element of the inverse of the matrix $\mathcal{I}$.

To construct the Fisher information matrix, we require second derivatives of the score function. Building off Eqs. (9.142) and (9.143), we obtain

$$\frac{\partial^2\ell}{\partial k_{+1}^2} = -\sum_{i=1}^{M}\frac{N_i c_{\mathrm{B},i}^2}{(\alpha + k_{+1}c_{\mathrm{B},i})^2}, \tag{9.148}$$

$$\frac{\partial^2\ell}{\partial\alpha^2} = -\sum_{i=1}^{M}\frac{N_i}{(\alpha + k_{+1}c_{\mathrm{B},i})^2}, \tag{9.149}$$

and

$$\frac{\partial\ell}{\partial k_{+1}\,\partial\alpha} = -\sum_{i=1}^{M}\frac{N_i c_{\mathrm{B},i}}{(\alpha + k_{+1}c_{\mathrm{B},i})^2}. \tag{9.150}$$

Inserting these results into Eq. (9.146), we obtain

$$\mathcal{I}(\hat{k}_{+1}, \hat{\alpha}) = -\left.\begin{bmatrix} \dfrac{\partial^2\ell}{\partial k_{+1}^2} & \dfrac{\partial\ell}{\partial k_{+1}\,\partial\alpha} \\[2ex] \dfrac{\partial\ell}{\partial k_{+1}\partial\alpha} & \dfrac{\partial^2\ell}{\partial\alpha^2} \end{bmatrix}\right|_{\alpha=\hat{\alpha},k_2=\hat{k}_2} \tag{9.151}$$

with the derivatives given by Eqs. (9.148)–(9.150). The inverse is

$$\mathscr{I}^{-1}(\hat{k}_{+1},\hat{\alpha}) = \frac{-1}{\det(\hat{k}_{+1},\hat{\alpha})}\begin{bmatrix} \dfrac{\partial^2 \ell}{\partial \alpha^2} & -\dfrac{\partial \ell}{\partial k_{+1}\,\partial \alpha} \\ -\dfrac{\partial \ell}{\partial k_{+1}\partial \alpha} & \dfrac{\partial^2 \ell}{\partial k_{+1}^2} \end{bmatrix}\Bigg|\alpha = \hat{\alpha}, k_2 = \hat{k}_2, \tag{9.152}$$

where

$$\det \mathscr{I}(\hat{k}_{+1},\hat{\alpha}) = \frac{\partial^2 \ell}{\partial k_{+1}^2}\frac{\partial^2 \ell}{\partial \alpha^2} - \left(\frac{\partial \ell}{\partial k_{+1}\,\partial \alpha}\right)^2\Bigg|_{\alpha=\hat{\alpha},k_2=\hat{k}_2}. \tag{9.153}$$

Collecting these results, we obtain expressions for the standard errors of the opening rate α, and blocker association rate $\hat{k}_{+1}$:

$$se(\hat{\alpha}) = \left(\det \mathscr{I}(\hat{k}_{+1},\hat{\alpha}) \sum_{i=1}^{M} N_i(\hat{\alpha} + \hat{k}_{+1}c_{\mathrm{B},i})^{-2}\right)^{1/2} \tag{9.154}$$

and

$$se(\hat{k}_{+1}) = \left(\det \mathscr{I}(\hat{k}_{+1},\hat{\alpha}) \sum_{i=1}^{M} N_i c_{\mathrm{B},i}^2(\hat{\alpha} + \hat{k}_{+1}c_{\mathrm{B},i})^{-2}\right)^{1/2}. \tag{9.155}$$

9.7.2.3 Confidence Intervals The procedure just discussed is based on the asymptotic properties of the score function, which under certain common conditions is normally distributed [55],

$$\ell(\theta) \sim N(\mathbf{0}, \mathscr{I}(\theta)). \tag{9.156}$$

In addition to generating the formula for the standard errors of the statistical point estimates $\hat{\theta}$ [e.g., Eqs. (9.154) and (9.155)], this result also informs us as to the appropriate method of calculating the confidence intervals for estimated rate constants

$$\left(\hat{\theta}_i + \Phi^{-1}\left(\frac{\alpha'}{2}\right) se(\hat{\theta}_i) \le \theta_i \le \hat{\theta}_i + \Phi^{-1}\left(1 - \frac{\alpha'}{2}\right) se(\hat{\theta}_i)\right), \tag{9.157}$$

where $\Phi^{-1}(\alpha')$ is the probit or quantile function for the normal distribution and, again, α' is the type I error rate. For the typical error rate of 0.05, $\Phi^{-1}(0.05/2) = -\Phi^{-1}(1-0.05/2) = 1.96$. Equations (9.157), (9.154), and (9.155) may be combined to obtain the confidence intervals of the opening rate α, and blocking rate k_{+1}.

9.8 SUMMARY

As we have illustrated, the fluctuations in molecular population are significant in small systems, including most processes legitimately on the nanoscale. Thus, scientific prudence requires that one account for these fluctuations in both theoretical and experimental investigations of such systems, especially when parametric estimates will be used to make decisions regarding health and safety.

The stochastic approach to chemical kinetics should be used for the characterization of all single-molecule experiments and should at least be considered when formulating experimental designs for investigations of small systems, depending on how small they are. Such considerations should not be limited to the accuracy of estimators or the correct calculation of standard errors or confidence intervals. The stochastic approach to chemical kinetics offers a way to utilize the randomness of such data sets to improve the robustness of experimental design.

We have considered two systems of biophysical relevance: receptor-mediated adhesion and ion transport. In so doing, we have demonstrated how one can construct mathematical models and test the hypotheses underlying those models. That is, we made no assumptions but, instead, formulated hypotheses that can be tested. We did not merely fit models. Being innately probabilistic, the stochastic approach to chemical kinetics tends to orient the experimentalist toward hypothesis testing.

When considering the development of hypotheses to test via a stochastic model, it is usually a good idea to consider what one can measure. For example, current is measured in patch-clamp experiments, whereas cellular distance and time are measured in the adhesion flow assay by way of video microscopy. The populations of tether bonds and open channels are not measured directly but, rather, are inferred. So too are the pause times and ionic flux intervals. However, these intervals can be inferred directly from experimental measurements.

It is for this reason that we have not discussed experiments in which biomolecular populations are measured directly. The direct quantification of individual populations is extraordinarily difficult. Fluorescence, scintillation, and so on, are all bulk quantification techniques which exhibit variances that exceed the intrinsic variance of molecular population. Moreover, one would require a unique signal for each species and a unique way to label the reactive species. These are not trivial experimental considerations. As a result, the grand probability distribution will rarely be useful for maximum likelihood estimation, even though related distributions—derived from the grand probability distribution—often offer concise estimates of rate constants.

REFERENCES

1. Doob JL. *Stochastic Processes*, Wiley, New York, 1953.
2. Gillespie DT. A general method for numerically simulating the stochastic time evolution of coupled chemical reactions. *J. Comput. Phys.* **22**:403–434, 1976.
3. Gillespie DT. Exact stochastic simulation of coupled chemical reactions. *J. Phys. Chem.* **81**(25):2340–2361, 1977.

4. Bharucha-Reid AT. *Elements of the Theory of Markov Processes and Their Applications*, McGraw-Hill, New York, 1960.
5. John PWM. Quadratic time homogeneous birth-and-death processes (abstract). *Ann. Math. Stat.* **27**:550, 1956.
6. Karlin S, McGregor JL. The classification of birth and death processes and the Stieltjes moment problem. *Trans. Am. Math. Soc.* **86**:366–400, 1957.
7. Delbrück M. Statistical fluctuations in autocatalytic kinetics. *J. Chem. Phys.* **8**:120–124, 1940.
8. Smoluchowski M. Grundriÿder koagulationskinetik kolloider lösungen. *Colloid. Polym. Sci.* **21**:1435–1536, 1917.
9. Bartholomay AF. On the linear birth and death processes of biology as Markoff chains. *Bull. Math. Biophys.* **20**:97–118, 1958.
10. Bartholomay AF. Stochastic models for chemical reactions: II. The unimolecular rate constant. *Bull. Math. Biophys.* **21**:363–373, 1959.
11. Bartholomay AF. A stochastic approach to statistical kinetics with application to enzyme kinetics. *Biochemistry* **1**(2):223–230, 1962.
12. Fredrickson AG. Stochastic models for sterilization. *Biotech. Bioeng.* **8**:167–182, 1966.
13. Fredrickson AG. Stochastic triangular reactions. *Chem. Eng. Sci.* **21**:687–691, 1966.
14. Rényi A. A discussion of chemical reactions using the theory of stochastic processes. *MTA Alk. Mater. Int. Kozl.* **2**:83–101, 1953.
15. McQuarrie DA, Jachimowski CJ, Russell ME. Kinetics of small systems: II. *J. Chem. Phys.* **40**(10):2914–2921, 1964.
16. Ishida K. Stochastic model for bimolecular reaction. *Bull. Chem. Soc. Jpn.* **41**:2472–2748, 1964.
17. Darvey IG, Ninham BW, Staff PJ. Stochastic models of second-order chemical reaction kinetics: the equilibrium state. *J. Chem. Phys.* **45**(6):2145–2155, 1966.
18. Laurenzi IJ. An analytical solution of the stochastic master equation for reversible bimolecular reaction kinetics. *J. Chem. Phys.* **113**(8):3315–3322, 2000.
19. Arslan E, Laurenzi IJ. Kinetics of autocatalysis in small systems. *J. Chem. Phys.* **128**:015101, 2008.
20. Laurenzi IJ, Bartels JD, Diamond SL. A general algorithm for exact simulation of multicomponent aggregation processes. *J. Comput. Phys.* **177**:418–449, 2002.
21. Lushnikov AA. Coagulation in finite systems. *J. Colloid. Interface Sci.* **65**:276–285, 1978.
22. Zhang X, De Cock K, Bugallo MF, Djuric PM. A general method for the computation of probabilities in systems of first order chemical reactions. *J. Chem. Phys.* **122**:104101–1 to 104101–8, 2005.
23. De Cock K, Zhang X, Bugallo MF, Djuric PM. Comment on "Stiffness in stochastic chemically reacting systems: the implicit tau-leaping method" [*J. Chem. Phys.* **119**:12784, 2003]. *J. Chem. Phys.* **121**:3347–3348, 2004.
24. Cozens-Roberts C, Lauffenburger DA, Quinn JA. Receptor-mediated cell attachment and detachment kinetics: I. Probabilistic model and analysis. *Biophys. J.* **58**:841–856, 1990.
25. Cozens-Roberts C, Quinn JA, Lauffenburger DA. Receptor-mediated cell attachment and detachment kinetics: II. Experimental model studies with the radial-flow detachment assay. *Biophys. J.* **58**:857–872, 1990.

26. Hammer DA, Apte SM. Simulation of cell rolling and adhesion on surfaces in shear-flow: general results and analysis of selectin-mediated neutrophil adhesion. *Biophys. J.* **63**:35–57, 1992.
27. Kaplanski G, Farnarier C, Tissot O, Pierres A, Alessi M-C, Kaplanski S, Bongrand P. Granulocyte–endothelium initial adhesion: analysis of transient binding events mediated by E-selectin in a laminar shear flow. *Biophys. J.* **64**:1922–1933, 1993.
28. Pierres A, Tissot O, Malissen B, Bongrand P. Dynamic adhesion of cd8-positive cells to antibody-coated surfaces: the initial step is independent of microfilaments and intracellular domains of cell-binding molecules. *J. Cell. Biol.* **125**:945–953, 1994.
29. Alon R, Chen S, Fuhlbridge R, Puri KD, Springer TA. The kinetics and shear threshold of transient and rolling interactions of L-selectin with its ligand on leukocytes. *Proc. Natl. Acad. Sci. USA* **95**:11631–11636, 1998.
30. Alon R, Chen S, Puri KD, Finger EB, Springer TA. The kinetics of L-selectin tethers and the mechanics of selectin-mediated rolling. *J. Cell. Biol.* **138**:1169–1180, 1997.
31. Alon R, Hammer DA, Springer TA. Lifetime of the P-selectin-carbohydrate bond and its response to tensile force in hydrodynamics flow. *Nature* **374**:539–542, 1995.
32. Chen S, Alon R, Fuhlbrigge RC, Springer TA. Rolling and transient tethering of leukocytes on antibodies reveal specializations of selectins. *Proc. Natl. Acad. Sci. USA* **94**:3172–3177, 1997.
33. Puri KD, Chen S, Springer TA. Modifying the mechanical property and shear threshold of L-selectin adhesion independently of equilibrium properties. *Nature* **392**:930–933, 1998.
34. Ramachandran V, Nollert MU, Qiu H, Liu W-J, Cummings RD, Zhu C, McEver RP. Tyrosine replacement in P-selectin glycoprotein ligand-1 affects distinct kinetic and mechanical properties of bonds with P- and L-selectin. *Proc. Natl. Acad. Sci. USA* **96**:13771–13776, 1999.
35. Smith MJ, Berg EL, Lawrence MB. A direct comparison of selectin-mediated transient, adhesive events using high temporal resolution. *Biophys. J.* **77**:3371–3383, 1999.
36. Doggett TA, Girdhar G, Lawshe A, Miller JL, Laurenzi IJ, Diamond SL, Diacovo TG. Alterations in the intrinsic properties of the gpibα–vwf tether bond define the kinetics of the platelet-type von willebrand disease mutation, gly233val. *Blood* **102**:152–160, 2003.
37. Doggett TA, Girdhar G, Lawshe A, Schmidtke DW, Laurenzi IJ, Diamond SL, Diacovo TG. Selectin-like kinetics and biomechanics promote rapid platelet adhesion in flow: the gpib α-vwf tether bond. *Biophys. J.* **83**:194–205, 2002.
38. Fukuda K, Doggett T, Laurenzi IJ, Liddington RC, Diacovo TG. The snake venom botrocetin acts as a biological brace to promote dysfunctional platelet aggregation. *Nat. Struct. Mol. Biol.* **12**(2):152–159, 2005.
39. Gillespie DT. A rigorous derivation of the stochastic master equation. *Physica A* **188**:402–425, 1992.
40. Neher E, Sakmann B. Single-channel currents recorded from membrane of denervated frog muscle fibres. *Nature* **260**:799–802, 1976.
41. Neher E, Sakmann B. Voltage-dependence of drug-induced conductance in frog neuromuscular junction. *Proc. Natl. Acad. Sci. USA* **72**(6):2140–2144, 2006.
42. Hodgkin AL, Huxley AF. Currents carried by sodium and potassium ions through the membrane of the giant axon of Loligo. *J. Physiol.* **116**:449–472, 1952.

43. Hodgkin AL, Huxley AF. A quantitative description of membrane current and its application to conduction and excitation in nerve. *J. Physiol.* **117**:500–544, 1952.

44. Hodgkin AL, Huxley AF, Katz B. Measurement of current-voltage relations in the membrane of the giant axon of Loligo. *J. Physiol.* **116**:424–448, 1952.

45. Colquhoun D, Hawkes AG. Relaxation and fluctuations of membrane currents that flow through drug-operated channels. *Proc. R. Soc. London B* **199**:231–262, 1977.

46. Colquhoun D, Hawkes AG. On the stochastic properties of single ion channels. *Proc. R. Soc. London B* **211**:205–235, 1981.

47. Colquhoun D, Hawkes AG. On the stochastic properties of bursts of single ion channel openings and of clusters of bursts. *Philos. Trans. R. Soc. London B* **300**:1–59, 1982.

48. McQuarrie DA. Kinetics of small systems. *J. Chem. Phys.* **38**(2):433–436, 1963.

49. McQuarrie DA. Stochastic approach to chemical kinetics. *J. Appl. Prob.* **4**:413–467, 1967.

50. Ishida K. The stochastic model for unimolecular gas reaction. *Bull. Chem. Soc. Jpn.* **33** (8):1030–1036, 1959.

51. Cox DR, Miller HD. *The Theory of Stochastic Processes*, Chapman & Hall, London, 1965.

52. Fisher RA. On an absolute criterion for fitting frequency curves. *Messenger Math.* **41**:155–160, 1912.

53. Fisher RA. On the mathematical foundations of theoretical statistics. *Philos. Trans. R. Soc. London A* **222**:309–368, 1922.

54. Abramowitz M, Stegun IA. *Handbook of Mathematical Functions with Formulas, Graphs, and Mathematical Table*, Dover Publications, New York, 1965.

55. Lawless JF. *Statistical Models and Methods for Lifetime Data*, Wiley Series in Probability and Statistics, Wiley, Hoboken, NJ, 2003.

10

IN SILICO MODELING OF ANGIOGENESIS AT MULTIPLE SCALES: FROM NANOSCALE TO ORGAN SYSTEM

AMINA A. QUTUB

Department of Bioengineering, Rice University, Houston, Texas

FEILIM MAC GABHANN

Institute for Computational Medicine and Department of Biomedical Engineering, Johns Hopkins University, Baltimore, Maryland

EMMANOUIL D. KARAGIANNIS

Department of Chemical Engineering, Massachusetts Institute of Technology, Cambridge, Massachusetts

ALEKSANDER S. POPEL

Department of Biomedical Engineering, School of Medicine, Johns Hopkins University, Baltimore, Maryland

10.1 INTRODUCTION

Angiogenesis research has already led to therapies for cancer, anemia, macular degeneration, and ischemia. In the coming decade, half a billion people worldwide are expected to benefit from drugs targeting angiogenesis in pathologies as diverse as obesity, arthritis, glioma, preeclampsia, peripheral artery disease, and neurodegeneration.

Multiscale Modeling of Particle Interactions: Applications in Biology and Nanotechnology,
Edited by Michael R. King and David J. Gee

Angiogenesis is the growth of capillaries from preexisting blood vessels. A detailed, mechanistic understanding of angiogenesis will improve fundamental scientific knowledge as well as lay the groundwork for drug discovery and patient-specific therapeutic regimens. An emerging tool in this process is multiscale *in silico* modeling.

We discuss quantitative approaches to study mechanisms of angiogenesis from the molecular to the tissue level. Among the methods reviewed are ordinary and partial differential equations, stochastic models, fractals, complex logical rules, agent-based architectures, and biological circuit analysis. Spatial scales range from the intracellular signaling level and ligand–receptor interactions, to cell-level movement and cell–matrix interactions, vessel branching and capillary network formation, tissue-level characteristics, and organ system response. Time scales correspond to these processes and span across seconds to many days.

Computational models of angiogenesis have already answered key biological questions and encouraged experimental follow-up on multiple biological levels. Models have predicted intracellular response to chronic hypoxia, sprout migration velocity based on local extracellular matrix composition, essential growth factor concentrations in skeletal muscle and tumors, and vessel growth from the proliferation and migration of individual cells, among other properties. Integration of molecular mechanisms with cell- and tissue-level models allows researchers to study angiogenesis from novel perspectives in time and space.

10.2 PARTICLE INTERACTIONS IN HYPOXIC RESPONSE AND ANGIOGENESIS

As a capillary grows from an existing blood vessel, a series of cellular events occur (Fig. 10.1). An endothelial cell from an existing vessel becomes activated by a stimulus such as hypoxia. In response to chemotactic gradients, the activated cell (called the *tip cell*) starts to migrate into the extracellular matrix by degrading it. Stalk cells adjacent to the tip cell proliferate and follow the tip cell. By this means, a sprout forms [1]. This capillary sprout continues to move toward a stimulus in response to chemical cues, mechanical factors, and a degree of persistence. Finally, the sprout joins an adjacent capillary. We divide the discussion of hypoxic response and angiogenesis by the size of the particles (scales) involved: from the nanoscale (molecular level) to the organ and organ systems levels (Fig. 10.2).

10.2.1 Molecules: O_2 Sensors, Growth Factors, Proteases, and Oxidation–Reduction Compounds

The growth of blood vessels can be stimulated and inhibited by many means, both chemical and mechanical. We focus on the chemical stimuli and, in particular, molecules that can be grouped into (1) oxygen-sensing compounds, (2) growth factors, (3) proteases, and (4) chemical entities involved in oxidation–reduction reactions.

10.2.1.1 O_2 Sensing Compound Tissues requiring oxygen drive the development of new blood vessels with O_2-carrying hemoglobin delivered via red blood cells. Cells

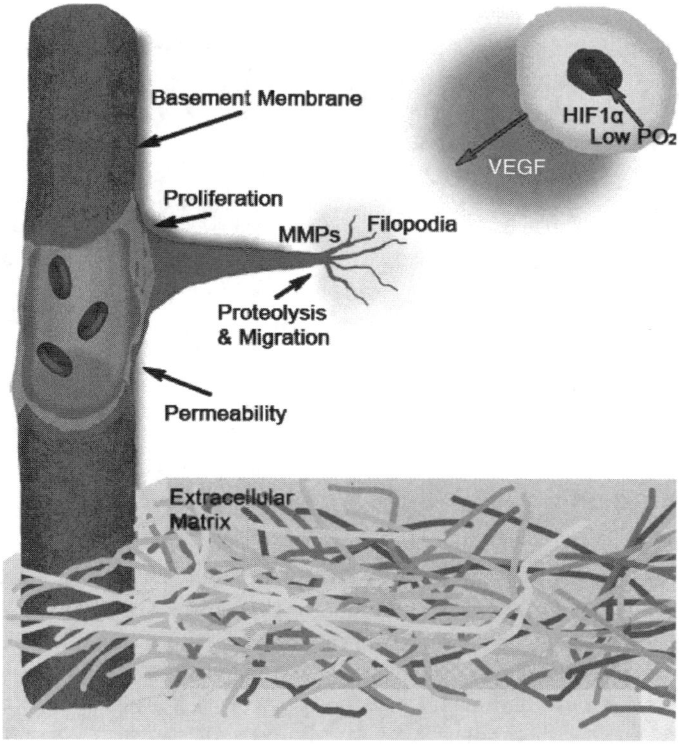

Figure 10.1 Initial steps in angiogenesis. Low oxygen in a neighboring cell (e.g., a cancer cell) triggers the expression of HIF1α, HIF1α up-regulates VEGF secretion, an adjacent vessel senses VEGF and a cell on the vessel becomes activated. This tip cell begins to migrate, using MMPs to proteolyze the surrounding matrix. Adjacent stalk cells proliferate behind the tip. HIF1α, hypoxia-inducible factor 1α. VEGF, vascular endothelial growth factor. MMPs, matrix metalloproteinases. (*See insert for color representation of figure.*)

sense an increase in oxygen demand and respond through several means, including transcriptional regulation. In response to hypoxia, the transcription factor hypoxia-inducible factor (HIF) activates hundreds of genes [2,3]. Various degrees and durations of hypoxia produce different activities of HIF degradation enzymes, HIF synthesis, and reactive oxygen species, and hence alter oxygen sensitivity. There are three known isoforms of HIF and three main isoforms of HIF prolyl hydroxylase enzymes. HIF1 (hypoxia-inducible factor 1), the isoform we focus on here, is hypothesized to be involved in the progression of angiogenesis and potentially, the initial activation of angiogenic response [4].

10.2.1.2 Growth Factors Genes activated by HIF1 include vascular endothelial growth factor (VEGF). VEGF proteins can induce a strong angiogenic response from cells, leading to cellular growth and movement. There are seven known human

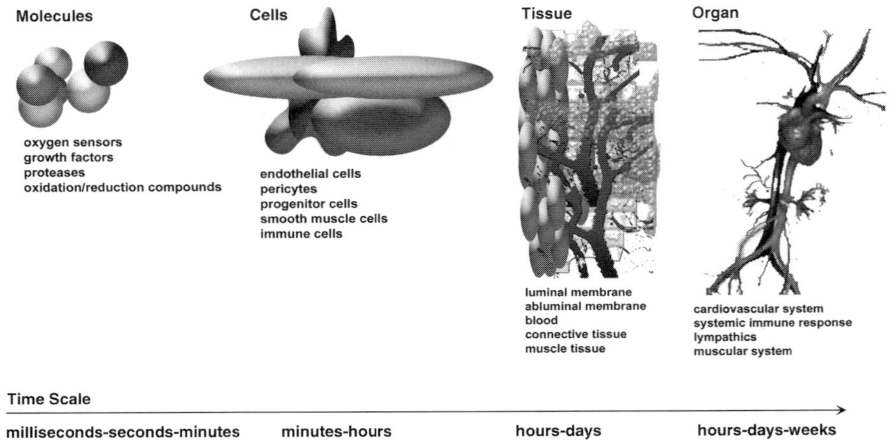

Figure 10.2 Time and spatial scales used in modeling processes in angiogenesis. (*See insert for color representation of figure.*)

isoforms of one critical VEGF gene (VEGF-A, not including $VEGF_{xxxb}$), five VEGF genes (VEGF-A, -B, -C, -D, and placental growth factor, PlGF), and five VEGF cell-surface receptors. Each receptor–ligand combination has a unique effect on cell behavior and, ultimately, on vascular pattern formation [5–7].

Although VEGF is a dominant factor in angiogenesis, it is just one of many growth factors associated with capillary growth. Microvascular growth is controlled by a balance of proangiogenic factors (e.g., VEGF, fibroblast growth factor, angiopoietin) with antiangiogenic factors (e.g., endostatin, thrombospondin-1, angiostatin). Local growth factor concentrations and gradients determine cell activation, cell migration, and cell proliferation at the onset of angiogenesis and throughout the process (Fig. 10.1). At the single- and multiple-cell level, other receptor–ligand combinations, such as notch and delta-like 4 ligand, change tip cell density and capillary branching properties [8].

10.2.1.3 Proteases Angiogenic sprouting also involves cell–matrix interactions by a combination of chemical and mechanical cues. Matrix metalloproteinases (MMPs) are molecules involved in the process of an activated cell proteolyzing its surrounding extracellular matrix (ECM); this activated cell forms a moving sprout tip and releases ECM-bound factors as it migrates. There are three known isoforms of the MMP2 protein, one of the 26 MMPs [9], and hundreds of peptides endogenous to the local matrix capable of altering cell migration or proliferation [10].

10.2.1.4 Oxidation–Reduction Compounds Compounds involved in oxidation–reduction are also central to angiogenesis. Local oxygen levels and oxygen consumption rates help determine the presence of reactive oxygen species and reactive nitrogen species [11,12]. Reactive oxygen and nitrogen species in turn modulate hypoxic response through transcriptional regulation, metabolic control, and vascular tone [13–15]. H_2O_2 can induce HIF1α expression by altering the enzymatic activities of the prolyl hydroxylases that degrade HIF1α [16]. The effects of NO are more controversial, if not more complex; both the level of NO and oxygen determine whether there is up- or down-regulation of HIF1α [17,18]. NO release by endothelial cells induces vascular relaxation and affects the levels of endothelin, a vasoconstrictor compound [19]. Metabolic changes due to oxygen deprivation or a genetic mutation also play a role in stimulating hypoxic and angiogenic response in the microvasculature. As one example, overproduction of succinate, a by-product of the TCA cycle, can lead to product inhibition of the prolyl hydroxylase reaction with HIF1α, thereby increasing HIF1α and HIF1-induced VEGF expression [20–22].

10.2.2 Cells, Tissue, and Organ Systems

10.2.2.1 Cells The different cell types involved in angiogenesis parallel the diverse categories of molecular compounds. Endothelial cells (ECs) are the cellular focus for this chapter. However, pericytes, precursor cells, stromal cells, and immune cells, among others, play important roles in cell signaling, growth factor production, EC movement, vessel stability, and capillary permeability [23]. Not only cell type but cell cycle, cell size, and cell position influence a cell's role in microvasculature growth. For example, the position of an endothelial cell on a capillary sprout determines what growth factor concentrations it sees, its activation and protein expression, and its locomotive properties [24,25].

10.2.2.2 Tissue and Organ Systems Once a capillary sprout develops, its fate is determined by the surrounding vascular network and tissue. A sprout can anastomose and attach to adjacent vessels, it can retract, it can split, it can branch, or it can become quiescent. As sprouts form and connect, a new capillary network emerges. The network's structure is determined by tissue composition and heterogeneity. Paracrine signaling from different organs and organ systems, shear force effects from blood flow, changing oxygen levels, inflammatory response, and lymphatics all can alter angiogenesis and the network's properties [23]. This developing network eventually becomes capable of carrying blood and bringing oxygen to hypoxic regions. With a process of such complexity across many biological levels, multiscale modeling becomes an essential tool in determining how these factors interact.

10.3 MODELING METHODS

Models of angiogenesis have been built using numerous mathematical and computational methods. We highlight a number of these methods here and describe how they relate to model properties across the molecular, cellular, tissue, and organ system levels (Tables 10.1 and 10.2).

Table 10.1 Model Methodologies with Examples of Existing Models and Their Characteristics Across Biological Levels

		Properties Explored by Models Categorized by Biological Level			
Model Methodology		Molecular	Cellular	Tissue	Organ System
Discrete	Monte Carlo [127]	VEGF molecular interactions VEGF-VEGFR interactions			
	2D Agent-based [40,47]	[VEGF] [FGF] [PDGF]	EC, SMC, pericytes, inflammatory cells Cell migration Cell proliferation Chemotaxis Cell cycle	Branching Capillary network formation	
	Potts model [37,42]	[VEGF] VEGF diffusion	EC–matrix interaction	Branching Capillary formation	
	2D Random walk-based models [41,43,128]	[FGF] [VEGF] VEGF diffusion, degradation, and production	Cell migration and persistence Chemotaxis Cell density	Branching Capillary network formation in tissue engineering construct	
Hybrid	3D Agent-based [119]	[VEGF] VEGF gradient Dll4 presence	EC activation EC migration and persistence EC proliferation EC matrix interactions Chemotaxis	Branching Capillary formation $Dll4^{-/+}$, $Dll4^{+/+}$	

	2D Discrete cell movement, continuous cell environment interactions [39,129–131]	Generic [TAF] [VEGF] Angiostatin	EC migration Pericytes, macrophages Cell–matrix interactions Matrix degradation Chemotaxis	Branching Capillary network formation Blood flow Vessel pruning	
Continuous	Reaction–diffusion [28,30–32,34,48, 68,76,80,83,90, 106,107,115, 132–135]	MMP interactions HIF1α degradation VEGF molecular interactions VEGF–VEGFR, VEGF–Neuropilin interactions Generic [TAF]	EC–matrix interactions MMP-driven release of VEGF Matrix heterogenity HIF1α autocrine synthesis	[VEGF] VEGF-VEGFR interactions Blood flow O_2 transport Vessel adaptation	[VEGF] VEGF organ system distribution
	Wave equations [33,34]	Generic [TAF]	EC–matrix interactions EC proliferation EC migration Cell-surface receptor binding	Capillary network formation	
	Fractals [35,36]			Capillary network formation	Tumor tissue vs. adjacent normal brain vascular network formation

Table 10.2 Highlights of Model Strengths by Methodology and Biological Level

Model Methodology		Example Model Strengths by Biological Level			
		Molecular	Cellular	Tissue	Organ System
Discrete	Monte Carlo	Single molecule interactions can be modeled			
	2D Agent-based		Multiple cell types and corresponding tissues can be modeled with the same superclass and overlapping methods/functions		
	Potts model	Minimizing energy principle is a concrete way to define biological processes, which limits possible particle interactions			
	2D Random walk-based models		Cell persistence is readily represented by weighting the random walk		

Hybrid	3D Agent-based		Can encompass both continuous equations and discrete processes Representations of in vivo vasculature structure Differences between in vivo and 2D in vitro vessel growth can be characterized	
	2D Discrete cell movement, continuous cell environment interactions		Cell processes such as proliferation can be a Boolean function of cell cycle, while diffusion and reaction of compounds and matrix interactions remain continuous	
Continuous	Reaction–diffusion	Detailed chemical-kinetics are captured		Pharmacokinetics and pharmacodynamics of angiogenic factors are readily modeled
	Wave Equations		Pattern recognition of cell movement and vessel formation	
	Fractals			Pattern recognition of vascular networks

10.3.1 Continuous Models

Reaction- and reaction–diffusion-based differential equation approaches using the continuum approximation have, so far, dominated angiogenesis modeling. They are the first methods to have been used at all four biological scales: molecular, cellular, tissue, and organ. A number of angiogenesis models used differential equations to represent a generic growth factor as a chemotactic stimulus (produced and released by a tumor mass), inducing growth of vessels into the tumor [26–28]. More recent models have included detailed equation-based network models of tumor-induced angiogenesis [29] and molecular-level interactions of VEGF and its receptors, coupled to vessel oxygenation [30].

Continuous angiogenesis models have been predominantly ordinary (ODE) or partial differential equations (PDE). The ODE-based models often explore chemical kinetic interactions between particles without considering spatial components. Examples include hydroxylation of HIF1 as a function of hypoxia (a process that occurs primarily in one location, the cell cytoplasm [31]), and VEGF–VEGFR binding dynamics at the cell membrane [32]. When spatial components are considered with concentration changes in time, PDEs are employed. This representation is necessary when including continuous particle movement (e.g., of cells, oxygen, or blood) where parameters such as velocity and changing vascular density are of interest. Wave equations (second-order PDEs used to describe phenomena that propagate in wave patterns) have also represented neocapillary formation in several studies [33,34].

Fractals—functions of recursive self-similarity that are everywhere continuous but nowhere differentiable—represent another continuous modeling methodology that has been used in the angiogenesis field. Recent implementations include representing microvascular trees as fractals and capillary changes in tumors as changes in fractal dimensions [35,36].

10.3.2 Discrete Models

Angiogenesis models have also been developed using discrete modeling methodology. Discrete models represent biological components as single, complete entities with guiding properties and rules that can be independent of adjacent compartments. This technique has been applied primarily at the cellular level, where cells are the discrete objects [37–41] and growth factor gradients may be discretized from continuous equations. Discrete models have included Potts models of angiogenic and vasculogenic growth [37,42], VEGF-driven angiogenic growth applied to a vascular engineering construct environment [43], and a cell-level rule-based model of network growth in mesenteric tissue [40].

10.3.3 Hybrid Models

In areas where differential equations and discrete models alone have limited utility, hybrid methodologies have been introduced. Emerging multiscale models are hybrids that can couple previously developed, detailed differential equations with discrete representations of growth factors, cells, extracellular matrix, and/or tissues.

Examples of hybrid models include a model of capillary growth in a corneal pocket assay where sprout formation is determined by a stochastic probability while diffusion, uptake, and degradation of the angiogenic factor bFGF in the corneal pocket is described by a PDE [44], and a deterministic model of network formation stemming from capillary movement through a heterogeneous extracellular matrix represented by a conductivity term including an anisotropy parameter that is varied randomly [45]. In the latter model, a *capillary indicator function* is introduced that is a binary function correlating to the absence or presence of capillaries. The hybrid approach is intrinsically designed with modularity whereas continuous models may need significant computer code changes (e.g., to introduce a new molecular compound involved in binding reactions), and discrete models may introduce error in their assumption that cells are in a single discrete state at every time step.

Logic-based models have also emerged in angiogenesis and cancer modeling as a means to represent biological processes as rules. Logic-based models can fall under discrete or hybrid categories. Rules can be of any form, expressed logically or mathematically; equations, Boolean rules, and probabilities are examples that have been employed [46,47]. Another type of model that, in cases, can be grouped into either discrete or hybrid methodology is the network-level model of angiogenic processes. Network models could be a series of differential equations, or they may employ Boolean logic to represent connectivity between compounds. A Boolean-based or thresholding approach to interactions can offer a means to characterize feedbacks and transcriptional regulation where detailed quantitative binding or signaling information is unavailable.

What methodology is optimal for angiogenesis modeling? The hypotheses being made, the availability and form of relevant experimental data, the biological levels of interest, and the required spatial resolution collectively determine the best method or combination of methods to use. Angiogenesis models can aid understanding of the biological system by simplifying interactions that are too complex to understand by experiments alone. Alternatively, the degree of complexity introduced in the models can be designed to parallel or extend beyond the biological resolution attainable by experiments. High model resolution comes at a cost, however; it may require exhaustive literature searches or extensive experimental comparisons. Furthermore, the complexity and uncertainties in the model can complicate parameter analysis.

Models have been developed to represent different stages in angiogenesis (Fig. 10.1): blood flow to oxygen transport, oxygen distribution to intracellular transcriptional response, hypoxic sensing to growth factor secretion, growth factor signaling to cell migration and proliferation, and capillary sprout formation to new capillary networks. Figure 10.3 illustrates the specific types of modules developed in our laboratory, across different biology levels. Below, we introduce these and related angiogenesis models grouped by their primary particle interactions and discuss routes for combining their predictive capabilities.

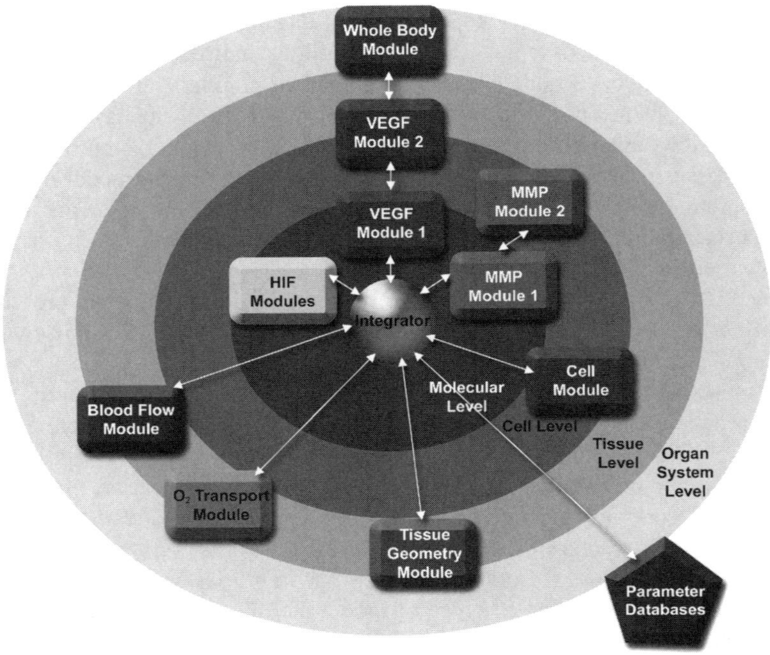

Figure 10.3 Angiogenesis modules (those shown were developed in our laboratory) organized by the dominant biological level they address.

10.4 OXYGEN SENSOR: HIF1α

When there is insufficient blood flow to meet a tissue's demand for oxygen, the body may compensate by developing a new network of capillaries (i.e., angiogenesis). To understand the mechanisms behind this process, it is logical to start by characterizing blood flow. Given a known microvascular geometry, blood flow can be calculated mathematically by solving a set of nonlinear algebraic equations for pressure at the network nodes (or bifurcation sites), and flow within segments [48,49]. From a given blood flow and estimated erythrocyte density, hematocrit distribution can be calculated throughout a capillary network [48]. From hematocrit, local oxygen levels can be calculated. There are many models of oxygen transport as a result of research in many different laboratories [50]. Once blood flow in the network is known, convection–diffusion–reaction PDEs governing oxygen transport are solved numerically, resulting in a three-dimensional distribution of oxygen in the microvascular network and surrounding tissue [51]. The resulting detailed distribution of oxygen provides a basis for modeling cellular oxygen sensing via hypoxia-inducible factor (HIF).

10.4.1 Hypoxia: Molecular Level

The three known HIF isoforms (HIF1, HIF2, and HIF3) are expressed in different ratios systemically and vary in cell-type expression and their influence on angiogenesis [52]. The chemical reactions involved in hydroxylation and ubiquitination of the isoforms are highly conserved. However, the reactions differ in hydroxylation protein affinity and, thereby, relative hydroxylation at different oxygen levels, across cell types [52,53].

HIF1, the first characterized member of the HIF family, is a heterodimer comprised of subunits HIF1α and HIF1β. HIF1α expression may be induced by a number of pathways and its degradation is highly sensitive to O_2 levels. Basal levels of HIF1α during normoxia are maintained by the prolyl hydroxylase enzymes (PHDs) [54]. PHDs hydroxylate HIF1α at Pro402 and Pro564 in the oxygen-dependent degradation domain of the protein. The isoforms of HIF PHDs are PHD1, PHD2, and PHD3. Each PHD isoform performs a unique function, with different kinetic properties and primary cellular locations [55].

The O_2 levels associated with normoxia can shift for an individual cell. If a cell is exposed to extended durations of low or high oxygen, PHDs may adjust the basal level of HIF1α and thus the cell responds to changes in O_2 with a new set point. Hence the absolute O_2 level associated with hypoxia depends strongly on microenvironmental conditions. A number of cofactors are necessary for the PHDs to be able to bind to HIF1α and hydroxylate it. Chemical-kinetic modeling has shown how iron, ascorbate (vitamin C), deoxoglutarate, and oxygen sequentially bind to the PHD2 enzyme and how this complex then binds and hydroxlyates HIF1α [31]. Once HIF1α is hydroxylated, it can then be tagged by the von Hippel Lindau protein, ubiquitinated, and degraded. The process all depends on the availability of oxygen.

When O_2 levels are insufficient, HIF1α rapidly accumulates in the cell nucleus. In the cell nucleus, HIF1α binds to its constitutively expressed dimerization partner, HIF1β, and triggers gene expression [2,56]. Hypoxic responsive genes triggered by HIF1α are involved in angiogenesis, energy metabolism, nutrient transport, cell cycle, and cell migration [2,57].

10.4.2 Hypoxia, HIF1, and Reactive Oxygen Species

How reactive oxygen species (ROSs) affect hypoxic response and the HIF1 pathway are questions steeped in debate and controversy [12,58]. Many studies have shown that ROSs, increase during hypoxia. The mechanism appears to be through electron transfer; at complex III of the mitochondrial electron chain, oxygen accepts electrons from ubisemiquinone [59]. Even if ROSs decrease during hypoxia, the resulting effects on HIF1 are still debated.

Modeling was used to show how ROSs may influence the regulation of HIF1 expression [22]. In the model, hydrogen peroxide was a representative ROS. During hydroxylation of HIF1α, hydrogen peroxide oxidizes ferrous iron (Fe^{2+}) to its ferric form (Fe^{3+}). This oxidation prevents ferrous iron from binding to the HIF1α hydroxylation enzymes prolyl hydroxylases (PHDs), and halts hydroxylation [60]. A separate means by which ROSs could affect HIF1 hydroxylation is by the recruitment of ascorbate (Asc) as a free-radical scavenger. This would prevent ascorbate from reducing ferric iron, and/or block ascorbate from binding directly to the PHDs. 2-oxoglutarate (2OG) and succinate (SC) are also cofactors in HIF1α

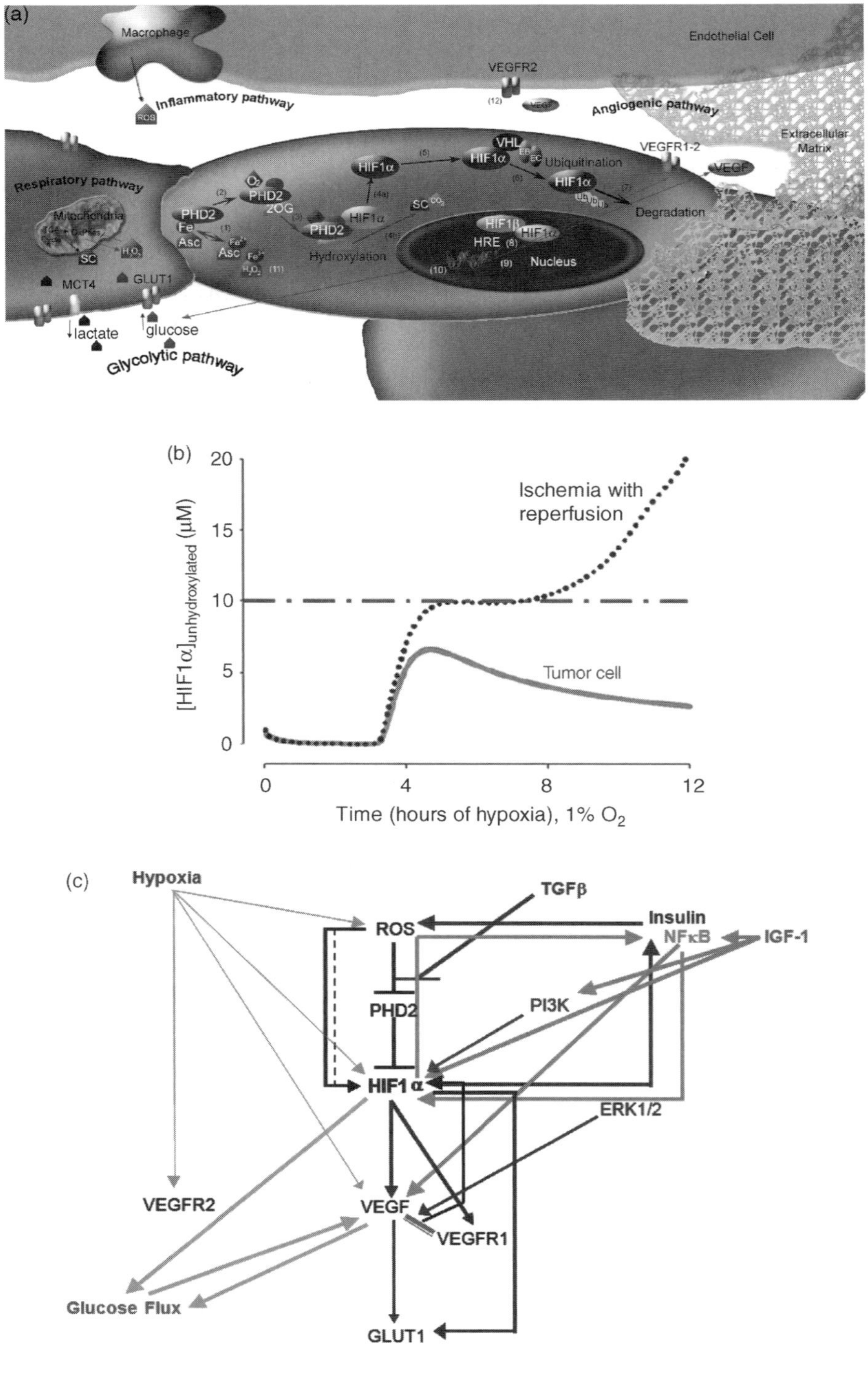
(a)
Macrophage
Endothelial Cell
VEGFR2
Inflammatory pathway
ROS
Angiogenic pathway
Extracellular Matrix
VEGFR1-2
VEGF
Respiratory pathway
Mitochondria
SC
VHL
HIF1α
Ubiquitination
Degradation
PHD2
ZOG
Fe
Asc
Hydroxylation
HIF1β
HRE
Nucleus
MCT4
GLUT1
lactate
glucose
Glycolytic pathway
(b)
Ischemia with reperfusion
Tumor cell
[HIF1α]unhydroxylated (μM)
Time (hours of hypoxia), 1% O2
0
5
10
15
20
0
4
8
12
(c)
Hypoxia
TGFβ
ROS
Insulin
NFκB
IGF-1
PHD2
PI3K
HIF1α
ERK1/2
VEGFR2
VEGF
VEGFR1
Glucose Flux
GLUT1

hydroxylation whose concentrations could be altered by free radicals and mitochondrial dysfunction [18,60,61]. Another method by which ROSs could influence the HIF1 pathway is through changing the availability of oxygen or changing PHD phosphorylation. Computational model results lend evidence to the hypothesis that ROSs work on the HIF1 system in two opposite ways, both up- and down-regulating HIF1 through modulation of PHD2 levels [22]. The model also demonstrated how ROSs have a differential effect in tumor cells compared to endothelial cells, in ischemic and ischemia–reperfusion conditions, leading to different HIF1α levels and apoptotic rates (Fig. 10.4b).

10.4.3 HIF1 Signaling

Up until now, HIF1 has been discussed in the context of intracellular molecular interactions. It is equally important to describe effects of HIF1 signaling in angiogenesis at the cell and tissue levels. HIF1α protein is regulated during angiogenesis by multiple pathways. These are known to include oxygen, reactive oxygen species, prolyl hydroxylases, and protein kinase signaling through PI3K. An interactome representation of HIF1 signaling as it relates to angiogenic pathways and VEGF is being developed (see Fig. 10.4c). The need for modeling to understand interactions in the HIF1 system becomes clear as connections are explored. The HIF1 dimer (composed of HIF1α and HIF1β) is known to bind to the hypoxic response element of VEGF, VEGFR1, and GLUT1 genes, and activate them under hypoxia. Independently, hypoxia can up-regulate VEGFR2. The ERK1/2 pathway can up-regulate VEGF-VEGFR1 binding and signaling, which leads to a positive feedback, up-regulating ERK1/2 and HIF1 [62]. Additional feedbacks at multiple levels (not shown) include HIF1 autocrine upregulation, PHD2 up-regulation by HIF1, and down-regulation of HIF1 when angiogenesis effectively increases oxygen supply. Furthermore, glycolysis and GLUT1 are likely to have effects on VEGF and its receptors, in terms of both protein concentrations and signaling. Signaling can be autocrine: for example, in a cancer cell, in skeletal muscle during exercise, or even in endothelial cells during angiogenesis. There could also be paracrine signaling where, for example, VEGF levels secreted from an adjacent tumor up-regulate the expression of HIF1 or kinases within endothelial cells. Computational modeling can highlight which HIF1 pathways dominate under different pathological conditions and predict dynamics of the interactions during neovascularization. As this network model is developed computationally and validated experimentally, this type of interactome network will provide a detailed link between molecular-based models of hypoxic sensing and VEGF secretion and inflammatory signaling during angiogenesis.

Figure 10.4 (a) HIF1 system and the pathways it affects; (b) relative levels of HIF1α over time at 1% O_2 as predicted by the model for cancer cells, and cells undergoing ischemia and reperfusion; (c) interactome network model of dominant pathways in HIF1 signaling as they relate to VEGF and inflammation during angiogenesis. [*See insert for color representation of part (a).*]

10.5 GROWTH FACTOR: VEGF

Vascular endothelial growth factor (VEGF) is a collective name for cytokines that are the products of a family of five genes: *vegfa*, *vegfb*, *vegfc*, *vegfd*, and *plgf*. Each of these genes encodes multiple, alternately spliced isoforms [63]. Although the translation of these cytokines is regulated by many microenvironmental stresses, one of the most important is hypoxia, particularly through the HIF pathway described above [64]. The cytokines serve as diffusible (shorter isoforms) or extracellular matrix-bound (longer isoforms) paracrine signal carriers. They bind to cell-surface VEGF receptors that are the products of five genes: three encoding receptor tyrosine kinases (*vegfr1*, *vegfr2*, *vegfr3*) and two encoding nonsignaling co-receptors, the Neuropilins (*nrp1*, *nrp2*) [65,66]. Neuropilins facilitate the binding of certain VEGF isoforms to VEGFR2 preferentially over VEGFR1 [67,68]. Cell-surface heparan sulfate proteoglycans also facilitate the receptor binding of a subset of VEGF isoforms [69–71].

10.5.1 VEGF Receptor–Ligand Interactions

Survival, migration, and proliferation are among the major cellular functions that are directly affected by VEGF receptor signaling. Therefore, cell behavior depends on the concentrations of each of the VEGF ligand family members, the cell-surface densities of the VEGF receptors, and the interaction network: that is, which ligands bind to which receptors. This last point is especially important because of the different, sometimes competing, signaling pathways downstream of each receptor. For example, VEGFR2 generally initiates pro-angiogenic signals while VEGFR1 has been shown to modulate them. As shown in Fig. 10.5a, VEGF ligands can bind multiple VEGF receptors (competitively, two receptor-binding sites per ligand plus two distal Neuropilin-binding sites for those ligands that can bind it) and receptors can bind multiple ligands (competitively, one ligand-binding domain per receptor monomer). The need for molecular-based computational models is apparent from this intricate network of parallel competitive receptor–ligand interactions (Fig. 10.5a), and we present here three examples of how mathematical models can aid in predicting the complex behavior of this system.

10.5.2 Example 1: VEGFR1—Signaling or Sequestering?

The placental growth factors (PlGF) — products of *plgf* — potentiate the effects of VEGF — products of *vegfa* — on cultured endothelial cells in vitro [72]. While VEGF binds VEGFR1 and VEGFR2, PlGF binds only VEGFR1 (Fig. 10.5a), so the working hypothesis had been that PlGF displaced VEGF from VEGFR1 to VEGFR2, increasing VEGFR2 signaling. There were two reasons that VEGFR1 was assumed not to transduce significant signals. First, it typically exhibits low kinase activity in substrate assays in vitro (relative to that of VEGFR2) [73]. Second, mice with a *vegfr1* gene lacking the kinase domain develop normally [74]. To test this working hypothesis, we created a computational model that simulated in vitro cell culture experiments and predicted the cell-surface receptor ligation following addition of

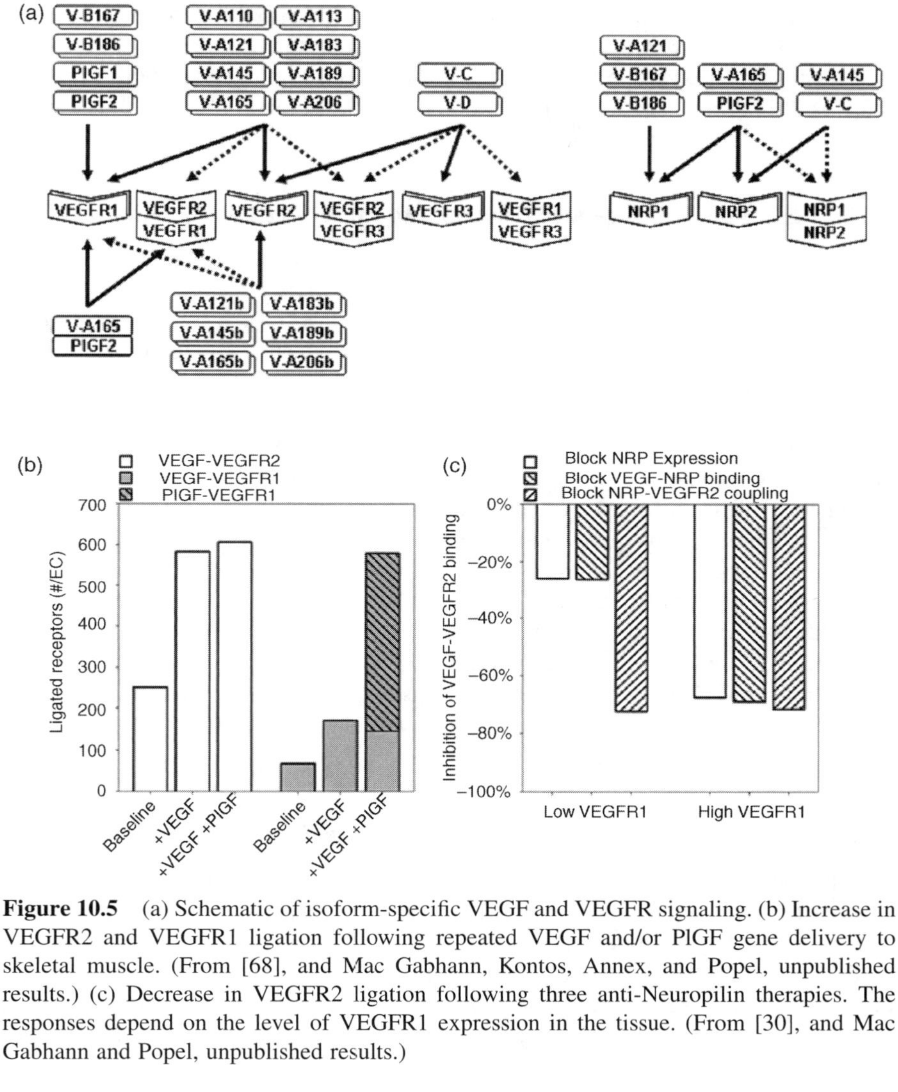

Figure 10.5 (a) Schematic of isoform-specific VEGF and VEGFR signaling. (b) Increase in VEGFR2 and VEGFR1 ligation following repeated VEGF and/or PlGF gene delivery to skeletal muscle. (From [68], and Mac Gabhann, Kontos, Annex, and Popel, unpublished results.) (c) Decrease in VEGFR2 ligation following three anti-Neuropilin therapies. The responses depend on the level of VEGFR1 expression in the tissue. (From [30], and Mac Gabhann and Popel, unpublished results.)

VEGF and PlGF to a dish of endothelial cells. Thus, we recreated, *in silico*, the experiments that had demonstrated PlGF–VEGF synergy [32,72]. The simulations, for the same conditions used in the experiment, predicted that while PlGF did displace some VEGF from VEGFR1, it was not sufficient to induce a significant change in VEGFR2 ligation. A large pool of VEGF was present in the cell culture media, maintaining the ligation level of VEGFR2. Thus, we produced an alternative hypothesis for PlGF–VEGF synergy: that PlGF and VEGF cause differential signaling of VEGFR1. This was later validated experimentally [75]. Of course, the simulations are specific to the culture conditions, cell types, and other microenvironmental conditions, and thus they do not rule out that PlGF may indeed "shift" VEGF from VEGFR1 to VEGFR2 in other situations (e.g., in vivo). However, our recent simulations have shown that in skeletal muscle, exogenous PlGF's primary effect is a significant change in the ligation of VEGFR1, and only a small change in VEGFR2 ligation is seen (Fig. 10.5b).

10.5.3 Example 2: Targeting Neuropilin to Inhibit VEGF Signaling

We expanded the above model, simulating in vitro experiments to include the co-receptor, Neuropilin-1, and two isoforms of VEGF: $VEGF_{121}$ (which does not appear to bind Neuropilin-1) and $VEGF_{165}$ (which does) [76]. Using a model of VEGF–Neuropilin–VEGFR2 binding in which $VEGF_{165}$ couples (i.e., acts as a bridge for) the two receptors through distal binding sites, we validated this model using seven independent sets of experiments from five groups of investigators, including the variations observed between cell types, some of which express more or fewer Neuropilin-1 compared to other VEGF receptors. We then used this model to test hypothetical mechanisms for an antibody to Neuropilin that had previously been shown to inhibit ocular neovascularization [77], and predicted that the antibody performed beyond what would be expected of a blockade of VEGF–Neuropilin-1 binding. Instead, simulation results based on an alternate hypothesis matched the data; the antibody appeared to be blocking the VEGF-induced bridging of VEGFR2 and Neuropilin-1. Building on these findings, we next built a model *in silico* of this interaction network applied to in vivo tumor tissue, and the simulations predicted that based on the mechanism above, the antibody would be an effective VEGF signaling inhibitor (Fig. 10.5b) [76,78].

10.5.4 Example 3: Three Receptors Equals Nine Signal Initiation Points

As mentioned above, the VEGF interactome (Fig. 10.5a) has at its core the building blocks of bivalent ligands and monovalent receptors. Thus, heterodimeric receptors can form and they have been shown to be important experimentally [79]. They are predicted by simulations to be prevalent, even to the exclusion of certain homodimers [80]. As a result, six receptor complexes (homodimers and heterodimers) could be formed from the three receptor tyrosine kinases by a ligand that can bind to each of the three (e.g., VEGFA ligands have been shown to bind to at least four of the six). This would result in nine distinct signaling initiation points for each such ligand because the phosphotyrosine profile of VEGFR1 following phosphorylation by the VEGFR2 kinase is different from that following phosphorylation by a second VEGFR1 [7,81]. Since this is repeated for each ligand (i.e., different ligands have different receptor phosphorylation profiles), the need for computational models is clear.

10.5.5 Tissue Heterogeneity Driving Oxygen-Dependent VEGF Gradients

The microenvironments within a tissue have a great bearing on the outcome of VEGF transport and signaling. Tissue microanatomy, in which there are cell types that secrete VEGF and others that express the VEGF receptors that transduce the signal and serve as sinks, results in gradients of VEGF concentration. Gradients of VEGF have often been hypothesized to be important for blood vessel guidance during angiogenic sprouting. Since each tissue has a different vascular structure, parenchymal cell architecture, and VEGF receptor expression profile, three-dimensional descriptions of each tissue based on experimentally measured parameters can give us insight into VEGF gradients that are believed to be present but are difficult to measure directly.

Using the same kinetic interaction network for VEGF and its receptors, we developed a three-dimensional model of a rat extensor digitorum longus (a skeletal muscle). Myocyte (muscle fiber), microvascular, and interstitial architecture are based on experimental measurements and the oxygen is calculated from the blood flowing through the vasculature and the consumptive needs of the parenchymal cells [30,82–84]. To approximate the oxygen-sensing and hypoxic response of HIF, the model calculates myocyte VEGF secretion as a function of the intracellular oxygen pressure, based on experimentally measured oxygen–HIF–VEGF relationships. The vasculature is representative of that observed in tissues and, as a result, the capillaries are not uniformly spaced within the muscle (as they are not in vivo). This leads to a nonhomogenous oxygen distribution and a nonhomogenous distribution of the VEGF receptors that serve as sinks for VEGF. As a result, the model predicts that there are 3% gradients in VEGF across 10 μm at baseline conditions; that is, there is an average 3% change in VEGF concentration across 10 μm [30]. As a result of this VEGF ligand gradient, there is a distribution of VEGFR2 activation across the capillary bed; each vessel in the model has a different density of ligated VEGF receptors. As a result, even if a tightly controlled, switchlike threshold of VEGFR2 signaling were required for angiogenic sprouting, some vessels could pass this threshold without activating the entire vessel bed. Simulating both hypothetical and experimentally-used therapeutics for increasing pro-angiogenic signaling, the model predicted that VEGF gradients are increased locally by the delivery of VEGF-expressing cells, but not by a more homogenous gene therapeutic approach. Exercise has been shown to upregulate both myocyte VEGF secretion and endothelial VEGF receptor expression and these, in combination, are predicted to significantly increase VEGF gradients.

10.5.6 From Tissue Expression to Whole-Body Distribution of VEGF

As mentioned above, while VEGF (also known as vascular permeability factor) is in all human tissues and in the blood, it is present at different levels in each. It may be communicated between tissues through transvascular permeability (a mechanism that it itself regulates) as well as through lymphatic transport. In both cases, the blood serves as an intermediary between tissues and thus it has become common to look to blood VEGF concentrations for information on the state of VEGF regulation in other tissues. While for normal breast tissue and skeletal muscle the experimentally measured VEGF concentration in the interstitium is similar to the plasma VEGF concentration [85–87], certain tumors overexpress VEGF and can raise plasma VEGF. However, these increases are seen, on average, across populations and interindividual variability is sufficiently high to hinder single-patient predictions [88]. The tissue-to-tissue communication of VEGF is also of particular interest in designing safe treatments for disease, allowing the prediction of possible deleterious side effects of systemic or regional VEGF-targeted therapeutics. To investigate these and other implications of multitissue VEGF transport, we have constructed compartment models that include the communication between a blood compartment and normal and diseased tissue compartments [89,90]. While VEGF production in a tissue was predicted to impact the blood concentration—though not as strongly as was expected—there did not appear to be significant tissue-to-tissue effects through the blood.

10.6 PROTEASES: MATRIX METALLOPROTEINASES

Matrix metalloproteinases (MMPs) are a growing family of zinc-binding enzymes. In vertebrates, they constitute a family of at least 26 proteins. MMPs play a significant role in angiogenesis; they modulate the extracellular matrix through proteolysis, cleave cytokines such as VEGF and bFGF [91–95], and, in part, regulate the activities of cell-adhesion molecules and signaling receptors [96–98]. Substrate specificity defines categories of MMPs including the collagenases (e.g., MMPs 1, 8, 13, MT1-MMP), gelatinases (e.g., MMPs 2, 9), or stromelysins (e.g., MMPs 3, 10, 11). MMPs are also categorized as diffusible or membrane-type MMPs (MT-MMPs) (e.g., MMP14 [99]). During angiogenesis, the MMPs collectively regulate sprout formation and tip cell movement [100], capillary tube formation and stability [101], and vessel regression.

10.6.1 Matrix Metalloproteinase Regulation

MMP proteolytic activity usually requires a bimolecular or multimolecular proteolytic processing event in which the MMP prodomain is cleaved [99]. Another, already-activated protease from the serine protease or MMP families can carry out this proteolytic event. MT1-MMP was the first of six MT-MMPs to be identified as a specific activator of pro-MMP2 (gelatinase A) at the cell surface [102]. Active MMPs 2, 3, and 13, as well as the serine protease, plasmin, are activators of pro-MMP9 [103].

Tissue inhibitors of metalloproteinases (or TIMPs) also regulate MMP activity. TIMPs proteins bind primarily to the active forms of MMPs, both surface localized and diffusible, and inhibit their proteolytic activity. Four tissue inhibitors of metalloproteinases have been discovered (TIMP 1 to 4) [104]. While they inhibit active MMPs on one hand, TIMPs may also assist in pro-MMP activation. TIMPs can facilitate interaction between a pro-MMP and an active MMP. For example, TIMP2 can be an activating adaptor for pro-MMP2 and an inhibitor of MT1-MMP. The MT1-MMP:TIMP2 complex acts as a cell surface receptor for pro-MMP2, allowing it to be activated [105]. Specifically, the carboxy-terminal domain of TIMP2 binds to the PEX domain of pro-MMP2 and the active amino-terminal domain of TIMP2 binds and inhibits the catalytic domain of a free MT1-MMP. A second TIMP2-free MT1-MMP molecule cleaves the pro-peptide of pro-MMP2 in the receptor complex and

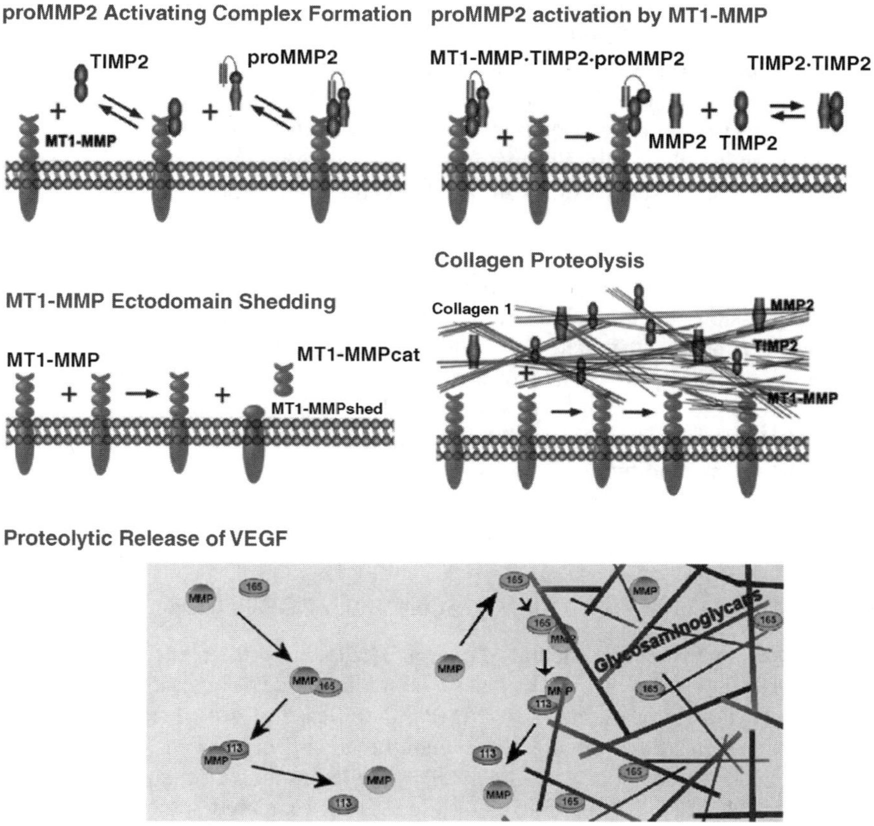

Figure 10.6 Schematic of proMMP2, MMP2, TIMP and MT-MMP interactions in the extracellular matrix (ECM), and subsequent proteolytic release of VEGF from the ECM. (*See insert for color representation of figure.*)

completes the pro-MMP2 activation [105]. Due to TIMP2's dual role, an increase in TIMP2 expression does not correlate directly with inhibition of protease activity [106]. TIMP2 regulates the localization of proteolytic activity from the cellular surface (MT1-MMP) to the diffusible forms (MMP2) [107]. Figure 10.6 illustrates these processes. Notably, TIMPs may also have MMP-independent functions, including interactions with VEGF–VEGFR2 binding.

10.6.2 Role of Matrix Metalloproteinases in Angiogenesis

MMPs collectively degrade most of the components of a basement membrane (BM) and the extracellular matrix (ECM). Both BM and ECM contain different types of collagens, either fibrillar (collagen I, II) or network forming (collagen IV), elastic fibers, proteoglycans, and glycosaminoglycans (GAGs). They also contain various glycoproteins; the biochemical "glue" for all of these components. It has been postulated that MMPs disengage the natural barriers of the BM and ECM through proteolytic digestion. Structural elements of the ECM potentially restrict the migration of the endothelial cells through the matrix [108].

Growth factor stimulation leads to the secretion of ECM-digesting enzymes on the surface of endothelial cells. Stromal cells also secrete proteases and in the case of tumor angiogenesis, surrounding tumor cells secrete these proteases. Cell production and secretion of proteases can profoundly alter ECM structure, reducing the restrictive physical barriers and allowing the migrating and proliferating ECs to penetrate avascular or poorly vascularized tissue [73,109,110]. In the case of the MMP expression by endothelial cells during sprouting angiogenesis, the proteolytic load has been mostly attributed to a single endothelial cell that is located at the tip of a proliferating endothelial sprout; thus called the tip cell. A sensory mechanism has been attributed to the tip cell which both guides the proliferating population of the endothelial sprout by sensing the underlying extracellular matrix, and proteolytically degrades the surrounding matrix, thus allowing the migration of the endothelial cells into the matrix. This mechanism allows the vectorial relocation of the proliferating cell population toward the oxygen-demanding regions.

10.6.3 Models of MMPs, Cell Migration, and VEGF Release

There is extensive experimental data on MMPs and their role in angiogenesis [73,105,111–114]. However, there is no integrated, quantitative model of extracellular matrix degradation by MMPs, the release of growth factors from the extracellular matrix, and the effects on endothelial cell movement.

To this end we developed biochemically detailed models of the post-secretional regulation of MMPs [106,107,115]. Our kinetic models include proMMP activation and MMP inhibition by TIMPs. The models capture the dynamics of the MT1-MMP/TIMP2/proMMP2 and MMP3/TIMP1/proMMP9 systems under in vitro conditions; they reproduce the biphasic dependence of MMP2 activation on TIMP2 and the biphasic inhibition of MMP9 by TIMP1 [106,115].

These biochemical models are incorporated into geometrically detailed models to provide spatial and temporal descriptions of collagen proteolysis as a tip cell moves through the ECM, as it does in angiogenesis. The model represents the MT1-MMP/TIMP2/proMMP2 system dynamics that determine an endothelial cells penetration of the basement membrane as well as its migration within the ECM.

Models are also being developed to study the role of MMPs in the release of VEGF. VEGF can be bound to matrix components, both heparan sulfate glycosaminoglycans in the basement membrane and ECM proteoglycans. Matrix-bound VEGF is biologically relevant. It is active either in its bound form or it can be converted into soluble molecules via proteolytic cleavage of VEGF [116]. A number of MMPs can directly cleave VEGF: MMP -1, -3, -7, -9, -16, and -19 [94]. The VEGF released regulates angiogenic sprout growth and direction as well as the degree of tissue oxygenation.

The exact role and cellular origin of MMPs under various physiological and pathological angiogenic processes are not completely understood. In skeletal muscle angiogenesis, MMPs are regulated by the nature of the mechanical stimuli and nature of the induced ischemia. Exercise-induced ischemia has different characteristics than those of injury-induced or artery ligation–induced ischemia. In injury, inflammation may cause the migration of neutrophils and macrophages. These immune cells secrete MMPs and alter the release of VEGF from the ECM [117]. Adding additional complexity, while MMPs can regulate VEGF levels in tissue, the products of MMP mediated proteolysis (e.g., angiostatin) can have an antiangiogenic role independent of cell migration [117].

10.7 CELLS AND TISSUES: CAPILLARY SPROUT FORMATION

As VEGF, MMPs, and the extracellular matrix interact at the molecular level, a series of events occur at the cell level during the onset of angiogenesis: tip cell activation, endothelial cell migration, chemotaxis and haptotaxis, and cell proliferation. Using an agent-based approach, we built a three-dimensional computational model to represent this cellular sprouting [46,118]. In the model the movement of cells is governed by rule-based logic designed from experimental work compiled from extensive literature research (Fig. 10.7). The model was first applied to conditions that might occur in a three-dimensional in vitro setting. We tested assumptions behind cell activation, migration, elongation, proliferation, and branching, and tested cell-level behavior in response to different stimuli: focusing, in the current model, on activation by a threshold change in VEGF and changes in Dll4 ligand presence. Single-cell behavior combines to produce a novel capillary network emerging out of combinatorially complex interactions of individual cells. Branching, multiple sprouts, anastomoses, looping, and regression are features used to characterize the resulting capillary network. The cell-based model presents a framework upon which to test hypotheses about the tissue-level effect of molecular and cell-level events (and vice versa). Rule- or event-based modeling allows us to easily produce *in silico* knockouts on multiple biological levels. Here we give an example, a molecular-level knockout of Dll4 ligand.

Delta-like ligand 4 (Dll4) is a ligand for Notch receptors. The transmembrane protein is critical to vascular development as haploinsufficiency of the Dll4 gene is embryonically lethal in many mouse strains [119–121]. Dll4 is expressed primarily in endothelial cells. Its expression is correlated to the local concentration of VEGF [122] as well as to VEGF receptor concentrations. Blocking VEGF decreases Dll4 [123] while Notch–Delta signaling down-regulates VEGFR2 [124]. The presence of Dll4 has been shown to reduce tip cell formation as a function of VEGF [125], and Notch suppresses branching and proliferation at the sprout tip [119]. While Dll4 deficiency increases sprout formation, the resulting vessels appear nonproductive (e.g., they are less capable of carrying blood or reducing hypoxia in surrounding tissue [123]). Using the computational approach described above, we showed the effects of VEGF protein concentrations and Dll4 haploinsufficiency on endothelial cells. Cell-level behavior alone contributed to differences in capillary network formation, and the capillary networks resulting from the model appear similar in vascular density, branching, and tortuosity to those found in Dll4 experiments (Fig. 10.7b).

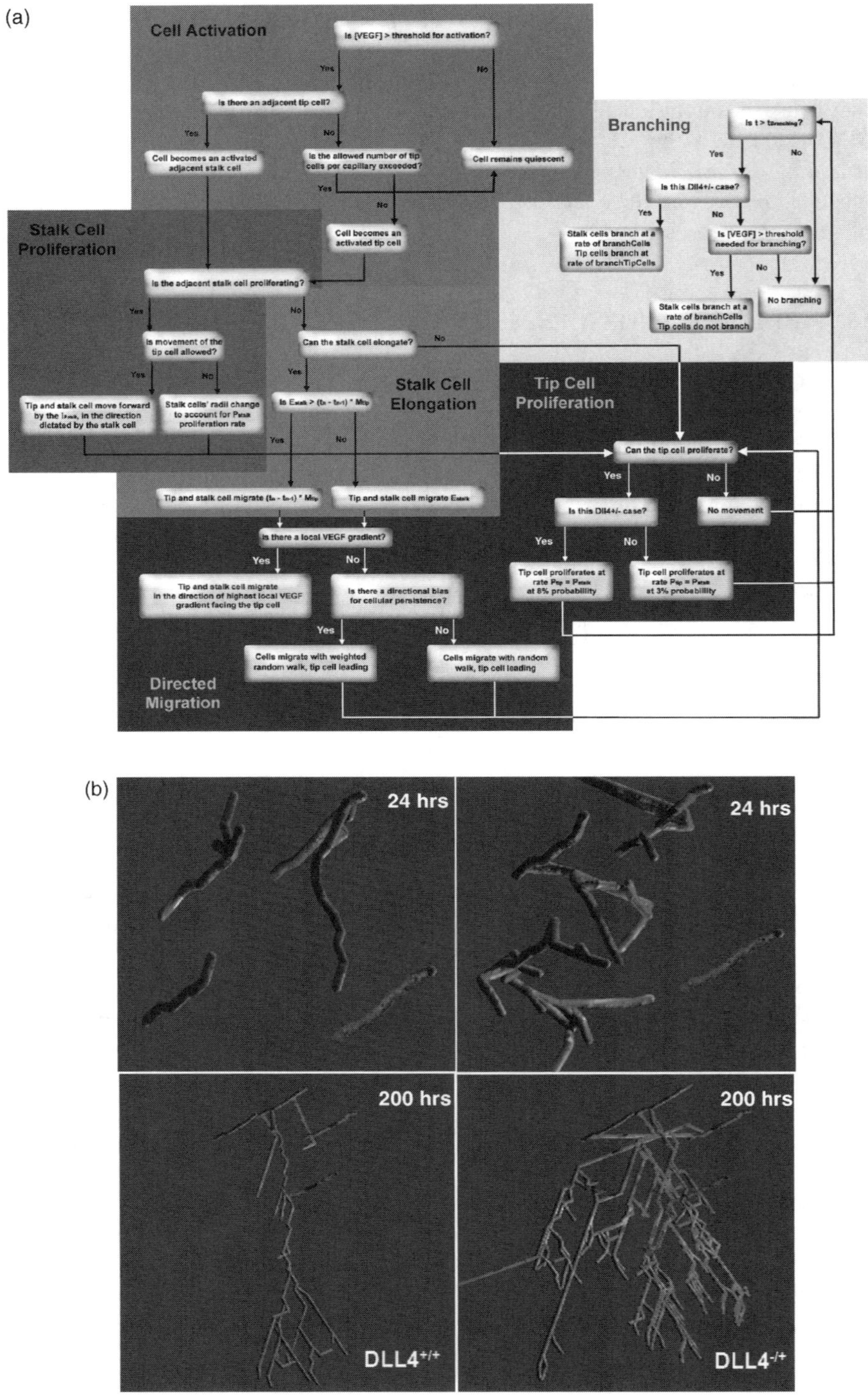
(a)
Cell Activation
Is [VEGF] > threshold for activation?
Yes
No
Is there an adjacent tip cell?
Yes
No
Cell becomes an activated adjacent stalk cell
Is the allowed number of tip cells per capillary exceeded?
Cell remains quiescent
Yes
No
Cell becomes an activated tip cell
Stalk Cell Proliferation
Is the adjacent stalk cell proliferating?
Yes
No
Is movement of the tip cell allowed?
Yes
No
dictated by the stalk cell
proliferation rate
Can the stalk cell elongate?
No
Yes
Stalk Cell Elongation
Yes
No
Branching
Yes
No
Is this Dll4+/- case?
Yes
No
Stalk cells branch at a rate of branchCells
Tip cells branch at rate of branchTipCells
Is [VEGF] > threshold needed for branching?
Yes
No
Stalk cells branch at a rate of branchCells
Tip cells do not branch
No branching
Tip Cell Proliferation
Can the tip cell proliferate?
Yes
No
Is this Dll4+/- case?
No movement
Yes
No
Tip cell proliferates at
at 8% probability
Tip cell proliferates at
at 3% probability
Is there a local VEGF gradient?
Yes
No
Tip and stalk cell migrate in the direction of highest local VEGF gradient facing the tip cell
Is there a directional bias for cellular persistence?
Yes
No
Cells migrate with weighted random walk, tip cell leading
Cells migrate with random walk, tip cell leading
Directed Migration
(b)
24 hrs
24 hrs
200 hrs
200 hrs
DLL4+/+
DLL4-/+

10.8 INTEGRATION

As other angiogenesis models are integrated with the molecular cell-level, rule-based model, its use will grow and its rules can be applied to specific in vivo physiological conditions (Fig. 10.2). The goal of combining models is also to coordinate feedbacks. Hence, blood flow leads to oxygen transport, which in turn leads to oxygen sensing via HIF, potential VEGF secretion, cell sensing, cell migration through the ECM, cell proliferation, sprout formation, and finally, new vessel building. Once the new vessel network is formed, blood flow is altered, and the models interact in a cycle. A first example of integration includes four models: blood flow, oxygen transport, VEGF, and cell modules. Model input is an initial three-dimensional vascular network structure obtained through quantitative descriptions or image analysis of microvessels in specific tissues. Blood flow and hematocrit in the vessels are calculated, from which oxygen distribution can be determined. Using an empirical correlation between oxygen, HIF expression, HIF activity, and local VEGF secretion levels, we can then predict VEGF distribution throughout the local tissue. VEGF is in turn sensed by endothelial cells. Cells can respond by proliferation and migration, leading to vessel growth. The formation of new capillaries starts the next round of calculations, and the new structure is fed back into the blood flow module.

A goal of this integration is to coordinate angiogenic models written in diverse languages and apply the process to dynamically represent healthy and diseased vascular conditions. A unit called the *controller* provides the integration platform. The controller starts, ends, and links modules written in different languages. It also passes parameters, including kinetic and physical parameters, between them. Individual modules will be validated for their specific application with available tissue-, cell-, and molecular-specific literature data, and the combined output of the integrated model will be compared to experiments. Integration allows predictions across temporal and spatial scales through an integrative hierarchy of molecular, cellular, tissue, and organ system components. This integration is essential in rational drug development and therapeutic regime design and will encourage related experiments in parallel.

10.9 CONCLUSIONS AND FUTURE DIRECTIONS

Multiscale angiogenesis modeling opens up numerous research paths, such as the integration of real-time modeling with experiments, personalized medicine, and design of combinatorial drug therapies. Two other paths we discuss here briefly are prediction of evolutionary patterns in the microvasculature and angiogenic synthetic biology.

Figure 10.7 (a) Experimentally based logical rules that govern cell processes in the agent-based angiogenesis model; (b) model output showing vessel growth in time for control conditions (left panel) and haploinsufficiency of the Notch ligand Dll4 (right panel). [*See insert for color representation of part (b).*]

As the angiogenesis models connect stimuli with responses, the evolutionary pattern of microvasculature and angiogenic-associated factors in species over time is an appealing application area. It is easy to envision modelers defining adaptive algorithms to govern interactions in the existing models and induce changes in protein function, cellular response, and eventually phenotype.

Programming these adaptive responses and molecular, cellular, and systems' fate in biological material is an equally intriguing avenue. Synthetic biology is a rapidly expanding field and applications in angiogenesis can be foreseen, with the huge advantage of preexisting multiscale models. Molecular and cellular models can be used as guides to the development of modified endothelial and precursor cells, while the tissue-level angiogenesis models can be compared to expected responses of these modified cells within tissue constructs. By this use of modeling, the connections between a person's cell genetic changes, proteomic changes, microenvironmental effects, and tissue-level changes have a good chance at being reverse engineered.

Computational models of angiogenesis have brought us a critical quantitative perspective on the field, predicted cell migratory speeds through different tissue matrices [107], allowed predictions of molecular- and cell-based therapies [30,31,84], and differentiated vasculature from cancerous brain tissue from noncancerous sites [36], among other contributions. Models have been developed using methods as diverse as ordinary differential equations to model biochemical reactions, partial differential equations to model blood flow and cell movement, agent- and rule-based models to represent biological events of proliferation, and network models to represent cell signaling (Fig. 10.8). As new tools develop for model validation, integration,

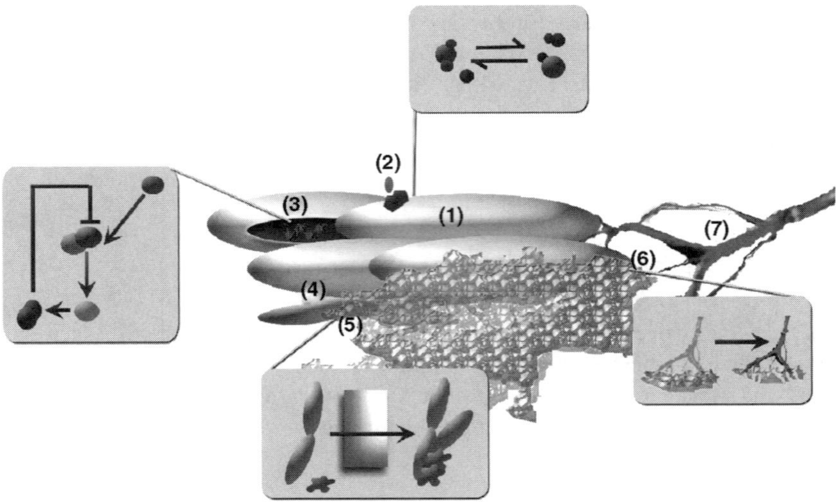

Figure 10.8 Examples of modeling areas and techniques that can be used to model angiogenesis at multiple levels, and advance angiogenic synthetic biology. (1) Intercellular and (2) membrane-based reactions modeled via chemical kinetic models; (3) transcriptional signaling modeled by biological networks; (4, 5) cell–cell and cell–matrix interactions modeled via agent-based programming techniques using rules; (6, 7) cell movement through the matrix and blood flow modeled via partial differential equations. (*See insert for color representation of figure.*)

visualization, and adaptation, the field of multiscale angiogenesis modeling becomes integral to rigorous experimental design and therapeutic advances. Design of an optimal therapy based on a person's microenvironment and the microenvironmental dynamics comes closer to reality with the use of angiogenesis modeling.

REFERENCES

1. Gerhardt H, Golding M, Fruttiger M, Ruhrberg C, Lundkvist A, Abramsson A, Jeltsch M, Mitchell C, Alitalo K, Shima D, et al. VEGF guides angiogenic sprouting utilizing endothelial tip cell filopodia. *J. Cell Biol.* **161**(6):1163–1177, 2003.
2. Semenza GL. Hydroxylation of HIF-1 oxygen sensing at the molecular level. *Physiology (Bethesda)* **19**:176–182, 2004.
3. Jiang BH, Semenza GL, Bauer C, Marti HH. Hypoxia-inducible factor 1 levels vary exponentially over a physiologically relevant range of O_2 tension. *Am. J. Physiol.* **271** (4 Pt. 1): C1172–C1180, 1996.
4. Dewhirst MW, Cao Y, Moeller B. Cycling hypoxia and free radicals regulate angiogenesis and radiotherapy response. *Nat. Rev. Cancer* **8**(6):425–437, 2008.
5. Yamazaki Y, Morita T. Molecular and functional diversity of vascular endothelial growth factors. *Mol. Divers.* **10**(4):515–527, 2006.
6. Roskoski R Jr. Vascular endothelial growth factor (VEGF) signaling in tumor progression. *Crit. Rev. Oncol. Hematol.* **62**(3):179–213, 2007.
7. Mac Gabhann F, Popel AS. Systems biology of vascular endothelial growth factors. *Microcirculation.* **15**(8):715–738, 2008.
8. Thurston G, Noguera-Troise I, Yancopoulos GD. The Delta paradox: DLL4 blockade leads to more tumour vessels but less tumour growth. *Nat. Rev. Cancer* **7**(5):327–331, 2007.
9. Amalinei C, Caruntu ID, Balan RA. Biology of metalloproteinases. *Rom. J. Morphol. Embryol.* **48**(4):323–334, 2007.
10. Karagiannis ED, Popel AS. A systematic methodology for proteome-wide identification of peptides inhibiting the proliferation and migration of endothelial cells. *Proc. Natl. Acad. Sci. USA* **105**(37):13775–13780, 2008.
11. White CW. Commentary on "Hypoxia, hypoxic signaling, tissue damage, and detection of reactive oxygen species (ROS)." *Free Radic. Biol. Med.* **40**(6):923–927, 2006.
12. Guzy RD, Schumacker PT. Oxygen sensing by mitochondria at complex: III. The paradox of increased reactive oxygen species during hypoxia. *Exp. Physiol.* **91**(5):807–819, 2006.
13. Brune B, Zhou J. The role of nitric oxide (NO) in stability regulation of hypoxia inducible factor-1alpha (HIF-1alpha). *Curr. Med. Chem.* **10**(10):845–855, 2003.
14. Guzy RD, Hoyos B, Robin E, Chen H, Liu L, Mansfield KD, Simon MC, Hammerling U, Schumacker PT. Mitochondrial complex III is required for hypoxia-induced ROS production and cellular oxygen sensing. *Cell Metab.* **1**(6):401–408, 2005.
15. Chandel NS, Budinger GR. The cellular basis for diverse responses to oxygen. *Free Radic. Biol. Med.* **42**(2):165–174, 2007.
16. Kohl R, Zhou J, Brune B. Reactive oxygen species attenuate nitric-oxide-mediated hypoxia-inducible factor-1alpha stabilization. *Free Radic. Biol. Med.* **40**(8): 1430–1442, 2006.

17. Forstermann U. Oxidative stress in vascular disease: causes, defense mechanisms and potential therapies. *Nat. Clin. Pract. Cardiovasc. Med.* **5**(6):338–349, 2008.
18. Kozhukhar AV, Yasinska IM, Sumbayev VV. Nitric oxide inhibits HIF-1alpha protein accumulation under hypoxic conditions: implication of 2-oxoglutarate and iron. *Biochimie* **88**(5):411–418, 2006.
19. Cardillo C, Kilcoyne CM. Cannon RO 3rd Panza JA. Interactions between nitric oxide and endothelin in the regulation of vascular tone of human resistance vessels in vivo. *Hypertension* **35**(6):1237–1241, 2000.
20. Koivunen P, Hirsila M, Remes AM, Hassinen IE, Kivirikko KI, Myllyharju J. Inhibition of hypoxia-inducible factor (HIF) hydroxylases by citric acid cycle intermediates: possible links between cell metabolism and stabilization of HIF. *J. Biol. Chem.* **282**(7):4524–4532, 2007.
21. Guzy RD, Sharma B, Bell E, Chandel NS, Schumacker PT. Loss of the SdhB, but not the SdhA, subunit of complex II triggers reactive oxygen species-dependent hypoxia-inducible factor activation and tumorigenesis. *Mol. Cell. Biol.* **28**(2):718–731, 2008.
22. Qutub A, Popel A. Reactive oxygen species regulate hypoxia-inducible factor HIF1alpha differentially in cancer and ischemia. *Mol. Cell. Biol.* **28**:5106–5119, 2008.
23. Adams RH, Alitalo K. Molecular regulation of angiogenesis and lymphangiogenesis. *Nat. Rev. Mol. Cell. Biol.* **8**(6):464–478, 2007.
24. Siekmann AF, Covassin L, Lawson ND. Modulation of VEGF signalling output by the Notch pathway. *Bioessays* **30**(4):303–313, 2008.
25. Holderfield MT, Hughes CC. Crosstalk between vascular endothelial growth factor, notch, and transforming growth factor-beta in vascular morphogenesis. *Circ. Res.* **102**(6): 637–652, 2008.
26. Deakin AS. Model for initial vascular patterns in melanoma transplants. *Growth* **40** (2):191–201, 1976.
27. Balding D, McElwain DL. A mathematical model of tumour-induced capillary growth. *J. Theor. Biol.* **114**(1):53–73, 1985.
28. Chaplain M, Anderson A. Mathematical modelling of tumour-induced angiogenesis: network growth and structure. *Cancer Treat. Res.* **117**:51–75, 2004.
29. Chaplain MA, McDougall SR, Anderson AR. Mathematical modeling of tumor-induced angiogenesis. *Annu. Rev. Biomed. Eng.* **8**:233–257, 2006.
30. Mac Gabhann F, Ji JW, Popel AS. Computational model of vascular endothelial growth factor spatial distribution in muscle and pro-angiogenic cell therapy. *PLoS Comput. Biol.* **2** (9):e127, 2006.
31. Qutub AA, Popel AS. A computational model of intracellular oxygen sensing by hypoxia-inducible factor HIF1 alpha. *J. Cell Sci.* **119**(Pt. 16): 3467–3480, 2006.
32. Mac Gabhann F, Popel AS. Model of competitive binding of vascular endothelial growth factor and placental growth factor to VEGF receptors on endothelial cells. *Am. J. Physiol. Heart Circ. Physiol.* **286**(1):H153–H164, 2004.
33. Olsen L, Sherratt JA, Maini PK, Arnold F. A mathematical model for the capillary endothelial cell-extracellular matrix interactions in wound-healing angiogenesis. *IMA J. Math. Appl. Med. Biol.* **14**(4):261–281, 1997.
34. Byrne HM, Chaplain MA. Mathematical models for tumour angiogenesis: numerical simulations and nonlinear wave solutions. *Bull. Math. Biol.* **57**(3):461–486, 1995.

35. Grizzi F, Russo C, Colombo P, Franceschini B, Frezza EE, Cobos E, Chiriva-Internati M. Quantitative evaluation and modeling of two-dimensional neovascular network complexity: the surface fractal dimension. *BMC Cancer* **5**:14, 2005.

36. Risser L, Plouraboue F, Steyer A, Cloetens P, Le Duc G, Fonta C. From homogeneous to fractal normal and tumorous microvascular networks in the brain. *J. Cereb. Blood Flow Metab.* **27**(2):293–303, 2007.

37. Bauer AL, Jackson TL, Jiang Y. A Cell-based model exhibiting branching and anastomosis during tumor-induced angiogenesis. *Biophys. J.* **92**(9):3105–3121, 2007.

38. Jabbarzadeh E, Abrams CF. Simulations of chemotaxis and random motility in 2D random porous domains. *Bull. Math. Biol.* **69**(2):747–764, 2007.

39. Anderson AR, Chaplain MA. Continuous and discrete mathematical models of tumor-induced angiogenesis. *Bull. Math. Biol.* **60**(5):857–899, 1998.

40. Peirce SM, Van Gieson EJ, Skalak TC. Multicellular simulation predicts microvascular patterning and *in silico* tissue assembly. *FASEB J.* **18**(6):731–733, 2004.

41. Stokes CL, Lauffenburger DA. Analysis of the roles of microvessel endothelial cell random motility and chemotaxis in angiogenesis. *J. Theor. Biol.* **152**(3):377–403, 1991.

42. Merks RM, Brodsky SV, Goligorksy MS, Newman SA, Glazier JA. Cell elongation is key to *in silico* replication of in vitro vasculogenesis and subsequent remodeling. *Dev. Biol.* **289**(1):44–54, 2006.

43. Jabbarzadeh E, Abrams CF. Strategies to enhance capillary formation inside biomaterials: a computational study. *Tissue Eng.* **13**(8):2073–2086, 2007.

44. Tong S, Yuan F. Numerical simulations of angiogenesis in the cornea. *Microvasc. Res.* **61**(1):14–27, 2001.

45. Sun S, Wheeler MF, Obeyesekere M, Patrick CW Jr. A deterministic model of growth factor-induced angiogenesis. *Bull. Math. Biol.* **67**(2):313–337, 2005.

46. Qutub A, Popel AS. Cell elongation, proliferation and migration differentiate endothelial cell phenotypes and determine capillary sprouting. *BMC Syst. Biol.* **33**:13, 2009.

47. Bailey AM, Thorne BC, Peirce SM. Multi-cell agent-based simulation of the microvasculature to study the dynamics of circulating inflammatory cell trafficking. *Ann. Biomed. Eng.* **35**(6):916–936, 2007.

48. Tsoukias NM, Goldman D, Vadapalli A, Pittman RN, Popel AS. A computational model of oxygen delivery by hemoglobin-based oxygen carriers in three-dimensional microvascular networks. *J. Theor. Biol.* **248**(4):657–674, 2007.

49. Pries AR, Secomb TW, Gaehtgens P, Gross JF. Blood flow in microvascular networks. Experiments and simulation. *Circ. Res.* **67**(4):826–834, 1990.

50. Goldman D. Theoretical models of microvascular oxygen transport to tissue. *Microcirculation.* **15**(8):795–811, 2008.

51. Ji JW, Tsoukias NM, Goldman D, Popel AS. A computational model of oxygen transport in skeletal muscle for sprouting and splitting modes of angiogenesis. *J. Theor. Biol.* **241** (1):94–108, 2006.

52. Shohet RV, Garcia JA. Keeping the engine primed: HIF factors as key regulators of cardiac metabolism and angiogenesis during ischemia. *J. Mol. Med.* **85**(12):1309–1315, 2007.

53. Maynard MA, Qi H, Chung J, Lee EH, Kondo Y, Hara S, Conaway RC, Conaway JW, Ohh M. Multiple splice variants of the human HIF-3 alpha locus are targets of the von

Hippel–Lindau E3 ubiquitin ligase complex. *J. Biol. Chem.* **278**(13):11032–11040, 2003.

54. Khanna S, Roy S, Maurer M, Ratan RR, Sen CK. Oxygen-sensitive reset of hypoxia-inducible factor transactivation response: prolyl hydroxylases tune the biological normoxic set point. *Free Radic. Biol. Med.* **40**(12):2147–2154, 2006.
55. Appelhoff RJ, Tian YM, Raval RR, Turley H, Harris AL, Pugh CW, Ratcliffe PJ, Gleadle JM. Differential function of the prolyl hydroxylases PHD1, PHD2, and PHD3 in the regulation of hypoxia-inducible factor. *J. Biol. Chem.* **279**(37):38458–38465, 2004.
56. Powell FL. Functional genomics and the comparative physiology of hypoxia. *Annu. Rev. Physiol.* **65**:203–230, 2003.
57. Wang GL, Jiang BH, Rue EA, Semenza GL. Hypoxia-inducible factor 1 is a basic-helix-loop-helix-PAS heterodimer regulated by cellular O_2 tension. *Proc. Natl. Acad. Sci. USA* **92**(12):5510–5514, 1995.
58. Pouyssegur J, Mechta-Grigoriou F. Redox regulation of the hypoxia-inducible factor. *Biol. Chem.* **387**(10–11):1337–1346, 2006.
59. Bell EL, Emerling BM, Chandel NS. Mitochondrial regulation of oxygen sensing. *Mitochondrion* **5**(5):322–332, 2005.
60. Pan Y, Mansfield KD, Bertozzi CC, Rudenko V, Chan DA, Giaccia AJ, Simon MC. Multiple factors affecting cellular redox status and energy metabolism modulate hypoxia-inducible factor prolyl hydroxylase activity in vivo and in vitro. *Mol. Cell. Biol.* **27**(3): 912–925, 2007.
61. Gottlieb E, Tomlinson IP. Mitochondrial tumour suppressors: a genetic and biochemical update. *Nat. Rev. Cancer* **5**(11):857–866, 2005.
62. Das B, Yeger H, Tsuchida R, Torkin R, Gee MF, Thorner PS, Shibuya M, Malkin D, Baruchel S. A hypoxia-driven vascular endothelial growth factor/Flt1 autocrine loop interacts with hypoxia-inducible factor-1alpha through mitogen-activated protein kinase/ extracellular signal-regulated kinase 1/2 pathway in neuroblastoma. *Cancer Res.* **65**(16): 7267–7275, 2005.
63. Ladomery MR, Harper SJ, Bates DO. Alternative splicing in angiogenesis: the vascular endothelial growth factor paradigm. *Cancer Lett.* **249**(2):133–142, 2007.
64. Semenza GL. HIF-1: using two hands to flip the angiogenic switch. *Cancer Metastasis Rev.* **19**(1–2):59–65, 2000.
65. Shibuya M. Vascular endothelial growth factor receptor-1 (VEGFR-1/Flt-1): a dual regulator for angiogenesis. *Angiogenesis* **9**(4):225–230 (discussion 231), 2006.
66. Staton CA, Kumar I, Reed MW, Brown NJ. Neuropilins in physiological and pathological angiogenesis. *J. Pathol.* **212**(3):237–248, 2007.
67. Soker S, Miao HQ, Nomi M, Takashima S, Klagsbrun M. VEGF165 mediates formation of complexes containing VEGFR-2 and neuropilin-1 that enhance VEGF165-receptor binding. *J. Cell. Biochem.* **85**(2):357–368, 2002.
68. Mac Gabhann F, Popel AS. Interactions of VEGF isoforms with VEGFR-1, VEGFR-2, and neuropilin in vivo: a computational model of human skeletal muscle. *Am. J. Physiol. Heart Circ. Physiol.* **292**(1):H459–H474, 2007.
69. Gitay-Goren H, Soker S, Vlodavsky I, Neufeld G. The binding of vascular endothelial growth factor to its receptors is dependent on cell surface–associated heparin-like molecules. *J. Biol. Chem.* **267**(9):6093–6098, 1992.

70. Cohen T, Gitay-Goren H, Sharon R, Shibuya M, Halaban R, Levi BZ, Neufeld G. VEGF121, a vascular endothelial growth factor (VEGF) isoform lacking heparin binding ability, requires cell-surface heparan sulfates for efficient binding to the VEGF receptors of human melanoma cells. *J. Biol. Chem.* **270**(19):11322–11326, 1995.

71. Vander Kooi CW, Jusino MA, Perman B, Neau DB, Bellamy HD, Leahy DJ. Structural basis for ligand and heparin binding to neuropilin B domains. *Proc. Natl. Acad. Sci. USA* **104**(15):6152–6157, 2007.

72. Carmeliet P, Moons L, Luttun A, Vincenti V, Compernolle V, De Mol M, Wu Y, Bono F, Devy L, Beck H, et al. Synergism between vascular endothelial growth factor and placental growth factor contributes to angiogenesis and plasma extravasation in pathological conditions. *Nat. Med.* **7**(5):575–583, 2001.

73. Hiratsuka S, Maru Y, Okada A, Seiki M, Noda T, Shibuya M. Involvement of Flt-1 tyrosine kinase (vascular endothelial growth factor receptor-1) in pathological angiogenesis. *Cancer Res.* **61**(3):1207–1213, 2001.

74. Hiratsuka S, Minowa O, Kuno J, Noda T, Shibuya M. Flt-1 lacking the tyrosine kinase domain is sufficient for normal development and angiogenesis in mice. *Proc. Natl. Acad. Sci. USA* **95**(16):9349–9354, 1998.

75. Autiero M, Waltenberger J, Communi D, Kranz A, Moons L, Lambrechts D, Kroll J, Plaisance S, De Mol M, Bono F, et al. Role of PlGF in the intra- and intermolecular cross talk between the VEGF receptors Flt1 and Flk1. *Nat. Med.* **9**(7):936–943, 2003.

76. Mac Gabhann F, Popel AS. Differential binding of VEGF isoforms to VEGF receptor 2 in the presence of neuropilin-1: a computational model. *Am. J. Physiol. Heart Circ. Physiol.* **288**(6):H2851–H2860, 2005.

77. Oh H, Takagi H, Otani A, Koyama S, Kemmochi S, Uemura A, Honda Y. Selective induction of neuropilin-1 by vascular endothelial growth factor (VEGF): a mechanism contributing to VEGF-induced angiogenesis. *Proc. Natl. Acad. Sci. USA* **99**(1):383–388, 2002.

78. Mac Gabhann F, Popel AS. Targeting neuropilin-1 to inhibit VEGF signaling in cancer: comparison of therapeutic approaches. *PLoS Comput. Biol.* **2**(12):e180, 2006.

79. Huang K, Andersson C, Roomans GM, Ito N, Claesson-Welsh L. Signaling properties of VEGF receptor-1 and -2 homo- and heterodimers. *Int. J. Biochem. Cell Biol.* **33**(4): 315–324, 2001.

80. Mac Gabhann F, Popel AS. Dimerization of VEGF receptors and implications for signal transduction: a computational study. *Biophys. Chem.* **128**(2–3):125–139, 2007.

81. Dixelius J, Makinen T, Wirzenius M, Karkkainen MJ, Wernstedt C, Alitalo K, Claesson-Welsh L. Ligand-induced vascular endothelial growth factor receptor-3 (VEGFR-3) heterodimerization with VEGFR-2 in primary lymphatic endothelial cells regulates tyrosine phosphorylation sites. *J. Biol. Chem.* **278**(42):40973–40979, 2003.

82. Ji JW, Mac Gabhann F, Popel AS. Skeletal muscle VEGF gradients in peripheral arterial disease: simulations of rest and exercise. *Am. J. Physiol. Heart Circ. Physiol.* **293**(6): H3740–H3749, 2007.

83. Mac Gabhann F, Ji JW, Popel AS. VEGF gradients, receptor activation, and sprout guidance in resting and exercising skeletal muscle. *J. Appl. Physiol.* **102**(2):722–734, 2007.

84. Mac Gabhann F, Ji JW, Popel AS. Multi-scale computational models of pro-angiogenic treatments in peripheral arterial disease. *Ann. Biomed. Eng.* **35**(6):982–994, 2007.

85. Dabrosin C, Chen J, Wang L, Thompson LU. Flaxseed inhibits metastasis and decreases extracellular vascular endothelial growth factor in human breast cancer xenografts. *Cancer Lett.* **185**(1):31–37, 2002.
86. Dabrosin C, Margetts PJ, Gauldie J. Estradiol increases extracellular levels of vascular endothelial growth factor in vivo in murine mammary cancer. *Int. J. Cancer* **107**(4): 535–540, 2003.
87. Hoffner L, Nielsen JJ, Langberg H, Hellsten Y. Exercise but not prostanoids enhance levels of vascular endothelial growth factor and other proliferative agents in human skeletal muscle interstitium. *J. Physiol.* **550**(Pt. 1): 217–225, 2003.
88. Kut C, Mac Gabhann F, Popel AS. Where is VEGF in the body? A meta-analysis of VEGF distribution in cancer. *Br. J. Cancer* **97**(7):978–985, 2007.
89. Wu FTH, Stefanini MO, Mac Gabhann F, Popel AS.A compartment model of VEGF distribution in humans in the presence of soluble VEGF receptor-1 acting as a ligand trap. *PLoS ONE,* **4**(4):e5108, 2009.
90. Stefanini MO, Wu FT, Mac Gabhann F, Popel AS. A compartment model of VEGF distribution in blood, healthy and diseased tissues. *BMC Syst. Biol.* **2**:77, 2008.
91. Houck KA, Leung DW, Rowland AM, Winer J, Ferrara N. Dual regulation of vascular endothelial growth factor bioavailability by genetic and proteolytic mechanisms. *J. Biol. Chem.* **267**(36):26031–26037, 1992.
92. Takeda N, Maemura K, Imai Y, Harada T, Kawanami D, Nojiri T, Manabe I, Nagai R. Endothelial PAS domain protein 1 gene promotes angiogenesis through the transactivation of both vascular endothelial growth factor and its receptor. *Flt-1. Circ. Res.* **95**(2):146–153,2004.
93. Keyt BA, Berleau LT, Nguyen HV, Chen H, Heinsohn H, Vandlen R, Ferrara N. The carboxyl-terminal domain (111-165) of vascular endothelial growth factor is critical for its mitogenic potency. *J. Biol.Chem.* **271**(13):7788–7795, 1996.
94. Lee S, Jilani SM, Nikolova GV, Carpizo D, Iruela-Arispe ML. Processing of VEGF-A by matrix metalloproteinases regulates bioavailability and vascular patterning in tumors. *J. Cell Biol.* **169**(4):681–691, 2005.
95. Whitelock JM, Murdoch AD, Iozzo RV, Underwood PA. The degradation of human endothelial cell-derived perlecan and release of bound basic fibroblast growth factor by stromelysin, collagenase, plasmin, and heparanases. *J. Biol.Chem.* **271**(17):10079–10086, 1996.
96. Cauwe B, Van den Steen PE, Opdenakker G. The biochemical, biological, and pathological kaleidoscope of cell surface substrates Processed by matrix metalloproteinases. *Crit. Rev. Biochem. Mol. Biol.* **42**(3):113–185, 2007.
97. Ichikawa Y, Ishikawa T, Momiyama N, Kamiyama M, Sakurada H, Matsuyama R, Hasegawa S, Chishima T, Hamaguchi Y, Fujii S, et al. Matrilysin (MMP-7) degrades VE-cadherin and accelerates accumulation of beta-catenin in the nucleus of human umbilical vein endothelial cells. *Oncol. Rep.* **15**(2):311–315, 2006.
98. Ratnikov BI, Rozanov DV, Postnova TI, Baciu PG, Zhang H, DiScipio RG, Chestukhina GG, Smith JW, Deryugina EI, Strongin AY. An alternative Processing of integrin alpha(v) subunit in tumor cells by membrane type-1 matrix metalloproteinase. *J. Biol.Chem.* **277**(9):7377–7385, 2002.
99. Visse R, Nagase H. Matrix metalloproteinases and tissue inhibitors of metalloproteinases: structure, function, and biochemistry. *Circ. Res.* **92**(8):827–839, 2003.

100. Haas TL, Milkiewicz M, Davis SJ, Zhou AL, Egginton S, Brown MD, Madri JA, Hudlicka O. Matrix metalloproteinase activity is required for activity-induced angiogenesis in rat skeletal muscle. *Am. J. Physiol. Heart Circ. Physiol.* **279**(4):H1540–H1547, 2000.

101. Chun TH, Sabeh F, Ota I, Murphy H, McDonagh KT, Holmbeck K, Birkedal-Hansen H, Allen ED, Weiss SJ. MT1-MMP-dependent neovessel formation within the confines of the three-dimensional extracellular matrix. *J. Cell Biol.* **167**(4):757–767, 2004.

102. Brinckerhoff CE, Matrisian LM. Matrix metalloproteinases: a tail of a frog that became a prince. *Nat. Rev. Mol. Cell Biol.* **3**(3):207–214, 2002.

103. Van den Steen PE, Dubois B, Nelissen I, Rudd PM, Dwek RA, Opdenakker G. Biochemistry and molecular biology of gelatinase B or matrix metalloproteinase-9 (MMP-9). *Crit. Rev. Biochem. Mol. Biol.* **37**(6):375–536, 2002.

104. Woessner JF Jr. MMPs and TIMPs: an historical perspective. *Mol. Biotechnol.* **22**(1): 33–49, 2002.

105. Butler GS, Butler MJ, Atkinson SJ, Will H, Tamura T, Schade van Westrum S, Crabbe T, Clements J, d'Ortho MP, Murphy G. The TIMP2 membrane type 1 metalloproteinase "receptor" regulates the concentration and efficient activation of progelatinase A: a kinetic study. *J. Biol. Chem.* **273**(2):871–880, 1998.

106. Karagiannis ED, Popel AS. A theoretical model of type I collagen proteolysis by matrix metalloproteinase (MMP) 2 and membrane type 1 MMP in the presence of tissue inhibitor of metalloproteinase 2. *J. Biol. Chem.* **279**(37):39105–39114, 2004.

107. Karagiannis ED, Popel AS. Distinct modes of collagen type I proteolysis by matrix metalloproteinase (MMP) 2 and membrane type I MMP during the migration of a tip endothelial cell: insights from a computational model. *J. Theor. Biol.* **238**(1):124–145, 2006.

108. Sottile J. Regulation of angiogenesis by extracellular matrix. *Biochim. Biophys. Acta Rev. Cancer* **1654**(1):13–22, 2004.

109. Murphy G, Gavrilovic J. Proteolysis and cell migration: creating a path? *Curr. Opin. Cell Biol.* **11**(5):614–621, 1999.

110. Horino K, Kindezelskii AL, Elner VM, Hughes BA, Petty HR. Tumor cell invasion of model 3-dimensional matrices: demonstration of migratory pathways, collagen disruption, and intercellular cooperation. *FASEB J.* **15**(6):932–939, 2001.

111. Koike T, Vernon RB, Hamner MA, Sadoun E, Reed MJ. MT1-MMP, but not secreted MMPs, influences the migration of human microvascular endothelial cells in 3-dimensional collagen gels. *J. Cell. Biochem.* **86**(4):748–758, 2002.

112. Galvez BG, Matias-Roman S, Albar JP, Sanchez-Madrid F, Arroyo AG. Membrane type 1-matrix metalloproteinase is activated during migration of human endothelial cells and modulates endothelial motility and matrix remodeling. *J. Biol. Chem.* **276** (40):37491–37500, 2001.

113. Hotary K, Allen E, Punturieri A, Yana I, Weiss SJ. Regulation of cell invasion and morphogenesis in a three-dimensional type I collagen matrix by membrane-type matrix metalloproteinases 1, 2, and 3. *J. Cell Biol.* **149**(6):1309–1323, 2000.

114. Koike T, Vernon RB, Hamner MA, Sadoun E, Reed MJ. MT1-MMP, but not secreted MMPs, influences the migration of human microvascular endothelial cells in 3-dimensional collagen gels. *J. Cell. Biochem.* **86**(4):748–758, 2002.

115. Vempati P, Karagiannis ED, Popel AS. A biochemical model of the matrix metalloproteinase 9 activation and inhibition. *J. Biol. Chem.* **282**(52):37585–37596, 2007.

116. Park JE, Keller GA, Ferrara N. The vascular endothelial growth factor (VEGF) isoforms: differential deposition into the subepithelial extracellular matrix and bioactivity of extracellular matrix-bound VEGF. *Mol. Biol.Cell* **4**(12):1317–1326, 1993.

117. Rundhaug JE. Matrix metalloproteinases and angiogenesis. *J. Cell. Mol. Med.* **9**(2): 267–285, 2005.

118. Qutub AA, Liu G, Vempati P, Popel AS. Integration of angiogenesis modules at multiple scales: from molecular to tissue. *Proceedings of the Pacific Symposium on Biocomputation,* 316–327, 2009.

119. Gale NW, Dominguez MG, Noguera I, Pan L, Hughes V, Valenzuela DM, Murphy AJ, Adams NC, Lin HC, Holash J, et al. Haploinsufficiency of delta-like 4 ligand results in embryonic lethality due to major defects in arterial and vascular development. *Proc. Natl. Acad. Sci. USA* **101**(45):15949–15954, 2004.

120. Krebs LT, Shutter JR, Tanigaki K, Honjo T, Stark KL, Gridley T. Haploinsufficient lethality and formation of arteriovenous malformations in Notch pathway mutants. *Genes Dev.* **18**(20):2469–2473, 2004.

121. Duarte A, Hirashima M, Benedito R, Trindade A, Diniz P, Bekman E, Costa L, Henrique D, Rossant J. Dosage-sensitive requirement for mouse Dll4 in artery development. *Genes Dev.* **18**(20):2474–2478, 2004.

122. Lobov IB, Renard RA, Papadopoulos N, Gale NW, Thurston G, Yancopoulos GD, Wiegand SJ. Delta-like ligand 4 (Dll4) is induced by VEGF as a negative regulator of angiogenic sprouting. *Proc. Natl. Acad. Sci. USA* **104**(9):3219–3224, 2007.

123. Noguera-Troise I, Daly C, Papadopoulos NJ, Coetzee S, Boland P, Gale NW, Lin HC, Yancopoulos GD, Thurston G. Blockade of Dll4 inhibits tumour growth by promoting non-productive angiogenesis. *Nature* **444**(7122):1032–1037, 2006.

124. Sainson RC, Aoto J, Nakatsu MN, Holderfield M, Conn E, Koller E, Hughes CC. Cell-autonomous notch signaling regulates endothelial cell branching and proliferation during vascular tubulogenesis. *FASEB J.* **19**(8):1027–1029, 2005.

125. Hellstrom M, Phng LK, Hofmann JJ, Wallgard E, Coultas L, Lindblom P, Alva J, Nilsson AK, Karlsson L, Gaiano N, et al. Dll4 signalling through Notch1 regulates formation of tip cells during angiogenesis. *Nature* **445**(7129):776–780, 2007.

11

LARGE-SCALE SIMULATION OF BLOOD FLOW IN MICROVESSELS

PROSENJIT BAGCHI

Department of Mechanical and Aerospace Engineering, Rutgers University, Piscataway, New Jersey

11.1 INTRODUCTION

The human circulatory system is composed of the heart, blood, and blood vessels, facilitating the delivery of oxygen and nutrients to every cell in the body. The *macrocirculation* refers to the flow of blood in large blood vessels, such as aorta, large arteries, and veins, having internal diameter greater than about 500 μm. The *microcirculation* refers to the flow of blood in vessels in the diameter range 5 to 500 μm, and it consists of arterioles, venules, and capillaries. Nearly 10^9 blood vessels have diameter in the range 5 to 500 μm. They provide the maximum resistance to the blood flow and thus determine a major fraction of the pumping power required by the heart.

Blood is a multiphase suspension that is composed primarily of the plasma, red blood cells, white blood cells, and platelets. The plasma is a liquid that constitutes about 55% of the total blood volume. It also contains many submicrometer and nanoscale particulates, such as proteins and lipids, apart from minerals, glucose, and enzymes. Nearly 90% of plasma is made of water; the viscosity of plasma (ca. 1.2 cP) is nearly equal to that of water (1 cP).

Red blood cells (RBCs) or erythrocytes are sacs of liquid protein (hemoglobin) that are enclosed by deformable membranes. They constitute 40 to 45% of the total blood

Multiscale Modeling of Particle Interactions: Applications in Biology and Nanotechnology,
Edited by Michael R. King and David J. Gee

volume. A relatively small fraction of blood volume is occupied by white blood cells and platelets. A normal healthy RBC has a biconcave discoid shape under resting conditions. The diameter of a typical RBC is about 8 μm, with a thickness of 2 μm. The viscosity of the hemoglobin is about five times that of the plasma. The membrane of an RBC is composed of a lipid bilayer with an underlying, nearly two-dimensional, protein network called a *cytoskeleton*. This unique structure of the membrane, along with the liquidlike nature of the hemoglobin, makes an erythrocyte a highly deformable particle. Normal erythrocytes, while in circulation, repeatedly flow through capillaries having internal diameters between 3 and 5 μm, much less than their undeformed size (8 μm).

In the microcirculation, the size of a blood cell is comparable to the diameter of a blood vessel. The particulate nature of blood becomes important in these vessels, unlike in the macrocirculation, where whole blood can be modeled as a single-phase liquid. Due to the deformable nature of the erythrocytes and their high volume fraction, the flow of blood in the microcirculation behaves as a semidense suspension of deformable particles. Rheology of blood in microcirculation is determined primarily by the deformability of the erythrocytes and the multiphase nature of blood. Experiments on blood flow in microvessels over a century have revealed complex rheological behavior such as the Fahraeus and Fahraeus–Lindqvist effects [1–6]. The physical mechanism giving rise to the Fahraeus–Lindqvist effect is believed to be a radial motion of the erythrocytes forming a cell-depleted region near the wall [7]. The radial motion arises due to extreme deformability of the erythrocytes [8]. Radial motion of normal erythrocytes in dilute suspension has also been observed by many, including Goldsmith and co-workers nearly 30 years ago [9]. In a dilute suspension, individual RBCs migrate continuously toward the center of the vessel. In a dense suspension, hydrodynamic interaction between adjacent cells also affects their motion. The deformation of individual cells, and cell–cell and cell–wall interactions together contribute to the microrheology of blood.

The motion of RBCs in narrow capillaries is often axisymmetric and hence amenable to theoretical analysis [10,11]. Extension to nonaxisymmetric single-file motion also has been made [12,13]. The main difficulty in the theoretical analysis of blood flow at small scales arises when the vessel diameter is in the approximate range 20 to 500 μm. Extension of models predicting single-cell dynamics to a large number of cells without sacrificing the detailed mechanics has remained a major computational challenge, as one must consider a large ensemble of cells in semidense suspension, as is the case for whole blood. The continuum models of whole blood flow in microvessels, such as the core-annular or two-phase models [14], rely on estimates of various physiological parameters, such as the width of the cell-free layer, which can be obtained only by high-resolution experiments [15] or by computational simulations in which the motion and deformation of individual cells are fully resolved. Three-dimensional simulations of bubbles, drops, and rigid particles have been pursued by various groups for both Stokes flow and inertial flows (see refs. [16–23], to name a few). Similar investigations with a large number of blood cells or capsules have been pursued only recently.

Computational modeling of blood flow in the microcirculation involves length scales ranging from nanometer to hundreds of micrometers. At the scale of the vessel, cell deformation is dictated by the local gradient of the flow, wall interaction, and cell–cell collision. At the nanoscale, the cytoskeleton and bilayer structures determine the mechanical properties of the cell. In the following, we first discuss some mesoscale models for cytoskeleton and bilayer deformation. This is followed by a discussion on a continuum model that can consider the flow of multiple, deformable, and interacting cells.

11.2 MESOSCALE MODELS

11.2.1 Spectrin-Level Models of Cytoskeleton

As mentioned above, an erythrocyte membrane is comprised of a lipid bilayer and an underlying, nearly two-dimensional skeletal network. Cholesterol and transmembrane proteins may be interspersed through the erythrocyte membrane. Well-resolved microscopic studies revealed that the skeletal network can be thought of as an ensemble of triangles, with the sides made of convoluted spectrin filaments. The contour length l_c of the spectrin filaments is about 200 nm, while their end-to-end distance l_o under zero stress is about 75 nm. The spectrins are connected head-on via short actin protofilaments at the vertices (called *junctional complexes*) of the triangles (Fig. 11.1). Most of the approximately 33,000 junctional complexes connect six neighboring spectrin filaments, while a relatively smaller number connect five and seven spectrins. Each spectrin is anchored to the lipid bilayer, approximately at its midlength, by a suspension complex made of ankyrin, band 3, and protein 4.2, preventing large relative motion between the bilayer and the cytoskeleton. The elastic property of the cytoskeleton arises due to thermal fluctuations of the spectrin chains.

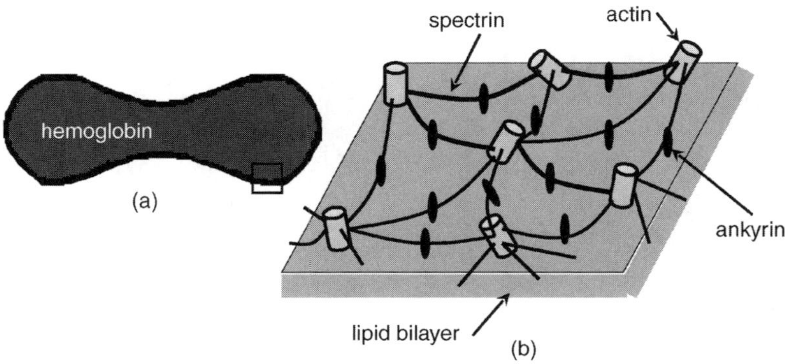

Figure 11.1 (a) Biconcave shape of an erythrocyte under resting conditions; (b) mesoscopic structure of a membrane patch showing the lipid bilayer and the cytoskeleton.

First, we consider the spectrin-level model developed by Boal [24]. Here the cytoskeleton is represented by a network of flexible polymer chains. Each spectrin filament is composed of a number of flexible tethers connected in series by hard beads of diameter a. The contour length of a spectrin chain is $l_c = 1.2n_s a$, where n_s is the number of tethers. The spectrin tetramers are connected at the sixfold junction vertices, which are free to move. The bilayer and the bilayer–cytoskeleton coupling are not explicitly modeled; rather, the former is assumed to be a flat surface (xy plane) with the midpoints of the spectrin filaments anchored to the xy plane. Due to computational limitations, a rectangular patch ($L_x \times L_y$) of the cytoskeleton of 0.3 μm length, instead of the entire cell surface, is simulated. In the simulations, the total number of junction vertices, N_j, considered was 16, and $4 \leq n_s \leq 30$. The Monte Carlo technique is used to generate a sequence of configurations of the spectrin network. At each finite step, beads are allowed to move freely, normal to the xy plane, with the constraint $|\mathbf{x}| \leq \sqrt{1.9a}$, where $\mathbf{x}$ is the distance between two adjacent beads in a chain. The constraint is equivalent to assuming a steric (short-range) repulsion for $0 < |\mathbf{x}| < a$ for all vertices, and a finite extensibility of the spectrin segments, unlike a Hookean spring. Each Monte Carlo step, when all bead positions are updated, results in new values of L_x and L_y. The new configuration is accepted or rejected as per a Boltzmann distribution $\exp[N_b \ln(1 + \Delta A_{xy}/A_{xy})]$, where $A_{xy} = L_x L_y$ is the area of the computational patch, $N_b = N_j(3n_s - 2)$ is the total number of beads, and ΔA_{xy} is the area change per Monte Carlo step.

The sequence of configurations generated is then used to predict the ensemble averaged area $\langle A_{xy} \rangle$ and the thickness $\langle z \rangle$ of the network. The in-plane shear and area compression moduli, E_s and E_a, respectively, are then extracted from the area fluctuations as

$$\frac{E_a}{K_B T} = \frac{\langle A_{xy} \rangle}{\langle A_{xy}^2 \rangle - \langle A_{xy} \rangle^2} \tag{11.1}$$

and

$$E_s = \frac{E_p E_a}{4E_a - E_p}, \tag{11.2}$$

where

$$E_p = \frac{K_B T}{\langle A_{xy} \rangle} \left[\frac{1}{\langle L_x^2 \rangle / \langle L_x \rangle^2 - 1} + \frac{1}{\langle L_y^2 \rangle / \langle L_y \rangle^2 - 1} \right], \tag{11.3}$$

K_B is the Boltzmann constant, and T is the temperature [24]. In their work, the quantities above are predicted from the simulations at zero stress as functions of n_s as follows:

$$\langle A/a^2 \rangle = 2.1\, n_s^{5/4}, \qquad \langle z \rangle / a = 0.17\, n_s^{4/5}, \tag{11.4}$$

where $A = A_{xy}/N_j$ is the area per junction vertex, and

$$E_s = 25\,K_BT/n_s^{1.7}a^2, \qquad E_a = 43\,K_BT/n_s^{1.7}a^2. \tag{11.5}$$

To apply these results to the erythrocyte membrane, we note that the contour length of a spectrin filament is about 2.67 times its end-to-end distance under zero stress. Then $A/A_c \sim 1/7$, where $A_c = \sqrt{3}\,l_c^2/2$. This value of A/A_c, referred to as the *in vivo value*, is obtained for $n_s \sim 26$ and $a \sim 6.4$ nm. Taking $T = 300$ K, we get $E_s \sim 0.01$ dyn/cm and $E_a \sim 0.017$ dyn/cm. The value predicted for E_s is comparable to the experimentally measured value (ca. 0.006 dyn/cm) obtained using the whole erythrocyte. In contrast, the predicted value of E_a is less than that obtained in whole cell measurements. The difference may be due to the area incompressibility condition of the actual erythrocyte that was not considered in the model.

Boey et al. [25] and Discher et al. [26] extended the foregoing model to study the stress-dependent deformation of the cytoskeleton. The simulations showed that at small deformation, the cytoskeleton behaves as a two-dimensional triangulated network of Hookean springs. Two non-Hookean characteristics are noted at large deformation. First, the network showed finite extensibility, with A/A_c approaching unity at large tensile stress. Second, A/A_c approached nonzero asymptotic values under larger compressive stress, implying that the steric repulsion prevents the collapse of the network. The area compression modulus E_a is minimum at zero stress and its value increases under tension or compression. The shear modulus E_s increases with increasing tension but remains constant at its zero-stress value when subject to compression.

11.2.2 Coarse-Grained Models

A direct extension of the spectrin chain model to whole cell deformation is computationally demanding, as it would require tracking nearly 33,000 junction vertices and hence $> 10^6$ tethers and beads. An alternative strategy is to use coarse-grained models that can faithfully reproduce the results of the tether/bead network at a smaller scale, yet can simulate whole cell deformation using modest computing resources. Discher et al. [26] developed a coarse-grained model in which the tether–bead structure is replaced by an interaction potential between the junction vertices, thereby effectively reducing the system size by a factor of $3n_s$. The net interaction potential U is written as

$$U = U_s + U_r + U_b + U_a + U_v, \tag{11.6}$$

where U_s is an attractive potential representing the springlike behavior of the spectrin, U_r is steric repulsion, and U_b is bending energy. The remaining two, U_a and U_v, incorporate cell surface area and volume incompressibility conditions, respectively.

The potential U_s is developed based on a DNA polymer chain model as

$$\frac{U_s}{K_B T} = \sum_{3N_j} \frac{(l_m/4b)\tilde{l}^2(3 - 2\tilde{l})}{1 - \tilde{l}}, \tag{11.7}$$

where l is the intervertex distance, l_m is the maximum distance, $\tilde{l} = l/l_m$, and b is the persistence length [27]. The potential gives a Hookean spring behavior at small l but prevents unbounded extension as $l \rightarrow l_m$, consistent with the observation of Boey et al. [25]. The steric repulsion is modeled as

$$U_r = \sum \frac{C}{A}, \quad \text{with} \quad \frac{C}{K_B T} = \frac{93\sqrt{3}\, l_m^3/16b)\tilde{l}_o^4(4\tilde{l}_o^2 - 9\tilde{l}_o + 6)}{(1 - \tilde{l}_o)^2} \tag{11.8}$$

where A is the area of each triangular surface element, $\tilde{l}_o = l_o/l_m$, and the summation is over the triangles. The bending energy is modeled as

$$U_b = k_b \sum (1 - \mathbf{n}_i \cdot \mathbf{n}_j), \tag{11.9}$$

where $\mathbf{n}$ is a unit vector normal to a triangular surface element, i and j are indices for adjacent surface elements, and $k_b \sim 69K_B T$. The surface area and volume conservation conditions are approximated as

$$U_a = \frac{k_{\text{area}}(A - A_{\text{ref}})^2}{2A_{\text{ref}}}, \tag{11.10}$$

$$U_v = \frac{k_{\text{vol}}(V - V_{\text{ref}})^2}{2V_{\text{ref}}}, \tag{11.11}$$

respectively, where A_{ref} and V_{ref} are the surface area and volume of the cell at zero stress. The parameters k_{area} and k_{vol} are set to high values so that A and V remain within a few percent of the reference values.

Using the model above, Discher et al. [26] simulated micropipette aspiration of erythrocytes. In the simulations, the cell surface was discretized with about 6,000 to 18,000 sixfold nodes (junction vertices). The simulation results for aspirated cell length agree reasonably well with the experimental measurements [28]. In agreement with fluorescence imaging measurements, the simulations show nonhomogeneous deformation of the cytoskeleton. The node density increases rapidly near the entrance and then decreases steadily to a minimum value at the tip of the cell. Further, the spectrin chains in the entrance region are compressed in the circumferential direction and elongated in the axial direction. Such anisotropy is not predicted by continuum models in which the assumption of axisymmetry is used in the modeling of the micropipette aspiration. In a later development, Li et al. [29] extended the coarse-grained model to consider a random network of spectrin tetramers and applied it to simulate the large deformation of erythrocytes during optical tweezer experiment.

11.2.3 Lipid Bilayer Models

The models described above do not explicitly consider the interaction between the lipid bilayer and the underlying cytoskeleton. The actin protofilament located at the center of the sixfold junction vertices is extended into the lipid bilayer and hence experiences a hydrodynamic drag. In addition to the random thermal fluctuation, the elastic force borne by the spectrins is transmitted to the bilayer via the actin protofilaments and the ankyrins. Further, steric repulsion exists between the bilayer and the actins and spectrins. A mesoscopic model of the bilayer attached to a network of spectrin filaments was developed by Lin and Brown [30]. In the model, a rectangular patch of the bilayer over a length scale of about 0.1 μm is considered and is represented by the xy plane. Local deformation of the bilayer caused by thermal fluctuations is represented by the height $h(\mathbf{r}, t)$, where $\mathbf{r}$ is the two-dimensional position vector on the xy plane. The evolution of h is governed by the Langevin equation as

$$\frac{\partial h(\mathbf{r}, t)}{\partial t} = \int \left[-K_c \nabla^4 h(\mathbf{r}', t) + \sigma \nabla^2 h(\mathbf{r}', t) + F_{\text{int}}(\mathbf{r}', t) + F_{\text{rand}}(\mathbf{r}', t) \right] \times \Lambda(\mathbf{r} - \mathbf{r}') d\mathbf{r}', \tag{11.12}$$

where $\Lambda(\zeta) = 1/8\pi\eta|\zeta|$ is the diagonal component of the Oseen tensor, η is the dynamic viscosity of the surrounding liquid, K_c is the bending modulus of the bilayer, and σ is the surface tension. In Eq. (11.12), the first two terms in square brackets represent the bending resistance and the preexisting tension in the bilayer. The fourth term represents thermal fluctuations and is modeled as a Gaussian white noise with zero mean, satisfying the fluctuation–dissipation theorem.

F_{int} represents the interaction force between the bilayer and the cytoskeleton and is composed of two components, F_{ank} and $F_{\text{rep}}^{\text{sp}}$. The first component represents the interaction force due to the attachment of the cytoskeleton to the bilayer at discrete locations by the ankyrins and is modeled as

$$F_{\text{ank}} = \gamma h(\mathbf{r}, t) \sum_i \exp\left[- \left| \frac{\mathbf{r} - \mathbf{R}_i}{c_1/4} \right|^2 \right], \tag{11.13}$$

where the $\mathbf{R}_i$ are the locations of the ankyrin attachments, labeled by the index i. The height h is essentially made to be zero at $\mathbf{R}_i$ by taking large values of the coefficient γ. The second component, $F_{\text{rep}}^{\text{sp}}$, represents the steric repulsion between the spectrins and the bilayer and is modeled as

$$F_{\text{rep}}^{\text{sp}}(\mathbf{r}) = \frac{\epsilon}{\lambda} \sum_i e^{-h(\mathbf{r})/\lambda} \exp\left[- \left(\frac{a_i x + b_i y + c_i}{c_1/4} \right)^2 \right], \tag{11.14}$$

where $a_i x + b_i y + c_i = 0$ represents a line connecting two adjacent attachment sites; $\lambda = 0.2$ nm and $\epsilon = K_B T/100c_1^2$ are so chosen that the force increases rapidly as $h \to 0$. Equation (11.12) is integrated numerically to evolve $h(\mathbf{r}, t)$ in time, starting from the initial configuration $h(\mathbf{r}, 0) = 0$. Since the bilayer patch is doubly periodic, it is easy to solve the equation in Fourier space since the Fourier modes are decoupled.

11.2.4 Bilayer–Cytoskeleton Interaction

In Lin and Brown [30], the cytoskeleton was assumed to be an immobile structure, and thus the deformation of the spectrin filaments and motion of the junction vertices were not considered. The dynamics of the bilayer is expected to be affected by the cytoskeletal deformation and fluctuation, and vice versa. This coupled dynamics was considered by Zhu et al. [31] by combining the cytoskeleton model and the bilayer model described in the two preceding sections. For the bilayer, in addition to F_{ank} and $F_{\text{rep}}^{\text{sp}}$, they introduced a repulsive force between the actins and the bilayer as

$$F_{\text{rep}}^{\text{ac}}(\mathbf{r}) = \frac{\epsilon}{\lambda} \sum_j e^{-h(\mathbf{r})/\lambda} \exp\left[- \left| \frac{\mathbf{r} - \mathbf{R}_j}{c_1/4} \right|^2 \right], \tag{11.15}$$

where the index j is used to label the actins [31]. Further, the motion of the actin protofilaments was modeled explicitly. For this, the force $\mathbf{f}$ acting on a protofilament was written as

$$\mathbf{f} = \mathbf{f}_{\text{s}} + \mathbf{f}_{\text{rep}} + \mathbf{f}_{\text{rand}}, \tag{11.16}$$

where $\mathbf{f}_s$ is the attractive force imparted by the spectrin chains and is derived from the potential given in Eq. (11.7) for the worklike chain model. The second term $\mathbf{f}_{\text{rep}}$ represents the steric repulsion between the bilayer and the actin protofilaments and hence is equal to $-F_{\text{rep}}^{\text{ac}}$. The last term $\mathbf{f}_{\text{rand}}$ is the Brownian force modeled as the Gaussian white noise, as mentioned above. The net force $\mathbf{f}$ is balanced by the hydrodynamic viscous drag on the protofilaments exerted by the surrounding fluid. Then the velocity $\mathbf{u}$ of the protofilaments is computed as $u_i = f_i/D_i, i = x, y, z$, where D_i are the drag coefficients in the respective directions. The expressions for D_i are obtained from the theory of the Stokes flow by assuming the protofilaments to be long, thin cylinders of circular cross section. The positions of the protofilaments are updated as $d\mathbf{R}/dt = \mathbf{u}$. The model was used to simulate a bilayer–cytoskeleton complex over a rectangular patch 0.4 μm in length.

11.3 CONTINUUM MODELS

One of the challenges in the computational study of blood flow in microcirculation is the consideration of a large ensemble of cells flowing as a dense suspension. This is an especially formidable problem when large deformations of multiple interacting cells are considered, especially for a geometry that is not triply periodic. At this level of description, the detailed molecular structure of the lipid bilayer and the cytoskeleton can be neglected and a continuum model of the cells can be very useful. The individual cell is then modeled as a capsule or vesicle (i.e., a liquid drop surrounded by an infinitesimally thin membrane). Typically, four quantities determine a cell's mechanical behavior: the ratio of the viscosity of the interior liquid (hemoglobin) to the exterior liquid (plasma), the extensional elastic modulus of the membrane, the shear modulus, and the bending resistance. Unlike a capsule, a vesicle does not have a shear resistance; rather, it satisfies the area incompressibility condition and often has a bending resistance, as in the case of the lipid bilayer.

11.3.1 Membrane Models

The elastic response of the membrane may be expressed in terms of a strain energy function W. The Mooney–Rivlin model (see, e.g., ref. [20]) states that

$$W = \frac{E_s}{6}(1-\psi)\left[2\Lambda_2 + e^{-2\Lambda_1} - 1\right] + \frac{E_s}{6}\psi\left[2\Lambda_2 e^{-2\Lambda_1} + 2e^{-2\Lambda_1} + e^{2\Lambda_1} - 3\right]. \quad (11.17)$$

In Eq. (11.17) ψ is a nondimensional parameter (varying from 0 to 1) which introduces nonlinear behavior in the model. Λ_1 $(= \log \epsilon_1\epsilon_2)$ and $\Lambda_2[= (1/2)(\epsilon_1^2 + \epsilon_2^2) - 1]$ are the strain invariants, where ϵ_1 and ϵ_2 are the principal extension ratios and E_s is the surface elastic modulus of the membrane. The neo-Hookean model, which represents a strain-softening material, is a special case of the Mooney–Rivlin law for $\psi \to 0$,

$$W = \frac{E_s}{6}(\epsilon_1^2 + \epsilon_2^2 + \epsilon_1^{-2}\epsilon_2^{-2} - 3). \quad (11.18)$$

Skalak et al. [32] proposed a strain energy function to describe a strain-hardening material as

$$W = \frac{E_s}{2}\left(\Lambda_2^2 + \Lambda_2 - \frac{e^{2\Lambda_1} - 1}{2}\right) + \frac{E_a}{8}(e^{2\Lambda_1} - 1)^2. \quad (11.19)$$

The first term on the right represents the shear deformation, and the second term represents the area extension. Another example is the model developed by Evans and Skalak [33],

$$W = E_a(\epsilon_1\epsilon_2 - 1)^2 + E_s\left(\frac{\epsilon_1^2 + \epsilon_2^2}{2\epsilon_1^2\epsilon_2^2} - 1\right). \quad (11.20)$$

Comparison with experimental measurements using human erythrocytes yields $E_a \approx 500$ dyn/cm and $E_s \approx 6 \times 10^{-3}$ dyn/cm. The high value of E_a and the low value of E_s imply large deformation of RBCs without a significant change in surface area.

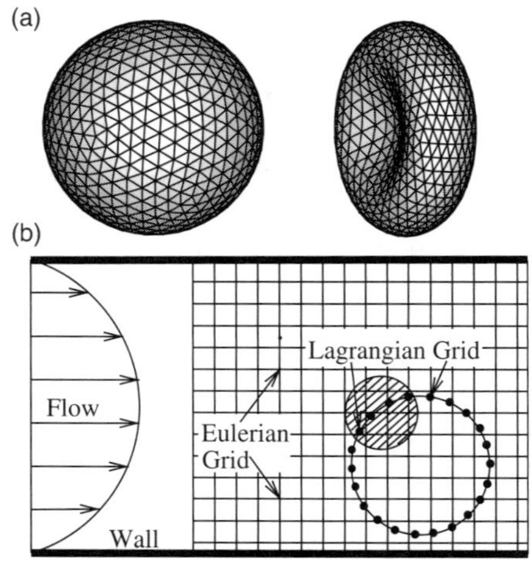

Figure 11.2 (a) Triangulation of the cell surface; (b) Eulerian and Lagrangian grids.

11.3.2 Membrane Force Calculation

The membrane force $\mathbf{f}(\mathbf{x}', t)$ generated can be obtained conveniently using a finite element approach [34,35]. The method has been adopted in several recent works on capsule deformation in a shear flow [36–40]. First, the cell surface is discretized using flat triangular elements (Fig. 11.2). The triangulated surface mesh can be obtained from the open-source GNU triangulated surface (GTS) library. In the process, an icosahedron is recursively divided and its vertices are projected onto a spherical or biconcave shape, resulting in an ordered network in which each Lagrangian node on the surface is surrounded by five or six triangular elements, similar to the spectrin network of the cytoskeleton [29].

The force distribution in the cell membrane is obtained from the deformation of individual surface triangles. Consider a general three-dimensional deformation of a capsule subject to an external shear. Due to deformation and tank treading of the capsule, a triangular surface element does not remain on the same plane after deformation. The displacement of the three vertices of an element arises due to deformation of the element as well as due to its rigid-body rotation. However, the rigid-body rotation does not result in any elastic force. Thus, one needs to obtain the displacements of the three vertices of the deformed element with respect to the undeformed element on a common plane by using a transformation rule,

$$\mathbf{x}_l^P = \mathbf{M}(\mathbf{x}_l - \mathbf{x}_0), \qquad \mathbf{X}_l^P = \mathbf{R}(\mathbf{X}_l - \mathbf{X}_0), \tag{11.21}$$

where $l = 0, 1, 2$ represent the three vertices, $\mathbf{x}_l^P$ and $\mathbf{X}_l^P$ are the coordinates of the vertices of the undeformed and deformed elements, respectively; on the common plane P; and $\mathbf{M}$ and $\mathbf{R}$ are the transformation matrices. Then the in-plane nodal displacements are $\mathbf{v}_l = \mathbf{X}_l^P - \mathbf{x}_l^P$ for $l = 0, 1, 2$. Now we can compute the deformation gradient tensor, $\mathbf{D}$, and the right Cauchy–Green tensor, $\mathbf{G} = \mathbf{D}^T D$, which relates to the extension ratios as

$$\epsilon_{1,2}^2 = \frac{1}{2}\left[G_{11} + G_{22} \pm \sqrt{(G_{11} - G_{22})^2 + 4G_{12}^2}\right]. \tag{11.22}$$

The principal of virtual work is used to evaluate the in-plane forces at the triangle vertices as

$$\mathbf{f}_l^P = \frac{\delta W}{\delta \mathbf{v}_l} = \frac{\partial W}{\partial \epsilon_1}\frac{\partial \epsilon_1}{\partial \mathbf{v}_l} + \frac{\partial W}{\partial \epsilon_2}\frac{\partial \epsilon_2}{\partial \mathbf{v}_l} \qquad \text{for} \quad l = 0, 1, 2. \tag{11.23}$$

Once the in-plane forces for an individual element are found, they are transformed to the global coordinates by using the transformation rule $\mathbf{f} = \mathbf{R}^{\mathrm{T}}\,\mathbf{f}^{\mathrm{P}}$, where $\mathbf{R}$ is the transformation matrix, as explained above. Since each node is shared by five or six surface triangles, the resultant force $\mathbf{f}(\mathbf{x}', t)$ is the vector addition of the forces contributed by all the shared elements.

11.3.3 Flow Solver

Once the membrane force is calculated, the next step is to couple the cell deformation with the surrounding flow. The problem belongs to the broader class of fluid–structure interaction problem for highly deformable surfaces. One convenient way of treating the problem is the immersed boundary/front-tracking method [41–43]. In the continuum description of interest here, the fluid flow, both exterior and interior to the cell, is governed by the Navier–Stokes equation. The principle of the front-tracking method is to use a single set of equations for the fluids both inside and outside the capsule. The fluid equations are solved on a fixed Eulerian grid, and the interface is tracked in a Lagrangian manner by a set of marker points (Fig. 11.2). The cell membrane is accounted for by introducing a body force $\mathbf{F}(\mathbf{x}, t)$ in the governing equations such that it is zero everywhere in the flow except at the interface:

$$\mathbf{F}(\mathbf{x}, t) = \int_{\partial S} \mathrm{f}(\mathbf{x}', t)\delta(\mathbf{x} - \mathbf{x}')\, d\mathbf{x}', \tag{11.24}$$

where $\mathbf{x}$ is the location of an arbitrary point in the flow domain, $\mathbf{x}'$ is any point on the interface, ∂S is the entire interface, and δ is the three-dimensional delta function, which vanishes everywhere except at the interface. We assume that both the plasma and the hemoglobin may be modeled as incompressible Newtonian fluids. For fluid motion under an externally applied pressure gradient, the governing equations are

$$\nabla \cdot \mathbf{u} = 0, \qquad \rho\left[\frac{\partial \mathbf{u}}{\partial t} + \mathbf{u} \cdot \nabla \mathbf{u}\right] = -\nabla p - \frac{d\bar{P}}{dx}\mathbf{e}_x + \nabla \cdot \tau + \mathbf{F}, \tag{11.25}$$

where $\mathbf{u}(\mathbf{x}, t)$ is the fluid velocity, ρ is the density, p is a local pressure, $\mathbf{e}_x$ is the unit vector along the x-axis, and

$$\tau = \mu\left[\nabla\mathbf{u} + (\nabla\mathbf{u})^T\right] \tag{11.26}$$

is the viscous stress tensor. $\mu(\mathbf{x}, t)$ is the viscosity in the entire fluid and may be written as

$$\mu(\mathbf{x}, t) = \mu_p[1 + (\lambda - 1)I(\mathbf{x}, t)], \tag{11.27}$$

where $\lambda\mu_p$ is the viscosity of the cytoplasmic fluid and $I(\mathbf{x}, t)$ is an indicator function, which is 1 in the interior of a cell and zero outside. For normal erythrocytes, the viscosity ratio $\lambda = 5$. Note that the net pressure gradient in Eq. (11.25) is the sum of the externally applied constant pressure gradient $-d\bar{P}/dx$, and the fluctuating local pressure gradient $-\nabla p$. The latter is generated due to the presence of the interacting cells.

The time marching of the flow solver is done using a two-step, time-split scheme. In the first step, known as the *predictor step*, an advection–diffusion equation is solved. The interfacial force is retained in the advection–diffusion equation. In the time integration, the nonlinear terms are treated explicitly using a second-order Adams–Bashforth scheme, and the viscous terms are treated semi-implicitly using the Crank–Nicholson scheme. Spatial derivatives are computed using a second-order finite-difference scheme. Since the diffusion terms are treated semi-implicitly, the advection–diffusion equation is solved using an alternative direction implicit (ADI) scheme. The second step involves solving a Poisson equation for the pressure, which must be solved fully implicitly to force the final velocity field to be divergence-free. If the problem at hand is periodic over some distance, the Poisson equation can be Fourier transformed, yielding a set of two-dimensional decoupled PDEs which is inverted by an efficient matrix diagonalization method, reducing computational expenses. The velocity field obtained from the predictor step is then corrected to be divergence-free using the pressure.

Once the fluid velocity is known, the cell surface velocity is computed as

$$\mathbf{u}(\mathbf{x}', t) = \int_S \mathbf{u}(\mathbf{x}, t)\delta(\mathbf{x} - \mathbf{x}')d\mathbf{x}, \tag{11.28}$$

where S indicates the entire flow domain. The Lagrangian points on the membrane are then advected as

$$\frac{d\mathbf{x}'}{dt} = \mathbf{u}(\mathbf{x}', t). \tag{11.29}$$

Numerically, Eq. (11.29) is treated explicitly using the second-order Adams–Bashforth scheme.

For numerical implementation, a smooth representation of the δ-function is used as

$$D(\mathbf{x}-\mathbf{x}') = \begin{cases} \dfrac{1}{64\,\Delta^3}\displaystyle\prod_{i=1}^{3}\left[1+\cos\frac{\pi}{2\Delta}(x_i - x_i')\right] & \text{for} \quad |x_i - x'_i| \leq 2\Delta, \quad i=1,2,3, \\ 0, & \text{otherwise}, \end{cases} \qquad (11.30)$$

where Δ is the Eulerian grid size. As the cells move and deform, $\mu(\mathbf{x},t)$ is updated by solving a Poisson equation for the indicator function $I(\mathbf{x},t)$ as

$$\nabla^2 I = \nabla \cdot \mathbf{G}, \qquad (11.31)$$

where $\mathbf{G} = \int_{\partial S} \delta(\mathbf{x}-\mathbf{x}')\mathbf{n}\, d\mathbf{x}'$ and $\mathbf{n}$ is the unit normal to the cell surface.

We have described a computational methodology for unsteady capsule dynamics in three dimensions using a front-tracking/immersed boundary method for the deformable interface, and a finite-difference/Fourier transform method for the flow solver. The methodology can address problems ranging from the dynamics of a single capsule, interaction between a pair of capsules, and the motion of multiple capsules under a variety of flows such as simple or parabolic shear, wall-bounded flows, and oscillatory flows. The methodology can also be applied to address adhesive rolling motion of a white blood cell [44,45], and interaction between a red blood cell and a white blood cell.

11.3.4 Simulation of Multiple Cells

Now we present some results on the motion of multiple deformable cells in a microvessel using the continuum method described above (see ref. [40] for details). We consider the cell motion in a pressure-driven flow in a microchannel bounded by two parallel plates that are separated by a distance H (Fig. 11.3). In the absence

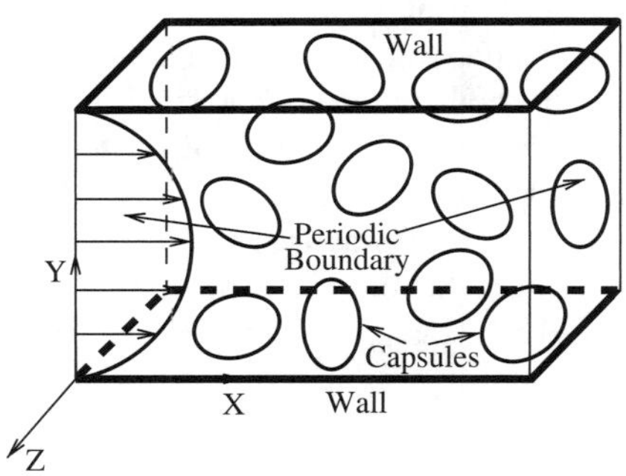

Figure 11.3 Three-dimensional computational domain for simulation of cell suspension in a microchannel.

of cells, the undisturbed flow is a fully developed parabolic (Poiseuille) flow of pure plasma. The computational domain is a cube of side length H. No-slip conditions are imposed at the top and bottom plates, and periodic conditions are imposed at the open boundaries in the x and z directions. We consider two types of cells, one having an initially spherical shape and the other having the biconcave shape representative of the erythrocytes in the unstressed state. For the spherical cells, the viscosity values of the interior and exterior fluids are assumed to be the same and equal to the plasma viscosity (i.e., 1.2 cP). For the biconcave cells, the viscosity values of the interior and exterior fluids are taken to be 6 and 1.2 cP, respectively, as in the case of normal erythrocytes and plasma. The cell membrane is assumed to follow the neo-Hookean law. Computations are performed in dimensionless form. The important dimensionless parameters are the capillary number $\mathrm{Ca} = \mu_p U_{cl}/E_s$, which is the ratio of the viscous force to the elastic force of the capsule membrane, the size ratio a/H, where a is the diameter of an undeformed spherical cell, and the viscosity ratio λ. The cell volume fraction (hematocrit) is 26%. For the biconcave cells, $H/a = 4.6$ yielding 113 cells in the computational domain. For the spherical cells, we take $H/a = 12.5$, which yields 1096 cells in the domain.

The Eulerian resolution is determined by the ratio a/H and the cell volume fraction. We ensure that there are about 20 Eulerian points per cell diameter and about six grid points between adjacent cells. For an ensemble of 1096 cells, we use a 288^3 grid, whereas for the biconcave cells we use 160^3 grid. The Lagrangian resolution used to discretize the cell surface is 1280 nodes for low Ca, and 5120 nodes for higher Ca. The dimensionless time step used in the simulation is 10^{-3}. The runtime is about 6 hours for unit dimensionless time using 8 1.7-MHz IBM p690 processors for the case of 1096 spherical capsules.

Figure 11.4 shows snapshots of the cell suspension for the two cases after the flow has reached a quasi-steady state starting from an initial random distribution of the cells. Deformation of the cells is evident in the figure. The amount of deformation depends on the capsule location with respect to the wall. Capsules near the wall deform more than those near the center due to the gradient in shear. Hydrodynamic interaction between adjacent cells results in the formation of a flat surface. For the biconcave cells, those near the wall align with the flow direction, whereas those near the center flow align with a nearly vertical alignment. Figure 11.5 shows the mean velocity profile over the channel cross section compared with the parabolic velocity. The well-known plug-flow profile is predicted in our simulations. Figure 11.6 shows the time-average distribution of cells across the channel height in terms of a local hematocrit $H_c(y,t)$. Once the suspension reaches a statistically stationary state, the cell distributions at frequent intervals are analyzed to obtain $H_c(y,t)$. As described earlier, the indicator function $I(x,y,z,t)$ can be used to calculate $H_c(y,t)$. Noting that I changes from zero to 1 across the cell membrane, we have

$$H_c(y,t) = \frac{\int_x \int_z I(x,y,z,t)\,dx\,dz}{\int_x \int_z dx\,dz}, \tag{11.32}$$

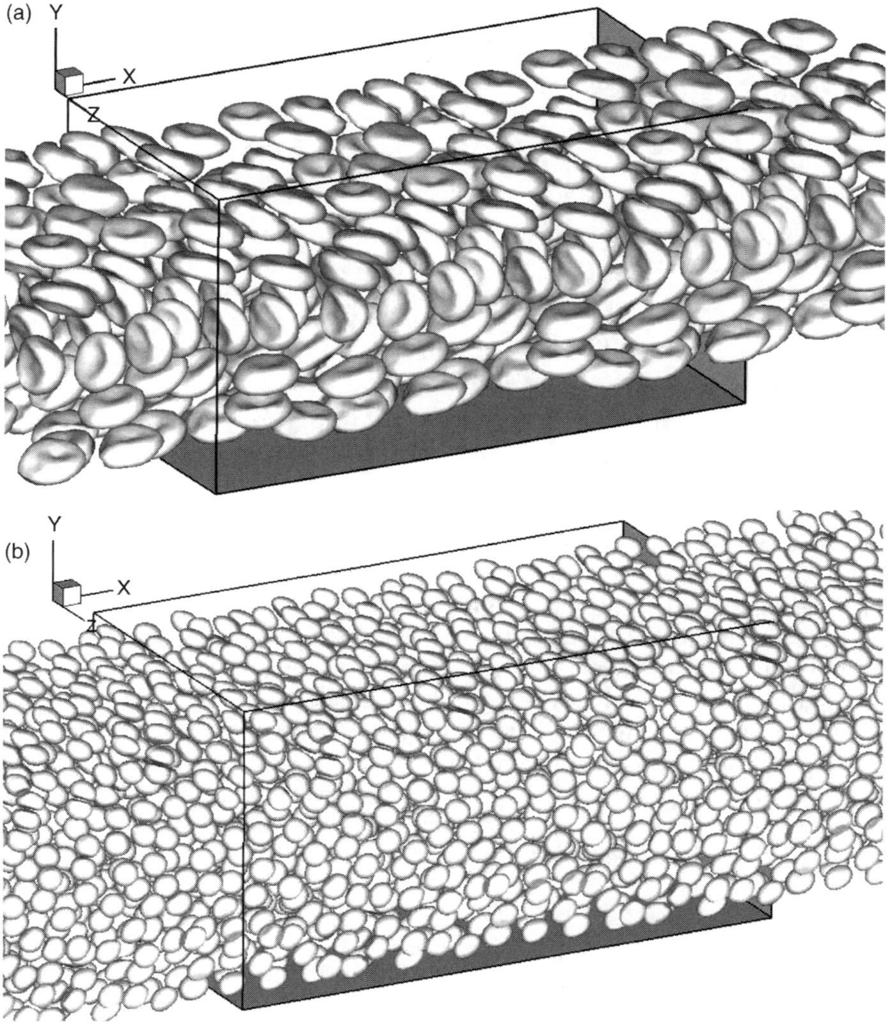

Figure 11.4 Instantaneous cell distribution for (a) 113 biconcave cells, and (b) 1096 initially spherical cells.

which is obtained over a time window and then time-averaged to obtain $H_c(y)$. Figure 11.6 shows the ratio $H_c(y)/H_t$, where H_t is the tube hematocrit for the case considered in Fig. 11.5. The ratio goes to zero near the wall, indicating the depletion of the cells, and is around 1 near the center. Thus, the formation of the cell-depleted layer near the wall can be predicted by the simulation. The oscillating nature of $H_c(y)/H_t$ near the center of vessel is due to cells flowing in layers in the channel.

The simulations presented above are computation and data intensive and probably the first of their type. They provide a wealth of information on the

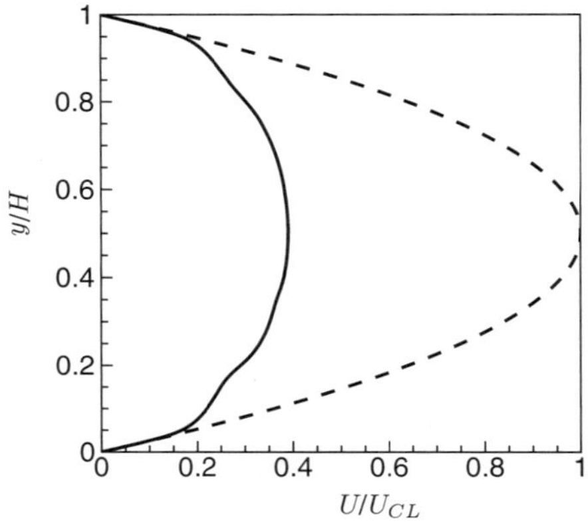

Figure 11.5 Mean velocity of spherical cell suspension for $H/a = 6.3$, $\mathrm{Ca} = 0.05$, and hematocrit 26% (solid line), compared with the parabolic velocity profile.

dynamics of semidense suspension of capsules and biological cells. The database from such simulations can be further postprocessed to gain deeper insight into the microrheology of microcirculatory blood flow. For example, the simulation results can be used to predict the Fahraeus and Fahraeus–Lindqvist effects, and the cell-free layer width under varying hematocrit, cell deformability, and vessel size. The collision frequency of cells as a function of cell deformability, hematocrit, and vessel size can also be obtained, which is of importance in modeling erythrocyte aggregation. The coefficient of hydrodynamic dispersion of cells and that of tracer particles can also be obtained. These are of importance in solute and drug transport in microvessels. Finally, these types of simulations may enable the formulation of better low-dimensional models of blood flow in small vessels by incorporating more physics.

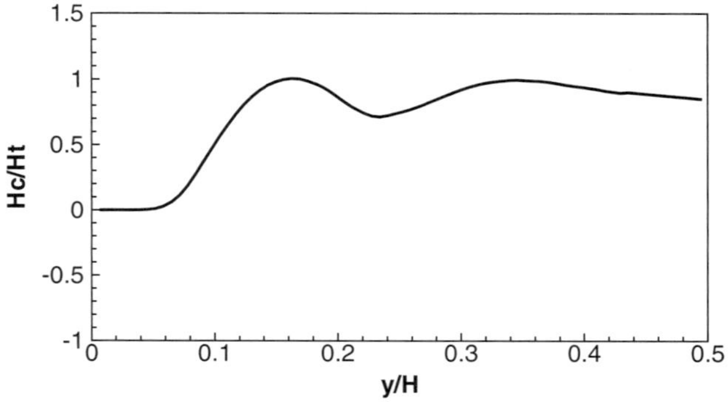

Figure 11.6 Mean particle distribution for the case considered in Fig.11.5.

11.4 CONCLUSION

In this chapter we discussed some mesoscale models for cytoskeleton and bilayer deformation to predict mechanical properties of the erythrocyte membrane, followed by a discussion on a continuum model that can consider the flow of multiple, deformable, and interacting cells. Progress has been made in using boundary integral and boundary element methods, immersed boundary and immersed interface methods, the lattice-Boltzmann technique, and dissipative particle dynamics techniques. These recent advances in numerical methods and computational resources have enabled researchers to consider detailed simulations of blood flow in microvessels in the presence of a large ensemble of deformable blood cells without sacrificing the detailed dynamics of the individual cell. Many research groups have recently succeeded in simulating hundreds to thousands of flowing cells in semidense suspension [40,46–49]. Based on this success, future research may consider a next generation of problems with direct relevance to physiological processes, such as detailed simulations of cellular motion in organ-specific microvascular networks, taking into consideration the geometric complexity and bifurcation of the vessels, analysis of drug deposition in microcirculation in the presence of interacting blood cells, and sickle cell polymerization.

REFERENCES

1. Cokelet GR, Goldsmith HL. Decreased hydrodynamic resistance in the two-phase flow of blood through small vertical tubes at low flow rates. *Circ. Res.* **68**:1–17, 1991.
2. Goldsmith HL, Cokelet GR, Gaehtgens P. Robin Fahraeus: evolution of his concepts in cardiovascular physiology. *Am. J. Physiol.* **257**:H1005, 1989.
3. Pries AR, Neuhaus D, Gaehtgens P. Blood viscosity in tube flow: dependence on diameter. *Am. J. Physiol.* **263**:H1770–H1778, 1992.
4. Popel AS, Johnson PC. Microcirculation and hemorheology. *Annu. Rev. Fluid Mech.* **37**:43–69, 2005.
5. Secomb TW. Mechanics of red blood cells and blood flow in microcirculation. In Modeling and Simulation of Capsules and Biological Cells, ed. C. Pozrikidis, Chapman & Hall/CRCPress,BocaRaton,FL,2003.
6. Fahraeus R, Lindqvist T. The viscosity of blood in narrow capillary tube. *Am. J. Physiol.* **96**:562–568, 1931.
7. Kim S, Kong RL, Popel AS, Intaglietta M, Johnson PC. Temporal and spatial variations of cell-free layer width in arterioles. *Am. J. Physiol. Heart Circ. Physiol.* **293**:H1526–H1535, 2007.
8. HappelJ,BrennerH,LowReynoldsNumberHydrodynamics,Kluwer,Hinghom,MA,1983.
9. Goldsmith HL. Red cell motions and wall interactions in tube flow. *Fed. Proc.* **30**:1578–1590, 1971.
10. Secomb TW, Skalak R, Ozkaya N, Gross JF. Flow of axisymmetric red blood cells in narrow capillaries. *J. Fluid Mech.* **163**:405–423, 1986.

11. Queguiner C, Barthes-Biesel D. Axisymmetric motion of capsules through cylindrical channels. *J. Fluid Mech.* **348**:349–376, 1997.
12. Hsu R, Secomb TW. Motion of nonaxisymmetric red blood cells in cylindrical capillaries. *J. Biomech. Eng.* **111**:147–151, 1989.
13. Coulliette C, Pozrikidis C. Motion of an array of drops through a cylindrical tube. *J. Fluid Mech.* **358**:1–28, 1998.
14. Sharan M, Popel AS. A two-phase model for flow of blood in narrow tubes with increased effective viscosity near the wall. *Biorheology* **38**:415–428, 2001.
15. Damiano ER, Long DS, Smith ML. Estimation of viscosity profiles using velocimetry data from parallel flows of linearly viscous fluids: application to microvascular hemodynamics. *J. Fluid Mech.* **512**:1–19, 2004.
16. Nott PR, Brady JF. Pressure-driven flow of suspensions: simulation and theory. *J. Fluid Mech.* **275**:157–199, 1994.
17. Loewenberg M, Hinch EJ. Numerical simulation of a concentrated emulsion in shear flow. *J. Fluid Mech.* **321**:395–419, 1996.
18. Zinchenko AZ, Davis RH. An efficient algorithm for hydrodynamical interaction of many deformable drops. *J. Comput. Phys.* **157**:539–587, 2000.
19. Mortazavi S, Tryggvason G. A numerical study of the motion of drops in Poiseuille flow: 1. Lateral migration of one drop. *J. Fluid Mech.* **411**:325–350, 2000.
20. Pozrikidis C. Numerical simulation of the flow-induced deformation of red blood cells. *Ann. Biomed. Eng.* **31**:1194–1205, 2003.
21. Pozrikidis C. Numerical simulation of cell motion in tube flow. *Ann. Biomed. Eng.* **33**(2):165–178, 2005.
22. Patankar NA, Ko T, Choi HG, Joseph DD. A correlation for the lift-off of many particles in plane Poiseuille flows of Newtonian fluids. *J. Fluid Mech.* **445**:55–76, 2001.
23. Zhang Z, Prosperetti A. A second-order method for three-dimensional particle simulations. *J. Comput. Phys.* **210**:292–324, 2005.
24. Boal DH. Computer simulation of a model network for the erythrocyte cytoskeleton. *Biophys. J.* **67**(2):521–529, 1994.
25. Boey SK, Boal DH, Discher DE. Simulations of the erythrocyte cytoskeleton at large deformation: I. Microscopic models. *Biophys. J.* **75**(3):1573–1583, 1998.
26. Discher DE, Boal DH, Boey SK. Simulations of the erythrocyte cytoskeleton at large deformation: II. Micropipette aspiration. *Biophys. J.* **75**(3):1584–1597, 1998.
27. Marko JF, Siggia ED. Statistical mechanics of supercoiled DNA. *Phys. Rev. E.* **52**(3):2912–2938, 1995.
28. Waugh R, Evans EA. Thermoelasticity of red blood cell membrane. *Biophys. J.* **26**(1):115–131, 1979.
29. Li J, Dao M, Lim CT, Suresh S. Spectrin-level modeling of the cytoskeleton and optical tweezers stretching of the erythrocyte. *Biophys. J.* **88**(5):3707–3719, 2005.
30. Lin LCL, Brown FLH. Dynamic simulations of membranes with cytoskeletal interactions. *Phys. Rev. E* **72**:011910, 2005.
31. Zhu Q, Vera C, Asaro RJ, Sche P, Sung LA. A hybrid model for erythrocyte membrane: a single unit of protein network coupled with lipid bilayer. *Biophys. J.* **93**(2):386–400, 2007.
32. Skalak R, Tozeren A, Zarda PR, Chien S. Strain energy function of red blood cell membranes. *Biophys. J.* **13**:245–264, 1973.

33. Evans EA, Skalak R, Mechanics and Thermodynamics of Biomembranes, CRC Press, Boca Raton, FL, 1980.

34. Charrier JM, Shrivastava S, Wu R. Free and constrained inflation of elastic membranes in relation to thermoforming-non-axisymmetric problems. *J. Strain Anal.* **24**(2):55–74, 1989.

35. Shrivastava S, Tang J. Large deformation finite element analysis of non-linear viscoelastic membranes with reference to thermoforming. *J. Strain Anal.* **28**:31–43, 1993.

36. Eggleton CD, Popel AS. Large deformation of red blood cell ghosts in a simple shear flow. *Phys. Fluids* **10**(8):1834–1845, 1998.

37. Li X, Sarkar K. Front-tracking simulation of deformation and buckling instability of a liquid capsule enclosed by an elastic membrane. *J. Comput. Phys.* **227**(10):4998–5018, 2008.

38. Doddi SK, Bagchi P. Effect of inertia on the hydrodynamic interaction between two liquid capsules in a shear flow. *Int. J. Multiphase Flow* **34**:375–392, 2008.

39. Doddi SK, Bagchi P. Lateral migration of a capsule in a plane Poiseuille flow in a channel. *Int. J. Multiphase Flow* **34**:966–986, 2008.

40. Doddi SK, Bagchi P. Three-dimensional computational modeling of multiple deformable cells flowing in microvessels. *Phys. Rev. E* **79**:046318, 2009.

41. Tryggvason G, Bunner B, Esmaeeli A, Al-Rawahi N, Tauber W, Han J, Nas S, Jan Y. A front tracking method for the computations of multiphase flow. *J. Comput. Phys.* **169**:708–759, 2001.

42. Peskin CS. Numerical analysis of blood flow in the heart. *J. Comput. Phys.* **25**:220–233, 1977.

43. Miller LA, Peskin CS. A computational fluid dynamics of 'clap and fling' in the smallest insects. *J. Exp. Biol.* **208**(Pt. 2): 195–212, 2005.

44. Pappu V, Doddi SK, Bagchi P. A computational study of leukocyte adhesion and its effect on flow pattern in microvessels. *J. Theor. Biol.* **254**:483–498, 2008.

45. Pappu V, Bagchi P. 3D computational modeling and simulation of leukocyte rolling adhesion and deformation. *Comp. Biol. Med.* **38**:738–753, 2008.

46. Ding EJ, Aidun CK. Cluster size distribution and scaling for spherical particles and red blood cells in pressure-driven flows at small Reynolds number. *Phys. Rev. Lett.* **96**(20):204502, 2006.

47. Dupin MM, Halliday I, Care CM, Alboul L, Munn LL. Modeling the flow of dense suspensions of deformable particles in three dimensions. *Phys. Rev. E* **75**:066707, 2007.

48. Freund JB. Leukocyte margination in a model microvessel. *Phys. Fluids* **19**(2):023301, 2007.

49. Pivkin IV, Richardson PD, Karniadakis G. Blood flow velocity effects and role of activation delay time on growth and form of platelet thrombi. *Proc. Natl. Acad. Sci. USA* **103**(46):17164–17169, 2006.

12

MOLECULAR TO MULTICELLULAR DEFORMATION DURING ADHESION OF IMMUNE CELLS UNDER FLOW

DAVID J. GEE

Department of Mechanical Engineering, Rochester Institute of Technology, Rochester, New York

MICHAEL R. KING

Department of Biomedical Engineering, Cornell University, Ithaca, New York

12.1 INTRODUCTION

The human immune system is triggered by internal and external factors. For example, when the immune system detects the presence of a bacterial or viral pathogen, it responds by mobilizing leukocytes to the site of infection. Similarly, when low-density lipoprotein (LDL) cholesterol invades the intraarterial wall, leukocytes are dispatched to consume or phagocytose the cholesterol particles. Research into immune system response in the microvasculature has progressed steadily over the past decade and is known to be mediated by multiple receptor–ligand adhesion molecules. For example, it is known that P-selectin is rapidly up-regulated to the plasma membrane of endothelial cells (ECs) upon stimulation by cytokines [1], and that selectin binding is a critical step in the inflammatory cascade [2]. This has been demonstrated in vitro for leukocytes rolling on artificial lipid bilayers containing purified P-selectin [3], for cell-free rolling under specific receptor–ligand presentation (Sialyl Lewis^x/selectin) [4], and in vivo in exposed rat mesentery [5]. Studies with

Multiscale Modeling of Particle Interactions: Applications in Biology and Nanotechnology, Edited by Michael R. King and David J. Gee

double-knockout mice have similarly demonstrated the critical role of selectins in inflammatory response [6]. Specifically, those mice exhibited compromised immune system response. In short, the selectin family of molecules is critical to the initial stages of leukocyte recruitment to sites of host infection or injury in the microvasculature.

During the initial stages of inflammatory response, surface-bound selectins and their carbohydrate ligands are primarily responsible for tethering and slow rolling of leukocytes over inflamed endothelium. It has been shown that exposure to inflammatory chemokines activates vascular endothelial cells and induces up-regulation of E- and P-selectin [1]. Initial adhesive contact also depends on molecules such as P-selectin glycoprotein ligand-1 (PSGL-1) and L-selectin, which are found on the tips of leukocyte microvilli [7]. For leukocytes in the near-wall environment, recruitment is initiated by tethering and rolling over activated endothelium under flow. Rolling, in which the leukocyte slows down with respect to the mean flow velocity, is caused by the coordinated formation (at the leading edge) and breakage (at the trailing edge) of receptor–ligand bonds.

Integrins and immunoglobulins are examples of other families of adhesion molecules that participate in the inflammatory response [8–10]. The ligation of β_2 integrins (including LFA-1 and Mac-1) with their EC surface ligands (e.g., ICAM-1, VCAM-1) is also known to play a vital role in the advanced stages of immune response by transitioning cell rolling to a firmly adherent cell capable of transmigrating across the endothelial cell layer and into tissue [3,11]. In vivo studies have shown that deficiency in E- and P-selectin can eliminate cellular inflammatory response even when integrins and their respective ligands are available [6]. Thus, selectin-mediated rolling is critical for inflammatory response [2].

Cell-free assays have reproduced the essential features of leukocyte rolling in vivo: the inherent noisiness of rolling velocity, the dependence of average velocity on adhesion molecule density on either surface, and shear rate dependence [4,12,13]. Thus, dilute cell-free systems can be utilized to study the biophysics of adhesion in a controlled parameter space. However, in reality, blood hematocrit severely complicates the physics of leukocyte recruitment. Melder et al. [14] demonstrated that for selectin- and integrin-mediated rolling in vitro, 32% hematocrit improved by twofold the cell capture efficiency of T-lymphocytes on an activated endothelial monolayer. Furthermore, it was demonstrated in vivo that 30% hematocrit produced a 20-fold increase in rolling when monocytes were perfused through an intact vessel in mouse mesentery [15]. In contrast, Mitchell and co-workers performed an in vitro study showing no measurable difference in the number of leukocytes rolling on E-selectin for whole blood or saline perfusion [16]. These competing claims highlight the challenges in trying to use the idealized results from parallel-plate flow chamber experiments in the interpretation of in vivo microcirculatory observations.

Computational modeling of blood flow and adhesion is an active area of research. Several models have been proposed to account for blood cell deformation. Bathe et al. [17] modeled the cell as a viscoelastic Maxwell material with constant cortical tension. Dong et al. [18] and Dong and Lei [19] modeled the cell membrane as an elastic two-dimensional ring in shear flow and predicted that the cell–substrate contact area would nearly double as the wall shear stress was increased from 0.5 to 20 dyn/cm^2. Observations of leukocyte rolling in vivo have shown that as the wall shear rate increased from 50 to 800 s^{-1}, leukocytes elongated by 140% of their undeformed diameter and the cell–substrate contact area increased 3.6-fold [20]. Both model and in vivo observations showed little variation in rolling velocity with increasing shear rate, suggesting that increased cell deformation resulted in greater adhesive forces acting to attenuate any increase in rolling velocity. Blood flow in the heart has been modeled with a technique known as the immersed boundary method (IBM) [21]. The novelty of the IBM was in the discretization of objects or cells by a cell-fixed Lagrangian grid within an immersive flow discretized by an Eulerian grid. This allowed for the modeling of geometrically complex free-boundary structures as force fields immersed in a larger fluid domain. The technique has been applied to other biological problems, such as flagellar swimming [22,23], platelet aggregation [24], and bacterial deposition [25]. Bottino [26] used the technique to model passive viscoelastic networks such as the actin cytoskeleton of ameboid cells, and Marella and Udaykumar [27] used the technique to model the flow interaction of an axisymmetric deformable leukocyte. Recently, several groups have developed three-dimensional, single-cell simulations of deformable leukocytes rolling on selectin surfaces, although these approaches are somewhat limited to shorter runtimes due to the need for supercomputing resources [28,29].

Hammer and Apte [30] developed the adhesive dynamics (AD) simulation to model selectin-mediated leukocyte rolling for low Reynolds number shear flow. This was later refined to include, for the first time, multiparticle cell–cell and cell–substrate hydrodynamic interactions [31]. This refined algorithm has been referred to as multiparticle adhesive dynamics (MAD). The method is based on a boundary integral equation (BIE) approach whereby the governing partial differential equations describing the particulate flow are recast as an integral equation for an unknown double-layer surface-density distribution and solved iteratively with a boundary element method (BEM). This technique has been used to simulate interacting cell adhesive events in viscous shear flow by treating the force-dependent, molecular binding kinetics using a Monte Carlo method. The technique can be extended to consider more complex flows (such as extensional or recirculating flows), nonspherical particles (such as platelets or biconcave disks [32]), and periodic systems representing large numbers of particles. Several extensions have been made to the multiparticle adhesive dynamics simulation in order to include cell deformation due to shear flow and adhesive cell–substrate interactions [33]. The new modules are based on a direct BIE approach to compute cell surface tractions and displacements. The solution procedure can be organized into three distinct components (termed the *mobility problem*, *exterior problem*, and *interior problem*) which are solved sequentially. In short, solution of the mobility problem leads to the cell's rigid body motion. Next, an exterior problem is solved for the cell's external stress distribution consistent with the mobility solution. The external surface traction drives the interior problem for cell deformation and completes one computational cycle.

Leukocyte trafficking in the microvasculature during inflammatory response is known to be mediated by multiple receptor–ligand adhesion molecules. Numerical modeling of this critical physiological response presents interesting challenges due to the nature of cell–cell interactions (large numbers of interactions, cell deformation, and a relatively large domain of influence of each cell over its neighbors). In this chapter we provide an overview of the physiological elements of the inflammatory cascade followed by a detailed discussion of an efficient computational technique that has been formulated and developed to model certain aspects of the inflammatory response. To model cell deformation, an elastic solid constitutive model for the leukocyte has been implemented. The computational technique is based on a boundary integral equation method (BIEM) and is extremely well suited to treat the half-space domain under consideration.

12.2 INNATE IMMUNE RESPONSE IS MEDIATED BY PATROLLING LEUKOCYTES

The human immune response is composed of multiple layers of defense against invading viruses, bacteria, and other pathogens and is broadly organized into two functional groups. An *innate immune response* is so-called because the host immediately recognizes the presence of the pathogen and mobilizes leukocytes in defense. This trafficking of leukocytes into extracellular tissue is organized to kill or phagocytose the pathogen.

An *adaptive immune response* is effected primarily by a special group of leukocytes known as *lymphocytes*: B cells and T cells. If the invading pathogen is able to circumvent the host innate immune response, the adaptive immune system is eventually able to recognize the foreign pathogen as nonself, thereby initiating the pathogen's destruction. The adaptive immune response has been described by Germain and Margulies [34] and Gray [35] and involves antigen presentation via a major histocompatibility complex (MHC) molecule; typically, class I or II MHC molecules.

12.2.1 Components of Whole Blood

Whole blood is composed of several components: red blood cells, white blood cells, platelets, and plasma. Red blood cells (RBCs) are responsible for transporting oxygen throughout the body and are devoid of DNA. White blood cells (WBCs), or leukocytes, effect immune response and are subdivided into five different classes: neutrophils, eosinophils, basophils, monocytes, and lymphocytes. Neutrophils, eosinophils, and basophils are also known as granulocytes, due to the presence of granules in their cytoplasm. For example, the MAC-1 integrin (important to inflammation) is stored in these granules. The neutrophil is sometimes referred to as polymorphonuclear (PMN), due to the multilobed appearance of the nucleus within the cell cytoplasm. Monocytes and lymphocytes are also known as mononuclear cells: again, a reference to the appearance of the nucleus. Normal human blood contains between approximately 4 to 10 × 10^9 leukocytes/L. This concentration is dominated by neutrophils and lymphocytes.

12.2.2 Functional Response of Leukocytes

The body responds to tissue damage or injury by the local production and release of cytokines and other inflammatory mediators. These signaling molecules initiate an *adhesion cascade* (Fig. 12.1) with the goal of attracting leukocytes to the region of injury, where they then execute their functional task to destroy the invading pathogen. For example, upon tissue injury, histamine may be released by mast cells in order to effect vasodilation. These and other signaling molecules are detected by specific counter-receptors on endothelial cells. The resulting ligation can then trigger multiple intercellular signaling pathways, resulting in the rapid up-regulation of P-selectin along the endothelium. This is important because P-selectin is known to play a key role in the initial stages of inflammatory response, and one ligand for P-selectin (P-selectin glycoprotein-1, PSGL-1) is known to be expressed constitutively on the surface of circulating neutrophils. Another important cytokine released by mast cells and macrophages, among others, is TNFα. In response to bacterial infection or other inflammatory mediators (such as IL-1), TNFα binds cell surface receptors, induces up-regulation of additional cell adhesion molecules, and aids the immune response by acting as a chemoattractant for neutrophils. In this way, neutrophils home to the site of injury or infection.

12.2.3 Cell Morphology

Scanning electron microscopy (SEM) images of a passive leukocyte shows that the surface is composed of multiple folds of excess membrane referred to as *microvilli*

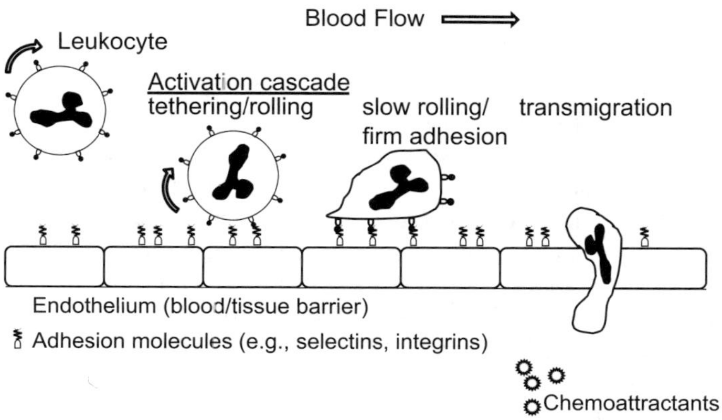

Figure 12.1 Schematic of adhesion cascade.

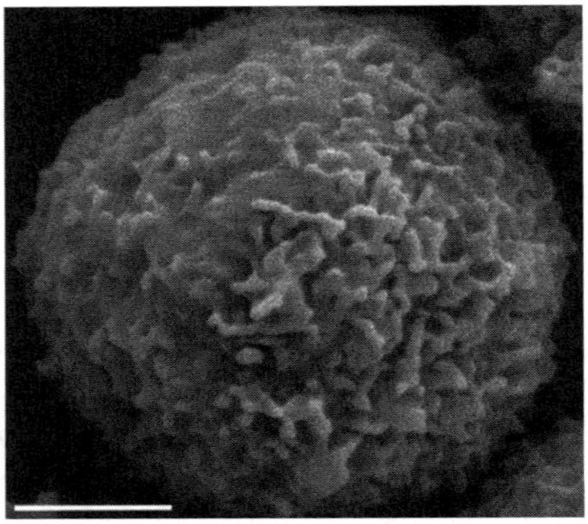

Figure 12.2 Scanning electron micrograph of human neutrophil. This research was originally published in *Blood*. [From Majstoravich S, Zhang J, Nicholson-Dykstra S, Linder S, Friedrich W, Siminovitch K, Higgs H. Lymphocyte microvilli are dynamic, actin-dependent structures that do not require Wiskott–Aldrich syndrome protein (WASp) for their morphology. *Blood* **104**:1396–1403. Copyright © 2004, American Society of Hematology.]

(Fig. 12.2). For a leukocyte in shear flow and under adhesive load, the microvilli will unfold and thereby increase the net surface contact area between the leukocyte and substrate. Compared to a smooth spherical cell of the same nominal diameter, various estimates give the unfolded surface area of a leukocyte as between 120 and 137% higher [27,36]. Beyond this important effect, it has been shown that certain surface receptors are preferentially located either near the tips of these membrane folds (PSGl-1, L-selectin, VLA-4) or in the valleys or troughs between the tips (Mac-1, LFA-1) [37]. This can have important physical implications relating to the phased response of the neutrophil (i.e., progressive activation).

12.3 PROBLEM FORMULATION

Blood flow in the microvasculature may be approximated as a Newtonian fluid and is described by a set of partial differential equations called the *Navier–Stokes equations*. These equations are nonlinear and are challenging to solve, even numerically. A great simplification is afforded in cases where viscous forces dominate rather than inertial forces. This is the case in the microcirculation. As a result, the nonlinear inertial terms in the fluid flow model are negligible compared to the viscous terms and the governing partial differential equations reduce to a linear set known as the *Stokes equations* (i.e., low Reynolds number flow). This simplification enables the application of standard methods of linear analysis in the solution of the differential equations. However, numerical modeling with the Stokes equations presents its own set of complications, including treatment of the relatively large domain of influence for each suspended particle with respect to its neighboring particles.

Particles or cells immersed in such a flow may be subjected to forces as a result of cell–cell collisions, cell–substrate adhesive interactions, and fluid shear forces. The previous version of MAD considered the cells as rigid spheres. Here the modeling is advanced by considering the cell to be deformable. As a first step, a linear elastic model has been implemented. For isotropic material response, this introduces two material (Lamé) parameters into the mathematical formulation. By combining the momentum equation with a linear isotropic elastic constitutive model, we arrive at the classical equation for small elastic deformations known as the *Navier equation.* This partial differential equation is also linear and has as a parameter the material Poisson's ratio, which, in turn, may be expressed in terms of the two elastic moduli.

12.3.1 Microcirculatory Flow Model

Tissue damage and infection occurs readily at the extremities of the body in regions vascularized by arterioles and venules. To describe the fluid flow through this circulatory path, we start with the Navier–Stokes equations for incompressible flow,

$$\rho \frac{D\mathbf{v}}{Dt} = \mu\nabla^2\mathbf{v} - \nabla p, \qquad \nabla \cdot \mathbf{v} = 0, \tag{12.1}$$

where ρ and μ are the fluid density and viscosity, respectively, and v_i and p are velocity and pressure, respectively. In the expression above, $D(\,\cdot\,)/Dt$ denotes the total derivative, $\nabla^2(\,\cdot\,)$ and $\nabla(\,\cdot\,)$ are the Laplace and gradient operator's, respectively, and $\nabla\cdot(\,\cdot\,)$ is the divergence operator. A Newtonian model for blood is a reasonable first approximation, and values for density and viscosity are taken for plasma. If we nondimensionalize the momentum equation [38] using characteristic velocity, U_0, length, L_0, and pressure p_0 $(= \mu U_0/L_0)$, the inertial terms are seen to be scaled by the Reynolds number, Re $(=\rho U_0 L_0/\mu)$:

$$\mathrm{Re}\left(\frac{\partial\tilde{\mathbf{v}}}{\partial\tilde{t}} + \tilde{\mathbf{v}}\cdot\tilde{\nabla}\tilde{\mathbf{v}}\right) = \tilde{\nabla}^2\tilde{\mathbf{v}} - \tilde{\nabla}\tilde{p}, \tag{12.2}$$

where, for example, $\tilde{p} = p/p_0$ and similarly for $\tilde{v}_i$ and $\tilde{t}$. Flows in the microvasculature are characterized by $\mathrm{Re} \ll 1$ due, primarily, to the extremely small characteristic values for U_0 and L_0. Typical blood vessel diameters in the microcirculation are between 10 and 300 μm, and the characteristic diameter of a neutrophil is between 8 and 10 μm. Hence, by neglecting the inertial terms in comparison to the viscous and pressure terms, the governing equations simplify as

$$\mu\nabla^2\mathbf{v} - \nabla p = \mathbf{0}, \qquad \nabla \cdot \mathbf{v} = 0, \tag{12.3}$$

where we have reverted back to dimensional variables. This set of linear differential equations, the Stokes equations, represent the flow in the region under consideration.

12.3.2 Tether Bond Dissociation Kinetics Model

Hammer and Apte [30] developed a simulation to model cell–substrate adhesive dynamics interactions for low Reynolds number linear shear flow. The neutrophil was modeled as a rigid sphere with random surface distribution of adhesion molecules, while the planar surface—extending to infinity—contained a random surface distribution of matched ligand. For cells flowing in the near-wall region, receptor–ligand pairs are tested for bond formation. Importantly, bond formation is not deterministic. Bonds that do form are modeled as springs connecting cell to substrate (Fig. 12.3). The length and orientation of each bond contributes to the adhesive force and torque on the cell. A summation over all bonds determines the net adhesive traction.

One model commonly used to describe the kinetics of biomolecular bond failure is due to Bell [39]. In the model, the rate of bond dissociation, k_r, depends on the magnitude of the bond force, F, as follows:

$$k_r = k_r^0 \exp\left(\frac{r_0 F}{k_b T}\right), \tag{12.4}$$

where the unstressed off-rate, k_r^0, and reactive compliance, r_0, have been determined experimentally [40,41]. Here k_b is Boltzmann's constant and T is the absolute temperature. Once the bond dissociation rate is known, the bond formation rate follows directly from the Boltzmann distribution for affinity [42]:

$$k_f = k_f^0 \exp\left[\frac{\sigma|\mathbf{x}_b - \boldsymbol{\lambda}|(r_0 - 0.5|\mathbf{x}_b - \boldsymbol{\lambda}|)}{k_b T}\right], \tag{12.5}$$

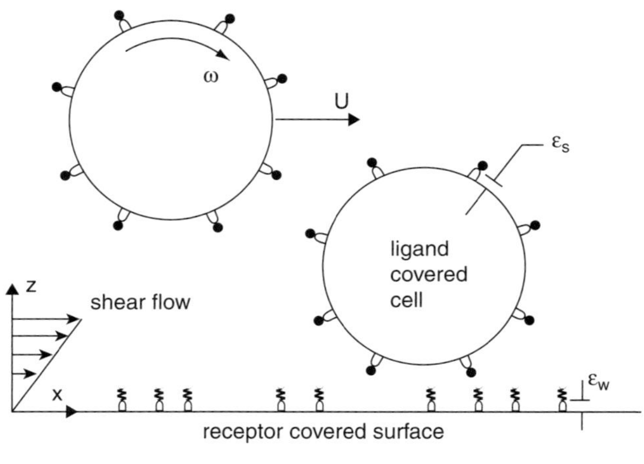

Figure 12.3 P-selectin/PSGL-1 tether bond model. Ligation of receptor and counter-receptor is modeled as a spring force connecting cell to substrate. Lifetime of an individual bond depends, in part, on the force on the bond. Bond formation is modeled as a stochastic process.

where σ is the spring constant and $|\mathbf{x}_b - \boldsymbol{\lambda}|$ is the deviation bond length. The AD algorithm was shown to reproduce the fine-scale dynamics of selectin-mediated rolling [43], but did not treat cell–cell interactions.

King and Hammer refined the AD algorithm to include cell–cell interactions [31,44]. Near and far-field hydrodynamic interactions between cells are included accurately via the boundary element solution of the Stokes equation. Contact interactions are treated analytically by including a short-range repulsive force, F_{rep} (representing contact forces between surfaces) of the form

$$F_{\text{rep}} = \frac{F_0 \exp(-\tau \cdot \varepsilon)}{1 - \exp(-\tau \cdot \varepsilon)} \tag{12.6}$$

where τ^{-1} is a length scale and ε denotes the surface-to-surface separation. In the simulations reported here, $\tau = 2000\,\mu\text{m}^{-1}$ and $F_0 = 10^{15}\,\text{fN}$. These forces, directed either along the line connecting the cell centers or normal to the plane, and gravitational forces are included in the summation for the cell's externally applied force vector.

12.3.3 Cell Deformation Model

In innate immune response, cell trafficking to sites of tissue injury or damage is directed by cell signaling events. These events proceed through multiple pathways and involve cytokines and other inflammatory mediators. The up-regulation of P-selectin on endothelial cells due to inflammatory stimuli is known to play an important role in the initial tethering and rolling (i.e., coordinated formation and breakage of bonds) of leukocytes on inflamed endothelium. As observed in vivo, leukocytes tend to remain approximately spherical during this phase [45]. As the immune cell becomes progressively activated, its firm adhesion to and subsequent transmigration across the endothelium into tissue involves greater and greater amounts of active shape change, as the leukocyte must eventually squeeze between endothelial cell–cell junctions (paracellular) or else pass directly through an endothelial cell (transcellular) [46,47]. We have chosen to model the small passive deformations associated with the initial tethering and rolling phase with a linear elastic model [48]. Thus, specification of the two Lamé parameters, λ_p and μ_p, is sufficient to characterize the small deformations due to imposed fluid shear or receptor-mediated adhesion.

For an elastic solid, the equilibrium equation assuming negligible body forces is given by [49]

$$\nabla \cdot \boldsymbol{\sigma} = \mathbf{0}, \tag{12.7}$$

where $\boldsymbol{\sigma}$ is the Cauchy stress tensor, not to be confused with the spring constant, σ. For Hooke's law and linear strain–displacement relation,

$$\sigma_{ij} = 2\mu_p \left[\frac{v_p}{1 - 2v_p} \delta_{ij} \varepsilon_{kk} + \varepsilon_{ij} \right], \qquad \varepsilon_{ij} = \frac{u_{i,j} + u_{j,i}}{2}, \tag{12.8}$$

where ε_{ij} is the strain tensor, u_i is displacement, μ_p is shear modulus, and $v_p = \lambda_p/2(\lambda_p + \mu_p)$ is Poisson's ratio. Taking the divergence of the stress tensor,

$$\sigma_{ij,\,j} = 2\mu_p \left[\frac{v_p}{1-2v_p} \frac{\partial}{\partial x_i} \varepsilon_{kk} + \varepsilon_{ij,\,j} \right], \tag{12.9}$$

where $\varepsilon_{kk} = u_{k,k}$ and $\varepsilon_{ij,\,j} = [\nabla^2 u_i + \nabla(u_{k,k})_i]/2$. Substituting for ε_{kk} and $\varepsilon_{ij,\,j}$, it can be shown that the equilibrium equation in terms of displacement may be given by

$$\frac{1}{1-2v_p} \nabla\nabla \cdot \mathbf{u} + \nabla^2 \mathbf{u} = \mathbf{0}. \tag{12.10}$$

This is known as *Navier's equation.*

12.3.4 Boundary Conditions

To complete the mathematical description, boundary conditions must be given. For the flow problem, we enforce the usual condition for a viscous fluid whereby the relative tangential velocity is zero at all fluid boundaries (i.e., the fluid adheres to the boundary). In addition, the transpiration velocity is also zero. For the displacement problem, the half-space boundary is rigid and therefore the displacement there is zero. Mathematically, these conditions are equivalent to

$$\mathbf{v}(\mathbf{x}) = \mathbf{0}, \quad \mathbf{u}(\mathbf{x}) = \mathbf{0}, \qquad \mathbf{x} \epsilon \Gamma_b, \tag{12.11}$$

where Γ_b denotes the bounding surface of the half-space (i.e., $\Gamma = \Gamma_b \cup \Gamma_p$, $p = 1, 2, \ldots, M$ for M particles). Despite the linearity of the governing flow equations, an analytical solution for particle velocities is still the exception rather than the norm, due to the slow decay of the influence of one particle over the motion of neighboring particles. Even in the case of an analytical solution, the particle must usually be represented by a specialized coordinate system such as a spherical coordinate system. For many-particle problems or for geometrically complex particle shapes, numerical solution is the only practical recourse.

12.4 NUMERICAL FORMULATION

We must resort to numerical methods to solve the set of equations outlined above. We choose a boundary element method due to the reduction in unknowns from a three-dimensional space to a three-dimensional surface. The complete problem can be organized into three distinct major components which are solved sequentially: the mobility, exterior, and interior problems. Solution of the mobility problem is based on the completed double-layer boundary integral equation method (CDL-BIEM) [50] and leads to the cell's rigid-body motion. The exterior problem is solved using a traction-based completed adjoint double-layer method (sometimes referred to as

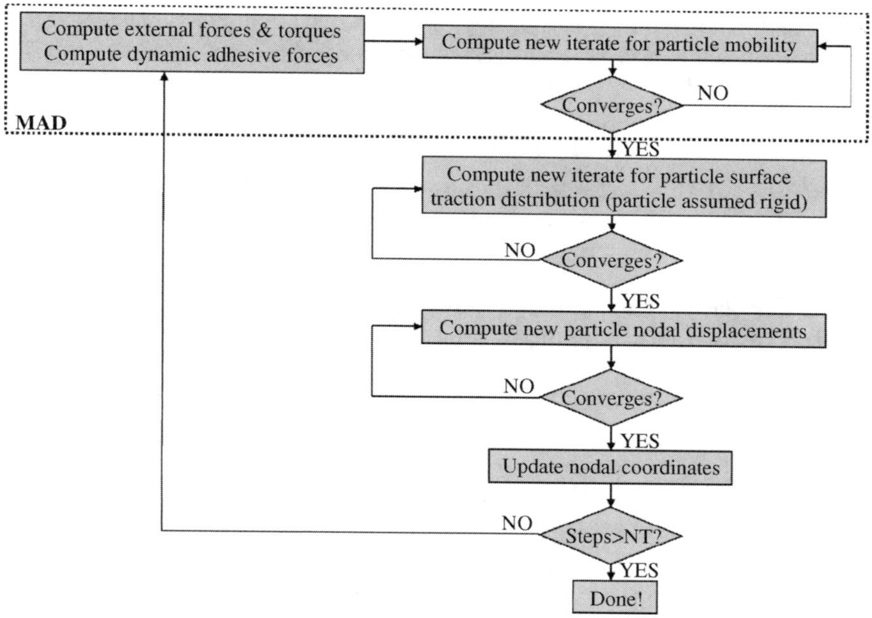

Figure 12.4 Flowchart for the deformable multiparticle adhesive dynamics algorithm.

CADL-BIEM) and yields the cell's surface traction distribution. The traction distribution drives the interior problem for cell deformation and the subsequent solution of the interior deformation problem then completes one computational cycle. The sequence for one complete time step is illustrated schematically in Fig. 12.4.

12.4.1 Indirect Boundary Integral Equation Method for Cell Mobility

As mentioned, viscous flow in the microcirculation is governed by the Stokes and continuity equations,

$$\mu\nabla^2\mathbf{v}-\nabla p = \mathbf{0}, \qquad \nabla\cdot\mathbf{v} = 0.$$

A domain-based numerical method (e.g., FEM, finite difference, etc.) for the solution of the mobility problem for the motion of multiple suspended particles would be a major computational task requiring the discretization of the entire flow domain. Since we are interested primarily in the motion of the particles, a numerical technique constructed for this task may be more efficient. The boundary element method is one such technique, and it proceeds from an integral equation restatement of the governing differential equations. There are several excellent books on numerical methods that describe the BEM (see, e.g., Banerjee and Butterfield [51], or Brebbia et al. [52]). Following Kim and Karilla [50], we give a brief overview of the formulation.

For a flow domain G (containing the immersed particles) bounded by Γ, let velocity fields $\mathbf{v}_1$ and $\mathbf{v}_2$ represent two solutions of the Stokes equations. Then for the associated stress fields $\boldsymbol{\sigma}_1$ and $\boldsymbol{\sigma}_2$, the Lorentz reciprocal theorem [53] states that

$$\int_\Gamma \mathbf{v}_1 \cdot (\boldsymbol{\sigma}_2 \cdot \mathbf{n})\, d\Gamma - \int_G \mathbf{v}_1 \cdot (\nabla \cdot \boldsymbol{\sigma}_2)\, dG = \int_\Gamma \mathbf{v}_2 \cdot (\boldsymbol{\sigma}_1 \cdot \mathbf{n})\, d\Gamma - \int_G \mathbf{v}_2 \cdot (\nabla \cdot \boldsymbol{\sigma}_1) dG. \tag{12.12}$$

Note that $\mathbf{v}_1$ and $\boldsymbol{\sigma}_1$ are associated with the solution to Eq. (12.3), while here, $\mathbf{v}_2$ and $\boldsymbol{\sigma}_2$ are associated with the solution of the related singular problem,

$$\mu\nabla^2\mathbf{v} - \nabla p = -\boldsymbol{f}\delta(\mathbf{x}), \qquad \nabla \cdot \mathbf{v} = 0, \tag{12.13}$$

where $\delta\,(\mathbf{x})$ is the Dirac delta function. Note that for $\mathbf{x} \neq \mathbf{0}$, we recover Eq. (12.3). The solution to this equation may be obtained, for example, with the Fourier transform technique. It can be shown that

$$\mathbf{v}_2(\mathbf{x}) = \boldsymbol{f} \cdot \frac{\mathcal{G}(\mathbf{x})}{8\pi\mu}, \qquad \mathcal{G}_{ij}(\mathbf{x}) = \frac{1}{r}\delta_{ij} + \frac{1}{r^3}x_i x_j, \qquad r = |\mathbf{x}|, \tag{12.14}$$

where $\mathcal{G}(\mathbf{x})$ is the Oseen tensor, and

$$p_2(\mathbf{x}) = \boldsymbol{f} \cdot \frac{\mathcal{P}(\mathbf{x})}{8\pi\mu}, \qquad \mathcal{P}_j(\mathbf{x}) = 2\mu\frac{x_j}{r^3} + \mathcal{P}_j^\infty \tag{12.15}$$

satisfy Eq. (12.13) and therefore represent the solution.

This solution is particularly useful when we notice that the divergence of the associated stress of the Oseen tensor is a delta function. That is,

$$\sum\nolimits_{ijk,k} = -\delta_{ij}\delta(\mathbf{x}), \tag{12.16}$$

where δ_{ij} is the Kronecker delta and Σ_{ijk} is given by

$$8\pi\mu\sum\nolimits_{ijk} = -\mathcal{P}_j\delta_{ik} + \mu(\mathcal{G}_{ij,k} + \mathcal{G}_{kj,i}) = -6\mu\frac{x_i x_j x_k}{r^5}. \tag{12.17}$$

Without loss of generality, we allow the point of application of the unit point force (i.e., $|\boldsymbol{f}| = 1$) to be anywhere within G (i.e., $\delta(\mathbf{x}-\mathbf{y}),\ \ \mathbf{y}\in G$). For a domain of integration including the singular point $\mathbf{x} = \mathbf{y}$ and recalling that $\nabla \cdot \boldsymbol{\sigma}_1 = \mathbf{0}$, Eq. (12.12) simplifies as

$$\mathbf{v}_1(\mathbf{y}) = -\int_\Gamma \mathbf{v}_1 \cdot (\boldsymbol{\sigma}_2 \cdot \mathbf{n})\, d\Gamma_x + \int_\Gamma \mathbf{v}_2 \cdot (\boldsymbol{\sigma}_1 \cdot \mathbf{n})\, d\Gamma_x, \qquad \mathbf{y}\in G \tag{12.18}$$

where $\mathbf{n}$ is the unit outward normal. Finally, substituting for the singular solution we arrive at an integral equation for velocity as a function of the boundary traction, $\mathbf{t}$ $(= \boldsymbol{\sigma} \cdot \hat{\mathbf{n}})$:

$$\mathbf{v}(\mathbf{y}) = \int_{\Gamma} \mathbf{v} \cdot \left(\sum \cdot \hat{\mathbf{n}} \right) d\Gamma_{\mathrm{x}} - \frac{1}{8\pi\mu} \int_{\Gamma} \mathcal{G} \cdot \mathbf{t} \, d\Gamma_{\mathrm{x}}, \qquad \mathbf{y} \in G, \tag{12.19}$$

where $\hat{\mathbf{n}} = -\mathbf{n}$ and we have dropped the 1,2 subscript notation. The first integral on the right-hand side (RHS) is known as the *double-layer potential*, while the second integral on the RHS is known as the *single-layer potential*.

For a rigid particle it can be shown that the velocity field $\mathbf{v}(\mathbf{y})$ may be computed with only the single-layer potential. However, this leads to a Fredholm integral equation of the first kind with its attendant issues for numerical solution of certain classes of problems (see Kim and Karilla [50]). An integral equation of the second type results when we switch to an indirect formulation. This latter type of integral equation is particularly amenable to numerical solution by iterative approach. Dropping the single-layer term, Power and Miranda [54] show that the velocity field may be represented by

$$v_j(\mathbf{y}) = \int_{\Gamma} K_{ij}(\mathbf{x}, \mathbf{y}) \varphi_i(\mathbf{x}) \, d\Gamma_{\mathrm{x}}, \qquad \mathbf{y} \in G, \tag{12.20}$$

where

$$K_{ij}(\mathbf{x}, \mathbf{y}) = 2\hat{n}_k(\mathbf{x}) \sum\nolimits_{ijk} (\mathbf{x}, \mathbf{y}) = -\frac{3}{2\pi} \frac{r_i r_j r_k \hat{n}_k}{r^5}, \qquad \mathbf{r} = \mathbf{x} - \mathbf{y} \tag{12.21}$$

is the associated traction of the Oseen tensor and $\boldsymbol{\varphi}$ is the unknown double-layer distribution. For a Liapunov-smooth surface (i.e., tangent plane everywhere), the double layer undergoes a jump discontinuity as $\mathbf{y}$ approaches the particle surface $\boldsymbol{\xi}$. This may be calculated exactly and results in

$$v_j(\boldsymbol{\xi}) = \varphi_j(\boldsymbol{\xi}) + \int_{\Gamma} K_{ij}(\mathbf{x}, \boldsymbol{\xi}) \varphi_i(\mathbf{x}) \, d\Gamma_{\mathrm{x}}, \quad \boldsymbol{\xi} \in \Gamma. \tag{12.22}$$

The final form of the integral equation for the unknown double-layer surface density has been given by Phan-Thien et al. [55],

$$\begin{aligned} \varphi_j(\boldsymbol{\xi}) + \int_{\Gamma} K_{ij} \varphi_i \, d\Gamma_x + \varphi_j^{(p,l)}(\boldsymbol{\xi}) \left\langle \boldsymbol{\varphi}^{(p,l)}, \boldsymbol{\varphi} \right\rangle - \psi_j^{(p)}(\boldsymbol{\xi}) \left\langle \boldsymbol{\psi}^{(p)}, \boldsymbol{\varphi} \right\rangle \\ = b_j(\boldsymbol{\xi}) - \frac{1}{2} \psi_j^{(p)}(\boldsymbol{\xi}) \left\langle \boldsymbol{\psi}^{(p)}, \boldsymbol{b} \right\rangle, \qquad \boldsymbol{\xi} \in \Gamma, \end{aligned} \tag{12.23}$$

where $\Gamma = \cup \Gamma_p$, $p = 1, \ldots, M$, denotes the collection of M particles which are assumed Lyapunov smooth, $\boldsymbol{\varphi}^{(p,l)}$ represents the translational motion ($l = 1,2,3$) and rotational motion ($l = 4,5,6$) of particle p, $\boldsymbol{\psi}^{(p)}$ denotes the eigenvector of the adjoint operator $\boldsymbol{K}^*$, and $\langle \cdot\ ,\ \cdot \rangle$ denotes the natural inner product

$$\langle \mathbf{a}, \mathbf{b} \rangle = \int_\Gamma \mathbf{a}(\mathbf{x}) \cdot \mathbf{b}(\mathbf{x})\, d\Gamma. \tag{12.24}$$

For spheres, $\boldsymbol{b}(\boldsymbol{\xi})$ is given by

$$b_j(\boldsymbol{\xi}) = -v_j^\infty + \sum_{p=1}^{M} \left[F_i^{(p)} - \tfrac{1}{2}(T^{(p)} \times \nabla)_i \right] G_{ji}(\boldsymbol{\xi}, \mathbf{x}_c^{(p)}) \qquad \boldsymbol{\xi} \in \Gamma, \tag{12.25}$$

where $\boldsymbol{F}^{(p)}$ and $\boldsymbol{T}^{(p)}$ are the external force and torque, respectively, acting on particle p, $\boldsymbol{G}_{ji}$ is the Stokeslet or singularity solution, and $\mathbf{v}^\infty$ is the far-field velocity (i.e., the field velocity far away from the influence of all particles). In Eq. (12.25), $\mathbf{x}_c^{(p)}$ has been taken to be the center of the particle. According to the current state of the solution algorithm, $v_j(\boldsymbol{\xi})$ is a rigid-body motion and Eq. (12.23) represents one possible approach to ensure uniqueness of a potential solution.

In Eq. (12.21) for the associated traction, container boundaries, which generally would affect the particle motion, are noticeably absent. Our interest, however, is to compute the motion for particles interacting not only with each other but with a surface as a representation of the endothelium (i.e., adhesive interactions). In the model, the endothelium has been approximated as an infinite, rigid planar surface where no-slip velocity boundary conditions are enforced at all times. The influence functions $\boldsymbol{G}$ and $\boldsymbol{K}$ in the presence of a planar boundary have been derived previously [55], and $\boldsymbol{\varphi}^{(p,l)}$ and $\boldsymbol{\psi}^{(p)}$ are also given there. Significantly, in the half-space formulation the singularity solution is supplemented with an image Stokeslet situated at $y_i = (y_1, y_2, -y_3)$ (Fig. 12.5) such that the velocity boundary condition is satisfied automatically. This implies that the planar surface need not be discretized. This indirect formulation for $\boldsymbol{\varphi}$ is solved iteratively with a boundary element method. The rigid-body motion (i.e., translational and rotational motion) may be computed from Eq. (12.22) once the double-layer density is known.

12.4.2 Direct Method for Surface Tractions and Displacements

12.4.2.1 Exterior Problem The cell-surface traction distribution may be found by applying a traction-based, completed adjoint double-layer boundary element method [55–57]. It can be shown that the traction, $\boldsymbol{t}$, at point $\boldsymbol{\xi} \in \Gamma$ with outward normal unit vector $\hat{\boldsymbol{n}}(\boldsymbol{\xi})$ is given by

$$\boldsymbol{t}(\boldsymbol{\xi}) + \int_\Gamma \mathbf{K}^*(\boldsymbol{\xi}, \boldsymbol{y}) \cdot \boldsymbol{t}(\boldsymbol{y})\, d\Gamma_y + \sum_{p,l} \boldsymbol{\varphi}^{(p,l)}(\boldsymbol{\xi}) \left\langle \boldsymbol{\varphi}^{(p,l)}, \boldsymbol{t} \right\rangle = \boldsymbol{b}(\boldsymbol{\xi}), \qquad \boldsymbol{\xi} \in \Gamma, \tag{12.26}$$

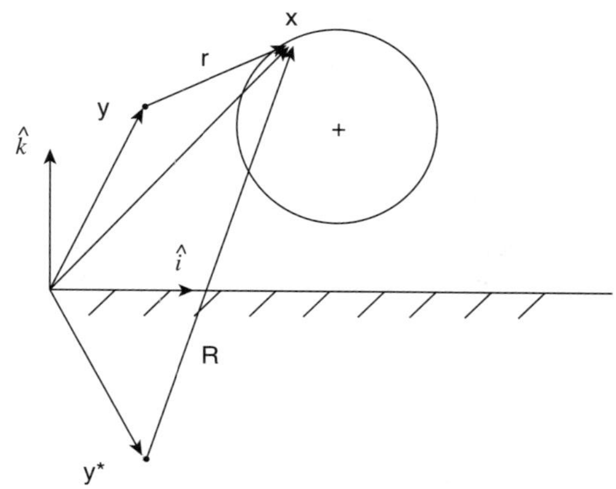

Figure 12.5 Semi-infinite formulation may be treated analytically by considering the effect of a point-force singularity in the mirror position below the $z = 0$ plane.

where

$$b_i(\boldsymbol{\xi}) = 2t_i^\infty + \sum_{p=1}^{M} \left(\frac{1}{S_p} F_i^{(p)} + \varepsilon_{ilk} \frac{T_l^{(p)}}{I_l^{(p)}} r_k^{(p)} \right). \tag{12.27}$$

Here $\boldsymbol{t}^\infty$ is the traction due to $\boldsymbol{v}^\infty$ (e.g., due to an imposed shear flow), ε_{ilk} is the permutation symbol, $\boldsymbol{r}^{(p)} \left(= \boldsymbol{\xi} - \mathbf{x}_c^{(p)} \right)$ is the vector from the particle center to the particle surface, and the $I_l^{(p)}$ are given by

$$I_l^{(p)} = -\int_{\Gamma_p} \mathbf{r}^{(p)} \cdot \mathbf{r}^{(p)} - \left(r_l^{(p)} \right)^2 d\Gamma. \tag{12.28}$$

$\boldsymbol{K}^*$ is the associated traction of $\boldsymbol{G}$, which, by definition, is the adjoint of the double-layer kernel. Similar to the case above for the mobility problem, the adjoint double-layer kernel $\boldsymbol{K}^*$ for a half-space is composed of several contributions: the associated traction for the Stokeslet, $K_{\mathbf{S}}^*$, and the associated traction for the image Stokeslet, $\boldsymbol{K}_{\mathbf{IS}}^*$,

$$\begin{aligned} \boldsymbol{K}_{\mathbf{S}}^* &= -\frac{3}{4\pi} \frac{r_i r_j r_m \hat{n}_m(\boldsymbol{\xi})}{r^5}, \quad \boldsymbol{r} = \mathbf{y} - \boldsymbol{\xi} \\ \boldsymbol{K}_{\mathbf{IS}}^* &= -\frac{3}{4\pi} \frac{R_i R_j R_m \hat{n}_m(\boldsymbol{\xi})}{R^5}, \quad \boldsymbol{R} = \mathbf{y} - \boldsymbol{\xi}^* \end{aligned} \tag{12.29}$$

and is likewise supplemented with additional terms, $\boldsymbol{K}_E^*$, such that $\boldsymbol{K}^* = \boldsymbol{K}_{\mathbf{S}}^* - \boldsymbol{K}_{\mathbf{IS}}^* + \boldsymbol{K}_E^*$, where $\boldsymbol{\xi}^*$ is the image of $\boldsymbol{\xi}$ with respect to the plane $z = 0$.

12.4.2.2 Interior Problem As noted previously, for infinitesimal elastic deformation the problem is governed by the Navier equation,

$$\frac{1}{1-2\nu_p}\nabla\nabla\cdot\mathbf{u}+\nabla^2\mathbf{u}=\mathbf{0}.$$

Using Betti's reciprocal theorem, this PDE is recast as an integral equation over the particle domain, D_p, resulting in the following integral equation for the displacement:

$$\mathbf{u}(\boldsymbol{\xi})+\int_{\Gamma_p}\boldsymbol{K}^{(p)}(\boldsymbol{\xi},\boldsymbol{y})\cdot\mathbf{u}(\boldsymbol{y})\,d\Gamma_y+\sum_l\boldsymbol{\varphi}^{(p,l)}(\boldsymbol{\xi})\left\langle\boldsymbol{\varphi}^{(p,l)},\mathbf{u}\right\rangle$$
$$=\int_{\Gamma_p}2\boldsymbol{G}^{(p)}(\boldsymbol{\xi},\boldsymbol{y})\cdot\boldsymbol{t}(\boldsymbol{y})\,d\Gamma_y,\qquad \boldsymbol{\xi}\in\Gamma_p, \tag{12.30}$$

where $\boldsymbol{G}$ is the Kelvinlet and $\boldsymbol{K}$ is the transpose of the associated traction of the Kelvinlet. To account for the half-space, the singularity solution is supplemented with an image term such that $\boldsymbol{G}^{(p)}=\boldsymbol{G}_{\mathrm{K}}^{(p)}-\boldsymbol{G}_{\mathrm{IK}}^{(p)}+\boldsymbol{G}_E^{(p)}$. The expressions for $\boldsymbol{G}$ on the right-hand side have been derived previously [57] and are repeated below for completeness:

$$\left(G_{ij}^{(p)}\right)_{\mathrm{K}}=\frac{1}{16\pi(1-\nu_p)\mu_p r}\left[(3-4\nu_p)\delta_{ij}+\frac{r_ir_j}{r^2}\right],$$

$$\left(G_{ij}^{(p)}\right)_{\mathrm{IK}}=\frac{1}{16\pi(1-\nu_p)\mu_p R}\left[\left(3-4\nu_p\right)\delta_{ij}+\frac{R_iR_j}{R^2}\right],$$

$$\left(G_{ij}^{(p)}\right)_E=\frac{\xi_3}{8\pi(1-\nu_p)\mu_p R^3}\left\{\delta_{j3}R_i+\delta_{i3}R_j-2\delta_{i3}\delta_{j3}R_3\right.$$
$$\left.+\frac{y_3}{3-4\nu_p}\left[2\delta_{i3}\delta_{j3}-\delta_{ij}+\frac{3R_i}{R^2}\left(R_j-2\delta_{j3}R_3\right)\right]\right\}. \tag{12.31}$$

The associated traction $\boldsymbol{K}^{(p)}\left(=\boldsymbol{K}_{\mathrm{K}}^{(p)}-\boldsymbol{K}_{\mathrm{IK}}^{(p)}+K_E^{(p)}\right)$ may be constructed similarly [57].

In the numerical procedure, **u** is solved for iteratively using the boundary element method. As mentioned previously, we have implemented a linear elastic constitutive model for the cell since rolling deformations at low to intermediate shear stresses ($\leq$5 dyn/cm^2) are relatively small in magnitude and thus may be approximated in this manner [58]. Additionally, Evans and co-workers have demonstrated that for compressive deformations on the order of hundreds of nanometers, the neutrophil response is well approximated as a linear elastic solid [59–61]. For higher stresses, it is necessary to alter the technique by using constitutive models capable of accurately representing large deformations. This can be accomplished, for example, with a liquid-drop model [62,63].

12.4.3 Surface Discretization

In the boundary element method, the solution of the integral equations given here yields the surface double-layer density, traction, and displacement. An additional calculation is required to solve for the particle's velocity. Field variables may also be calculated via postprocessing once the boundary solution is obtained. In this application, the surfaces include the elastic inclusions Γ_p, $p = 1, \ldots, M$, the planar surface Γ_b, and a hemispherical surface Γ_h with radius $h \rightarrow \infty$. It can be shown that the disturbance induced by the elastic inclusions decays at infinity. Therefore, the contribution from these surface integrals is zero. At the planar surface, zero-slip and zero-displacement boundary conditions are enforced at all times. In this way, the method reduces to computing values for the unknowns only over the surfaces of the particles.

The surface of each sphere has been discretized using 96 Quad9 elements: quadrilateral elements with three nodes per edge, one center node, and interpolated with second-order polynomials (Fig. 12.6). The unknown variable is considered uniform over each element and is evaluated using Gaussian quadrature. A 1×1 quadrature with a single collocation point/element is sufficient for capturing the effect of distant elements. For nearby elements, 2×2 quadrature with four collocation points/element is used. The solution is obtained in an iterative fashion using the value at the previous iteration as a guess for the new one. The convergence criterion used requires the iterates to achieve a relative error of $O(10^{-4})$. In the case for elastic inclusions, the piecewise constant nature of the element solutions may lead to poor surface representation. Hence, nodal values for displacements have been calculated using simple averages of the displacements of neighboring elements. For interacting particles, small time steps of order $O(10^{-7}\text{s})$ have been used and the calculations have been performed on a single 1.8-GHz AMD Opteron processor. The single- and double-layer kernels are singular (strong singularity for the elasticity problem, and weak singularity otherwise). The technique for treating singular contributions has been described previously [64].

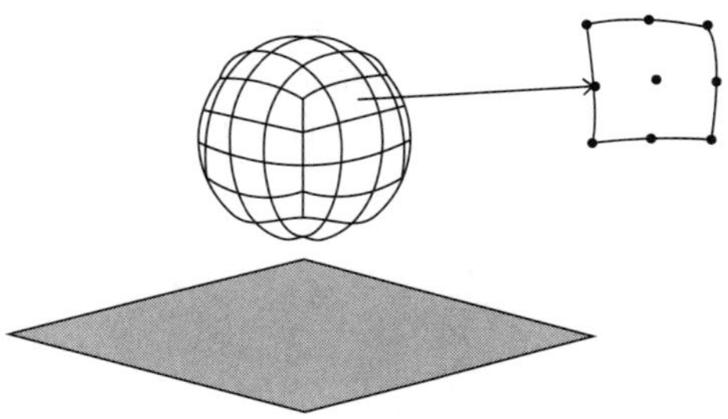

Figure 12.6 Boundary element surface discretization; Quad9 Lagrangian element.

12.5 SIMULATION RESULTS

It has been demonstrated that a firmly adherent rigid spherical cell will recruit additional circulating cells via far-field hydrodynamic interactions through a mechanism called *hydrodynamic recruitment*. This has been shown experimentally in cell-free assays, computed numerically with the multiparticle adhesive dynamics simulation (MAD), and is supported by measurements of leukocyte capture in vivo using the hamster cheek pouch model [31,65]. Furthermore, it has been shown that two or more leukocytes rolling in close proximity to each other will decrease their rolling velocity, thereby providing a nucleation site for a cluster of adherent or slow-rolling cells [66,67]. Cell deformation, as part of the rolling interaction, is anticipated to contribute to this type of recruitment by stabilizing rolling via enhanced presentation of ligand.

12.5.1 Uniform Translation and Sedimentation

Results from the new simulation were first compared against several benchmark solutions. The first case examined was the hydrodynamic drag on a single rigid sphere undergoing uniform translation in unbounded space. For quiescent flow far from the particle, the solution is given by the expression $6\pi\mu a U_0$ where a, U_0 are the particle radius and velocity, respectively, and μ is the viscosity. By solving for and then taking the integral of the surface traction over the cell surface, it can be shown that the computed drag is in excellent agreement with the theoretical value.

The drag on a particle in translational motion parallel to a planar wall was calculated as a function of distance from the wall. Proximity to the boundary is defined by the nondimensional parameter *h/a*, where *h* is the perpendicular distance from the wall to the particle center. This problem has been solved analytically by Lee and Leal [68] using a bipolar coordinate system. Using the new simulation, it is apparent that both numerical and analytical results are in good agreement, as indicated in Fig. 12.7a, as a function of the nondimensional separation. Incidentally, these two test cases highlight one of the key features of the BEM. Namely, when the fluid bounding surface is approached, no additional surface discretization is required since the kernel functions account for the wall's influence.

Next, for the case of two sedimenting spheres (denoted A and B) under constant external force, the sedimentation velocity has been shown to depend on the cell–cell separation distance d and the inclination angle θ of the line connecting the cell centers [69]. For the special case in which $\theta = 90°$ (i.e., cells aligned in the direction of gravity), the sedimentation velocity has been shown to be given by [69]

$$\mathrm{U_A} = \mathrm{U_B} = -\mathrm{U_0}\left(1 + \frac{3}{2}\frac{a}{d}\right)\hat{i}_z. \tag{12.32}$$

This result is plotted in Fig. 12.7b, along with simulation results (square symbols). Again, agreement with the theoretical result as a function of cell–cell separation

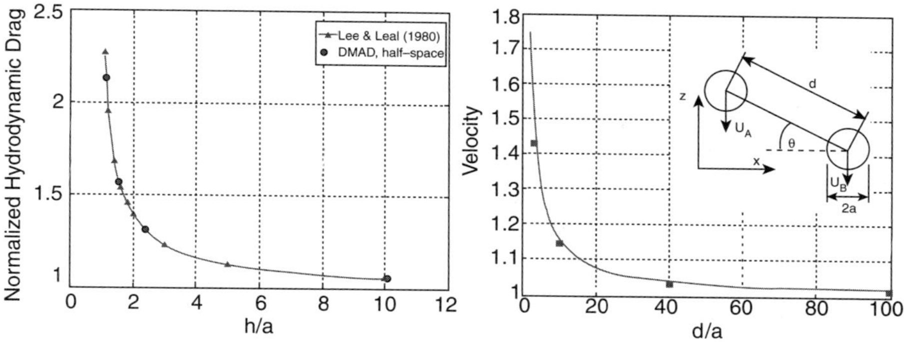

(a) (b)

Figure 12.7 (a) Normalized hydrodynamic drag for a single rigid sphere in motion parallel to a planar wall. The numerical result is denoted as DMAD (deformable multiparticle adhesive dynamics). (b) Sedimentation velocity for two spheres in unbounded flow. Simulation results (square symbols) are in excellent agreement with analytical results [Eq. (12.32)].

distance is excellent, degrading slightly as the cells come in very close proximity to each other.

12.5.2 Simple Shear Flow

As mentioned previously, in vivo data suggest that both circulating and rolling neutrophils generally retain their spherical shape. However, to visually demonstrate the cell deformation capability of the current model, cell deformation due to a simple shear flow may be accentuated by decreasing the cell's shear modulus relative to the imposed fluid shear stress. Here we consider a single value for the strain rate ($\dot{\gamma} = 100\ \mathrm{s}^{-1}$) and have taken $a = 4\ \mu$m. For simple shear flow, the characteristic fluid stress is given by the product of the fluid viscosity and the shear rate (i.e., $\mu\dot{\gamma}$). Then for a ratio of fluid stress-to-shear modulus equal to 0.1 (denoted as the viscous number, $V_i = \mu\dot{\gamma}/\mu_p$, in ref. [64]), an initially spherical elastic solid will deform into the ellipsoidal shape indicated in Fig. 12.8a. As the cell translates and rotates (i.e., tumbles), the directions of its major and minor axes are seen to remain fixed in orientation (Fig. 12.8b), exhibiting *tank-treading motion*. That is, for long-time motion the cell is seen to deform continuously even as it maintains the orientation of its principal axes. This result is in agreement with previous numerical simulations [64]. In the presence of a nearby planar boundary, the simulation indicates that the surface traction distribution is only moderately affected (Fig. 12.8c). Here we see only slight deviations in the overall deformed shape for cells with gap flows (i.e., the distance between the lowest point on the cell and the planar wall) on the order of 1 μm. Finally, it should be pointed out that, in general, the number of iterations required to achieve a converged solution increases as the wall is approached.

For physiological shear rates, we may increase the cell's shear modulus in order that the cell deformation due to fluid shear is infinitesimal. This conforms to earlier

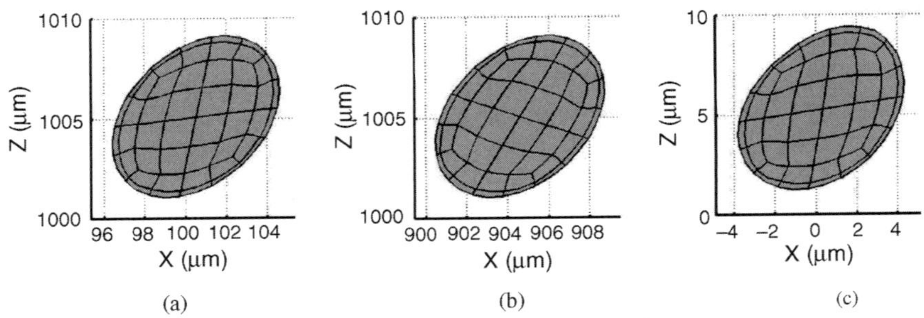

Figure 12.8 Tank-treading motion for elastically deformable cell in unbounded linear shear flow. Deformation has been accentuated by considering small values for the cell's shear modulus. The cell tumbles as it translates in flow direction (i.e., the x-direction) (a), but maintains the orientation of its principal axes (b). In the near-wall environment, the surface tractions are affected only moderately (c).

observations which indicated that rolling cells early in the activation cascade generally retain their spherical shape [45]. As a model of cell–cell collisions in the near-wall environment, we simulated the interaction of two deformable cells for $V_i = 0.01$ (Fig. 12.9). To model the neutrophil surface roughness, surfaces were covered with a distribution of small bumps assumed to have negligible influence on the fluid flow. The surface roughness value was taken to be 175 nm [44]. If the cell-cell separation distance becomes too small, a short-range repulsive force representing van der Waals and electrostatic interactions keeps the two interacting surfaces from touching [Eq. (12.6)]. Then as a result of the imposed simple shear flow, the elevated upper cell eventually overtakes the lower cell. By plotting position histories for both cells during the interaction (Fig. 12.10), it can be seen that the lower cell is displaced transiently toward the reactive boundary as a result of the interaction. The positions of the cell centers of gravity are seen to recover once the two-particle interaction is completed. However, the cumulative effect of many such repeated interactions—due to high hematocrit and the high density of white blood cells in circulation—is likely to increase the susceptibility of the cell to the formation of transient adhesive interactions, and thereby to induce rolling. This idea of cell collisions driving cells toward the wall has been shown previously, where it was demonstrated that collisions with erythrocytes can drive white cells toward the wall [14,15].

We have seen evidence for this type of interaction in experiments for cells flowing and rolling over functionalized surfaces. In these flow experiments, the lower surface of a parallel-plate flow chamber was coated with 2 μg/mL multivalent sialyl Lewis^x-PAA-biotin using precoated NeutrAvidin biotin-binding protein [70]. The surface was then blocked for nonspecific adhesion with 2% bovine serum albumin in Dulbecco's phosphate-buffered saline for 1 hour. Next, isolated neutrophils ($10^6\,\mathrm{mL}^{-1}$) were introduced into the chamber using a syringe pump at a flow rate corresponding to a wall shear rate of $100\,\mathrm{s}^{-1}$. The flow chamber was mounted on an inverted microscope and cell–cell collision events were monitored using a microscope-linked CCD camera and recorded on high-quality VCR tapes for cell tracking analyses.

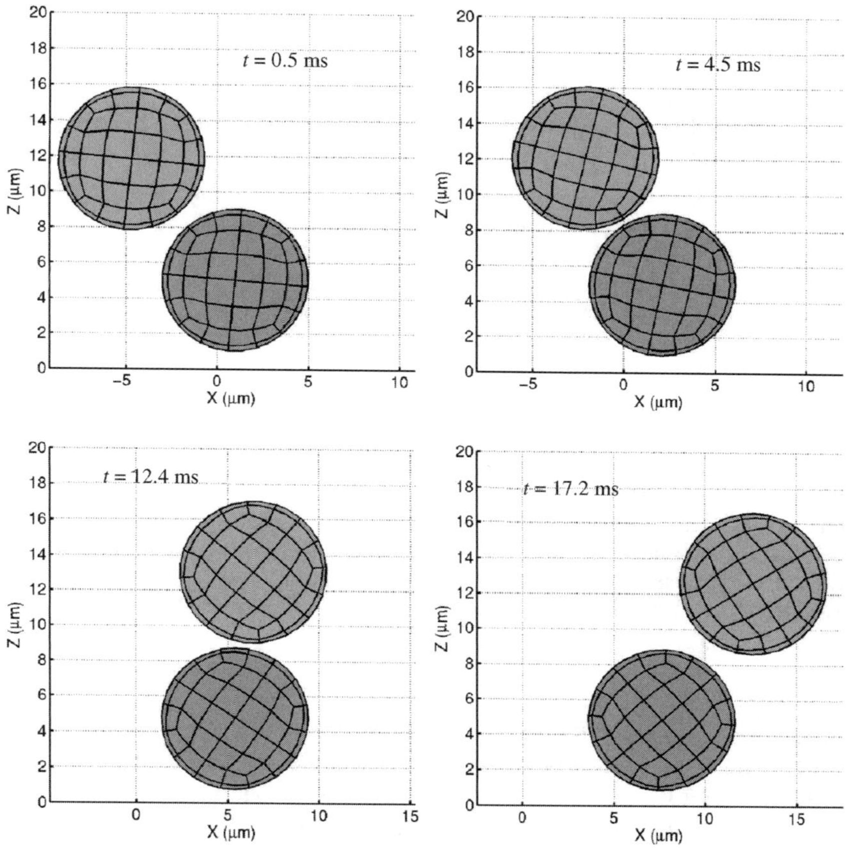

Figure 12.9 Near-wall binary collision for two elastically deformable cells ($\dot{\gamma} = 100\,s^{-1}$, $V_i = 0.01$).

By analyzing the video data and tracking the position coordinates of each neutrophil on every frame, cell–cell collision data were seen to conform to the numerical simulations.

We are most interested in simulations with adhesive interactions since this mimics several features of cell rolling over inflamed endothelium. P-selectin (CD62) is rapidly up-regulated on stimulated endothelium and plays an important role in early immune response. Furthermore, due to the general ruffled nature of the neutrophil surface and the clustering of CD62 counter-receptors on these microvilli, it is believed that CD62 density is an important parameter for immune response. For the model of the endothelium as a planar surface, a physiologically relevant distribution of adhesion molecules was used [3]. To model the roughness of the planar surface, it was covered with a distribution of small bumps assumed to have negligible influence on the fluid flow. The surface roughness value was taken to be 50 nm [44]. As before,

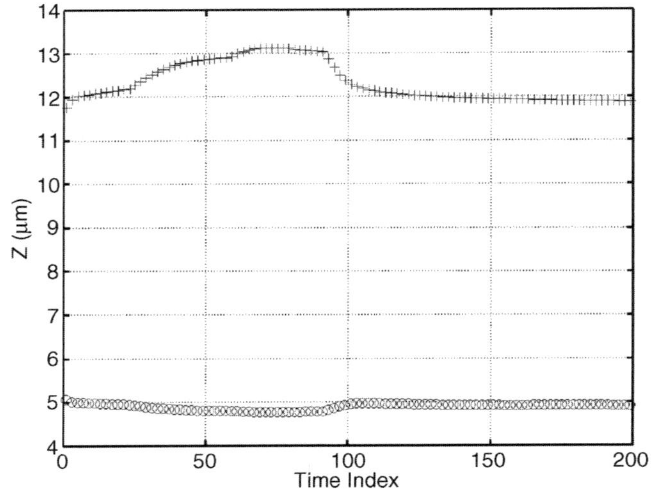

Figure 12.10 Trace of z-position history above the planar wall for binary collision. Both cells are seen to recover eventually to their precollision heights above the plane. However, the lower cell is temporarily displaced closer to the reactive substrate as a result of the binary collision. The result of many such collisions is expected to enhance rolling interactions for the near-wall cell.

short-range repulsive forces keep the two interacting surfaces from making mathematical contact. As mentioned above, we have previously described a dissociation kinetics model for P-selectin/PSGL-1 transient bond formation. With the unstressed rate of bond formation as the only parameter, adhesive simulations were performed by placing the cell within a reactive distance to the wall. Upon bond formation, bond coordinates are tracked and the bond lifetime, which depends in part on the force on the bond, is checked at each cycle. By summing the total number of bonds and the total force for each element of the surface, we may compute the localized adhesive traction for each element. This contribution is then added to Eq. (12.27) in solving for the surface traction distribution, Eq. (12.26).

A series of images illustrating the cell adhesion algorithm is given in Fig. 12.11. In this example for a linear shear flow ($\dot{\gamma} = 100\,\mathrm{s}^{-1}$), the bond spring constant was set to $10^6\,\mathrm{fN}/\mu\mathrm{m}$ and a small time step ($dt = 10^{-7}\,\mathrm{s}$) was used when the cell sediments or is placed within the bond reactive region. In Fig. 12.11a, the cell lower surface is just within the bond interaction region. In the ensuing frames, several P-selectin/PSGL-1 bonds form between the surfaces, and these induce transient adhesion to the surface, with resulting deformation to the cell, Fig. 12.11b and c. Cell rotation and translation eventually induces these bonds to rupture. In this case, the cell translates downstream due to the fluid flow (Fig. 12.11d) and may eventually be recaptured. For long-lived bonds, fluid shear may induce a much more significant deformation to the cell, with resulting alteration to the fluid streamlines.

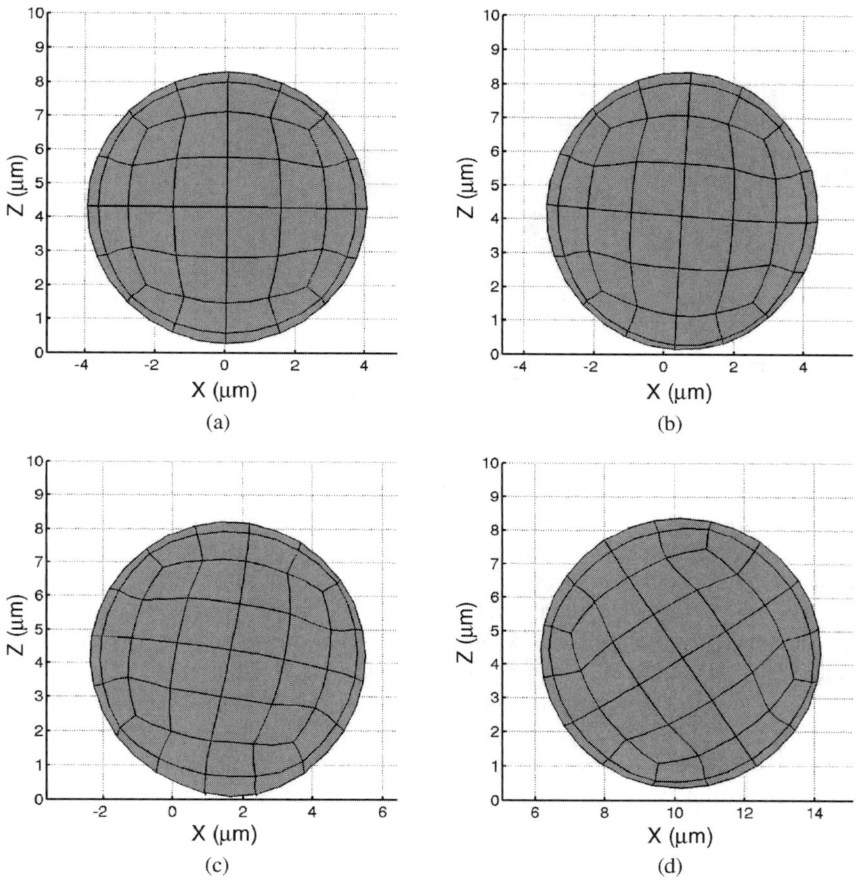

Figure 12.11 (a) Simulation of rolling interaction for deformable leukocyte over activated endothelium. Substrate ligand density controls potential adhesive interactions. (b,c) Adhesive forces alter the cell-surface traction distribution and induce cell deformation. As a result of these adhesive forces and cell deformation, the cell adhesive contact area increases. (d) Fluid shear forces may induce bond rupture with the result that the cell may reenter circulation. The cell may be captured again at a downstream location ($\dot{\gamma} = 100\ s^{-1}$, $V_i = 0.01$).

12.6 FUTURE DIRECTIONS

A new computer simulation has been formulated and implemented to study the role of cell deformation during the initial stages of inflammatory response. The numerical simulation is based on a boundary element method and is capable of modeling elastically deformable, interacting particles immersed in a semi-infinite, viscous, incompressible external Stokes flow. Deformations due to fluid stresses (both with and without the influence of a planar boundary) and adhesive stresses (for a cell rolling on a model of inflamed endothelium) have been demonstrated. One strength of the model in relation to domain-type numerical techniques is the reduction of unknowns from a three-dimensional space to the three-dimensional surface of the particles. These types of calculations are currently being performed on inexpensive, single-processor workstations and will, in the future, be performed in parallel on inexpensive computing clusters.

The model extends an earlier simulation for the multiparticle adhesive dynamics of rigid cells rolling over a model of inflamed endothelium. That model was shown to reproduce several of the phenomena characteristic of rolling adhesion. For example, it has been demonstrated that a firmly adherent rigid spherical cell will recruit additional circulating cells through a mechanism termed hydrodynamic recruitment. This has been demonstrated experimentally in cell-free assays, computed numerically with the multiparticle adhesive dynamics model, and is supported by measurements of leukocyte capture in vivo using the hamster cheek pouch model [31,65]. Furthermore, it has been shown that two or more leukocytes rolling in close proximity to each other will decrease their rolling velocity, thereby providing a nucleation site for a cluster of adherent or slow rolling cells [66,67]. However, the fact that rolling is much more stable in vivo than in cell-free assays points to fundamental shortcomings in our current understanding of selectin- and integrin-mediated adhesion based on results, to date, obtained in model systems. Cell deformation leading to dynamic cell flattening is widely considered to play an important role in inflammation by modulating cell rolling velocity. With the new simulation, we are able to compare experimental results with theoretical predictions of the relationships among shear rate, adhesion molecule density, and contact area between surfaces, and their resulting influence on the adhesive strength between neutrophils, neighboring cells, and the endothelium. Thus, multiscale simulations of this type will stimulate additional research into the effect of cell deformation on these recruitment mechanisms.

Acknowledgments

The authors gratefully acknowledge financial support from the National Institutes of Health.

REFERENCES

1. Kansas GS. Selectins and their ligands: current concepts and controversies. *Blood* **88**:3259–3287, 1996.
2. Arfors KE, et al. A monoclonal antibody to the membrane glycoprotein complex CD18 inhibits polymorphonuclear leukocyte accumulation and plasma leakage in vivo. *Blood* **69**:338–340, 1987.
3. Lawrence MB, Springer TA. Leukocytes roll on a selectin at physiologic flow rates: distinction from and prerequisite for adhesion through integrins. *Cell* **65**:859–873, 1991.

4. Brunk DK, Goetz DJ, Hammer DA. Sialyl Lewis^x/E-selectin mediated rolling in a cell-free system. *Biophys. J.* **71**:2902–2907, 1996.
5. Ley K, et al. Lectin-like cell adhesion molecule 1 mediates leukocyte rolling in mesenteric venules in vivo. *Blood* **77**:2553–2555, 1991.
6. Bullard D, Kunkel E, Kubo H, Hicks M, Lorenzo I, Doyle N, Doerschuk C, Ley K, Beaudet A. Infectious susceptibility and severe deficiency of leukocyte rolling and recruitment in E-selectin and P-selectin double mutant mice. *J. Exp. Med.* **18**:2329–2236, 1996.
7. Springer TA. Traffic signals on endothelium for lymphocyte recirculation and leukocyte emigration. *Annu. Rev. Physiol.* **57**:827–872, 1995.
8. Anderson DC, Springer TA. Leukocyte adhesion deficiency: an inherited defect in the Mac-1, LFA-1, and p150, 95 glycoproteins. *Annu. Rev. Med.* **38**:175–194, 1987.
9. Hemler ME. VLA proteins in the integrin family: structures, functions, and their role on leukocytes. *Annu. Rev. Immunol.* **8**:365–400, 1990.
10. Springer TA. Traffic signals for lymphocyte recirculation and leukocyte emigration: the multistep paradigm. *Cell* **74**:301–314, 1994.
11. Ley K, Laudanna C, Cybulsky MI, Nourshargh S. Getting to the site of inflammation: the leukocyte adhesion cascade updated. *Nat. Rev. Immunol.* **7**:678–689, 2007.
12. Greenberg AW, Brunk DK, Hammer DA. Cell-free rolling mediated by L-selectin and sialyl Lewis^x reveals the shear threshold effect. *Biophys. J.* **79**:2391–2402, 2000.
13. Rodgers SD, Camphausen RT, Hammer DA, Sialyl Lewis^x-mediated, PSGL-1-independent rolling adhesion on P-selectin. *Biophys. J.* **79**:694–706, 2000.
14. Melder RJ, Munn LL, Yamada S, Ohkubo C, Jain RK. Selectin- and integrin-mediated T-lymphocyte rolling and arrest on TNF-α-activated endothelium: augmentation by erythrocytes. *Biophys. J.* **69**:2131–2138, 1995.
15. Melder RJ, Yuan J, Munn LL, Jain RK. Erythrocytes enhance lymphocyte rolling and arrest in vivo. *Microvasc. Res.* **59**:316–322, 2000.
16. Mitchell DJ, Li P, Reinhardt PH, Kubes P. Importance of L-selectin-dependent leukocyte–leukocyte interactions in human whole blood. *Blood* **95**:2954–2959, 2000.
17. Bathe M, Shirai A, Doerschuk CM, Kamm RD. Neutrophil transit times through pulmonary capillaries: the effects of capillary geometry and fMLP-stimulation. *Biophys. J.* **83**:1917–1933, 2002.
18. Dong C, Cao J, Struble EJ, Lipowsky HH. Mechanics of leukocyte deformation and adhesion to endothelium in shear flow. *Ann. Biomed. Eng.* **27**:298–312, 1999.
19. Dong C, Lei X. Biomechanics of cell rolling: shear flow, cell-surface adhesion, and cell deformability. *J. Biomech.* **33**:35–43, 2000.
20. Firrell JC, Lipowsky HH. Leukocyte margination and deformation in mesenteric venules of rat. *Am. J. Physiol.* **256**:H1667–H1674, 1989.
21. Peskin CS. Numerical analysis of blood flow in the heart. *J. Comput. Phys.* **25**:220, 1977.
22. Fauci LJ, Peskin CS. A computational model of aquatic animal locomotion. *J. Comp. Phys.* **77**(1):85, 1988.
23. Fauci LJ. Interaction of oscillating filaments: a computational study. *J. Comput. Phys.* **86**(2):294, 1990.
24. Fogelson AL, Fauci LJ. Truncated newton methods and the modeling of complex immersed elastic structures. *Comm. Pure Appl. Math.* **46**:787, 1993.

25. Dillon R, Fauci LJ, Fogelson AL, Gaver D. Modeling biofilm processes using the immersed boundary method. *J. Comput. Phys.* **129**:57, 1996.

26. Bottino DC. Modeling viscoelastic networks and cell deformation in the context of the immersed boundary method. *J. Comput. Phys.* **147**:86, 1998.

27. Marella SV, Udaykumar HS. Computational analysis of the deformability of leukocytes modeled with viscous and elastic structural components. *Phys. Fluids* **16**:244–264, 2004.

28. Jadhav S, Eggleton CD, Konstantopoulos K. A 3-D computational model predicts that cell deformation affects selectin-mediated leukocyte rolling. *Biophys. J.* **88**:96–104, 2005.

29. Khismatullin DB, Truskey GA. 3D numerical simulation of receptor-mediated leukocyte adhesion to surfaces: effects of cell deformability and viscoelasticity. *Phys. Fluids* **17**:031505, 2005.

30. Hammer DA, Apte SM. Simulation of cell rolling and adhesion on surfaces in shear flow: general results and analysis of selectin-mediated neutrophil adhesion. *Biophys. J.* **63**:35–57, 1992.

31. King MR, Hammer DA. Multiparticle adhesive dynamics: hydrodynamic recruitment of rolling leukocytes. *Proc Natl Acad Sci USA* **98**:14919–14924, 2001.

32. Mody NA, King MR. Three-dimensional simulations of a platelet-shaped spheroid near a wall in shear flow. *Phys. Fluids* **17**:113302, 2005.

33. Gee DJ, King MR. Multiparticle adhesion of deformable cells. Presented at the 15th U.S. National Congress on Theoretical and Applied Mechanics, Boulder, CO, June 2006.

34. Germain RN, Margulies DH. The biochemistry and cell biology of antigen processing and presentation. *Annu. Rev. Immunol.* **11**:403–450, 1993.

35. Gray D. Immunological memory. *Annu. Rev. Immunol.* **11**:49–77, 1993.

36. Chien S, et al. Viscoelastic properties of leukocytes. In: *White Cell Mechanics: Basic Science and Clinical Aspects*, ed. HJ Meiselman, MA Lichtman, PL LaCelle, A.R. Liss, New York, 1984.

37. Gaborski TR, Clark A, Waugh RE, McGrath JL. Membrane mobility of β_2 integrins and rolling associated adhesion molecules in resting neutrophils. *Biophys. J.* **95**:4934–4947, 2008.

38. Panton RL. *Incompressible Flow*, Wiley, New York, 1984.

39. Bell GI. Models for the specific adhesion of cells to cells. *Science* **200**:618–627, 1978.

40. Alon R, Hammer DA, Springer TA. Lifetime of the P-selectin: carbohydrate bond and its response to tensile force in hydrodynamic flow. *Nature* **374**:539–542, 1995.

41. Smith MJ, Berg EL, Lawrence MB. A direct comparison of selectin-mediated transient, adhesive events using high temporal resolution. *Biophys. J.* **77**:3371–3383, 1999.

42. Bell GI, Dembo M, Bongrand P. Competition between non-specific repulsion and specific bonding. *Biophys. J.* **45**:1051–1064, 1984.

43. Chang KC, Hammer DA. Adhesive dynamics simulations of sialyl-Lewis^x/E-selectin-mediated rolling in a cell-free system. *Biophys. J.* **79**:1891–1902, 2000.

44. King MR, Hammer DA. Multiparticle adhesive dynamics. Interactions between stably rolling cells. *Biophys. J.* **81**:799–813, 2001.

45. Wojciechowski JC, Sarelius IH. Preferential binding of leukocytes to the endothelial junction region in venules in situ. *Microcirculation* **12**:349–359, 2005.

46. Carman CV, Springer TA. A transmigratory cup in leukocyte diapedesis both through individual vascular endothelial cells and between them. *J. Cell. Biol.* **167**:377–388, 2004.

47. Yang L, Froio RM, Sciuto TE, Dvorak AM, Alon R, Luscinskas FW. ICAM-1 regulates neutrophil adhesion and transcellular migration of TNF-α-activated vascular endothelium under flow. *Blood* **106**:584–592, 2005.

48. Gee DJ, Lee D, King MR. Evaluation of a linear elastic model for inclusion in a multiscale model of deformable leukocytes. Presented at the Fall Meeting, Biomedical Engineering Society, Los Angeles, CA, Sept. 2007.

49. Malvern LE. *Introduction to the Mechanics of a Continuous Medium*, Prentice-Hall, Englewood Cliffs, NJ, 1969.

50. Kim S, Karrila SJ. *Microhydrodynamics: Principles and Selected Applications*, Butterworth–Heinemann, Stoneham, MA, 1991.

51. Banerjee PK, Butterfield R. *Boundary Element Methods in Engineering Science*, McGraw-Hill, New York, 1981.

52. Brebbia CA, Telles JCF, Wrobel LC. *Boundary Element Techniques*, Springer-Verlag, New York, 1984.

53. Lorentz HA. Ein allgemeiner Satz die Bewegung einer reibenden Flüssigkeit betreffend, nebst einigen Anwendungen desselben [A general theorem concerning the motion of a viscous fluid and a few applications from it]. *Versl. Kon. Akad. Wetensch. Amsterdam* **5**:168–174, 1896.

54. Power H, Miranda G. Second kind integral equation formulation of Stokes flow past a particle of arbitrary shape. *SIAM J. Appl. Math.* **47**:689–698, 1987.

55. Phan-Thien N, Tullock D, Kim S. Completed double layer in half-space: a boundary element method. *Comput. Mech.* **9**:121–135, 1992.

56. Phan-Thien N, Fan XJ. Traction-based completed adjoint double layer boundary element method in elasticity. *Comput. Mech.* **16**:360–367, 1995.

57. Tran-Cong T, Phan-Thien N. Boundary element solution for half-space elasticity or Stokes problem with a no-slip boundary. *Comput. Mech.* **1**:259–268, 1986.

58. Schmid-Schonbein GW, Sung KLP, Tozeren H, Skalak R, Chien S. Passive mechanical properties of human leukocytes. *Biophys. J.* **36**:243, 1981.

59. Evans EA, Heinrich V, Yeung A, Kinoshita K. Nano-to-micro scale dynamics of P-selectin detachment from leukocyte interfaces: I. Membrane separation from the cytoskeleton. *Biophys. J.* **88**:2288–2298, 2005.

60. Heinrich V, Leung A, Evans E. Nano-to-micro scale dynamics of P-selectin detachment from leukocyte interfaces: II. Tether flow terminated by P-selectin dissociation from PSGL-1. *Biophys. J.* **88**:2299–2308, 2005.

61. King MR, Heinrich V, Evans E, Hammer DA. Nano-to-micro scale dynamics of P-selectin detachment from leukocyte interfaces: III. Numerical simulation of tethering under flow. *Biophys. J.* **88**:1676–1683, 2005.

62. Evans EA, Yeung A. Apparent viscosity and cortical tension of blood granulocytes determined by micropipette aspiration. *Biophys. J.* **56**:151, 1989.

63. Yeung A, Evans EA. Cortical shell–liquid core model for passive flow of liquid like spherical cells into micropipettes. *Biophys. J.* **56**:139, 1989.

64. Phan-Thien N, Fan XJ. A boundary integral formulation for elastically deformable particles in a viscous fluid. *Z. Angew. Math. Phys.* **47**:672–694, 1996.

65. Lee D, King MR, Hydrodynamic interactions between cells on reactive surfaces. In: *Principles of Cellular Engineering: Understanding the Biomolecular Interface*, ed. MR King, Academic Press, San Diego, CA, 2006.
66. King MR, Kim MB, Sarelius IH, Hammer DA. Hydrodynamic interactions between rolling leukocytes in vivo. *Microcirculation* **10**:401–409, 2003.
67. King MR, Ruscio AD, Kim MB, Sarelius IH. Interactions between stably rolling leukocytes in vivo. *Phys. Fluids* **17**:031501, 2005.
68. Lee SH, Leal G. Motion of a sphere in the presence of a plane interface: 2. An exact solution in bipolar co-ordinates. *J. Fluid Mech.* **98**:193–224, 1980.
69. Leal G. *Laminar Flow and Convective Transport Processes*, Butterworth–Heinemann, Stoneham, MA, 1992.
70. Lee D, Schultz JB, Knauf PA, King MR. Mechanical shedding of L-selectin from the neutrophil surface during rolling on Sialyl Lewis^x under flow. *J. Biol. Chem.* **282**:4812–4820, 2007.

INDEX

Multiscale Modeling of Particle Interactions: Applications in Biology and Nanotechnology,
Edited by Michael R. King and David J. Gee